International Series in Operations Research & Management Science

Volume 149

Series Editor

Frederick S. Hillier
Stanford University, CA, USA

Special Editorial Consultant

Camille C. Price
Stephen F. Austin State University, TX, USA

For further volumes:
http://www.springer.com/series/6161

Eric V. Denardo

Linear Programming and Generalizations

A Problem-based Introduction with Spreadsheets

Springer

Eric V. Denardo
Yale University
P.O. Box 208267
New Haven CT 06520-8267
USA
eric.denardo@yale.edu

Additional material to this book can be downloaded from http://extra.springer.com.

ISSN 0884-8289
ISBN 978-1-4419-6490-8 e-ISBN 978-1-4419-6491-5
DOI 10.1007/978-1-4419-6491-5
Springer New York Dordrecht Heidelberg London

Library of Congress Control Number: 2011920997

Printed on acid-free paper

Springer is part of Springer Science+Business Media (www.springer.com)

Preface

The title of this book adheres to a well-established tradition, but "linear programming and generalizations" might be less descriptive than "models of constrained optimization." This book surveys models that optimize something, subject to constraints. The simplest such models are linear, and the ideas used to analyze linear models generalize easily.

Over the past half century, dozens of excellent books have appeared on this subject. Why another? This book fuses five components:

- It uses examples to introduce general ideas.
- It engages the student in spreadsheet computation.
- It surveys the uses of constrained optimization.
- It presents the mathematics that relates to constrained optimization.
- It links the subject to economic reasoning.

Each of these components can be found in other books. Their fusion makes constrained optimization more accessible and more valuable. It stimulates the student's interest, it quickens the learning process, it helps students to achieve mastery, and it prepares them to make effective use of the material.

A well-designed example provides context. It can illustrate the applicability of the model. It can reveal a concept that holds in general. It can introduce the notation that will be needed for a more general discussion.

Examples mesh naturally with spreadsheet computation. To compute on a spreadsheet is to learn interactively – the spreadsheet gives instant feedback. Spreadsheet computation also takes advantage of the revolution that has occurred in computer hardware and software. Decades ago, constrained optimization required specialized knowledge and access to huge computers. It was a subject for experts. That is no longer the case. Constrained optimization has become vastly easier to learn and to use. Spreadsheets help the student to become facile with the subject, and it helps them use it to shape their professional identities.

Constrained optimization draws upon several branches of mathematics. Linear programming builds upon linear algebra. Its generalizations draw upon analysis, differential calculus, and convexity. Including the relevant math in a course on constrained optimization helps the student to master the math and to use it effectively.

Nearly every facet of constrained optimization has a close link to economic reasoning. I cite two examples, among many: A central theme of economics is the efficient allocation of scarce resources, and *the* canonical model for allocating scarce resources is the linear program. Marginal analysis is a key concept in economics, and it is exactly what the simplex method accomplishes. Emphasizing the links between constrained optimization and economics makes both subjects more comprehensible, and more germane.

The scope of this book reflects its components. Spreadsheet computation is used throughout as a teaching-and-learning aide. Uses of constrained optimization are surveyed. The theory is dovetailed with the relevant mathematics. The links to economics are emphasized.

The book is designed for use in courses that focus on the applications of constrained optimization, in courses that emphasize the theory, and in courses that link the subject to economics. A "use's guide" is provided; it takes the form of a brief preview of each of the six Parts that comprise this book.

Acknowledgement

This book's style and content have been shaped by decades of interaction with Yale students. Their insights, reactions and critiques have led me toward a problem-based approach to teaching and writing. With enthusiasm, I acknowledge their contribution. This book also benefits from interactions with my colleagues on the faculty. I am deeply indebted to Uriel G. Rothblum, Kurt Anstreicher, Ludo Van der Heyden, Harvey M. Wagner, Arthur J. Swersey, Herbert E. Scarf and Donald J. Brown, whose influences are evident here.

Contents

Part V – Game Theory

Part VI – Nonlinear Optimization

Part I–Prelude

This book introduces you, the reader, to constrained optimization. This subject consists primarily of linear programs, their generalizations, and their uses. Part I prepares you for what is coming.

Chapter 1. Introduction to Linear Programs

In this chapter, a linear program is described, and a simple linear program is solved graphically. Glimpses are provided of the uses to which linear programs can be put. The limitations that seem to be inherent in linear programs are identified, each with a pointer to the place in this book where it is skirted.

Chapter 2. Spreadsheet Computation

Chapter 2 contains the facets of Excel that are used in this book. Also discussed in Chapter 2 is the software that accompanies this text. All of the information in it is helpful, and some of it is vital.

Chapter 3. Mathematical Preliminaries

Presented in Chapter 3 is the mathematics on which an introductory account of linear programming rests. A familiar method for solving a system of linear equations is described as a sequence of "pivots." An Excel Add-In can be used to execute these pivots.

Chapter 1: Introduction to Linear Programs

1. Preview

The goals of this chapter are to introduce you to linear programming and its generalizations and to preview what's coming. The chapter itself is organized into six main sections:

- In the first of these sections, the terminology that describes linear programs is introduced and a simple linear program is solved graphically.
- In the next section, several limitations of linear programs are discussed, and pointers are provided to places in this book where these limitations are skirted.
- The third section describes optimization problems that seem not to be linear programs, but can be converted into linear programs.
- The fourth section introduces four themes that pervade this book.

E. V. Denardo, *Linear Programming and Generalizations,* International Series in Operations Research & Management Science 149,
DOI 10.1007/978-1-4419-6491-5_1, © Springer Science+Business Media, LLC 2011

- The fifth section introduces the computer codes that are used in this text.
- The sixth section consists of a brief account of the origins of the field.

Linear programming and its generalizations is a broad subject. It has a wide variety of uses. It has links to several academic fields. It is united by themes that are introduced here and are developed in later chapters.

2. An Example

A "linear program" is a disarmingly simple object. Its definition entails the terms, "linear expression" and "linear constraint." A **linear expression** appears below; its variables are x, y and z, and the dependence of this expression on x, y and z is linear.

$$3x - 2.5y + \sqrt{2}z$$

A **linear constraint** requires a linear inequality to take any one of the three forms that are illustrated below:

$$\begin{aligned} 3x - 2.5y + \sqrt{2}z &\leq 6, \\ 2x - 5y + z &= 3, \\ x &\geq 0. \end{aligned}$$

In other words, a linear constraint requires a linear expression to be less than or equal to a number, to be equal to a number, or to be greater than or equal to a number. The linear constraint $x \geq 0$ requires the number x to be nonnegative, for instance.

A **linear program** either maximizes or minimizes a linear expression subject to finitely many linear constraints. An example of a linear program is:

Program 1.1. $z^* = \text{Maximize } \{2x + 2y\}$ subject to the constraints

$$\begin{aligned} 1x + 2y &\leq 4, \\ 3x + 2y &\leq 6, \\ x \qquad &\geq 0, \\ y &\geq 0. \end{aligned}$$

The **decision variables** in a linear program are the quantities whose values are to be determined. Program 1.1 has two decision variables, which are x and y. Program 1.1 has four constraints, each of which is a linear inequality.

A big deal?

A linear program seems rather simple. Can something this simple be important? Yes! Listed below are three reasons why this is so.

- A staggeringly diverse array of problems can be posed as linear programs.
- A family of algorithms that are known as the simplex method solves nearly all linear programs with blinding speed.
- The ideas that underlie the simplex method generalize readily to situations that are far from linear and to settings that entail several decision makers, rather than one.

Linear programming describes the family of mathematical tools that are used to analyze linear programs. In tandem with the digital computer, linear programming has made mathematics vastly more useful. Linear programming also provides insight into a number of academic disciplines, which include mathematics, economics, computer science, engineering, and operations research. These insights are glimpsed in this chapter and are developed in later chapters.

Feasible solutions

Like any field, linear programming has its own specialized terminology (jargon). Most of these terms are easy to remember because they are suggested by normal English usage. A **feasible solution** to a linear program is a set of values of its decision variables that satisfies each of its constraints. Program 1.1 has many feasible solutions, one of which $x=1$ and $y=0$. The **feasible region** of a linear program is its set of feasible solutions. Program 1.1 has only two decision variables, so its feasible region can be represented on the plane. Figure 1.1 does so.

Figure 1.1. Feasible region for Program 1.1.

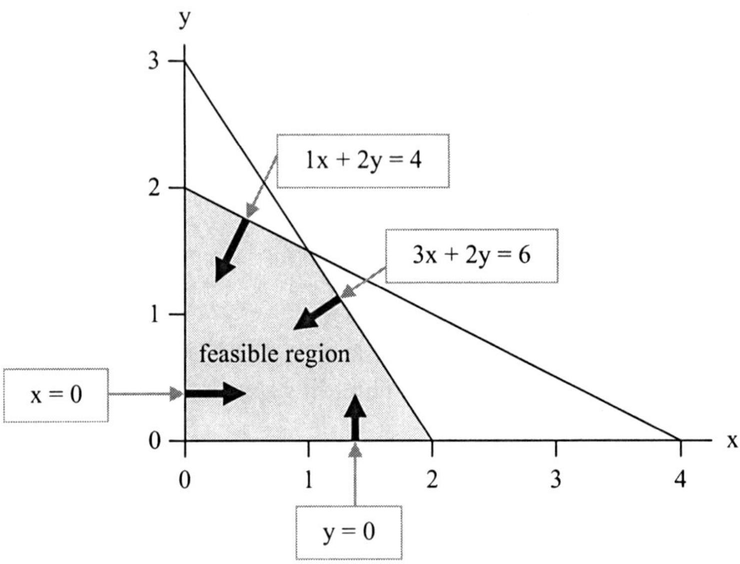

Figure 1.1 is easy to construct because the pairs (x, y) that satisfy a particular linear constraint form a "half-plane" whose boundary is the line on which this constraint holds as an equation. For example:

- The constraint $1x + 2y \leq 4$ is satisfied as an equation by the pairs (x, y) on the line $1x + 2y = 4$.
- Two points determine a line, and the line $1x + 2y = 4$ includes the points (pairs) (0, 2) and (4, 0).
- Since (0, 0) satisfies the constraint $1x + 2y \leq 4$ as a strict inequality, this constraint is satisfied by the half plane in which (0, 0) lies.
- In Figure 1.1, a thick arrow points from the line $1x + 2y = 4$ into the half-plane that satisfies the inequality $1x + 2y \leq 4$.

The feasible region for Program 1.1 is the intersection of four half-planes, one per constraint. In Figure 1.1, the feasible region is the area into which the thick arrows point, and it is shaded.

Optimal solutions

Each feasible solution assigns an **objective value** to the quantity that is being maximized or minimized. The feasible solution x=1, y=0 has 2 as its objective value, for instance. An **optimal solution** to a linear program is a feasible solution whose objective value is largest in the case of a maximization problem, smallest in the case of a minimization problem. The **optimal value** of a linear program is the objective value of an optimal solution to it.

An optimal solution to Program 1.1 is x=1 and y=1.5, and its optimal value $z^* = 2x + 2y = (2)(1) + (2)(1.5) = 5$. To convince yourself that this is the optimal solution to Program 1.1, consider Figure 1.2. It augments Figure 1.1 by including two "iso-profit" lines, each of which is dashed. One of these lines contains the points (x, y) whose objective value equals 4, the other contains the pairs (x, y) whose objective value that equals 5. It is clear, visually, that the unique optimal solution to Program 1.1 has x=1 and y=1.5.

Figure 1.2. Feasible region for Program 1.1, with two iso-profit lines.

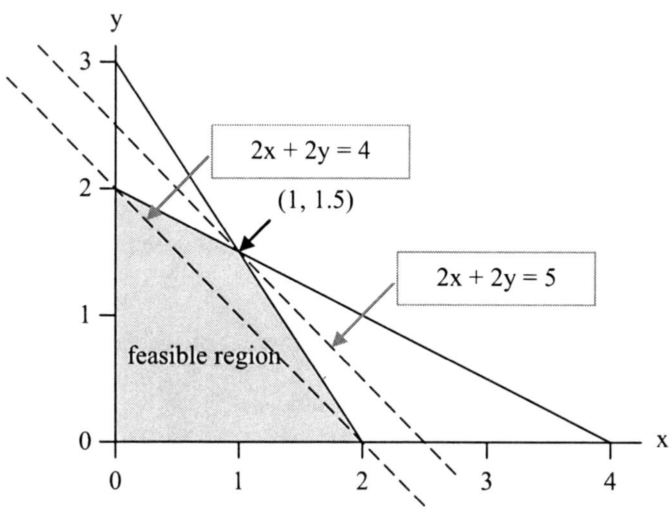

A linear program can have only one optimal value, but it can have more than one optimal solution. If the objective of Program 1.1 were to maximize (x + 2y), its optimal value would be 4, and every point on the line segment connecting (0, 2) and (1, 1.5) would be optimal.

A taxonomy

Linear programs divide themselves into categories. A linear program is **feasible** if it has at least one feasible solution, and it is said to be **infeasible** if it has no feasible solution. Program 1.1 is feasible, but it would become infeasible if the constraint $x + y \geq 3$ were added to it. Infeasible linear programs do arise in practice. They model situations that are so tightly restricted as to have no solution.

A linear program is said to be **unbounded** if it is feasible and if the objective value of its feasible solutions can be improved without limit. An example of an unbounded linear program is:

Max $\{x\}$, subject to $x \geq 2$.

An unbounded linear program is almost invariably a signal of an incorrect formulation: it is virtually never possible to obtain an infinite amount of anything that is worthwhile.

A linear program is **feasible and bounded** if it is feasible and if its objective cannot be improved without limit. Highlighted below is a property of linear programs that are feasible and bounded:

> Each linear program that is feasible and bounded has at least one optimal solution.

This property is not quite self-evident. It should be proved. The simplex method will provide a proof.

Each linear program falls into one of these three categories:

- The linear program may be infeasible.
- It may be feasible and bounded.
- It may be unbounded.

To **solve** a linear program is to determine which of these three categories it lies in and, if it is feasible and bounded, to find an optimal solution to it.

Bounded feasible regions

A linear program is said to have a **bounded feasible region** if some number K exists such that each feasible solution equates every decision variable to a number whose absolute value does not exceed K. Program 1.1 has a bounded feasible region because each feasible solution equates each decision variable to a number between 0 and 2.

If a linear program is unbounded, it must have an unbounded feasible region. The converse is not true, however. A linear program that has an unbounded feasible region can be feasible and bounded. To see that this is so, consider Program 1.2.

Program 1.2. $z^* =$ Minimize $\{4u + 6v\}$ subject to the constraints

$$\begin{aligned} 1u + 3v &\geq 2, \\ 2u + 2v &\geq 2, \\ u \quad &\geq 0, \\ v &\geq 0. \end{aligned}$$

Figure 1.3 plots the feasible region for Program 1.2. This feasible region is clearly unbounded. Program 1.2 is bounded, nonetheless; every feasible solution has objective value that exceeds 0.

Figure 1.3. Feasible region for Program 1.2.

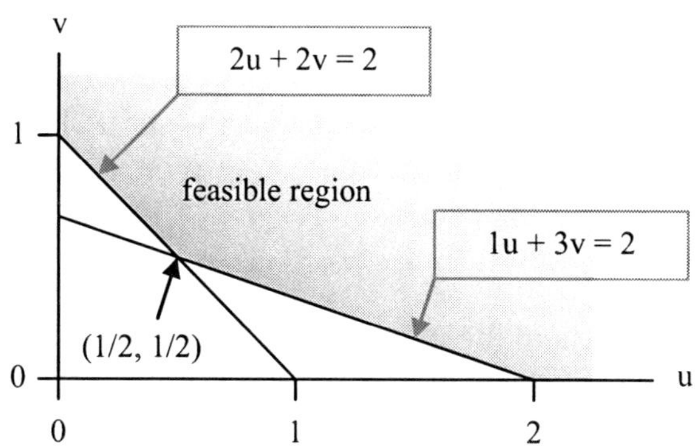

You might suspect that unbounded feasible reasons do not arise in practice, but that is not quite accurate. In a later chapter, we'll see that every linear program is paired with another, which is called its "dual." We will see that if a linear program is feasible and bounded, then so is its dual, in which case both linear programs have the same optimal value, and at least one of them has an unbounded feasible region. Programs 1.1 and 1.2 are each other's duals, by the way. One of their feasible regions is unbounded, as must be.

3. Generalizations

A linear program is an optimization problem that fits a particular format: A linear expression is maximized or minimized subject to finitely many linear constraints. Discussed in this section are the limitations imposed by this format, along with the parts of this book where most of them are circumvented.

Constraints that hold strictly

A linear program requires each constraints to take one of three forms; a linear expression can be "≥" a number, it can be "=" a number, or it can be "≤" a number. Strict inequalities are *not allowed.* One reason why is illustrated by this optimization problem:

Minimize $\{3y\}$, subject to $y>2$.

This problem does not have an optimal solution. The "infimum" of its objective equals 6, and setting y slightly above 2 comes "close" to 6, but an objective value of 6 is not achievable. Ruling out strict inequalities eliminates this difficulty.

On the other hand, the simplex method can – and will – be used to find solutions to linear systems that include one or more strict inequalities. To illustrate, suppose a feasible solution to Program 1.1 is sought for which the variables x and y are positive. To construct one, use the linear program:

Maximize θ, subject to the constraints of Program 1.1 and

$$\theta \le x,$$

$$\theta \le y.$$

In Chapter 12, strict inequalities emerge in a second way, as a facet of a subject called "duality."

Integer-valued variables

A linear program lets us impose constraints that require the decision variable x to lie between 0 and 1, inclusive. On the other hand, linear programs *do not* allow us to impose the constraint that restrict a decision variable x to the values 0 and 1. This would seem to be a major restriction. Lots of entities (people, airplanes, and so forth) are integer-valued.

An **integer program** is an optimization problem that would become a linear program if we suppressed the requirement that its decision variables be integer-valued. The simplex method is so fast that it is used in as a subroutine in algorithms that solve integer programs. How that occurs is described in Chapter 13.

In addition, an important class of integer programs can be solved by a single application of the simplex method. That's because applying the simplex method to these integer programs can be guaranteed to produce an optimal solution that is integer-valued. These integer programs are "network flow" models whose data are "integer-valued." They are studied in Chapter 9.

Competition

A linear program models a situation in which a single decision maker strives to select the course of action that maximizes the benefit received. At first glance, the subject seems to have nothing to do with game theory, that is, with models of situations in which multiple decision makers can elect to cooperate or compete. But it does! Chapters 14, 15 and 16 of this book adapt the ideas and algorithms of linear programming to models of competitive behavior.

Non-linear functions

Linear programs require that the objective and constraints have a particular form, that they be linear. A **nonlinear program** is an optimization problem whose objective and/or constraints are described by functions that fail to be linear. The ideas used to solve linear programs generalize to handle a variety of nonlinear programs. How that occurs is probed in Chapter 20.

4. Linearization

Surveyed in this section are some optimization problems that do not present themselves as linear programs but that can be converted into linear programs.

A "maximin" objective

Suppose we wish to find a solution to a set of linear constraints that maximizes the smaller of two measures of benefit, for instance, to solve:

Program 1.3. $z^* =$ Maximize the smaller of $(2x+2y)$ and $(x-3y)$, subject to

$$\begin{aligned} 1x + 2y &\le 4, \\ 3x + 2y &\le 6, \\ x \qquad &\ge 0, \\ y &\ge 0. \end{aligned}$$

The object of Program 1.3 is to maximize the smaller of two linear expressions. This is not a linear program because its objective is not a linear expression. To convert Program 1.3 into an equivalent linear program, we maximize the quantity t subject to constraints that keep t from exceeding the linear expressions $(2x+2y)$ and $(x-3y)$. In other words, we replace Program 1.3 by

Program 1.3′. $z^* =$ Maximize $\{t\}$, subject to

$$\begin{aligned} t &\le 2x + 2y, \\ t &\le 1x - 3y, \\ 1x + 2y &\le 4, \\ 3x + 2y &\le 6, \\ x \qquad &\ge 0, \\ y &\ge 0. \end{aligned}$$

Program 1.3′ picks the feasible solution to Program 1.3 that maximizes the smaller of the linear expressions $2x+2y$ and $1x-3y$, exactly as desired.

A "minimax" objective

Suppose we wish to find a solution to a set of linear constraints that minimizes the larger of two linear expressions, e.g., that minimizes the larger of

$(2x + 2y)$ and $(x - 3y)$, subject to the constraints of Program 1.3. The same trick works, as is suggested by:

Program 1.4. Minimize {t}, subject to

$$t \geq 2x + 2y,$$

$$t \geq 1x - 3y,$$

and the constraints of Program 1.3.

Evidently, it is easy to recast a "maximin" or a "minimax" objective in the format of a linear program. This conversion enhances the utility of linear programs. Its role in John von Neumann's celebrated minimax theorem is discussed in Chapter 14.

"Maximax" and "minimin" objectives?

Suppose we seek to maximize the *larger* of the linear expressions $(2x + 2y)$ and $(1x - 3y)$, subject to the constraints of Program 1.3. It does *not* suffice to maximize {t}, subject to the original constraints and $t \geq 2x + 2y$ and $t \geq 1x - 3y$. This linear program is unbounded; t can be made arbitrarily large. For the same reason, we cannot use a linear program to minimize the smaller of two linear expressions.

The problem of maximizing the larger of two or more linear expressions can be posed as an integer program, as can the problem of minimizing the smaller of two or and more linear expressions. How to do this will be illustrated in Chapter 7.

Decreasing marginal benefit

A linear program seems to require that the objective vary linearly with the level of a decision variable. In Program 1.1, the objective is to maximize the linear expression, $2x + 2y$. Let us replace the addend 2y in this objective by the (nonlinear) function p(y) that is exhibited in Figure 1.4. This function illustrates the case of **decreasing marginal benefit**, in which the (profit) function p(y) has slope that decreases as the quantity y increases.

Figure 1.4. A function p(y) that illustrates decreasing marginal profit.

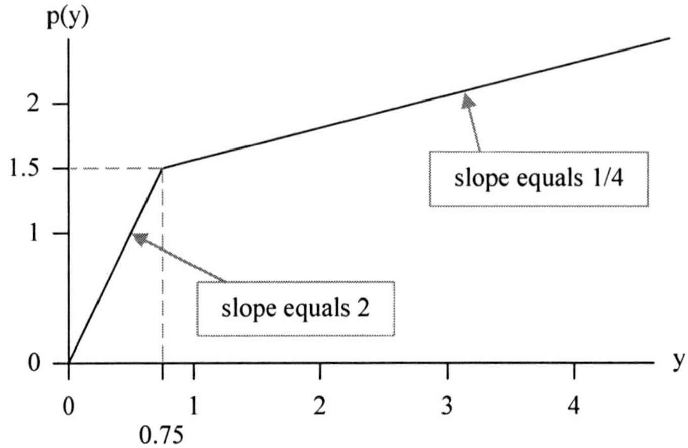

Decreasing marginal benefit occurs when production above a certain level requires extra expense, for instance, by the use of overtime labor. The profit function p(y) in Figure 1.4 can be accommodated by introducing two new decision variables, y_1 and y_2, along with the constraints

$$y_1 \geq 0, \qquad y_1 \leq 0.75, \qquad y_2 \geq 0, \qquad y = y_1 + y_2,$$

and replacing the addend 2y in the objective by $(2y_1 + 0.25y_2)$. This results in:

Program 1.5. $z^* = \text{Maximize } \{2x + 2y_1 + 0.25y_2\}$ subject to the constraints

$$1x + 2y \leq 4,$$
$$3x + 2y \leq 6,$$
$$y = y_1 + y_2,$$
$$y_1 \leq 0.75,$$
$$x \geq 0, \quad y \geq 0, \quad y_1 \geq 0, \quad y_2 \geq 0.$$

To verify that Program 1.5 accounts correctly for the profit function p(y) in Figure 1.4, we consider two cases. First, if the total quantity y does not exceed 0.75, it is optimal to set $y_1 = y$ and $y_2 = 0$. Second, if the total quantity y does exceed 0.75, it is optimal to set $y_1 = 0.75$ and $y_2 = y - 0.75$.

An unintended option

Program 1.5 is a bit more subtle than it might seem. Its constraints allow an unintended option, which is to set $y_2 > 0$ while $y_1 < 1$. This option is ruled out by optimization, however. In this case and in general:

> Linear models of decreasing marginal benefit introduce unintended options that are ruled out by optimization.

The point is that a linear program will not engage in a more costly way to do something if a less expensive method of doing the same thing is available.

Increasing marginal cost

Net profit is the negative of net cost: A net profit of \$6.29 is identical to a net cost of −\$6.29, for instance. Maximizing net profit is precisely equivalent to minimizing net cost. Because of this, the same trick that handles the case of decreasing marginal profit also handles the case of increasing marginal cost. One or more unintended options are introduced, but they are ruled out by optimization. Again, the more costly way of doing something is avoided.

Increasing marginal return?

A profit function exhibits **increasing marginal return** if its slope increases with quantity. One such function is exhibited in Figure 1.5. Its slope equals 1/2 for quantities below 1 and equals 2 for quantities above 1.

Figure 1.5. A profit function that exhibits increasing marginal return.

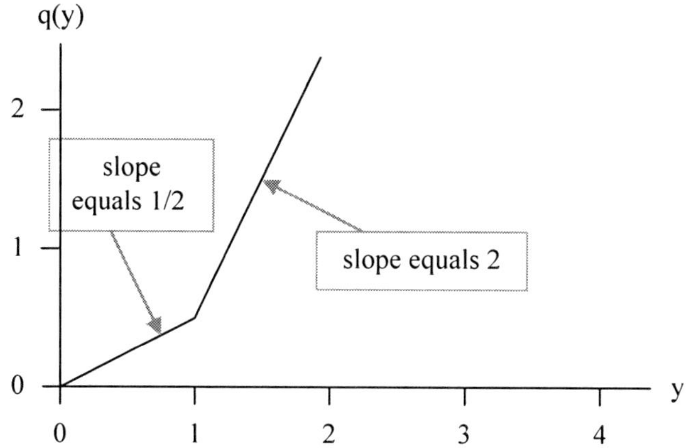

Let us turn our attention to the variant of Program 1.1 whose object is to maximize the nonlinear expression, $\{2x + q(y)\}$. Proceeding as before would lead to:

Program 1.6. $z^* = \text{Maximize } \{2x + 0.5y_1 + 2y_2\}$ subject to the constraints

$$1x + 2y \le 4,$$
$$3x + 2y \le 6,$$
$$y = y_1 + y_2,$$
$$y_1 \le 1,$$
$$x \ge 0, \quad y \ge 0, \quad y_1 \ge 0, \quad y_2 \ge 0.$$

Program 1.6 introduces an unintended option, which is to set y_2 positive while y_1 is below 1, and this option is *selected* by optimization. Indeed, in Program 1.6, it cannot be optimal to set y_1 positive. Given the option, the linear program chooses the more profitable way to do something. In this case and in general:

Linear models of increasing marginal return introduce unintended options that are *selected* by optimization.

Increasing marginal return (equivalently, decreasing marginal cost) cannot be handled by a linear program. It requires an integer program. Chapter 7 includes a discussion of how to use binary variables (whose values are either 0 or 1) to handle increasing marginal return.

Absolute value in the performance measure

Our attention now turns to an optimization whose constraints are linear but whose objective weighs the absolute value of one or more linear expressions. To illustrate, let a and b be fixed *positive* numbers, and consider:

Program 1.7. Minimize $\{a|x - 1| + b|y - 2|\}$, subject to the constraints of Program 1.1.

Program 1.7 is easily converted into an equivalent linear program. To do so, we introduce two new decision variables, t and u, and consider:

Program 1.7′. Minimize $\{at + bu\}$, subject to the constraints of Program 1.1 and

(1) $$x - 1 \le t, \quad -t \le x - 1, \quad y - 2 \le u, \quad -u \le y - 2.$$

The decision variables t and u appear in no constraints other than (1). To see what value is taken by the decision variable t, we consider two cases:

- If x exceeds 1, the first two constraints in (1) are satisfied by any value of t that has $t \geq x - 1$, and the fact that a is positive guarantees that the objective is minimized by setting $t = x - 1$.
- If 1 exceeds x, the first two constraints in (1) are satisfied by any value of t that has $t \geq 1 - x$, and the fact that a is positive guarantees that the objective is minimized by setting $t = 1 - x$.

A similar observation applies to y. Programs 1.7 and 1.7′ have the same optimal value, and the optimal solution to Program 1.7′ specifies values of x and y that are optimal for Program 1.7.

An alternative to least-squares regression

To illustrate a least-squares regression model, we again let a and b be fixed *positive* numbers (data), and consider:

Program 1.8. Minimize $\{a(x - 1)^2 + b(y - 2)^2\}$, subject to the constraints in Program 1.1.

By squaring the difference, these models place higher weights on observations that are further from the norm, e.g., on outliers. Should you wish to weigh the observations proportionally to their distance from the norm, substitute the criterion

$$\{a|x - 1| + b|y - 2|\},$$

and convert the model to a linear program, exactly as was done for Program 1.7.

An alternative to variance minimization

The justly-famous Markowitz model of portfolio theory allocates a budget among investments so as to minimize the variance of the return, subject to the constraint that the expectation of the return is at least as large as some preset value. This optimization takes the form of Program 1.8 with a and b being nonnegative numbers that sum to 1. This model is an easily-solved nonlinear program.

On the other hand, the variance *squares* the difference between the outcome and its expectation, and it weighs upside and downside differences equally. Substituting a "mean absolute deviation" for the variance produces a linear program that may make better sense. Also, removing two of the constraints in (1) minimizes the expected *downside* variability, which might make still better sense.

Constraints on ratios

A **ratio constraint** places an upper bound or a lower bound on the ratio of two linear expressions. To illustrate, we append to Program 1.1 the ratio constraint.

$$\frac{x}{y} \le 0.8.$$

This constraint is not linear, so it cannot be part of a linear program.

But the other constraints in Program 1.1 guarantee $y \ge 0$, and multiplying an inequality by a nonnegative number preserves its sense. In particular, multiplying the ratio constraint that is displayed above by the nonnegative number y produces the linear constraint

$$x \le 0.8y.$$

This conversion must be qualified, slightly, because ratios are not defined when their denominators equal zero. *If* the constraint $x/y \le 0.8$ is intended to mean that x cannot be positive when $y = 0$, it is equivalent to $x \le 0.8y$. In general:

> Multiplying a ratio constraint by its denominator converts it to a linear constraint *if* its denominator can be guaranteed to be positive.

If the denominator of a ratio is guaranteed to be nonnegative (rather than positive), one needs to take care when it equals zero, as is suggested above.

Optimizing a ratio*

The next three subsections concern a linear program whose objective function is a ratio of linear expressions. These subsections are starred. They cover a specialized topic that can be skipped or deferred with no loss of continuity. Readers who are facile with matrix notation may wish to read them now, however.

Program 1.9, below, maximizes the ratio of two linear expressions, subject to linear constraints. Its data form the $m \times n$ matrix A, the $m \times 1$ vector b, the $1 \times n$ vector c and the $1 \times n$ vector d. Its decision variables are the entries in the $n \times 1$ vector x.

Program 1.9. $z^* = \text{Maximize} \left\{ \frac{cx}{dx} \right\}$, subject to the constraints

$$(2) \qquad Ax = b, \quad x \geq 0.$$

Program 1.9 will be analyzed under

Hypothesis A:

1. Every vector x that satisfies $Ax=b$ and $x \geq 0$ has $dx>0$.

2. No vector x satisfies $Ax=0$, $x \geq 0$ and $dx>0$.

It was A. Charnes and W. W. Cooper who showed that Program 1.9, which they dubbed a **linear fractional program**, can be converted into an equivalent linear program[1].

Interpretation of Hypothesis A*

Before converting Program 1.9 into an equivalent linear program, we pause to ask ourselves: How can we tell whether or not a particular model satisfies Hypothesis A? A characterization of Hypothesis A appears below:

> Hypothesis A is satisfied if and only if there exist positive numbers L and U such that every feasible solution x to Program 1.9 has
>
> $$L \leq dx \leq U.$$

In applications, it is often evident that every vector x that satisfies (2) assigns a value to d x that is bounded away from 0 and from $+\infty$.

As a point of logic, we can demonstrate that Program 1.9 is equivalent to a linear program without verifying the characterization of Hypothesis A that is highlighted above. And we shall.

[1] A. Charnes and W. W. Cooper, "Programming with linear fractional functionals," *Naval Research Logistics Quarterly*, V. 9, pp.181-186, 1962.

A change of variables*

A change of variables will convert Program 1.9 into an equivalent linear program. The decision variables in this equivalent linear program are the number t and the $n \times 1$ vector $\hat{x}$ that will be related to x via

$$(3) \qquad t = \frac{1}{dx} \qquad \text{and} \qquad \hat{x} = xt.$$

This change of variables converts the objective of Program 1.9 to $cxt = c\hat{x}$. Also, multiplying the constraints in (2) by the positive number t produces the constraints $A\hat{x} = bt$ and $\hat{x} \geq 0$ that appear in:

Program 1.10. $z^* =$ Maximize $c\hat{x}$, subject to

$$(4) \qquad A\hat{x} = bt, \quad d\hat{x} = 1, \quad \hat{x} \geq 0, \quad t \geq 0.$$

Programs 9 and 10 have the same data, namely, the matrix A and the vectors b, c and d. They have different decision variables. Feasible solutions to these two optimization problems are related to each other by:

Proposition 1.1. Suppose Hypothesis A is satisfied. Equation (3) relates each solution x to (2) to a solution $(\hat{x}, t)$ to (4), and conversely, with objective values

$$(5) \qquad \frac{cx}{dx} = c\hat{x}.$$

Proof. First, consider any solution x to (2). Part 1 of Hypothesis A guarantees that dx is positive, hence that t, as defined by (3), is positive. Thus, $1 = dxt = d\hat{x}$. Also, multiplying (2) by t and using $\hat{x} = xt$ verifies that $A\hat{x} = bt$ and that $\hat{x} \geq 0$, so (4) is satisfied. In addition, $(cx)/(dx) = (cx)t = c(xt) = c\hat{x}$, so (5) is satisfied.

Next, consider any solution $(\hat{x}, t)$ to (4). Part 2 of Hypothesis A guarantees $t>0$. This allows us to define x by $x = \hat{x}/t$. Dividing $A\hat{x} = bt$ and $\hat{x} \geq 0$ by the positive number t verifies (2). Also, since $x = \hat{x}/t$ and $d\,\hat{x} = 1$, we have

$$\frac{cx}{dx} = \frac{cx}{d\hat{x}} \times t = \frac{cx}{1} \times t = c\hat{x},$$

which completes the proof. ∎

Proposition 1.1 shows how every feasible solution to Program 1.9 corresponds to a feasible solution to Program 1.10 that has the same objective value. Thus, rather than solving Program 1.9 (which is nonlinear), we can solve Program 1.10 (which is linear) and use (3) to construct an optimal solution to Program 1.9.

5. Themes

Discussed in this section are several themes that are developed in later chapters. These themes are:

- The central role played by the simplex pivot.
- The contributions made by linear programming to mathematics.
- The insights provided by linear programming into economics.
- The broad array of situations that can be modeled and solved as linear programs and their generalizations.

A subsection is devoted to each theme.

Pivoting

At the heart of nearly every software package that solves linear programs lies the simplex method. The simplex method was devised by George B. Dantzig in 1947. An enormous number of person-years have been invested in attempts to improve on the simplex method. Algorithms that compete with it in specialized situations have been devised, but nothing beats it for general-purpose use, especially when integer-valued solutions are sought. Dantzig's simplex method remains the best general-purpose solver six decades after he proposed it.

At the core of the simplex method lies the *pivot*, which plays a central role in Gauss-Jordan elimination. In Chapter 3, we will see how Gauss-Jordan elimination pivots in search of a solution to a system of linear equations. In Chapter 4, we will see that the simplex method keeps on pivoting, in search of an optimal solution to a linear program. In Chapter 15, we'll see how a slightly different pivot rule (called *complementary pivoting*) finds the solution to a non-zero sum matrix game. And in Chapter 16, we'll see how complementary pivots find an approximation to a Brouwer fixed-point. That's a remarkable progression – variants of a simple idea solve a system of linear equations, a linear program, and a fixed-point equation.

The simplex method presents a dilemma for theoreticians. It is the best general-purpose solver of linear programs, but its worst-case behavior is abysmal. It solves practical problems with blazing speed, *but* there exist classes of specially-constructed linear programs for which the number of pivots required by the simplex method grows exponentially with the size of the problem. Many researchers have attempted to explain why these "bad" problems do not arrive in practice. Chapter 4 includes a thumb-nail discussion of that issue.

Impact on mathematics

The analysis of linear programs and their generalizations have had a profound impact on mathematics. Three facets of this impact are noted here.

People, commodities and a great many other items exist in nonnegative quantities. But, prior to the development of linear programming, linear algebra was nearly bereft of results concerning inequalities. The simplex method changed that. Linear algebra is now rife with results that concern inequalities. Some of these results appear in Chapter 12. Additionally, the simplex method is the main technique for solving linear systems some of whose decision variables are required to be nonnegative.

The simplex method actually solves a pair of linear programs – the one under attack *and* its dual. That it does so is an important – and largely unanticipated – facet of linear algebra whose implications are probed in Chapter 12. Duality is an important addition to the mathematician's tool kit; it facilitates the proof of many theorems, as is evident in nearly every issue of the journals, *Mathematics of Operations Research* and *Mathematical Programming.*

Finally, as noted above, a generalization of the simplex pivot computes approximate solutions to Brouwer's fixed-point equation, thereby making a deep contribution to nonlinear mathematics.

The overarching impact of linear programming on mathematics may have been to emphasize the value of problem-based research.

Economic reasoning

This book includes several insights provided by linear programming and its generalizations into economic reasoning. Two such insights are noted here.

To prepare for a discussion of one of these insights, it is observed that nearly every list of the most important concepts in economic reasoning includes at least two of the following:

- The break-even price (a.k.a. shadow price) of a scarce resource.
- The opportunity cost of engaging in an activity.
- The importance of thinking at the margin, of assessing the incremental benefit of doing something.

Curiously, throughout much of the economics literature, no clear link is drawn between these three concepts. It will be seen in Chapter 5 that these concepts are intimately related *if* one substitutes for opportunity cost the notion of *relative opportunity cost* of doing something, this being the reduction in benefit due to setting aside the resources needed to do that thing.

Within economics, these three concepts are usually described in the context of an *optimal* allocation of resources. In Chapter 12, however, it will be seen that these three concepts apply to each step of the simplex method, that it uses them to pivot from one "basis" to another as it seeks an optimal solution.

It was mentioned earlier that every linear program is paired with another, in particular, that Programs 1.1 and 1.2 are each duals. This duality provides economic insight at several different levels. Three illustrations of its impact are listed below.

- In Chapter 5, a duality between production quantities and prices is established: Specifically, the dual of the problem of producing so as to make the most profitable use of a bundle of resources is the problem of setting least costly prices on those resources such that no activity earns an "excess profit."
- In Chapter 14, duality is used to construct a general equilibrium for a stylized model of an economy. One linear program in this pair sets production and consumption quantities that maximize the consumer's

well-being while requiring the market for each good to "clear." The dual linear program sets prices that maximize the producers' profits. Their optimal solutions satisfy the consumer's budget constraint, thereby constructing a general equilibrium.

- In Chapter 14, duality is also seen to be a simple and natural way in which to analyze and solve von Neumann's celebrated matrix game.

Linear programming also provides a number of insights into financial economics.

Areas of application

Several chapters of this book are devoted to the situations that can be modeled as linear programs and their generalizations.

- Models of the allocation of scarce resources are surveyed in Chapter 7.
- Network-based optimization problems are the subject of Chapters 8 and 9.
- Applications that entail strict inequalities are discussed in Chapter 12
- Methods for solving integer programs are included in Chapter 13.
- Models of competitive behavior are studied in Chapters 14-16.
- Optimality conditions for nonlinear programs are presented in Chapter 20.

The applications in the above list are of a linear program, without regard to its dual. Models of competition can be analyzed by a linear program *and* its dual. These include the aforementioned model of an economy in general equilibrium.

6. Software

An enormous number of different software packages have been constructed that solve linear programs and their generalizations. Many of these packages are available for classroom use, either at nominal charge or at no charge. Each package has advantages and disadvantages. Several of them dovetail nicely with spreadsheet computation. You – and your instructor – have a choice. You may find it convenient to use any of a variety of software packages.

One choice

To keep the exposition simple, this book is keyed to a pair of software packages. They two are:

- *Solver*, which comes with Excel. The original version of Solver was written by Frontline Systems. Solver is now maintained by Microsoft.
- *Premium Solver*, which is written and distributed by Frontline Systems. An educational version of Premium Solver is available, free of charge.

These packages are introduced in Chapter 2, and their uses are elaborated upon in subsequent chapters. These packages (and many others) have user interfaces that are amazingly user-friendly.

Large problems

Solver and Premium Solver for Education can handle all of the linear and nonlinear optimization problems that appear in this text. These codes fail on problems that are "really big" or "really messy" – those with a great many variables, with a great many constraints, with a large number of integer-valued variables, or with nonlinear functions that are not differentiable. For big problems, you will need to switch to one of the many commercially available packages, and you may need to consult an expert.

7. The Beginnings

Presented in this section is a brief account of the genesis of linear programming. It began just before World War II in the U.S.S.R and just after World War II in the United States.

Leoinid V. Kantorovich

In Leningrad (now St. Petersburg) in 1939, a gifted mathematician and economist named L. V. Kantorovich (1912-1986) published a monograph on the best way to plan for production[2]. This monograph included a linear program, and it recognized the importance of duality, but it seemed to omit a systematic method of solution. In 1942, Kantorovich published a paper that included a complete description of a network flow problem, including duality, again without a systematic solution method[3].

For the next twenty years, Kantorovich's work went unnoticed in the West. Nor was it applauded within the U. S. S. R., where planning was centralized and break-even prices were anathema. It was eventually recognized that Kantorovich was the first to explore linear programming and that he probed it deeply. Leonid V. Kantorovich richly deserved his share of the 1975 Nobel Prize in Economics, awarded for work on the optimal allocation of resources.

George B. Dantzig

George B. Dantzig (1914-2005) spent the years 1941 to 1945 in Washington, D.C., working on planning problems for the Air Force. To understand why this might be excellent preparation for the invention of linear programming, contemplate even a simple planning problem, such as organizing the activities needed to produce parachutes at the rate of 5,000 per month.

After war's end, Dantzig returned to Berkeley for a few months to complete his Ph. D. degree. By the summer of 1946, Dantzig was back in Washington as the lead mathematician in a group whose assignment was to mechanize the planning problems faced by the Air Force. By the spring of 1947, Dantzig had observed that a variety of Air Force planning problems could be posed as linear programs. By the summer of 1947 he had developed the simplex method. These and a string of subsequent accomplishments have cemented his stature as the preeminent figure in linear programming.

Tjalling C. Koopmans

Tjalling C. Koopmans (1910-1985) developed an interest in economics while earning a Ph. D. in theoretical physics from the University of Leyden. In 1940, he immigrated to the United States with his wife and six-week old

---

[2] Kantorovich, L. V., *The mathematical method of production planning and organization*, Leningrad University Press, Leningrad, 1939. Translated in *Management Science*, V. 6, pp. 366-422, 1960.

[3] Kantorovich, L. V., "On the translocation of masses," *Dokl. Akad. SSSR*, V. 37, pp. 227–229.

daughter. During the war, while serving as a statistician for the British Merchant Shipping Mission in Washington, D.C., he built a model of optimal routing of ships, with the attendant shadow costs. Koopmans shared the 1975 Nobel Prize in economics with Kantorovich for his contributions to the optimal allocation of resources.

An historic conference

A conference on activity analysis was held from June 20-24, 1949, at the Cowles Foundation, then located at the University of Chicago. This conference was organized by Tjalling Koopmans, who had become very excited about the potential for linear programming during a visit by Dantzig in the spring of 1947. The volume that emerged from this conference was the first published compendium of results related to linear programming[4]. The participants in this conference included six future Nobel Laureates (Kenneth Arrow, Robert Dorfman, Tjalling Koopmans, Paul Samuelson, Herbert Simon and Robert Solow) and five future winners of the von Neumann Theory Prize in Operations Research (George Dantzig, David Gale, Harold Kuhn, Herbert Simon and Albert Tucker).

Military applications and the digital computer

Dantzig's simplex method made possible the solution of a host of industrial and military planning problems – in theory. Solving these problems called for vastly more computational power than could be achieved by scores of operators of desk calculators. It was an impetus for the development of the digital computer.

With amazing foresight, the Air Force organized Project SCOOP (scientific computation of optimal programs) and funded the development and acquisition of digital computers that could implement the simplex method. These computers included:

- The SEAC (short for Standards Eastern Automatic Computer), which, in 1951, solved a 48-equation 71-variable linear program in 18 hours.
- UNIVAC I, installed in 1952, which solved linear programs as large as 250 equations and 500 variables.

---

[4] *Activity analysis of production and allocation: Proceedings of a conference*, Tjalling C. Koopmans, ed., John Wiley & Sons, New York, 1951.

It is difficult for a person who is not elderly to appreciate what clunkers these early computers were – how hard it was to get them to do anything. But Moore's law may help: If computer power doubles every two years, accomplishing anything was more difficult by a factor of *one billion* (roughly $2^{60/2}$) sixty years ago.

Industrial applications

In a characteristically gracious memoir, William W. Cooper discussed the atmosphere in the early days[5]. In the late 1940s at Carnegie Institute of Technology (now Carnegie Mellon University), a group that he directed wrestled with the efficient blending of aviation fuels. Cooper describes the extant state of linear programming as "embryonic … no publications were available." He reports that his group's attempt to adapt activity analysis to the blending problem was "fraught with difficulties." He acknowledges failing to recognize fully the significance of Dantzig's work. His group quickly produced two seminal papers, one on blending aviation fuels,[6] another on the resolution of degeneracy.[7]

In the same memoir, Cooper recounted his surprise at the response to these papers. A large number of firms contacted him to express an eagerness to learn more about these new methods for planning and control of their operations. Within the oil industry, he received inquiries from the Soviet Bloc. The oil industry would quickly become a major user of linear programming and its generalizations.

8. Review

This chapter is designed to introduce you to linear programming and to provide you with a feel for what is coming.

---

[5] Cooper. W. W., "Abraham Charnes and W. W. Cooper (et al): A brief history of a long collaboration in developing industrial uses of linear programming," *Operations Research*, V. 50, pp. 35-41.

[6] Charnes, A., W. W. Cooper and B. Mellon, "Blending aviation gasolines – a study of programming interdependent activities in an integrated oil company, *Econometrica*, V. 20, pp 135-159, 1952.

[7] Charnes, A., "Optimality and degeneracy in linear programming," *Econometrica*, V. 20, pp 160-170, 1952.

Terminology

The terminology that appears in Section 2 is used throughout this book, indeed, throughout the literature on linear programming and its generalizations. Before proceeding, you should be familiar with each of the terms that appear in boldface in that section – linear constraint, linear program, feasible solution, objective value, infeasible linear program, and so forth.

Utility

It is hoped that you now have a feel for the value of studying linear programming and its generalizations. Within this chapter, it has been observed that:

- The basic model is flexible – some optimization problems that appear to be nonlinear can be converted into equivalent problems that are linear.

- The methods have broad applicability – they adapt to handle strict inequalities, integer-valued variables, nonlinearities, and competition.

- Pivots are potent – with them, we can tackle systems of linear equations, linear programs, and fixed-point problems.

- Duality is central – it plays key roles in models of competition, in economics, and in the mathematics that relates to optimization.

- Applications are ubiquitous – problems from many fields can be formulated as linear programs and their generalizations.

- The subject provides insight into several academic disciplines – these include computer science, economics, engineering, operations research, and mathematics.

- Modern computer packages are user friendly – they solve a variety of optimization problems with little effort on the part of the user, and they are quick.

Its breadth, insight and usefulness may make linear programming the most important development in applicable mathematics to have occurred during the last 100 years.

9. Homework and Discussion Problems

1. *(subway cars)* The Transit Authority must repair 100 subway cars per month, and it must refurnish 50 subway cars per month. Both tasks can be done by the Transit Authority, and both can be contracted to private shops, but at a higher cost. Private contracting increases the cost by $2000 per car repaired and $2500 per car refurnished.

   The Transit Authority repairs and refurnishes subway cars in four shops. Repairing each car consumes $1/150^{\text{th}}$ of the monthly capacity of its Evaluation shop, $1/60^{\text{th}}$ of the capacity of its Assembly shop, none of the capacity of its Paint shop, and $1/60^{\text{th}}$ of the capacity of its Machine shop. Refurnishing each car requires $1/100$ of the monthly capacity if its Evaluation shop, $1/120^{\text{th}}$ of the monthly capacity of its Assembly shop, $1/40^{\text{th}}$ of the monthly capacity of its Paint shop, and none of the capacity of its Machine shop.

   (a) Formulate the problem of minimizing the monthly expense for private contracting as a linear program. Solve it graphically.

   (b) Formulate the problem of maximizing the monthly saving for repairing and refurbishing in the Authority's own shops as a linear program. Does this linear program have the same solution as the one in part (a)? If so, why? If not, why?

2. *(A woodworking shop)* A woodworking shop makes cabinets and tables. The profit earned from each cabinet equals $700. The profit earned by each table equals $500. The company's carpentry shop has a capacity of 120 hours per week. Its finishing shop has a capacity of 80 hours per week. Making each cabinet requires 20 hours of carpentry and 15 hours of finishing. Making each table requires 10 hours of carpentry and 10 hours of finishing. The company wishes to determine the rates of production (numbers of cabinets and tables per week) that maximize profit.

   (a) Write down a linear program whose optimal solution accomplishes this.

   (b) Solve your linear program graphically.

3. *(a fire drill)* The principal of a new elementary school seeks an allocation of students to exit doors that empties the school as quickly as possible

in the case of a fire. On a normal school day, there are 450 people in the building. It has three exterior doors. With a bit of experimentation, she learned that about 1.5 minutes elapse between the sounding of a fire alarm and the emergence of people from door A, after which people can emerge at the rate of 60 per minute. The comparable data for doors B and C are delay of 1.25 minutes and 1.0 minutes, and rates of 40 per minute and 50 per minute, respectively.

(a) Write a linear program whose optimal solution allocates people to doors in a way that empties the school as quickly as possible.

(b) Can you "eyeball" the optimal solution to this linear program? *Hint:* After the first 1.5 minutes, are people filing out at the rate of 150 per minute?

4. *(deadheading)* SW airline uses a single type of aircraft. Its service has been disrupted by a major winter storm. A total of 20 aircraft, each with its crew, must be *deadheaded* (flown without passengers) in order to resume its normal schedule. To the right of the table that appears below are the excess supplies at each of three airports. (These total 20). To the bottom are the excess demands at five other airports. (They also total 20). Within the table are the deadheading costs. For instance, the airline has 9 aircraft too many at airport A, it has 5 aircraft too few at airport V, and the cost of deadheading each aircraft from airport A to airport V is 25 thousand dollars. The airline wishes to resume its normal schedule with the least possible expense on deadheading.

(a) Suppose 10 is subtracted from each cost in the right-most column. Does this subtract $30,000 (which equals $10 \times 3 \times \$1{,}000$) from the cost of every plan that restores 3 planes to airport Z?

(b) Subtract the smallest cost in each column from every cost in that column. Did this alter the relative desirability of different plans?

(c) With respect to the costs obtained from part (b), "eyeball" a shipping plan whose cost is close to zero. How far from optimum can it be? Have you established a lower bound on the cost of resuming SW airline's normal schedule?

	V	W	X	Y	Z	supply
A	25	10	20	25	20	9
B	5	10	80	20	40	4
C	10	40	75	10	10	7
demand	5	2	4	6	3	

5. *(linear fractional program)** Suppose Program 1.9 has a bounded feasible region. Can there be a nonzero solution to the constraints $A\ x = 0$ and $x \geq 0$? Is part 2 of Hypothesis A guaranteed? Explain your answers.

6. *(cotton tents)* During WW II, Dantzig's group used mechanical calculators to help them plan and organize the production of all items as complicated as aircraft. Imagine something relatively simple, specifically, the job of organizing the production of standard-issue cotton military tents at the rate of 15,000 per month. Describe a (triangular) "goes-into" matrix whose entries would determine what goods would need to be produced, each at a prescribed monthly rate. (You may wish to check the web to see what a standard-issue military tent might have looked like.) Do production capacities come into play? Can you conceive of a role for a linear program? If so, might it be necessary to account for decreasing marginal benefit and/or ratio constraints? If so, why?

Chapter 2: Spreadsheet Computation

1. Preview

Spreadsheets make linear programming easier to learn. This chapter contains the information about spreadsheets that will prove useful. Not all of that information is required immediately. To prepare for Chapters 3 and 4, you should understand:

- a bit about Excel functions, especially the *sumproduct* function;
- what a *circular reference* is;

E. V. Denardo, *Linear Programming and Generalizations*, International Series in Operations Research & Management Science 149,
DOI 10.1007/978-1-4419-6491-5_2,

- how to download from the Springer website and activate a group of Excel Add-Ins called OP_TOOLS;
- how to use Solver to find solutions to systems of linear equations.

Excel has evolved, and it continues to evolve. The same is true of Solver. Several versions of Excel and Solver are currently in use. A goal of this chapter is to provide you with the information that that is needed to make effective use of the software with which your computer is equipped.

Excel for PCs

If your computer is a PC, you could be using Excel 2003, 2007 or 2010. Excel 2003 remains popular. Excel 2007 and Excel 2010 have different file structures. To ease access, each topic is introduced in the context of Excel 2003 and is adapted to more recent versions of Excel in later subsections. Needless to say, perhaps, some subsections are more relevant to you than others.

Excel for Macs

If your computer is a Mac that is equipped with a version of Excel that is dated prior to 2008, focus on the discussion of Excel 2003, which is quite similar. If your computer is a Mac that is equipped with Excel 2011, focus on the discussion of Excel 2010, which is similar.

But if your computer is equipped with Excel 2008 (for Macs only), its software has a serious limitation. Excel 2008 does not support Visual Basic. This makes it less than ideal for scientific and business uses. You will *not* be able to use your computer to take the grunt-work out of the calculations in Chapters 3 and 4, for instance. Upgrade to Excel 2011 as soon as possible. It does support Visual Basic. Alternatively, use a different version of Excel, either on your computer or on some other.

2. The Basics

This section contains basic information about Excel. If you are familiar with Excel, scan it or skip it.

At first glance, a spreadsheet is a pretty dull object – a rectangular array of cells. Into each cell, you can place a number, or some text, or a function. The function you place in a cell can call upon the values of functions in other cells. And *that* makes a spreadsheet a potent programming language, one that has revolutionized desktop computing.

Cells

Table 2.1 displays the upper left-hand corner of a spreadsheet. In spreadsheet lingo, each rectangle in a spreadsheet is called a **cell**. Evidently, the columns are labeled by letters, the rows by numbers. When you refer to a cell, the column (letter) *must* come first; cell B5 is in the second column, fifth row.

Table 2.1. A spreadsheet

	A	B	C	D
1				
2	0.3			
3	24	←	=2^3 * 3	
4	2.718281828	←	= EXP(1)	
5	1.414213562	←	= SQRT(2)	
6	mean			
7	= mean	←	'= mean	
8	1/3			
9				
10				

You **select** a cell by putting the cursor in that cell and then clicking it. When you select a cell, it is outlined in heavy lines, and a **fill handle** appears in the lower right-hand corner of the outline. In Table 2.1, cell C9 has been selected. Note the fill handle – it will prove to be very handy.

Entering numbers

Excel allows you to enter about a dozen different types of information into a cell. Table 2.1 illustrates this capability. To enter a **number** into a cell, select that cell, then type the number, and then depress either the Enter key or any one of the arrow keys. To make cell A2 look as it does, select cell A2, type 0.3 and then hit the Enter key.

Entering functions

In Excel, functions (and only functions) begin with the "=" sign. To enter a **function** into a cell, select that cell, depress the "=" key, then type the function, and then depress the Enter key. The function you enter in a cell will *not* appear there. Instead, the cell will display the **value** that the function has been assigned.

It Table 2.1, cell A3 displays the value 24, but it is clear (from column C) that cell A3 contains the function $2^3 \times 3$, rather than the number 24. Similarly, cell A5 displays the number 1.414…, which is the value of the function $\sqrt{2}$, evaluated to ten significant digits.

Excel includes over 100 functions, many of which are self-explanatory. We will use only a few of them. To explore its functions, on the Excel Insert menu, click on Functions.

Entering text

To enter **text** into a cell, select that cell, then type the text, and then depress either the Enter key or any one of the arrow keys. To make cell A6 look as it does, select cell A6 and type *mean*. Then hit the Enter key. If the text you wish to place in a cell could be misinterpreted, begin with an *apostrophe*, which will not appear. To make cell A7 appear as it does in Table 2.1, select cell A7, type '= *mean*, and hit the Enter key. The leading apostrophe tells Excel that what follows is text, not a function.

Formatting a cell

In Table 2.1, cell A8 displays the fraction 1/3. Making that happen looks easy. But suppose you select cell A8, type 1/3 and then press the Enter key. What will appear in cell A8 is "3-Jan." Excel has decided that you wish to put a date in cell A8. And Excel will interpret *everything* that you subsequently enter into cell A8 as a date. Yuck!

With Excel 2003 and earlier, the way out of this mess is to click on the Format menu, then click on Cells, then click on the Number tab, and then select either General format or a Type of Fraction.

Format Cells with Excel 2007

With Excel 2007, the Format menu disappeared. To get to the Format Cells box, double-click on the Home tab. In the menu that appears, click on the Format icon, and then select Format Cells from the list that appears. From here on, proceed as in the prior subsection.

Format Cells with Excel 2010

With Excel 2010, the Format Cells box has moved again. To get at it, click on the Home tab. A horizontal "ribbon" will appear. One block on that ribbon is labeled Number. The lower-right hand corner of the Number block has a tiny icon. Click on it. The Format Cells dialog box will appear.

Entering Fractions

How can you get the fraction 1/3 to appear in cell A8 of Table 2.1? Here is one way. First, enter the function =1/3 in that cell. At this point, 0.333333333 will appear there. Next, with cell A8 still selected, bring the Format Cells box into view. Click on its Number tab, select Fraction and the Type labeled Up to one digit. This will round the number 0.333333333 off to the nearest one-digit fraction and report it in cell A8.

The formula bar

If you select a cell, its content appears in the **formula bar**, which is the blank rectangle just above the spreadsheet's column headings. If you select cell A5 of Table 2.1, the formula =SQRT(2) will appear in the formula bar, for instance. What good is the formula bar? It is a nice place to edit your functions. If you want to change the number in cell A5 to $\sqrt{3}$, select cell A5, move the cursor onto the formula bar, and change the 2 to a 3.

Arrays

In Excel lingo, an **array** is a rectangular block of cells. Three arrays are displayed below. The array B3:E3 (note the colon) consists of a row of 4 cells, which are B3, C3, D3 and E3. The array B3:B7 consists of a column of 5 cells. The array B3:E7 consists of 20 cells.

B3:E3 B3:B7 B3:E7

Absolute and relative addresses

Every cell in a spreadsheet can be described in four different ways because a "$" sign can be included or excluded before its row and/or column. The came cell is specified by:

B3 B$3 $B3 B3

In Excel jargon, a **relative** reference to a column or row omits the "$" sign, and an **absolute** (or **fixed**) reference to a column or row includes the "$" sign.

Copy and Paste

Absolute and relative addressing is a clever feature of spreadsheet programs. It lets you repeat a pattern and compute recursively. In this subsection, you will see what happens when you Copy the content of a cell (or of an array) onto the Clipboard and then Paste it somewhere else.

With Excel 2003 and earlier, select the cell or array you want to reproduce. Then move the cursor to the **Copy** icon (it is just to the right of the scissors), and then click it. This puts a copy of the cell or array you selected on the **Clipboard**. Next, select the cell (or array) in which you want the information to appear, and click on the **Paste** icon. What was on the clipboard will appear where you put it *except for* any cell addresses in functions that you copied onto the Clipboard. They will change as follows:

- The relative addresses will shift the number rows and/or columns that separate the place where you got it and the place where you put it.
- By contrast, the absolute addresses will not shift.

This may seem abstruse, but its uses will soon be evident.

Copy and Paste with Excel 2007 and Excel 2010

With Excel 2007, the Copy and Paste icons have been moved. To make them appear, double-click on the Home tab. The Copy icon will appear just below the scissors. The Paste icon appears just to the left of the Copy icon, and it has the word "Paste" written below it. With Excel 2010, the Copy and Paste icons are back in view – on the Home tab, at the extreme left.

3. Expository Conventions

An effort has been made to present material about Excel in a way that is easy to grasp. As concerns keystroke sequences, from this point on:

> This text displays each Excel keystroke sequence in boldface type, omitting both:
>
> - The **Enter** keystroke that finishes the keystroke sequence.
> - Any English punctuation that is not part of the keystroke sequence.

For instance, cells A3, A4 and A5 of Table 2.1 contain, respectively,

=2^3*3 **=EXP(1)** **=SQRT(2)**

Punctuation is *omitted* from keystroke sequences, even when it leaves off the period at the end of the sentence!

The spreadsheets that appear in this text display the *values* that have been assigned to functions, rather than the functions themselves. The convention that is highlighted below can help you to identify the functions.

> When a spreadsheet is displayed in this book:
>
> - If a cell is outlined in dotted lines, it displays the value of a function, and that function is displayed in some other cell.
> - The "$" signs in a function's specification suggest what other cells contain similar functions.

In Table 2.1, for instance, cells A3, A4 and A5 are outlined in dotted lines, and column C specifies the functions whose values they contain. Finally:

> The Springer website contains two items that are intended for use with this book. They can be downloaded from http://extras.springer.com/2011/978-1-4419-6490-8.

One of the items at the Springer website is a folder that is labeled, "Excel spreadsheets – one per chapter." You are encouraged to download that folder now, open its spreadsheet for Chapter 2, note that this spreadsheet contains sheets labeled Table 2.1, Table 2.2, …, and experiment with these sheets as you proceed.

4. The Sumproduct Function

Excel's SUMPRODUCT function is extremely handy. It will be introduced in the context of

Problem 2.A. For the random variable X that is described in Table 2.2, compute the mean, the variance, the standard deviation, and the mean absolute deviation.

Table 2.2. A random variable, X.

	A	B	C	D
1				
2		A random variable X		
3				
4		outcome	probability	value of X
5		a	0.30	-6
6		b	0.55	3.2
7		c	0.12	10
8		d	0.03	22

The sumproduct function will make short work of Problem 2.A. Before discussing how, we interject a brief discussion of discrete probability models. If you are facile with discrete probability, it is safe to skip to the subsection entitled "Risk and Return."

A discrete probability model

The random variable X in Table 2.2 is described in the context of a *discrete probability model*, which consists of "outcomes" and "probabilities:"

- The *outcomes* are mutually exclusive and collectively exhaustive. Exactly one of the outcomes will occur.
- Each outcome is assigned a nonnegative number, which is interpreted as the *probability* that the outcome will occur. The sum of the probabilities of the outcomes must equal 1.0.

A *random variable* assigns a number to each outcome.

The probability model in Table 2.2 has four outcomes, and the sum of their probabilities does equal 1.0. Outcome b will occur with probability 0.55, and the random variable X will take the value 3.2 if outcome b occurs.

A measure of the center

The random variable X in Table 2.2 takes values between –6 and +22. The *mean* (a.k.a. *expectation*) of a random variable represents the "center" of its probability distribution. The mean of a random variable X is denoted as μ or E(X), and it is found by multiplying the probability of each outcome by the value that the random variable takes when that outcome occurs and taking the sum. For the data in Table 2.2, we have

$$\begin{aligned}\mu = E(X) &= (0.30) \times (-6) + (0.55) \times (3.2) + (0.12) \times (10) + (0.03) \times (22) \\ &= 1.82.\end{aligned}$$

The mean of a random variable has the same unit of measure as does the random variable itself. If X is measured in dollars, so is its mean. The mean is a *weighted* average; each value that X can take is weighed (multiplied) by its probability.

Measures of the spread

There are several measures of the *spread* of a random variable, that is, of the difference (X – μ) between the random variable X and its mean. The most famous of these measures of spread is known as the variance. The variance of a random variable X is denoted as σ^2 or Var(X) and is the expectation of the square of (X – μ). For the data in Table 2.2, we have

$$\begin{aligned}\sigma^2 = \text{Var}(X) &= (0.30) \times (-6 - 1.82)^2 + (0.55) \times (3.2 - 1.82)^2 \\ &\quad + (0.12) \times (10 - 1.82)^2 + (0.03) \times (22 - 1.82)^2, \\ &= 39.64.\end{aligned}$$

The unit of measure of the variance is the *square* of the unit of measure of the random variable. If X is measured in dollars, Var(X) is measured in (dollars) × (dollars), which is a bit weird.

The standard deviation of a random variable X is denoted as σ or StDev(X) and is the square root of its variance. For the data in Table 2.2,

$$\sigma = \text{StDev}(X) = 6.296.$$

The standard deviation of a random variable has the same unit of measure as does the random variable itself.

A less popular measure of the spread of a random variable is known as its mean absolute deviation. The mean absolute deviation of a random variable X is denoted MAD(X) and it is the expectation of the absolute value of $(X - \mu)$. For the data in Table 2.2,

$$\begin{aligned} \text{MAD}(X) &= (0.30) \times |-6 - 1.82| + (0.55) \times |3.2 - 1.82| \\ &\quad + (0.12) \times |10 - 1.82| + (0.03) \times |22 - 1.82|, \\ &= 4.692 \end{aligned}$$

Taking the square (in the variance) and then the square root (in the standard deviation) seems a bit contrived, and it emphasizes values that are far from the mean. For many purposes, the mean absolute deviation may be a more natural measure of the spread in a distribution.

Risk and return

Interpret the random variable X as the profit that will be earned from a portfolio of investments. A tenet of financial economics is that in order to obtain a higher *return* one must accept a higher *risk*. In this context, E(X) is taken as the measure of return, and StDev(X) as the measure of risk. It can make sense to substitute MAD(X) as the measure of risk. Also, as suggested in Chapter 1, a portfolio X that minimizes MAD(X) subject to the requirement that E(X) be at least as large as a given threshold can be found by solving a linear program.

Using the sumproduct function

The arguments in the sumproduct function must be arrays that have the same number of rows and columns. Let us suppose we have two arrays of the same size. The sumproduct function multiplies each element in one of these arrays by the corresponding element in the other and takes the sum. The same is true for three arrays of the same size. That makes it easy to compute the mean, the variance and the standard deviation, as is illustrated in Table 2.3

Table 2.3. A spreadsheet for Problem 2.A.

	B	C	D	E	F	G	H
1							
2				A random variable X			
3							
4	outcome	probability	X	X - μ	|X - μ|		
5	a	0.30	-6	-7.82	7.82		
6	b	0.55	3.2	1.38	1.38		
7	c	0.12	10	8.18	8.18		
8	d	0.03	22	20.18	20.18		
9							
10					= ABS(D8 - C$13)		
11				= D8 - C$13			
12							
13	E(X) =	1.82		= SUMPRODUCT(C5:C8 , D5:D8)			
14	Var(X) =	39.64		= SUMPRODUCT(C5:C8 , E5:E8, E5:E8)			
15	StDev(X) =	6.296		= SQRT(C14)			
16	MAD(X) =	4.692		=SUMPRODUCT(C5:C8 , F5:F8)			

Note that:

- The function in cell C13 multiplies each entry in the array C5:C8 by the corresponding entry in the array D5:D8 and takes the sum, thereby computing $\mu = E(X)$.
- The functions in cells E5 through E8 subtract 1.82 from the values in cells D5 through D8, respectively.
- The function in cell D13 sums the product of corresponding entries in the *three* arrays C5:C8 and E5:E8 and E5:E8, thereby computing Var(X).

The arrays in a sumproduct function must have the same number of rows and the same number of columns. In particular, a sumproduct function will *not* multiply each element in a row by the corresponding element in a column of the same length.

Dragging

The functions in cells E5 through E8 of Table 2.3 could be entered separately, but there is a better way. Suppose we enter just *one* of these functions, in particular, that we enter the function **=D5 – C$13** in cell E5. To **drag** this function downward, proceed as follows:

- Move the cursor to the lower right-hand corner of cell E5. The fill handle (a small rectangle in the lower right-hand corner of cell E5) will change to a Greek cross ("+" sign).

- While this Greek cross appears, depress the mouse, slide it down to cell E8 and then release it. The functions **=D6 – C$13** through **=D8 – C$13** will appear in cells E6 through E8. Nice!

Dragging downward increments the *relative* row numbers, but not the *fixed* row numbers. Similarly, dragging to the right increases the *relative* column numbers, but leaves the *fixed* column numbers unchanged. Dragging is an especially handy way to repeat a pattern and to execute a recursion.

5. Array Functions and Matrices

As mentioned earlier, in Excel lingo, an *array* is any rectangular block of cells. Similarly, an **array function** is an Excel function that places values in an array, rather than in a single cell. To have Excel execute an array function, you must follow this protocol:

- Select the array (block) of cells whose values this array function will determine.

- Type the name of the array function, but *do not hit* the Enter key. Instead, hit **Ctrl+Shift+Enter** (In other words, depress the Ctrl and Shift keys and, while they are depressed, hit the Enter key).

Matrix multiplication

A **matrix** is a rectangular array of numbers. Three matrices are exhibited below, where they have been assigned the names (labels) A, B and C.

$$A = \begin{bmatrix} 0 & 1 & 2 \\ -1 & 1 & -1 \end{bmatrix}, \quad B = \begin{bmatrix} 3 & 2 \\ 2 & 0 \\ 1 & 1 \end{bmatrix}, \quad C = \begin{bmatrix} 4 & 2 \\ 1 & 3 \end{bmatrix},$$

The product A B of two matrices is defined if – and only if – the number of columns in A equals the number of rows in B. If A is an $m \times n$ matrix and B is an $n \times p$ matrix, the **matrix product** A B is the $m \times p$ matrix whose ij^{th}

element is found by multiplying each element in the ith row of A by the corresponding element in the jth column of B and taking the sum.

It is easy to check that matrix multiplication is **associative**, specifically, that (A B) C = A (B C) if the number of columns in A equals the number of rows in B and if the number of columns in B equals the number of rows in C.

A spreadsheet

Doing matrix multiplication by hand is tedious and error-prone. Excel makes it easy. The matrices A, B and C appear as arrays in Table 2.4. That table also displays the matrix product A B and the matrix product A B C. To create the matrix product A B that appears as the array C10:D11 of Table 2.4, we took these steps:

- Select the array C10:D11.
- Type **=mmult(C2:E3, C6:D8)**
- Hit **Ctrl+Shift+Enter**

Table 2.4. Matrix multiplication and matrix inversion.

	B	C	D	E	F	G	H	I	J
1									
2	A =	0	1	2					
3		-1	1	-1					
4									
5					C =	4	2		
6		3	2			1	3		
7	B =	2	0						
8		1	1						
9									
10	A B =	4	2		= mmult(C2:E3 , C6:D8)				
11		-2	-3						
12									
13	A B C =	18	14		= mmult(C2:E3 , mmult(C6:D8 , G5:H6))				
14		-11	-13						
15									
16	C^{-1} =	0.3	-0.2		= minverse(G5:H6)				
17		-0.1	0.4						

The matrix product A B C can be computed in either of two ways. One way is to multiply the array A B in cells C10:D11 by the array C. The other way is by using the **=mmult(array, array)** function *recursively*, as has been done in Table 2.4. Also computed in Table 2.4 is the inverse of the matrix C.

Quirks

Excel computes array functions with ease, but it has its quirks. One of them has been mentioned – you need to remember to end each array function by hitting **Ctrl+Shift+Enter** rather than by hitting **Enter** alone.

A second quirk concerns 0's. With non-array functions, Excel (wisely) interprets a "blank" as a "0." When you are using array functions, it does not; you *must enter the 0's*. If your array function refers to a cell containing a blank, the cells in which the array is to appear will contain an (inscrutable) error message, such as ##### or #Value.

The third quirk occurs when you decide to alter an array function or to eliminate an array. To do so, you must begin by selecting *all* of the cells in which its output appears. Should you inadvertently attempt to change a portion of the output, Excel will proclaim, "You cannot change part of an Array." If you then move the cursor – or do most anything – Excel will repeat its proclamation. A loop! To get out of this loop, hit the **Esc** key.

6. A Circular Reference

An elementary problem in algebra is now used to bring into view an important limitation of Excel. Let us consider

Problem 2.B. Find values of x and y that satisfy the equations

$$x = 6 - 0.5y,$$

$$y = 2 + 0.5x.$$

This is easy. Substituting $(2 + 0.5x)$ for y in the first equation gives $x=4$ and hence $y=4$.

Let us see what happens when we set this problem up in a naïve way for solution on a spreadsheet. In Table 2.5, formulas for x and y have been placed in cells B4 and B5. The formula in each of these cells refers to the value in the other. A *loop* has been created. Excel insists on being able to evaluate the functions on a spreadsheet in some sequence. When Excel is presented with Table 2.5, it issues a **circular reference** warning.

Table 2.5. Something to avoid.

	A	B	C	D
1				
2		a circular reference		
3				
4	value of x =	2	←	=2 - 0.5*B5
5	value of y =	0	←	= 1 + 1*B4

You can make a circular reference warning disappear. If you do make it disappear, your spreadsheet is all but certain to be gibberish. It is emphasized:

> **Danger:** Do not ignore a "circular reference" warning. You can make it go away. If you do, you will probably wreck your spreadsheet.

This seems ominous. Excel cannot solve a system of equations. But it can, with a bit of help.

7. Linear Equations

To see how to get around the circular reference problem, we turn our attention to an example that is slightly more complicated than Problem 2.B. This example is

Problem 2.C. Find values of the variables A, B and C that satisfy the equations

$$\begin{aligned} 2A + 3B + 4C &= 10, \\ 2A - 2B - C &= 6, \\ A + B + C &= 1. \end{aligned}$$

You probably recall how to solve Problem 2.C, and you probably recall that it requires some grunt-work. We will soon see how to do it on a spreadsheet, without the grunt-work.

An ambiguity

Problem 2.C exhibits an ambiguity. The letters A, B and C are the *names* of the variables, and Problem 2.C asks us to find *values* of the variables A, B and C that satisfy the three equations. You and I have no trouble with this ambiguity. Computers do. On a spreadsheet, the name of the variable A will be placed in one cell, and its value will be placed in another cell.

A spreadsheet for Problem 2.C

Table 2.6 presents the data for Problem 2.C. Cells B2, C2 and D2 contain the labels of the three decision variables, which are A, B and C. Cells B6, C6 and D6 have been set aside to record the values of the variables A, B and C. The data in the three constraints appear in rows 3, 4 and 5, respectively.

Table 2.6. The data for Problem 2.C.

	A	B	C	D	E	F	G
1							
2	label	A	B	C			RHS
3	equation (1)	2	3	4	9	=	10
4	equation (2)	2	-2	-1	-1	=	6
5	equation (3)	1	1	1	3	=	1
6	value	1	1	1			
7							
8			= SUMPRODUCT(B5:D5 , B$6:D$6)				

Note that:

- Trial values of the decision variables have been inserted in cells B6, C6 and D6.
- The "=" signs in cells F3, F4 and F5 are memory aides; they remind us that we want to arrange for the numbers to their left to equal the numbers to their right, but they have nothing to do with the computation.

- The **sumproduct** function in E5 multiplies each entry in the array B$6:D$6 by the corresponding entry in the array B5:D5 and reports their sum.

- The "$" signs in cell E5 suggest – correctly – that this function has been dragged upward onto cells E4 and E3. For instance, cell E3 contains the value assigned to the function

  =SUMPRODUCT(B3:D3, B$6:D$6)

  and the number 9 appears in cell E3 because Excel assigns this function the value $9 = 2 \times 1 + 3 \times 1 + 4 \times 1$.

The standard format

The pattern in Table 2.6 works for any number of linear equations in any number of variables. This pattern is dubbed the "standard format" for linear systems, and it will be used throughout this book. A linear system is expressed in **standard format** if the columns of its array identify the variables and the rows identify the equations, like so:

- One row is reserved for the values of the variables (row 6, above).

- The entries in an equation's row are:
  - The equation's coefficient of each variable (as in cells B3:D3, above).
  - A sumproduct function that multiplies the equation's coefficient of each variable by the value of that variable and takes the sum (as in cell E3).
  - An "=" sign that serves (only) as a memory aid (as in cell F3).
  - The equation's right-hand-side value (as in cell G3).

What is missing?

Our goal is to place numbers in cells B6:D6 for which the values of the functions in cells E3:E5 equal the numbers in cells G3:G5, respectively. Excel cannot do that, by itself. We will see how to do it with Solver and then with Premium Solver for Education.

8. Introducing Solver

This section is focused on the simplest of Solver's many uses, which is to find a solution to a system of linear equations. The details depend, slightly, on the version of Excel with which your computer is equipped.

A bit of the history

Let us begin with a bit of the history. Solver was written by Frontline Systems for inclusion in an early version of Excel. Shortly thereafter, Microsoft took over the maintenance of Solver, and Frontline Systems introduced Premium Solver. Over the intervening years, Frontline Systems has improved its Premium Solver repeatedly. Recently, Microsoft and Frontline Systems worked together in the design of Excel 2010 (for PCs) and Excel 2011 (for Macs). As a consequence:

- If your computer is equipped with Excel 2003 or Excel 2007, Solver is perfectly adequate, but Premium Solver has added features and fewer bugs.
- If your computer is equipped with Excel 2010 (for PCs) or with Excel 2011 (for Macs), a great many of the features that Frontline Systems introduced in Premium Solver have been incorporated in Solver itself, and many bugs have been eliminated.
- If your computer is equipped with Excel 2008 for Macs, it does not support Visual Basic. Solver is written in Visual Basic. The **=pivot(cell, array)** function, which is used extensively in this book, is also written in Visual Basic. You will not be able to use Solver or the "pivot" function until you upgrade to Excel 2011 (for Macs). Until then, use some other version of Excel as a stopgap.

Preview

This section begins with a discussion of the version of Solver with which Excel 2000, 2003 and 2007 are equipped. The discussion is then adapted to Excel 2010 and 2011. Premium Solver is introduced in the next section.

Finding Solver

When you purchased Excel (with the exception of Excel 2008 for Macs), you got Solver. But Solver is an "Add-In," which means that it may not be ready to use. To see whether Solver is up and running, open a spreadsheet.

With Excel 2003 or earlier, click on the Tools menu. If Solver appears there, you are all set; Solver is installed and activated. If Solver does not appear on the Tools menu, it may have been installed but not activated, and it may not have been installed. Proceed as follows:

- Click again on the Tools menu, and then click on Add-Ins. If Solver is listed as an Add-In but is not checked off, check it off. This activates Solver. The next time you click on the Tools menu, Solver will appear and will be ready to use.
- If Solver does not appear on the list of Add-Ins, you will need to find the disc on which Excel came, drag Solver into your Library, and then activate it.

Finding Solver with Excel 2007

If your computer is equipped with Excel 2007, Solver is not on the Tools menu. To access Solver, click on the Data tab and then go to the Analysis box. You will see a button labeled Solver if it is installed and active. If the Solver button is missing:

- Click on the Office Button that is located at the top left of the spreadsheet.
- In the bottom right of the window that appears, select the Excel Options button.
- Next, click on the Add-Ins button on the left and look for Solver Add-In in the list that appears.
- If it is in the inactive section of this list, then select Manage: Excel Add-Ins, then click Go…, and then select the box next to Solver Add-in and click OK.
- If Solver Add-in is not listed in the Add-Ins available box, click Browse to locate the add-in. If you get prompted that the Solver Add-in is not currently installed on your computer, click Yes to install it.

Finding Solver with Excel 2010

To find Solver with Excel 2010, click on the Data tab. If Solver appears (probably at the extreme right), you are all set. If Solver does not appear, you will need to activate it, and you may need to install it. To do so, open an Excel spreadsheet and then follow this protocol:

- Click on the File menu, which is located near the top left of the spreadsheet.
- Click on the Options tab (it is near the bottom of the list) that appeared when you clicked on the File menu.
- A dialog box named Excel Options will pop up. On the side-bar to its left, click on Add-Ins. Two lists of Add-Ins will appear – "Active Application Add-Ins" and "Inactive Application Add-Ins."
  - If Solver is on the "Inactive" list, find the window labeled "Manage: Excel Add-Ins," click on it, and then click on the "Go" button to its right. A small menu entitled Add-Ins will appear. Solver will be on it, but it will not be checked off. Check it off, and then click on OK.
  - If Solver is not on the "Inactive" list, click on Browse, and use it to locate Solver. If you get a prompt that the Solver Add-In is not currently installed on your computer, click "Yes" to install it. After installing it, you will need to activate it; see above.

Using Solver with Excel 2007 and earlier

Having located Solver, we return to Problem 2.C. Our goal is to have Solver find values of the decision variables A, B and C that satisfy the equations that are represented by Table 2.6. With Excel 2007 and earlier, the first step is to make the Solver dialog box look like Figure 2.1. (The Solver dialog box for Excel 2010 differs in ways that are described in the next subsection.)

To make your Solver dialog box look like that in Figure 2.1, proceed as follows:

- With Excel 2003, on the Tools menu, click on Solver. With Excel 2007, go to the Analysis box of the Data tab, and click on Solver.
- Leave the Target Cell blank.
- Move the cursor to the By Changing Cells window, then select cells B6:D6, and then click.
- Next, click on the Add button.

Figure 2.1. A Solver dialog box for Problem 2.C.

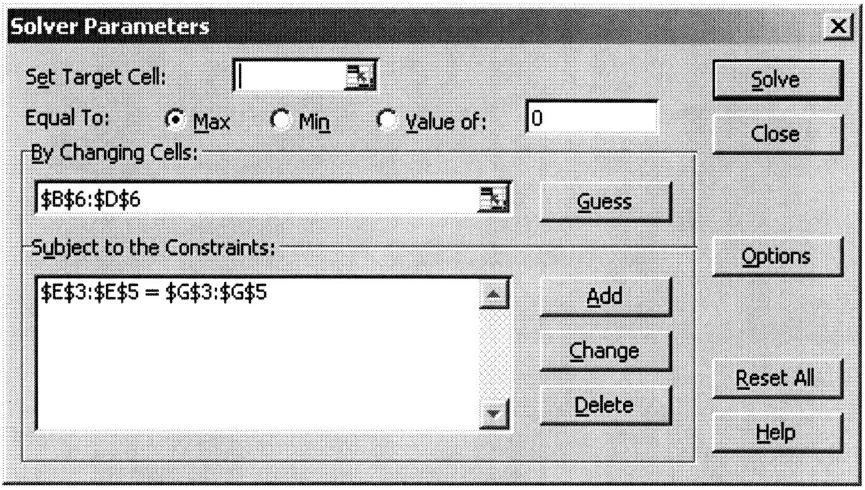

- An Add Constraint dialog box will appear. Proceed as follows:
  - Click on the Cell Reference window, then select cells E3:E5 and click.
  - Click on the triangular button on the middle window. On the drop-down menu that appears click on "=".
  - Click on the Constraint window. Then select cells G3:G5 and click. This will cause the Add Constraint dialog box to look like:

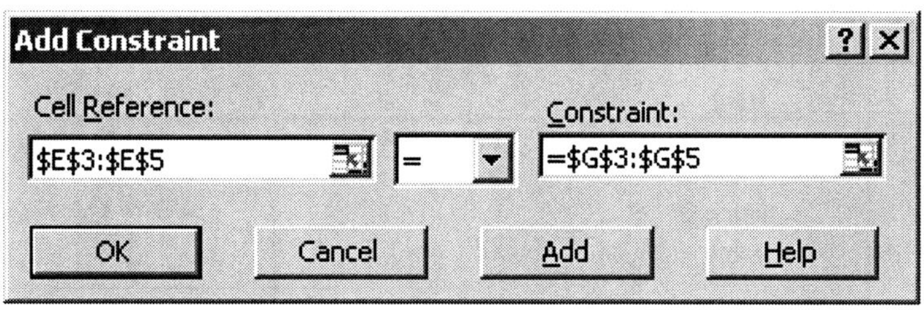

  - Click on OK. This will close the Add Constraint dialog box and return you to the Solver dialog box, which will now look exactly like Figure 2.1.

- In the Solver dialog box, *do not* click on the Solve button. Instead, click on the Options button and, on the Solver Options menu that appears (see below) click on the Assume Linear Model window. Then click on the OK button. And then click on Solve.

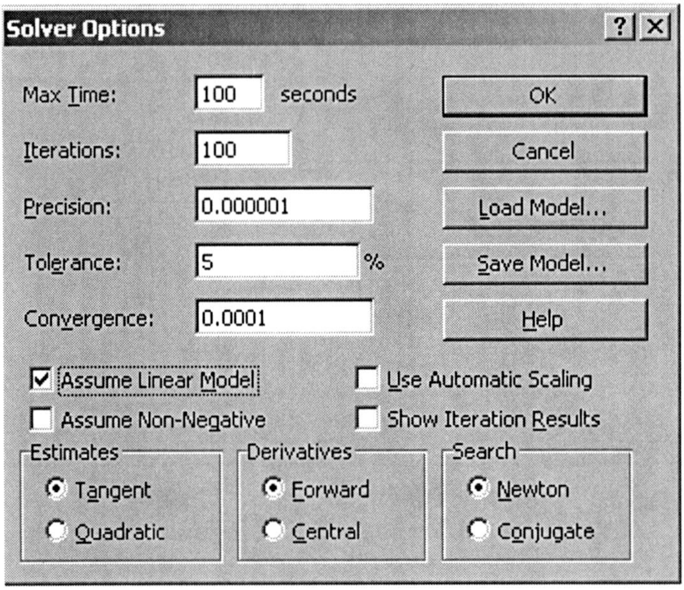

In a flash, your spreadsheet will look like that in Table 2.7. Solver has succeeded; the values it has placed in cells B6:D6 that enforce the constraints E3:E5=G3:G5; evidently, setting A=0.2, B=–6.4 and C=7.2 which solves Problem 2.C.

Table 2.7. A solution to Problem 2.C.

	A	B	C	D	E	F	G
1							
2	label	A	B	C			RHS
3	equation (1)	2	3	4	10	=	10
4	equation (2)	2	-2	-1	6	=	6
5	equation (3)	1	1	1	1	=	1
6	value	0.2	-6.4	7.2			
7							
8			= SUMPRODUCT(B5:D5 , B$6:D$6)				

Using Solver with Excel 2010

Presented as Figure 2.2 is a Solver dialog box for Excel 2010. It differs from the dialog box for earlier versions of Excel in the ways that are listed below:

- The cell for which the value is to be maximized or minimized in an optimization problem is labeled *Set Objective*, rather than *Target Cell.*
- The method of solution is selected on the main dialog box rather than on the Options page.

Figure 2.2. An Excel 2010 Solver dialog box.

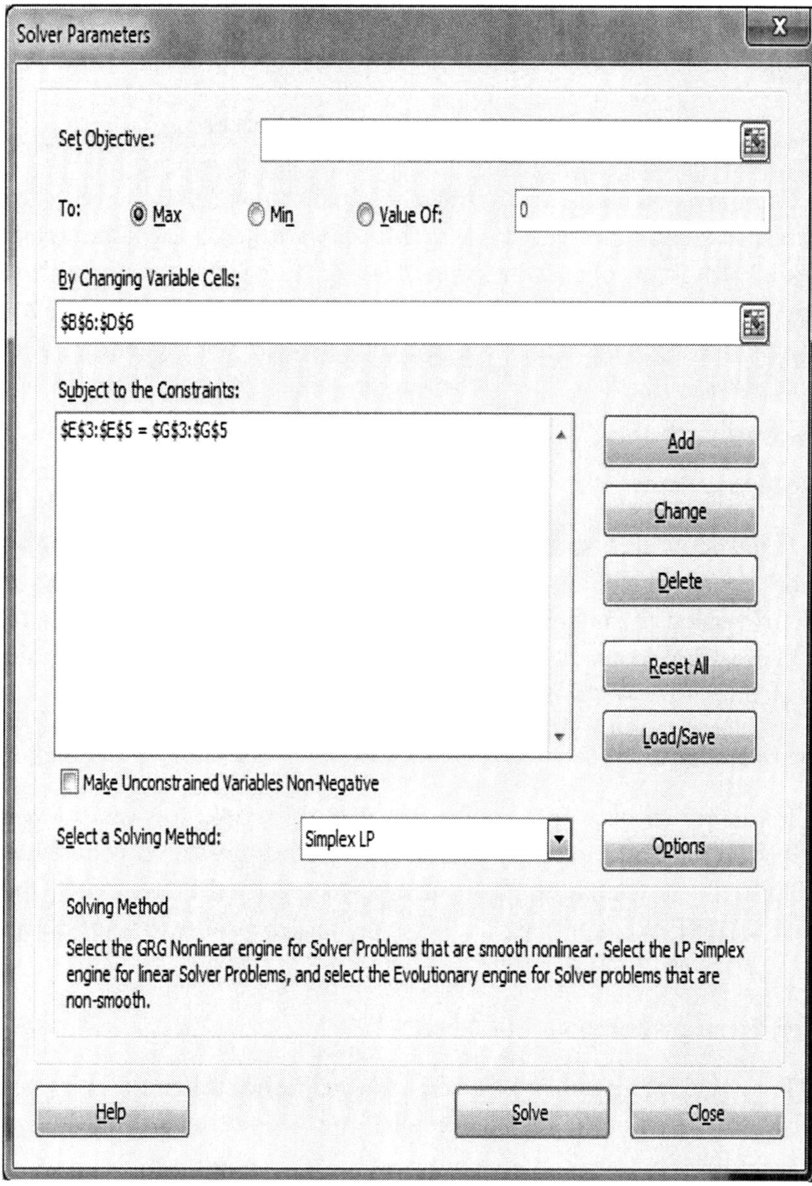

- The capability to constrain the decision variables to be nonnegative appears on the main dialog box, rather than on the Options page.
- A description of the "Solving Method" that you have selected appears at the bottom of the dialog box.

Fill this dialog box out as you would for Excel 2007, but remember to select the option you want in the "nonnegative variables" box.

9. Introducing Premium Solver

Frontline Systems has made available for educational use a bundle of software called the Risk Solver Platform. This software bundle includes Premium Solver, which is an enhanced version of Solver. This software bundle also includes the capability to formulate and run simulations and the capability to draw and roll back decision trees. Sketched here are the capabilities of Premium Solver. This sketch is couched in the context of Excel 2010. If you are using a different version of Excel, your may need to adapt it somewhat.

Note to instructors

If you adopt this book for a course, you can arrange for the participants in your course (including yourself, of course) to have free access to the educational version of the Risk Solver Platform. To do so, call Frontline Systems at 755 831-0300 (country code 01) and press 0 or email them at academics@ solver.com.

Note to students

If you are enrolled in a course that uses this book, you can download the Risk Solver Platform by clicking on the website http://solver.com/student/ and following instructions. You will need to specify the "Textbook Code," which is DLPEPAE, and the "Course code," which your instructor can provide.

Using Premium Solver as an Add-In

Premium Solver can be accessed and used in two different ways – as an Add-In or as part of the Risk Solver Platform. Using it as an Add-In is discussed in this subsection. Using it as part of the Risk Solver Platform is discussed a bit later.

To illustrate the use of Premium Solver as an Add-In, begin by reproducing Table 2.6 on a spreadsheet. Then, in Excel 2010, click on the File button. An Add-Ins button will appear well to the right of the File button. Click on the Add-Ins button. After you do so, you will see a rectangle at the left with a light bulb and the phrase “Premium Solver Vxx.x” (currently V11.0). Click on it. A Solver Parameters dialog box will appear. You will need to make it look like that in Figure 2.3.

Figure 2.3. A dialog box for using Premium Solver as an Add-In.

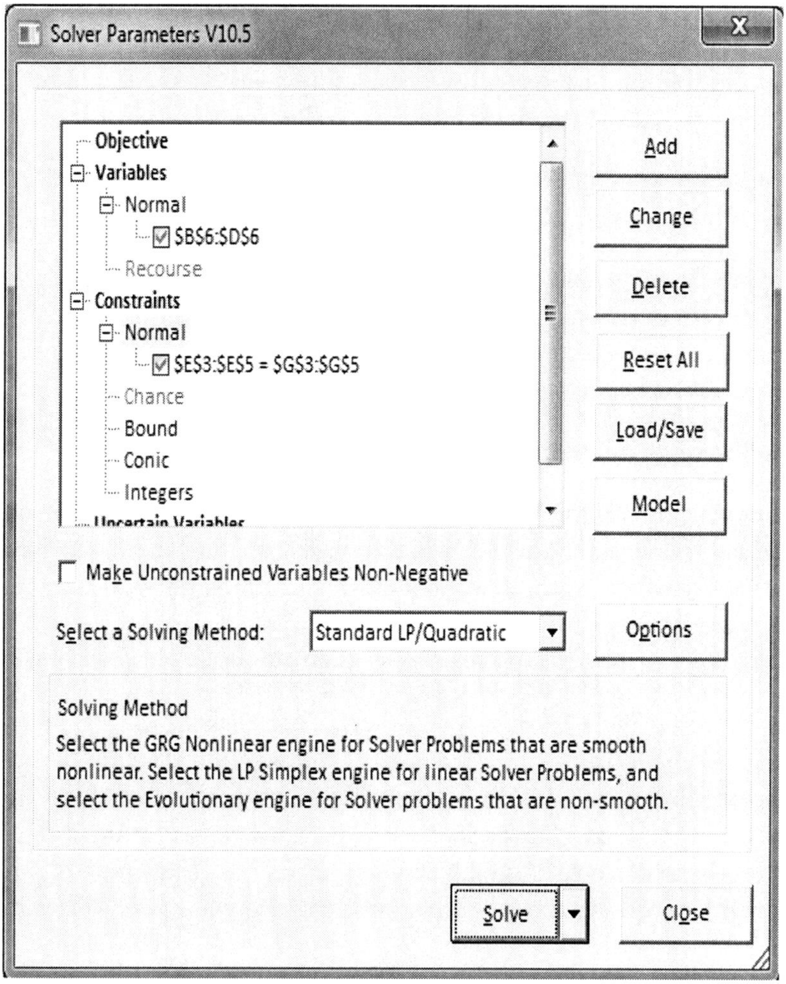

Filling in this dialog box is easy:

- In the window to the left of the Options button, click on Standard LP/Quadratic.
- Next, in the large window, click on Normal Variables. Then click on the Add button. A dialog box will appear. Use it to identify B6:D6 as the cells whose values Premium Solver is to determine. Then click on OK. This returns you to the dialog box in Figure 2.3, with the variables identified.
- In the large window, click on Normal Constraints. Then click on the Add button. Use the (familiar) dialog box to insert the constraints E3:E5 = G3:G5. Then click on OK.
- If the button that makes the variables nonnegative is checked off, click on it to remove the check mark. Then click on Solve.

In a flash, your spreadsheet will look like that in Table 2.7. It will report values of 0.2, –6.4 and 7.2 in cells B7, C7, and D7.

When Premium Solver is operated as an Add-In, it is **modal**, which means that you cannot do anything outside its dialog box while that dialog box is open. Should you wish to change a datum on your spreadsheet, you need to close the dialog box, temporarily, make the change, and then reopen it.

Using Premium Solver from the Risk Solver Platform

But when Premium Solver is operated from the Risk Solver Platform, it is **modeless**, which means that you can move back and forth between Premium Solver and your spreadsheet without closing anything down. The modeless version can be very advantageous.

To see how to use Premium Solver from the Risk Solver Platform, begin by reproducing Table 2.6 on a spreadsheet. Then click on the File button. A Risk Solver Platform button will appear at the far right. Click on it. A menu will appear. Just below the File button will be a button labeled Model. If that button is *not* colored, click on it. A dialog box will appear at the right; in it, click on the icon labeled Optimization. A dialog box identical to Figure 2.4 will appear, except that neither the variables nor the constraints will be identified.

Figure 2.4. A Risk Solver Platform dialog box.

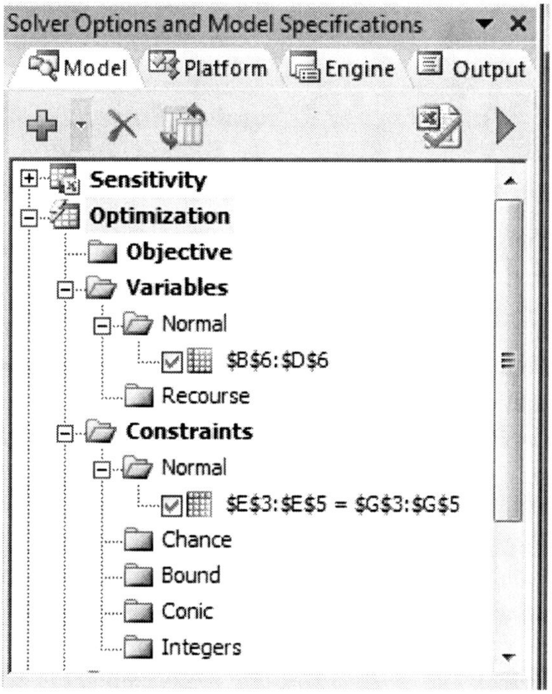

Making this dialog box look exactly like Figure 2.4 is not difficult. The green Plus sign (Greek cross) just below the word "Model" is used to add information. The red "X" to its right is used to delete information. Proceed as follows:

- Select cells B6:D6, then click on Normal Variables, and then click on Plus.
- Click on Normal Constraints and then click on Plus. Use the dialog box that appears to impose the constraints E3:E5 = G3:G5.

It remains to specify the solution method you will use and to execute the computation. To accomplish this:

- Click on Engine, which is to the right of the Model button, and select Standard LP/Quadratic Engine.
- Click on Output, which is to the right of the Engine button. Then click on the green triangle that points to the right.

In an instant, your spreadsheet will look exactly like Table 2.7. It will exhibit the solution A = 0.2, B = –6.4 and C = 7.2.

10. What Solver and Premium Solver Can Do

The user interfaces in Solver and in Premium Solver are so "friendly" that it is hard to appreciate the 800-pound gorillas (software packages) that lie behind them. The names and capabilities of these software packages have evolved. Three of these packages are identified below:

1. The package whose name includes "LP" finds solutions to systems of linear equations, to linear programs, *and* to integer programs. In newer versions of Premium Solver, it also finds solutions to certain quadratic programs.
2. The package whose name includes "GRG" is somewhat slower, but it can find solutions to systems of nonlinear constraints and to nonlinear programs, with or without integer-valued variables.
3. The package whose name includes "Evolutionary" is markedly slower, but it can find solutions to problems that elude the other two.

Premium Solver and the versions of Solver that are in Excel 2010 and Excel 2011 include all three packages. Earlier editions of Excel include the first two of these packages. A subsection is devoted to each.

The LP software

When solving linear programs and integer programs, use the LP software. It is quickest, and it is guaranteed to work. If you use it with earlier versions of Solver, remember to shift to the Options sheet and check off Assume Linear Model. To use it with Premium Solver as an Add-In, check off Standard LP/Quadratic in a window on the main dialog box. The advantages of this package are listed below:

- Its software checks that the system you claim to be linear actually *is* linear – and this is a debugging aid. (Excel 2010 is equipped with a version of Solver that can tell you what, if anything, violates the linearity assumptions.)

- It uses an algorithm that is virtually foolproof.
- For technical reasons, it is more likely to find an integer-valued optimal solution if one exists.

The GRG software

When you seek a solution to a system of nonlinear constraints or to an optimization problem that includes a nonlinear objective and/or nonlinear constraints, try the GRG (short for generalized reduced gradient) solver. It *may* work. Neither it nor any other computer program can be guaranteed to work in all nonlinear systems. To make good use of the GRG solver, you need to be aware of an important difference between the it and the LP software:

- When you use the LP software, you can place any values you want in the changing cells before you click on the *Solve* button. The values you have placed in these cells will be ignored.
- On the other hand, when you use the GRG software, the values you place in the changing cells are important. The software *starts* with the values you place in the changing cells and attempts to improve on them.

The closer you start, the more likely the GRG software is to obtain a solution. It is emphasized:

When using the GRG software, try to "start close" by putting reasonable numbers in the changing cells.

The multi-start feature

Premium Solver's GRG code includes (on its options menu) a "multi-start" feature that is designed to find solutions to problems that are not convex. If you are having trouble with the GRG code, give it a try.

A quirk

The GRG Solver may attempt to evaluate a function outside the range for which it is defined. It can attempt to evaluate the function **=LN(cell)** with a negative number in that cell, for instance. Excel's **=ISERROR(cell)** function can help you to work around this. To see how, please refer to the discussion on page 643 of Chapter 20.

Numerical differentiation

It is also the case that the GRG Solver differentiates numerically; it *approximates* the derivative of a function by evaluating that function at a variety of points. It is safe to use any function that is differentiable and whose derivative is continuous. Here are two examples of functions that should be avoided:

- The function =**MIN(x, 6)** which is not differentiable at x=6.
- The function =**ABS(x)** which is not differentiable at x=0.

If you use a function that is not differentiable, you may get lucky. And you may not. It is emphasized:

Avoid functions that are not differentiable.

Needless to say, perhaps, it is a *very* good idea to avoid functions that are not continuous when you use the GRG Solver.

The Evolutionary software

This software package is markedly slower, but it does solve problems that elude the simplex method and the generalized reduced gradient method. Use it when the GRG solver does not work.

The Gurobi and the SOCP software

The Risk Solver Platform includes other optimization packages. The Gurobi package solves linear, quadratic, and mixed-integer programs very effectively. Its name is an amalgam of the last names of the founders of Gurobi Optimization, who are Robert Bixby, Zonghao Gu, and Edward Rothberg. The SOCP engine quickly solves a generalization of linear programs whose constraints are cones.

11. An Important Add-In

The array function =**PIVOT(cell, array)** executes pivots. This function is used again and again, starting in Chapter 3. The function =**NL(q, μ, σ)** computes the expectation of the amount, if any, by which a normally distributed random variable having μ as its mean and σ as its standard deviation exceeds the number q. That function sees action in Chapter 7.

Neither of these functions comes with Excel. They are included in an Add-In called OP_TOOLS. This Add-In is available at the Springer website. You are urged to download this addend and install it in your Library before you tackle Chapter 3. This section tells how to do that.

Begin by clicking on the Springer website for this book, which is specified on page 39. On that website, click on the icon labeled OP_TOOLS, copy it, and paste it into a convenient folder on your computer, such as My Documents. Alternatively, drag it onto your Desktop.

What remains is to insert this Add-In in your Library and to activate it. How to do so depends on which version of Excel you are using.

With Excel 2003

With Excel 2003, the Start button provides a convenient way to find and open your Library folder (or any other). To accomplish this:

- Click on the Start button. A menu will pop up. On that menu, click on Search. Then click on For Files and Folders. A window will appear. In it, type Library. Then click on Search Now.
- After a few seconds, the large window to the right will display an icon for a folder named Library. Click on that icon. A path to the folder that contains your Library will appear toward the top of the screen. Click on that path.
- You will have opened the folder that contains your library. An icon for your Library is in that folder. Click on the icon for your Library. This opens your Library.

With your library folder opened, drag OP_TOOLS into it. Finally, activate OP_TOOLS, as described earlier.

With Excel 2007 and Excel 2010

With Excel 2007 and Excel 2010, clicking on the Start button is *not* the best way to locate your Library. Instead, open Excel. If you are using Excel

2007, click on the Microsoft Office button. If you are using Excel 2010, click on File.

Next, with Excel 2007 or 2010, click on Options. Then click on the Add-Ins tab. In the Manage drop-down, choose Add-Ins and then click Go. Use Browse to locate OP_TOOLS and then click on OK. Verify that OP_TOOLS is on the Active Add-Ins list, and then click on OK at the bottom of the window.

To make certain that OP_TOOLS is up and running, select a cell, enter = **NL(0, 0, 1)** and observe that the number 0.398942 appears in that cell.

12. Maxims for Spreadsheet Computation

It can be convenient to hide data within functions, as has been done in Table 2.1 and Table 2.5. This can make the functions easier to read, but it is dangerous. The functions do not appear on your spreadsheet. If you return to modify your spreadsheet at a later time, you may not remember where you put the data. It is emphasized:

> **Maxim on data:** Avoid hiding data within functions. Better practice is to place each element of data in a cell and refer to that cell.

A useful feature of spreadsheet programming is that the spreadsheet gives instant feedback. It displays the value taken by a function as soon as you enter it. Whenever you enter a function, use test values to check that you constructed it properly. This is especially true of functions that get dragged – it is easy to leave off a "$" sign. It is emphasized:

> **Maxim on debugging:** Test each function as soon as you create it. If you drag a function, check that you inserted the "$" signs where they are needed.

The fact that Excel gives instant feedback can help you to "debug as you go."

13. Review

All of the information in this chapter will be needed, sooner or later. You need not master all of it now. You can refer back to this chapter as needed. Before tackling Chapters 3 and 4, you should be facile with the use of spreadsheets to solve systems of linear equations via the "standard format." You should also prepare to use the software on the Springer website for this book.

A final word about Excel: When you change *any* cell on a spreadsheet, Excel automatically re-computes the value of each function on that sheet. This happens fast – so fast that you may not notice that it has occurred.

14. Homework and Discussion Problems

1. Use Excel to determine whether or not 989 is a prime number. Do the same for *991*. (*Hint:* Use a "drag" to divide each of these numbers by 1, 3, 5, …, 35.)

2. Use Solver to find a number x that satisfies the equation $x = e^{-x^2}$. (*Hint:* With a trial value of x in one cell, place the function e^{-x^2} in another, and ask Solver to find the value of x for which the numbers in the two cells are equal.)

3. (*the famous birthday problem*) Suppose that each child born in 2007 (not a leap year) was equally likely to be born on any day, independent of the others. A group of n such children has been assembled. None of these children are related to each other. Denote as Q(n) the probability that at least two of these children share a birthday. Find the smallest value of n for which $Q(n)>0.5$. *Hints:* Perhaps the probability P(n) that these n children were born on n different days be found (on a spreadsheet) from the recursion $P(n)=P(n-1)\,(365-n)/365$. If so, a "drag" will show how quickly P(n) decreases as n increases.

4. For the matrices A and B in Table 2.4, compute the matrix product BA. What happens when you ask Excel to compute $(BA)^{-1}$? Can you guess why?

5. Use Solver or Premium Solver to find a solution to the system of three equations that appears below. *Hint:* Use 3 changing cells and the Excel function =**LN(cell)** that computes the natural logarithm of a number.

$$\begin{aligned} 3A + 2B + 1C + 5\ln(A) &= 6 \\ 2A + 3B + 2C + 4\ln(B) &= 5 \\ 1A + 2B + 3C + 3\ln(C) &= 4 \end{aligned}$$

6. Recreate Table 2.4. Replace the "0" in matrix A with a blank. What happens?

7. The spreadsheet that appears below computes $1 + 2^n$ and 2^n for various values of n, takes the difference, and gets 1 for $n \le 49$ and gets 0 for $n \ge 50$. Why? *Hint:* Modern versions of Excel work with 64 bit words.

	A	B	C	D	E
1					
2		14	1E+14	1E+14	1
3		15	1E+15	1E+15	1
4		16	1E+16	1E+16	0
5		17	1E+17	1E+17	0
6					
7					
8			=1+10^B5	= 10^B5	= C5 - D5

Chapter 3: Mathematical Preliminaries

1. Preview

Presented in this chapter is the mathematics on which an introductory account of the simplex method rests. This consists principally of:

- A method for solving systems of linear equations that is known as Gauss-Jordan elimination.

E. V. Denardo, *Linear Programming and Generalizations*, International Series in Operations Research & Management Science 149,
DOI 10.1007/978-1-4419-6491-5_3, © Springer Science+Business Media, LLC 2011

- A discussion of vector spaces and their bases.
- An introduction to terminology that is used throughout this book.

Much of the information in this chapter is familiar. Gauss-Jordan elimination plays a key role in linear algebra, as do vector spaces. In this chapter, Gauss-Jordan elimination is described as a sequence of "pivots" that seek a solution to a system of equations. In Chapter 4, you will see that the simplex method keeps on pivoting, as it seeks an optimal solution to a linear program. Later in this chapter, it is shown that Gauss-Jordan elimination constructs a basis for a vector space.

One section of this chapter is starred. That section touches lightly on efficient numerical computation, an advanced topic on which this book does not dwell.

2. Gaussian Operations

Gauss-Jordan elimination wrestles a system of linear equations into a form for which a solution is obvious. This is accomplished by repeated and systematic use of two operations that now bear Gauss's name. These **Gaussian operations** are:

- To replace an equation by a non-zero constant c times itself.
- To replace an equation by the sum of itself and a constant d times another equation.

To replace an equation by a constant c times itself, multiply each addend in that equation by the constant c. Suppose, for example, that the equation $2x - 3y = 6$ is replaced by the constant -4 times itself. This yields the equation, $-8x + 12y = -24$. Every solution to the former equation is a solution to the latter, and conversely. In fact, the former equation can be recreated by replacing the latter by the constant $-1/4$ times itself.

Both of these Gaussian operations are **reversible** because their effects can be undone (reversed). To undo the effect of the first Gaussian operation, replace the equation that it produced by the constant $(1/c)$ times itself. To undo the effect of the second Gaussian operation, replace the equation that it produced by the sum of itself and the constant $-d$ times the other equation. Because Gaussian operations are reversible, they preserve the set of solutions to an equation system. It is emphasized:

Each Gaussian operation preserves the set of solutions to the equation system; it creates no new solutions, and it destroys no existing solutions.

Gauss-Jordan elimination will be introduced in the context of system (1), below. System (1) consists of four linear equations, which have been numbered (1.1) through (1.4). These equations have four variables or "unknowns," which are x_1, x_2, x_3, and x_4. The number p that appears on the right-hand side of equation (1.3) is a datum, not a decision variable.

(1.1) $$2x_1 + 4x_2 - 1x_3 + 8x_4 = 4$$

(1.2) $$1x_1 + 2x_2 + 1x_3 + 1x_4 = 1$$

(1.3) $$2x_3 - 4x_4 = p$$

(1.4) $$-1x_1 + 1x_2 - 1x_3 + 1x_4 = 0$$

An attempt will be made to solve system (1) for particular values of p. Pause to ask yourself: How many solutions are there are to system (1)? Has it no solutions? One? Many? Does the number of solutions depend on p? If so, how? We will find out.

3. A Pivot

At the heart of Gauss-Jordan elimination – and at the heart of the simplex method – lies the "pivot," which is designed to give a variable a coefficient of +1 in a particular equation and to give that variable a coefficient of 0 in each of the other equations. This pivot "eliminates" the variable from all but one of the equations. To **pivot** on a nonzero coefficient c of a variable x in equation (j), execute these Gaussian operations:

- First, replace equation (j) by the constant (1/c) times itself.
- Then, for each k other than j, replace equation (k) by itself minus equation (j) times the coefficient of x in equation (k).

This definition may seem awkward, but applying it to system (1) will make everything clear. This will be done twice – first by hand, then on a spreadsheet.

Let us begin by pivoting on the coefficient of x_1 in equation (1.1). This coefficient equals 2. This pivot executes the following *sequence* of Gaussian operations:

- Replace equation (1.1) with the constant (1/2) times itself.
- Replace equation (1.2) with itself minus 1 times equation (1.1).
- Replace equation (1.3) with itself minus 0 times equation (1.1).
- Replace equation (1.4) with itself minus −1 times equation (1.1).

The first of these Gaussian operations changes the coefficient of x_1 in equation (1.1) from 2 to 1. The second of these operations changes the coefficient of x_1 in equation (1.2) from 1 to 0. The third operation keeps the coefficient of x_1 in equation (1.3) equal to 0. The fourth changes the coefficient of x_1 in equation (1.4) from −1 to 0.

This pivot transforms system (1) into system (2), below. This pivot consists of Gaussian operations, so it preserves the set of solutions to system (1). In other words, each set of values of the variables x_1, x_2, x_3, and x_4 that satisfies system (1) also satisfies system (2), and conversely.

(2.1) $$1x_1 + 2x_2 - 0.5x_3 + 4x_4 = 2$$

(2.2) $$1.5x_3 - 3x_4 = -1$$

(2.3) $$2x_3 - 4x_4 = p$$

(2.4) $$3x_2 - 1.5x_3 + 5x_4 = 2$$

This pivot has *eliminated* the variable x_1 from equations (2.2), (2.3) and (2.4) because its coefficients in these equations equal zero.

4. A Basic Variable

A variable is said to be **basic** for an equation if its coefficient in that equation equals 1 and if its coefficients in the other equations equal zero. The pivot that has just been executed made x_1 basic for equation (2.1), exactly as planned. It is emphasized:

> A pivot on a nonzero coefficient of a variable in an equation makes that variable basic for that equation.

The next pivot will occur on a nonzero coefficient in equation (2.2). The variables x_3 and x_4 have nonzero coefficients in this equation. We could pivot on either. Let's pivot on the coefficient of x_3 in equation (2.2). This pivot consists of the following sequence of Gaussian operations:

- Replace equation (2.2) by itself divided by 1.5.
- Replace equation (2.1) by itself minus −0.5 times equation (2.2).
- Replace equation (2.3) by itself minus 2 times equation (2.2).
- Replace equation (2.4) by itself minus −1.5 times equation (2.2).

These Gaussian operations transform system (2) into system (3), below. They create no solutions and destroy none.

(3.1) $1x_1 + 2x_2 \qquad + 3x_4 = 5/3$

(3.2) $+ 1x_3 - 2x_4 = -2/3$

(3.3) $0x_4 = p + 4/3$

(3.4) $3x_2 \qquad + 2x_4 = 1$

This pivot made x_3 basic for equation (3.2). It *kept* x_1 basic for equation (3.1). That is no accident. Why? The coefficient of x_1 in equation (2.2) had been set equal to zero, so replacing another equation by itself less some constant times equation (2.2) cannot change its coefficient of x_1. The property that this illustrates holds in general. It is emphasized:

> Pivoting on a nonzero coefficient of a variable x in an equation has these effects:
>
> - The variable x becomes basic for the equation that has the coefficient on which the pivot occurred.
> - Any variable that had been basic for another equation remains basic for that equation.

5. Trite and Inconsistent Equations

The idea that motivates Gauss-Jordan elimination is to keep pivoting until a basic variable has been created for each equation. There is a complication, however, and it is now within view. Equation (3.3) is

$$0x_1 + 0x_2 + 0x_3 + 0x_4 = p + 4/3.$$

Let us recall that p is a datum (number), not a decision variable. It is clear that equation (3.3) has a solution if $p = -4/3$ and that it has no solution if $p \neq -4/3$. This motivates a pair of definitions. The equation

$$0x_1 + 0x_2 + \cdots + 0x_n = d$$

is said to be **trite** if $d = 0$. The same equation is said to be **inconsistent** if $d \neq 0$. A trite equation poses no restriction on the values taken by the variables. An inconsistent equation has no solution.

Gauss-Jordan elimination creates no solutions and destroys none. Thus, if Gauss-Jordan elimination produces an inconsistent equation, the original equation system can have no solution. In particular, system (1) has no solution if $p \neq -4/3$.

For the remainder of this section, it is assumed that $p = -4/3$. In this case, equations (3.1) and (3.2) have basic variables, and equation (3.3) is trite. Gauss-Jordan elimination continues to pivot, aiming for a basic variable for each non-trite equation. Equation (3.4) lacks a basic variable. The variables x_2

and x_4 have nonzero coefficients in equation (3.4). Either of these variables could be made basic for that equation. Let's make x_2 basic for equation (3.4). That is accomplished by executing this sequence of Gaussian operations:

- Replace equation (3.4) by itself divided by 3.
- Replace equation (3.1) by itself minus 2 times equation (3.4).
- Replace equation (3.2) by itself minus 0 times equation (3.4).
- Replace equation (3.3) by itself minus 0 times equation (3.4).

This pivot transforms system (3) into system (4).

$$
\begin{array}{llllll}
(4.1) & 1x_1 & & & + (5/3)x_4 = 1 \\
(4.2) & & & + 1x_3 & - \quad 2x_4 = -2/3 \\
(4.3) & & & & 0x_4 = 0 \\
(4.4) & & + 1x_2 & & + (2/3)x_4 = 1/3
\end{array}
$$

In system (4), each non-trite equation has been given a basic variable. A solution to system (4) is evident. Equate each basic variable to the right-hand-side value of the equation for which it is basic, and equate any other variables to zero. That is, set:

$$x_1 = 1, \qquad x_3 = -2/3, \qquad x_2 = 1/3, \qquad x_4 = 0.$$

These values of the variables satisfy system (4), hence must satisfy system (1).

More can be said. Shifting the non-basic variable x_4 to the right-hand side of system (4) expresses *every* solution to system (4) as a function of x_4. Specifically, for each value of x_4, setting

$$
\begin{array}{ll}
(5.1) & x_1 = \quad 1 - (5/3)x_4, \\
(5.2) & x_3 = -2/3 + \quad 2x_4, \\
(5.4) & x_2 = \quad 1/3 - (2/3)x_4,
\end{array}
$$

satisfies system (4) and, consequently, satisfies system (1). By the way, the question posed earlier can now be answered: If $p \neq -4/3$. system (1) has no solution, and if $p = -4/3$, system (1) has infinitely many solutions, one for each value of x_4.

The dictionary

System (5) has been written in a format that is dubbed the **dictionary** because:

- Each equation has a basic variable, and that basic variable is the sole item on the left-hand side of the equation for which it is basic.
- The nonbasic variables appear only on the right-hand sides of the equations.

In Chapter 4, the dictionary will help us to understand the simplex method.

Consistent equation systems

An equation system is said to be **consistent** if it has at least one solution and to be **inconsistent** if it has no solution. It has been demonstrated that if an equation system is consistent, Gauss-Jordan elimination constructs a solution. And if an equation system is inconsistent, Gauss-Jordan elimination constructs an inconsistent equation.

6. A Basic System

With system (4) in view, a key definition is introduced. A system of linear equations is said to be **basic** if each equation either is trite or has a basic variable. System (4) is basic because equation (4.3) is trite and the remaining three equations have basic variables.

Basic solution

A basic equation system's **basic solution** equates each non-basic variable to zero and equates each basic variable to the right-hand-side value of the equation for which it is basic. The basic solution to system (4) is:

$$x_1 = 1, \quad x_3 = -2/3, \quad x_2 = 1/3, \quad x_4 = 0.$$

Recap of Gauss-Jordan elimination

Gauss-Jordan elimination pivots in search of a basic system, like so:

> **Gauss-Jordan elimination** While at least one non-trite equation lacks a basic variable:
>
> 1. Select a non-trite equation that lacks a basic variable. Stop if this equation is inconsistent.
>
> 2. Else select any variable whose coefficient in this equation is nonzero, and pivot on it.

When Gauss-Jordan elimination is executed, each pivot creates a basic variable for an equation that lacked one. If Gauss-Jordan elimination stops at Step 1, an inconsistent equation has been identified, and the original equation system can have no solution. Otherwise, Gauss-Jordan elimination constructs a basic system, one whose basic solution satisfies the original equation system.

A coarse measure of work

A coarse measure of the effort needed to execute an algorithm is the number of multiplications and divisions that it entails. Let's count the number of multiplications and divisions needed to execute Gauss-Jordan elimination on a system of m linear equations in n decision variables.

Each equation has n + 1 data elements, including its right-hand side value. The first Gaussian operation in a pivot divides an equation by the coefficient of one of its variables. This requires n divisions (not n + 1) because it is not necessary to divide a number by itself. Each of the remaining Gaussian operations in a pivot replaces an equation by itself less a particular constant d times another equation. This requires n multiplications (not n+1) because it is not necessary to compute $d - d = 0$. We've seen that each Gaussian operation in a pivot requires n multiplications or divisions. Evidently:

- Each Gaussian operation entails n multiplications or divisions.

- Each pivot entails m Gaussian operations, one per equation, for a total of m n multiplications and divisions per pivot.

- Gauss-Jordan elimination requires as many as m pivots, for a total of m^2 n multiplications and divisions.

In brief:

> **Worst-case work:** Executing Gauss-Jordan elimination on a system of m linear equations in n unknowns requires as many as $m^2 n$ multiplications and divisions.

Doubling m and n multiplies the work bound $m^2 n$ by $2^3 = 8$. Evidently, the worst-case work bound grows as the *cube* of the problem size. That is not good news. Fortunately, as linear programs get larger, they tend to get sparser (have a higher percentage of 0's), and sparse-matrix techniques help to make large problems tractable. How that occurs is discussed, briefly, in the starred section of this chapter.

7. Identical Columns

A minor complication has been glossed over: a basic system can have more than one basic solution. To indicate how this can occur, consider system (6), below. It differs from system (1) in that p equals −4/3 and in that it has a fifth decision variable, x_5, whose coefficient in each equation equals that of x_2.

$$(6.1) \qquad 2x_1 + 4x_2 - 1x_3 + 8x_4 + 4x_5 = 4$$

$$(6.2) \qquad 1x_1 + 2x_2 + 1x_3 + 1x_4 + 2x_5 = 1$$

$$(6.3) \qquad 2x_3 - 4x_4 = -4/3$$

$$(6.4) \qquad -1x_1 + 1x_2 - 1x_3 + 1x_4 + 1x_5 = 0$$

From a practical viewpoint, the variables x_2 and x_5 are indistinguishable; either can substitute for the other, and either can be eliminated. But let's see what happens if we leave both columns in and pivot as before. The first pivot makes x_1 basic for equation (6.1). This pivot begins by replacing equation (6.1) by itself times (1/2). Note that the coefficients of x_2 and x_5 in this equation remain equal; they started equal, and both were multiplied by (1/2). The next step in this pivot replaces equation (6.2) by itself less equation (6.1).

The coefficients of x_2 and x_5 in equation (6.2) remain equal. And so forth. A general principle is evident. It is that:

> Identical columns stay identical after executing any number of Gaussian operations.

As a consequence, applying to system (6) the same sequence of Gaussian operations that transformed system (1) into system (4) produces system (7), below. System (7) is identical to system (4), except that the coefficient of x_5 in each equation equals the coefficient of x_2 in that equation.

(7.1) $1x_1 + (5/3)x_4 = 1$

(7.2) $+ 1x_3 - 2x_4 = -2/3$

(7.3) $0x_4 = 0$

(7.4) $+ 1x_2 + (2/3)x_4 + 1x_5 = 1/3$

The variables x_2 *and* x_5 are basic for equation (7.4). When x_2 became basic, x_5 also became basic. Do you see why?

System (7) has two basic solutions. One basic solution corresponds to selecting x_2 as the basic variable for equation (7.4), and it sets

(8) $x_1 = 1,\quad x_2 = 1/3,\quad x_3 = -2/3,\quad x_4 = 0,\quad x_5 = 0.$

The other basic solution corresponds to selecting x_5 as the basic variable for equation (7.4), and it sets

(9) $x_1 = 1,\quad x_2 = 0,\quad x_3 = -2/3,\quad x_4 = 0,\quad x_5 = 1/3.$

This ambiguity is due to identical columns. Gaussian operations are reversible, so columns that are identical *after* a Gaussian operation occurred must have been identical before it occurred. Hence, distinct columns stay distinct. In brief:

> If an equation has more than one basic variable, two or more variables in the original system had identical columns of coefficients, and all of them became basic for that equation.

The fact that identical columns stay identical is handy – in later chapters, it will help us to understand the simplex method.

8. A Basis and its Basic Solution

Consider any basic system. A set of variables is called a **basis** if this set consists of one basic variable for each equation that has a basic variable. System (4) has one basis, which is the set $\{x_1, x_3, x_2\}$ of variables. System (7) has two bases. One of these bases is the set $\{x_1, x_3, x_2\}$ of variables. The other basis is $\{x_1, x_3, x_5\}$.

Again, consider any basic system. Each basis for it has a unique **basic solution**, namely, the solution to the equation system in which each nonbasic variable is equated to zero and each basic variable is equated to the right-hand-side value of the equation for which it is basic. System (7) is basic. It has two bases and two basic solutions; equation (8) gives the basic solution for the basis $\{x_1, x_3, x_2\}$, and (9) gives the basic solution for the basis $\{x_1, x_3, x_5\}$.

The terms "basic variable," "basis," and "basic solution" suggest that a basis for a vector space lurks nearby. That vector space is identified later in this chapter.

9. Pivoting on a Spreadsheet

Pivoting by hand gets old fast. Excel can do the job flawlessly and painlessly. This section tells how.

A detached-coefficient tableau

The spreadsheet in Table 3.1 will be used to solve system (1) for the case in which $p=-4/3$. Rows 1 through 5 of Table 3.1 are a **detached-coefficient tableau** for system (1). *Note that:*

- Each variable has a column heading, which is recorded in row 1.
- Rows 2 through 5 contain the coefficients of the equations in system (1), as well as their right-hand-side values.
- The "=" signs have been omitted.

Table 3.1. Detached coefficient tableau for system (1) and the first pivot.

	A	B	C	D	E	F
1	equation	x1	x2	x3	x4	RHS
2	(1.1)	2	4	-1	8	4
3	(1.2)	1	2	1	1	1
4	(1.3)	0	0	2	-4	-1.3333333
5	(1.4)	-1	1	-1	1	0
6						
7	(2.1)	1	2	-0.5	4	2
8	(2.2)	0	0	1.5	-3	-1
9	(2.3)	0	0	2	-4	-1.3333333
10	(2.4)	0	3	-1.5	5	2
11						
12						
13			= pivot(B2 , B2:F5)			

The first pivot

This spreadsheet will be used to execute the same sequence of pivots as before. The first of these pivots will occur on the coefficient of x_1 in equation (1.1). This coefficient is in cell B2 of Table 3.1. Rows 7 though 11 display the result of that pivot. *Note that:*

- Row 7 equals row 2 multiplied by (1/2).
- Row 8 equals row 3 less 1 times row 7.
- Row 9 equals row 4 less 0 times row 7.
- Row 10 equals row 5 less −1 times row 7.

Excel functions could be used to create rows 7-10. For instance, row 7 could be obtained by inserting in cell B7 the function **=B2/$B2** and dragging it across the row. Similarly, row 8 could be obtained by inserting in cell B8 the function **=B3 – $B3 * B$7** and dragging it across the row. But there is an easier way.

An Add-In

As Table 3.1 suggests, the array function **=pivot(cell, array)** executes this pivot. The easy way to replicate rows 7-10 of Table 3.1 is as follows:

- Select the array **B7:F10** (This causes the result of the pivot to appear in cells B7 through F10.)
- Type =**pivot(B2, B2:F5)** to identify B2 as the pivot element and B2:F5 as the array of coefficients on which the pivot is to occur.
- Type **Ctrl+Shift+Enter** to remind Excel that this is an array function. (It is an array function because it places values in a block (array) of cells, rather than in a single cell.)

The function =**pivot(cell, array)** makes short work of pivoting. This function does not come with Excel, however. It is an Add-In. It is included in the software that accompanies this text, where it is one of the functions in *Optimization Tools*. Before you can use it, you must install it in your Excel Library and activate it. Chapter 2 tells how to do that.

The second and third pivots

Table 3.2 reports the result of executing two more pivots with the same array function.

Table 3.2. Two further pivots on system (1).

	A	B	C	D	E	F
1	equation	x1	x2	x3	x4	RHS
7	(2.1)	1	2	-0.5	4	2
8	(2.2)	0	0	1.5	-3	-1
9	(2.3)	0	0	2	-4	-1.3333333
10	(2.4)	0	3	-1.5	5	2
11						
12	(3.1)	1	2	0	3	1.6666667
13	(3.2)	0	0	1	-2	-0.6666667
14	(3.3)	0	0	0	0	0
15	(3.4)	0	3	0	2	1
16						
17	(4.1)	1	0	0	1.6666667	1
18	(4.2)	0	0	1	-2	-0.6666667
19	(4.3)	0	0	0	0	0
20	(4.4)	0	1	0	0.6666667	0.3333333
21						
22						
23			= pivot(C15, B12:F15)			

To execute these two pivots:

- Select the block B12:F15 of cells, type **=pivot(D8, B7:F10)** and then hit **Ctrl+Shift+Enter**
- Select the block B17:F20 of cells, type =**pivot(C15, B12:F15)** and then hit **Ctrl+Shift+Enter**

Rows 17-20 report the result of these pivots. The data in rows 17-20 are identical to those in system (4) with $p = -4/3$. In particular:

- The variable x_1 has been made basic for equation (4.1).
- The variable x_3 has been made basic for equation (4.2).
- Equation (4.3) has become trite.
- The variable x_2 has been made basic for equation (4.4).

Pivoting with an Add-In is easy and is error-proof. It has an added advantage – it *re-executes* the pivot sequence after each change in a datum. The moment you change a value in cells B2:F5 of the spreadsheet in Table 3.1, Excel re-executes the pivot sequence, and it does so with blazing speed.

10. Exchange Operations

Many presentations of Gauss-Jordan elimination include four Gaussian operations, of which only two have been presented. The other Gaussian operations are called **exchange operations**, and they appear below:

- Exchange the positions of a pair of equations.
- Exchange the positions of a pair of variables.

Like the others, these exchange operations can be undone. To recover the original equation system after doing an exchange operation, simply repeat it.

The exchange operations do not help us to construct a basis. They do serve a "cosmetic" purpose. They let us state results in simple language. For instance, the exchange operations let us place the basic variables on the diagonal and the trite equations at the bottom. To illustrate, reconsider Table 3.2.

Exchanging rows 19 and 20 shifts the trite equation to the bottom. Then, exchanging columns C and D puts the basic variables on the diagonal.

In linear algebra, the two Gaussian operations that were introduced earlier and the first of the above two exchange operations are known as **elementary row operations**. Most texts on linear algebra *begin* with a discussion of elementary row operations and their properties. That's because Gaussian operations are fundamental to linear algebra.

11. Vectors and Convex Sets

Modern computer codes solve linear systems that have of hundreds or thousands of equations, as does the simplex method. These systems are impossible to visualize. Luckily, the intuition obtained from 2-dimensional and 3-dimensional geometry holds up in higher dimensions. It provides insight as to what's going on. This section probes the relevant geometry, as it applies to vectors and convex sets. Much of this section may be familiar, but you might welcome a review.

Vectors

A linear program has some number n of decision variables, and n may be large. An ordered set $x = (x_1, x_2, \ldots, x_n)$ of values of these decision variables is called a **vector** or an **n-vector**, the latter if we wish to record the number of entries in it.

Similarly, the symbol $\Re^n$ denotes the set of all n-vectors, namely, the set that consists of each vector $x = (x_1, x_2, \ldots, x_n)$ as x_1 through x_n vary, independently, over the set $\Re$ of all real numbers. This set $\Re^n$ of all n-vectors is known as **n-dimensional space** or, more succinctly, as **n-space**. The n-vector $x=(0, 0, \ldots, 0)$ is called the **origin** of $\Re^n$.

Relax! There will be no need to visualize higher-dimensional spaces because we can proceed by analogy with plane and solid geometry. Figure 3.1 is a two-dimensional example. In it, the ordered pair $x=(5, 1)$ of real numbers is located five units to the right of the origin and 1 unit above it. Also, the ordered pair $y=(-2, 3)$ is located two units to the left of the origin and three units above it.

Figure 3.1. The vectors x=(5, 1) and y=(−2, 3) and their sum.

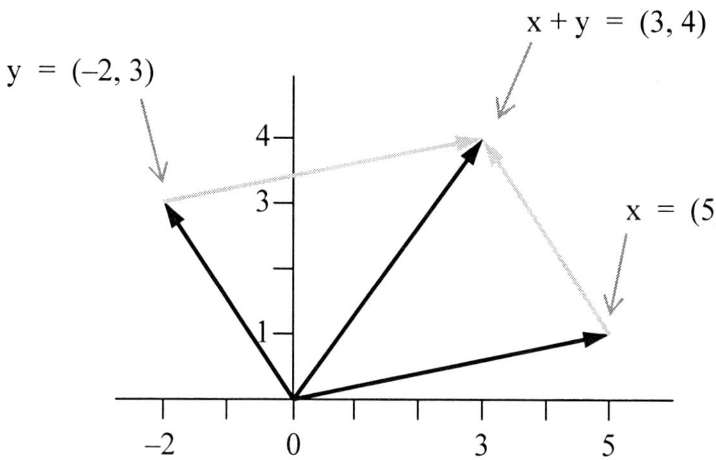

Vector addition

Let $x = (x_1, x_2, \ldots, x_n)$ and $y = (y_1, y_2, \ldots, y_n)$ be two n-vectors. The **sum**, x + y, of the vectors x and y is defined by

(10) $$x + y = (x_1 + y_1,\ x_2 + y_2,\ \ldots,\ x_n + y_n).$$

Vector addition is no mystery: simply add the components. This is true of vectors in $\Re^2$, in $\Re^3$, and in higher-dimensional spaces. Figure 3.1 depicts the sum x + y of the vectors x=(5, 1) and y=(−2, 3). Evidently,

$$(5, 1) + (-2, 3) = (5 - 2, 1 + 3) = (3, 4).$$

The gray lines in Figure 3.1 indicate that, graphically, to take the sum of the vectors (5, 1) and (−2, 3), we can shift the "tail" of either vector to the head of the other, while preserving the "length" and "direction" of the vector that is being shifted.

Scalar multiplication

If $x = (x_1, x_2, \ldots, x_n)$ is a vector and if c is a real number, the **scalar multiple** of x and c is defined by

(11) $$cx = (cx_1, cx_2, \ldots, cx_n).$$

Evidently, to multiply a vector x by a scalar c is to multiply each component of x by c. This scalar c can be any real number – positive, negative or zero.

What happens when the vector x in Figure 3.1 is multiplied by the scalar c=0.75? Each entry in x is multiplied by 0.75. This reduces the length of the vector x without changing the direction in which it points.

What happens when the vector x is multiplied by the scalar c=−1? Each entry in x is multiplied by −1. This reverses the direction in which x points, but does not change its length.

With y as a vector, the scalar product (−1) y is abbreviated as – y. With x and y as two vectors that have the same number n of components, the **difference**, x − y is given by

$$x - y = x + (-1)y = (x_1 - y_1, x_2 - y_2, \ldots, x_n - y_n).$$

Displayed in Figure 3.2 is the difference x − y of the vectors x=(5, 1) and y=(−2, 3). These two vectors have x − y=(5, 1) − (−2, 3)=(7, −2).

Figure 3.2. The vectors x=(5, 1) and y=(−2, 3) and their difference.

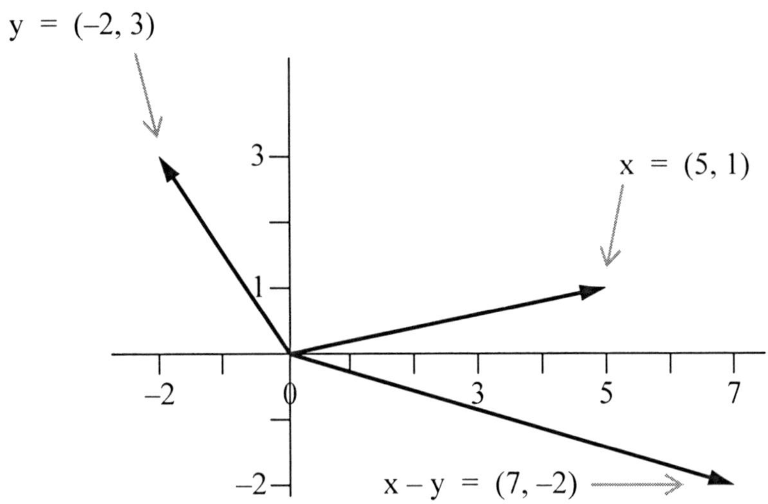

Convex combinations and intervals

Let $x=(x_1, x_2, \ldots, x_n)$ and $y=(y_1, y_2, \ldots, y_n)$ be two n-vectors and let c be a number that satisfies $0 \le c \le 1$. The vector

$$cx + (1 - c)y$$

is said to be a **convex combination** of the vectors x and y. Similarly, the **interval** between x and y is the set S of n-vectors that is given by

$$S = \{cx + (1 - c)y : 0 \leq c \leq 1\}. \tag{12}$$

Here and hereafter, a colon within a mathematical expression is read as "such that." Equation (12) defines the interval S as the set of all convex combinations of x and y. Figure 3.3 illustrates these definitions.

Figure 3.3. The thick gray line segment is the interval between x=(5, 1) and y=(−2, 3).

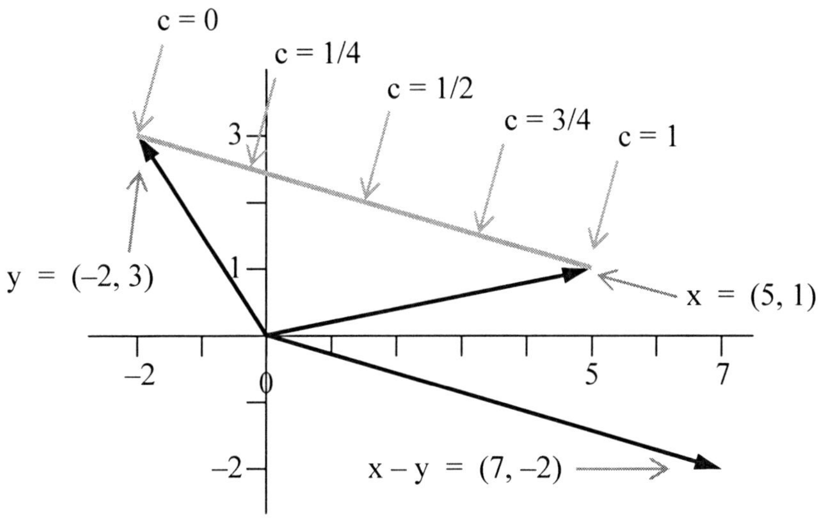

Each convex combination of the vectors x and y that are depicted in Figure 3.3 can be written as

$$cx + (1 - c)y = cx + y - cy = y + c(x - y), \tag{13}$$

where c is a number that lies between 0 and 1, inclusive. Evidently, the interval between x and y consists of each vector y + c(x − y) obtained by adding y and the vector c(x − y) as c varies from 0 to 1. Figure 3.3 depicts y + c(x − y) for the values c = 0, 1/4, 1/2, 3/4 and 1.

By the way, if x and y are distinct n-vectors, the **line** that includes x and y is the set L that is given by

$$L = \{cx + (1 - c)y : c \in \Re\}. \tag{14}$$

This line includes x (take c=1) and y (take c=0), it contains the interval between x and y, and it extends without limit in both directions.

Convex sets

A set C of n-vectors is said to be **convex** if C contains the interval between each pair of vectors in C. Figure 3.4 displays eight shaded subsets of $\Re^2$ (the plane). The top four are convex. The bottom four are not. Can you see why?

Figure 3.4. Eight subsets of the plane.

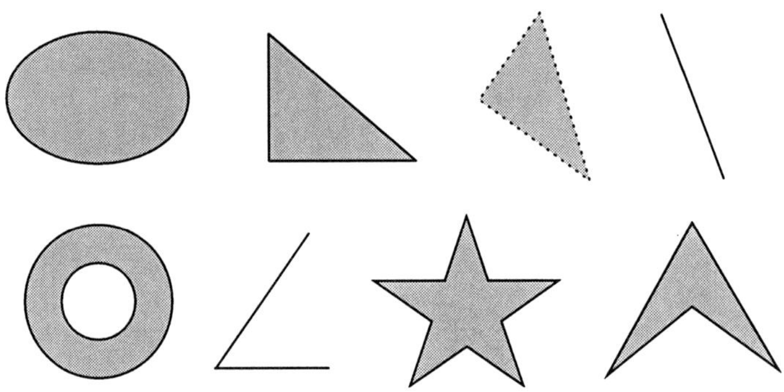

Convex sets will play a key role in linear programs and in their generalizations. A vector x that is a member of a convex set C is said to be an **extreme point** of C if x is not a convex combination of two other vectors in C. Reading from left to right, the four convex sets in Figure 3.4 have infinitely many extreme points, three extreme points, no extreme points, and two extreme points. Do you see why?

Unions and intersections

Let S and T be subsets of $\Re^n$. The **union** $S \cup T$ is the set of n-vectors that consists of each vector that is in S, or is in T or is in both. The **intersection** $S \cap T$ is the subset of $\Re^n$ that consists of each vector that is in S and is in T. It's easy to convince oneself visually (and to prove) that:

- The union $S \cup T$ of convex sets need not be convex.
- The intersection $S \cap T$ of convex sets must be convex.

Linear constraints

Let us recall from Chapter 1 that each constraint in a linear program requires a linear expression to bear one of three relationships to a number, these three being "=", "≤", and "≥." In other words, with a_0 through a_n as fixed numbers and x_1 through x_n as decision variables, each constraint takes one of these forms:

$$a_1 x_1 + a_1 x_2 + \cdots + a_n x_n = a_0$$

$$a_1 x_1 + a_1 x_2 + \cdots + a_n x_n \leq a_0$$

$$a_1 x_1 + a_1 x_2 + \cdots + a_n x_n \geq a_0$$

It's easy to check that the set of n-vectors $x=(x_1, x_2, \ldots, x_n)$ that satisfy a particular linear constraint is convex. As noted above, the intersection of convex sets is convex. Hence, the set of vectors $x = (x_1, x_2, \cdots, x_n)$ that satisfy *all* of the constraints of a linear program is convex. It is emphasized:

> The set of vectors that satisfy all of the constraints of a linear program is convex.

Convex sets play a crucial role in linear programs and in nonlinear programs.

12. Vector Spaces

The introduction to linear programming in Chapter 4 does not require an encyclopedic knowledge of vector spaces. It does use the information that is presented in this section and in the next two. A set V of n-vectors is called a **vector space** if:

- V is not empty.
- The sum of any two vectors in V is also in V.
- Each scalar multiple of each vector in V is also in V.

Each vector space V must contain the origin; that is so because V must contain at least one vector x and because it must also contain the scalar 0 times x, which is the origin. Each vector space is a convex set. Not every convex set is a vector space, however.

Geometric insight

It's clear, visually, that the subsets V of $\Re^2$ (the plane) that are vector spaces come in these three varieties:

- The set V whose only member is the origin is a vector space.
- Any line that passes through the origin is a vector space.
- The plane is itself a vector space.

Ask yourself: Which subsets of $\Re^3$ are vector spaces?

Linear combinations

Let v_1 through v_K be n-vectors, and let c_1 through c_K be scalars (numbers); the sum,

$$c_1 v_1 + c_2 v_2 + \cdots + c_K v_K, \tag{15}$$

is said to be a **linear combination** of the vectors v_1 through v_K. Evidently, a linear combination of K vectors multiplies each of them by a scalar and takes the sum.

Linearly independent vectors

The set $\{v_1, v_2, \ldots, v_K\}$ of n-vectors are said to be **linearly independent** if the only solution to

$$0 = c_1 v_1 + c_2 v_2 + \cdots + c_K v_K \tag{16}$$

is $0 = c_1 = c_2 = \cdots = c_K$. In other words, the n-vectors v_1 through v_K are linearly independent if the only way to obtain the vector 0 as a linear combination of these vectors is to multiply each of them by the scalar 0 and then add them up.

Similarly, the set $\{v_1, v_2, \ldots, v_K\}$ of n-vectors is said to be **linearly dependent** if these vectors are *not* linearly independent, equivalently, if a solution to (16) exists in which not all of the scalars equal zero. Convince yourself, visually, that:

- Any set of n-vectors that includes the origin is linearly dependent.
- Two n-vectors are linearly independent if neither is a scalar multiple of the other.
- In the plane, $\Re^2$, every set of three vectors is linearly dependent.

A set $\{v_1, v_2, \ldots, v_K\}$ of vectors in a vector space V is said to **span** V if every vector in V is a linear combination of these vectors.

A basis

Similarly, a set $\{v_1, v_2, \ldots, v_K\}$ of vectors in a vector space V is said to be a **basis** for V if the vectors v_1 through v_K are linearly independent and if every element of V is a linear combination of this set $\{v_1, v_2, \ldots, v_K\}$ of vectors.

Trouble?

A basis has just been defined as a set of *vectors*. Earlier, in our discussion of Gauss-Jordan elimination, a basis had been defined as a set of *decision variables*. That looks to be incongruous, but a correspondence will soon be established.

13. Matrix Notation

It will soon be seen that Gauss-Jordan elimination constructs a basis for the "column space" of a matrix. Before verifying that this is so, we interject a brief discussion of matrix notation.

In the prior section, the entries in the n-vector $x=(x_1, x_2, \ldots, x_n)$ could have been arranged in a row or in a column. When doing matrix arithmetic, it is necessary to distinguish between rows and columns.

Matrices

A "matrix" is a rectangular array of numbers. Whenever possible, capital letters are used to represent matrices. Depicted below is an $m \times n$ matrix A. Evidently, the integer m is the number of rows in A, the integer n is the number of columns, and A_{ij} is the number at the intersection of the ith row and j^{th} column of A.

(17)
$$A = \begin{bmatrix} A_{11} & A_{12} & \cdots & A_{1n} \\ A_{21} & A_{22} & \cdots & A_{2n} \\ \vdots & \vdots & \vdots & \\ A_{m1} & A_{m2} & \cdots & A_{mn} \end{bmatrix}$$

Throughout, when A is an $m \times n$ matrix, A_j denotes the jth column of A and A^i denotes the ith row of A.

$$A_j = \begin{bmatrix} A_{1j} \\ A_{2j} \\ \vdots \\ A_{mj} \end{bmatrix}, \qquad A^i = [A_{i1}\ A_{i2}\ \cdots\ A_{in}]$$

Matrix multiplication

This notation helps us to describe the product of two matrices. To see how, let E be a matrix that has r columns and let F be a matrix that has r rows. The matrix product E F can be taken, and the ij^{th} element $(EF)_{ij}$ of this matrix product equals the sum over k of $E_{ik} F_{kj}$. In other words,

(18)
$$(EF)_{ij} = \sum\nolimits_{k=1}^{r} E_{ik} F_{kj} = E^i F_j \qquad \text{for each i and j.}$$

Thus, the ij^{th} element of the matrix (EF) equals the product $E^i F_j$ of the ith row of E and the jth column of F.

Similarly, the ith row $(EF)^i$ of EF and j^{th} column $(EF)_j$ of EF are given by

(19)
$$(EF)^i = E^i F,$$

(20)
$$(EF)_j = EF_j.$$

It is emphasized:

> The ith row of the matrix product (EF) equals E^iF and the jth column of this matrix product equals EF_j

Vectors

In this context, a vector is a matrix that has only one row or only one column. Whenever possible, lower-case letters are used to represent vectors. Displayed below are an $n \times 1$ vector x and an $m \times 1$ vector b.

$$x = \begin{bmatrix} x_1 \\ x_2 \\ \vdots \\ x_n \end{bmatrix}, \qquad b = \begin{bmatrix} b_1 \\ b_2 \\ \vdots \\ b_m \end{bmatrix}$$

Evidently, a single subscript identifies an entry in a vector; for instance, x_j is entry in row j of x.

The equation Ax=b

A system of m linear equations in n unknowns is written succinctly as $Ax=b$. Here, the decision variables are x_1 through x_n, the number A_{ij} is the coefficient of x_j in the ith equation, and the number b_i is the right-hand side value of the ith equation.

The matrix equation $Ax=b$ appears repeatedly in this book. As a memory aide, the following conventions are employed:

- The data in the equation $Ax=b$ are the $m \times n$ matrix A and the $m \times 1$ vector b.
- The decision variables (unknowns) in this equation are arrayed into the $n \times 1$ vector x.

In brief, the integer m is the number of rows in the matrix A, and the integer n is the number of columns. Put another way, the matrix equation $Ax=b$ is a system of m equations in n unknowns.

The matrix product Ax

When the equation Ax=b is studied, the matrix product Ax is of particular importance. Evidently, Ax is an m × 1 vector. Expression (19) with E=A and F=x confirms that the ith element of Ax equals $A^i x$, indeed, that

(21) $$Ax = \begin{bmatrix} A_{11}x_1 + A_{12}x_2 + \cdots + A_{1n}x_n \\ A_{21}x_1 + A_{22}x_2 + \cdots + A_{2n}x_n \\ \vdots \qquad \vdots \qquad \vdots \\ A_{m1}x_1 + A_{m2}x_2 + \cdots + A_{mn}x_n \end{bmatrix}.$$

Note in expression (21) that the number (scalar) x_1 multiplies each entry in A_1 (the 1st column of A), that the scalar x_2 multiplies each entry in A_2, and so forth. In other words,

(22) $$Ax = A_1x_1 + A_2x_2 + \cdots + A_nx_n.$$

Equation (22) interprets Ax as a linear combination of the columns of A. It is emphasized:

The matrix product Ax is a linear combination of the columns of A. In particular, the scalar x_j multiplies A_j.

You may recall that the "column space" of a matrix A is the set of all linear combinations of the columns of A; we will get to that shortly.

The matrix product yA

Let y be any 1 × m vector. Since A is an m × n matrix, the matrix product yA can be taken. Equation (20) shows that the jth entry in yA equals yA_j. In other words,

(23) $$yA = (yA_1, yA_2 \ldots, yA_n).$$

Just as Ax is a linear combination of the columns of A, the matrix product yA is a linear combination of the rows of A, one in which y_1 multiplies each element of A^1, in which y_2 multiplies each element of A^2, and so forth.

(24) $$yA = y_1A^1 + y_2A^2 + \cdots + y_mA^m$$

In brief:

> The matrix product yA is a linear combination of the rows of A. In particular, the scalar y_i multiplies A^i.

An ambiguity

When A is a matrix, two subscripts denote an entry, a single subscript denotes a column, and a single superscript denotes a row. The last of these conventions must be taken with a grain of salt; "T" abbreviates "transpose," and A^T denotes the **transpose** of the matrix A, not its Tth row.

14. The Row and Column Spaces

Let A be an $m \times n$ matrix. For each $n \times 1$ vector x, equation (22) interprets the matrix product Ax as a linear combination of the columns of A. The set V_c that is specified by the equation,

$$(25) \qquad V_c = \{Ax : x \in \Re^{n\times 1}\},$$

is called the **column space** of the matrix A. Equation (25) reads, "V_c equals the set that contains Ax for every $n \times 1$ vector x." It is clear from equation (22) that V_c is the set of all linear combinations of the columns of the matrix A, moreover, that V_c is a vector space.

With A as an $m \times n$ matrix and with y as a $1 \times m$ vector, equation (24) interprets yA as a linear combination of the rows of A. The set V_r that is specified by the equation,

$$(26) \qquad V_r = \{yA : y \in \Re^{1\times m}\},$$

is called the **row space** of A. Evidently, V_r is the set of all linear combinations of the rows of the matrix A, and it too is a vector space.

A basis for the column space

Gauss-Jordan elimination can be used to construct a basis for the column space of a matrix. In fact, Gauss-Jordan elimination *has* been used to construct a basis for the column space of the 4×4 matrix A that is given by

(27) $$A = \begin{bmatrix} 2 & 4 & -1 & 8 \\ 1 & 2 & 1 & 1 \\ 0 & 0 & 2 & -4 \\ -1 & 1 & -1 & 1 \end{bmatrix}.$$

Let us see how. With A given by (27) and with x as a 4×1 vector, equation (22) shows the matrix product A x is this linear combination of the columns of A.

(28) $$\mathrm{A\,x} = \begin{bmatrix} 2 \\ 1 \\ 0 \\ -1 \end{bmatrix} x_1 + \begin{bmatrix} 4 \\ 2 \\ 0 \\ 1 \end{bmatrix} x_2 + \begin{bmatrix} -1 \\ 1 \\ 2 \\ -1 \end{bmatrix} x_3 + \begin{bmatrix} 8 \\ 1 \\ -4 \\ 1 \end{bmatrix} x_4.$$

Please observe that (28) is identical to the left-hand side of system (1).

A homogeneous equation

The matrix equation Ax=b is said to be **homogeneous** if its right-hand-side vector b consists entirely of 0's. With Ax given by (28), let us study solutions x to the (homogeneous) equation Ax=0. This equation appears below as

(29) $$\begin{bmatrix} 2 \\ 1 \\ 0 \\ -1 \end{bmatrix} x_1 + \begin{bmatrix} 4 \\ 2 \\ 0 \\ 1 \end{bmatrix} x_2 + \begin{bmatrix} -1 \\ 1 \\ 2 \\ -1 \end{bmatrix} x_3 + \begin{bmatrix} 8 \\ 1 \\ -4 \\ 1 \end{bmatrix} x_4 = \begin{bmatrix} 0 \\ 0 \\ 0 \\ 0 \end{bmatrix},$$

No new work is needed to identify the solutions to (29). To see why, replace the right-hand side values of system (1) by 0's and *repeat* the Gaussian operations that transformed system (1) into system (4), getting:

(30) $$\begin{bmatrix} 1 \\ 0 \\ 0 \\ 0 \end{bmatrix} x_1 + \begin{bmatrix} 0 \\ 0 \\ 0 \\ 1 \end{bmatrix} x_2 + \begin{bmatrix} 0 \\ 1 \\ 0 \\ 0 \end{bmatrix} x_3 + \begin{bmatrix} 5/3 \\ -2 \\ 0 \\ 2/3 \end{bmatrix} x_4 = \begin{bmatrix} 0 \\ 0 \\ 0 \\ 0 \end{bmatrix}.$$

These Gaussian operations preserve the set of solutions to the equation system; the scalars x_1 through x_4 satisfy (29) if and only if they satisfy (30). From this fact, we conclude that:

- The columns of A are linearly dependent because (30) is satisfied by equating x_4 to any nonzero number and setting

$$x_1 = -(5/3)x_4, \qquad x_3 = 2x_4, \qquad x_2 = -(2/3)x_4.$$

- The columns A_1, A_2 and A_3 are linearly independent because setting $x_4 = 0$ in (29) and (30) shows that the only solution to $A_1x_1 + A_2x_2 + A_3x_3 = 0$ is $x_1 = x_2 = x_3 = 0$
- The vector A_4 a linear combination of A_1, A_2 and A_3 because applying the same sequence of Gaussian operations to the system

$$\begin{bmatrix} 2 \\ 1 \\ 0 \\ -1 \end{bmatrix} x_1 + \begin{bmatrix} 4 \\ 2 \\ 0 \\ 1 \end{bmatrix} x_2 + \begin{bmatrix} -1 \\ 1 \\ 2 \\ -1 \end{bmatrix} x_3 = \begin{bmatrix} 8 \\ 1 \\ -4 \\ 1 \end{bmatrix}$$

transforms it into

$$\begin{bmatrix} 1 \\ 0 \\ 0 \\ 0 \end{bmatrix} x_1 + \begin{bmatrix} 0 \\ 0 \\ 0 \\ 1 \end{bmatrix} x_2 + \begin{bmatrix} 0 \\ 1 \\ 0 \\ 0 \end{bmatrix} x_3 = \begin{bmatrix} 5/3 \\ -2 \\ 0 \\ 2/3 \end{bmatrix},$$

which demonstrates that $A_4 = (5/3)A_1 + (2/3)A_2 - 2A_3$.

These observations imply that the set $\{A_1, A_2, A_3\}$ of vectors is a basis for the column space of A. This is so because the vectors A_1, A_2 and A_3 are linearly independent and because A_4 is a linear combination of them, which guarantees that every linear combination of A_1 through A_4 can be expressed as a linear combination of A_1, A_2 and A_3. The same line of reasoning works for every matrix. It is presented as:

Proposition 3.1 *(basis finder)*. Consider any matrix A. Apply Gauss-Jordan elimination to the equation Ax=0 and, at termination, denote as

$\{A_j : j \in C\}$ the set of columns on which pivots have occurred. This set $\{A_j : j \in C\}$ of columns is a basis for the column space of A.

Proof. This application of Gauss-Jordan elimination cannot terminate with an inconsistent equation because setting $x=0$ produces a solution to $Ax=0$. It must terminate with a basic solution. Denote as $\{A_j : j \in C\}$ the set of columns that on which pivots have occurred.

The analog of (30) indicates that the set $\{A_j : j \in C\}$ of columns must be linearly independent and that each of the remaining columns must be a linear combination of these columns. Thus, the set $\{A_j : j \in C\}$ of columns span the column space of A, which completes a proof. ■

Reconciliation

Early in this chapter, Gauss-Jordan elimination had been used to transform system (1) into system (4). Let us recall that system (4) is basic, specifically, that the set $\{x_1, x_2, x_3\}$ of decision variables is a basis for system (4).

In the current section, the same Gauss-Jordan procedure has been used to identify the set $\{A_1, A_2, A_3\}$ of columns as a basis for the column space of A. These are two different ways of making the same statement. It is emphasized:

> The statement that a set of *variables* is a basis for the equation system $Ax = b$ means that their *columns* of coefficients are a basis for the column space of A and that b lies in the column space of A.

A third way to describe a basis

When the variables in the equation system $Ax=b$ are labeled x_1 through x_n, a basis can also be described as a subset β of the first n integers. A subset β of the first n integers is now said to be a **basis** if $\{A_j : j \in \beta\}$ is a basis for the column space of A . In brief, the *same* basis for the column space of the 4×4 matrix A in equation (27) is identified in these three ways:

- As the set $\{A_1, A_2, A_3\}$ of columns of A.
- As the set $\{x_1, x_2, x_3\}$ of decision variables.
- As the set $\beta=\{1, 2, 3\}$ of integers.

Each way in which to describe a basis has its advantages: Describing a basis as a set of columns is precise. Describing a basis as a set of decision variables will prove to be particularly convenient in the context of a linear program. Describing a basis as a set of integers is succinct.

What about the row space?

A basis for the row space of a matrix A could be found by applying Gauss-Jordan elimination to the equation $A^T x = 0$, where A^T denotes the transpose of A. A second application of Gauss-Jordan elimination is not necessary, however.

Three important results

Three key results about vector spaces are stated and illustrated in this subsection. These three results are highlighted below:

> Three results:
>
> - Every basis for a vector space contains the same number of elements, and that number is called the **rank** of the vector space.
> - The row space and the column spaces of a matrix A have the same rank.
> - If the equation $Ax = b$ has a solution, execution of Gauss–Jordan elimination constructs a basic system, and the set of rows on which pivots occur is a basis for the row space of A.

All three of these results are important. Their proofs are postponed, however, to Chapter 10, which sets the stage for a deeper understanding of linear programming.

To illustrate these results, we recall that the coefficients of the decision variables in system (1) array themselves into the 4×4 matrix A in equation (27). A sequence of pivots transformed system (1) into system (4). These pivots occurred on coefficients in rows 1, 2 and 4, and they produced a basic tableau whose basis is the set $\{x_1, x_2, x_3\}$ of decision variables. Proposition 3.1 and the above results show that:

- The set $\{A_1, A_2, A_3\}$ of columns is a basis for the column space of the matrix A in (27).
- This matrix A has 3 as the rank of its column space.

- This matrix has 3 as the rank of its row space.
- The set $\{A^1, A^2, A^4\}$ of rows is a basis for the row space of A.

The rank of a vector space is also called its **dimension**; these terms are synonyms. "Dimension" jives better with our intuition. In 3-space, every plane through the origin has 2 as its dimension (or rank), for instance.

15. Efficient Computation*

Efficient computation is vital to codes that solve large linear programs, e.g., those having thousands of decision variables. Efficient computation is not essential to a basic grasp of linear programming, however. For that reason, it is touched upon lightly in this starred section.

Pivots make the simplex method easy to understand, but they are relatively inefficient. Gaussian elimination substitutes "lower pivots" for pivots. It solves an equation system with roughly half the work. Or less.

Lower pivots

To describe lower pivots, we identify the set S of equations on which lower pivots have not yet occurred. Initially, S consists of all of the equations in the system that is being solved. Each lower pivot selects an equation in S, removes it, and executes certain Gaussian operations on the equations that remain in S. Specifically, each **lower pivot** consists of these steps:

- Select an equation (j) in S and a variable x whose coefficient in equation (j) is not zero.
- Remove equation (j) from S.
- For each equation (k) that remains in S, replace equation (k) by itself less the multiple of equation (j) that equates the coefficient of x in equation (k) to zero.

This verbal description of lower pivots is cumbersome. But, as was the case for full pivots, an example will make everything clear.

A familiar example

To illustrate lower pivots, we return to system (1). This system will be solved a second time, with each "full" pivot replaced by the comparable lower pivot. For convenient reference, system (1) is reproduced here as system (31).

(31.1) $$2x_1 + 4x_2 - 1x_3 + 8x_4 = 4$$

(31.2) $$1x_1 + 2x_2 + 1x_3 + 1x_4 = 1$$

(31.3) $$2x_3 - 4x_4 = p$$

(31.4) $$-1x_1 + 1x_2 - 1x_3 + 1x_4 = 0$$

Initially, before any lower pivots have occurred, the set S consists of equations (31.1) through (31.4).

The first lower pivot

In this illustration, the same pivot elements will be selected as before. The first lower pivot will occur on the coefficient of x_1 in equation (31.1). This lower pivot eliminates (drives to zero) the coefficient of x_1 in equations (31.2), (31.3) and (31.4). This lower pivot is executed by removing equation (31.1) from S and then:

- Replacing equation (31.2) by itself minus (1/2) times equation (31.1).
- Replacing equation (31.3) by itself minus (0/2) times equation (31.1).
- Replacing equation (31.4) by itself minus (−1/2) times equation (31.1).

The three equations that remain in S become:

(32.2) $$1.5x_3 - 3x_4 = -1$$

(32.3) $$2x_3 - 4x_4 = p$$

(32.4) $$3x_2 - 1.5x_3 + 5x_4 = 2$$

The variable x_1 does not appear in equations (32.2), (32.3) and (32.4). These three equations are *identical* to equations (2.2), (2.3) and (2.4), as must be.

Equation (31.1) has been set aside, temporarily. After equations (32.2) through (32.4) have been solved for values of the variables x_2, x_3 and x_4, equation (31.1) will be solved for the value of x_1 that is prescribed by these values of x_2, x_3 and x_4.

The second lower pivot

As was the case in the initial presentation of Gauss-Jordan elimination, the second pivot element will be the coefficient of x_3 in equation (32.2). A lower pivot on this coefficient will drive to zero the coefficient of x_3 in equations (32.3) and (32.4). This lower pivot is executed by removing equation (32.2) from S and then:

- Replacing equation (32.3) by itself minus (2/1.5) times equation (32.2)
- Replacing equation (32.4) by itself minus (–1.5/1.5) times equation (32.2).

This lower pivot replaces (32.3) and (32.4) by equations (33.3) and (33.4).

(33.3) $$0x_4 = p + 4/3$$

(33.4) $$3x_2 \qquad + 2x_4 = 1$$

The variable x_3 has been eliminated from equations (33.3) and (33.4). These two equations are identical to equations (3.3) and (3.4), exactly as in the case for the first lower pivot.

Equation (32.2), on which this pivot occurred, is set aside. After solving equations (33.3) and (33.4) for values of the variables x_2 and x_4, equation (32.2) will be solved for the variable x_3 on which the lower pivot has occurred.

The next lower pivot is slated to occur on equation (33.3). Again, there are two cases to consider. If p is unequal to –4/3, equation (33.3) is inconsistent, so no solution can exist to the original equation system. Alternatively, if p=–4/3, equation (33.3) is trite, and it has nothing to pivot upon.

Let us proceed on the assumption that p=−4/3. In this case, equation (33.3) is trite, so it is removed from S, which reduces to S to equation (34.4), below.

$$(34.4) \qquad 3x_2 \qquad + 2x_4 = 1$$

The final lower pivot

Only equation (34.4) remains in S. The next step calls for a lower pivot on equation (34.4). The variables x_2 and x_4 have nonzero coefficients in equation (34.4), so a lower pivot could occur on either of them. As before, we pivot on the coefficient of x_2 in this equation. But no equations remain in S after equation (34.4) is removed. Hence, this lower pivot entails no arithmetic. As concerns lower pivots, we are finished.

Back-substitution

It remains to construct a solution to system (31). This is accomplished by equating to zero each variable on which no lower pivot has occurred and then solving the equations on which lower pivots have occurred in "reverse" order. In our example, no lower pivot has occurred on a the variable x_4. With $x_4 = 0$, the three equations on which lower pivots have occurred are:

$$2x_1 + 4x_2 - 1x_3 = 4$$

$$1.5x_3 = -1$$

$$3x_2 \qquad = 1$$

The first lower pivot eliminated x_1 from the bottom two equations. The second lower pivot eliminated x_3 from the bottom equation. Thus, these equations can be solved for the variables on which their lower pivots have occurred by working from the bottom up. This process is aptly called **back-substitution**. For our example, back-substitution first solves the bottom equation for x_2, then solves the middle equation for x_3, and then solves the top equation for x_1. This computation gives $x_2 = 1/3$ and $x_3 = -2/3$ and $x_1 = 1$, exactly as before.

Solving an equation system by lower pivots and back-substitution is known as **Gaussian elimination** and by the fancier label, **L-U decomposition.** By either name, it requires roughly half as many multiplications and divisions as does Gauss-Jordan elimination. This suggests that lower pivots are twice as good. Actually, lower pivots are a bit better; they allow us to take better advantage of "sparsity" and help us to control "round-off" error.

Sparsity and fill-in

Typically, a large system of linear equations is **sparse**, which means that all but a tiny fraction of its coefficients are zeros. As pivoting proceeds, a sparse equation system tends to "fill in" as nonzero entries replace zeros.

An adroit sequence of pivots can reduce the rate at which fill-in occurs. A simple method for retarding fill-in counts the number of nonzero elements that might be created by each pivot and select a pivot element that minimizes this number. This method works with full pivots, and it works a bit better with lower pivots, for which it is now described. Specifically:

- Keep track of the set R of rows on which lower pivots have not yet occurred and the set C of columns of for which variables have not yet been made basic.
- For each $k \in C$, denote as c_k the number of equations in R in which x_k has a nonzero coefficient.
- For each $j \in R$, denote as r_j the number of variables whose coefficients in equation (j) are nonzero.

Take a moment to convince yourself that a lower pivot on the coefficient of the variable x_k in equation (j) will fill in (render non-zero) at most $(r_j - 1)(c_k - 1)$ zeros. This motivates the rule that's displayed below.

Myopic pivoter (initialized as indicated above).

While R is nonempty:

Among the pairs (j, k) with $j \in R$ and $k \in C$ for which the coefficient of x_k in row j of the current tableau is nonzero, pick a pair that minimizes $(r_j - 1)(c_k - 1)$.

Execute a lower pivot on the coefficient of x_k in row j of the current tableau.

Remove k from C and j from R.

Update r_j for each equation $j \in R$, and update c_k for each $k \in C$, .

This rule is **myopic** (near-sighted) in the sense that it aims to minimize the amount of fill–in at the moment, without looking ahead.

Gaussian elimination with back-substitution requires roughly half as many multiplications and divisions, but the worst-case work count still grows as the cube of the problem size. As the problem size increases, the coefficient matrix tends to become increasingly sparse (have a larger fraction of zeros), and the work bound grows less rapidly if care is taken to pivot in a way that retards fill-in.

Pivoting on very small numbers

Modern implementations of Excel do floating-point arithmetic with a 64-bit word length. This allows about 16 digits of accuracy. In small or moderate-sized problems, round-off error is not a problem, provided we avoid pivoting on very small numbers.

To see what can go awry, consider a matrix (array) whose nonzero entries are between 1 and 100, except for a few that are approximately 10^{-6}. Pivoting on one of these tiny entries multiplies everything in its row by 10^6 and shifts the information in some of the other rows about 6 digits to the right. Doing that once may be OK. Doing two or three times can bury the information in the other rows. And that's without worrying about the round-off error in the pivot element. In brief:

> When executing Gauss-Jordan elimination, try not to pivot on coefficients that are several orders of magnitude below the norm.

16. Review

Gauss-Jordan elimination makes repeated and systematic use of two Gaussian operations. These operations are organized into pivots. Each pivot creates a basic variable for an equation that lacked one. Each pivot keeps the variables that had been basic for the other equations basic for those equations. Gauss-Jordan elimination keeps pivoting until:

- Either it constructs an inconsistent equation.
- Or it creates a basic system, specifically, a basic variable for each non-trite equation.

If Gauss-Jordan elimination constructs an inconsistent equation, the original equation system can have no solution. If Gauss-Jordan elimination constructs a basic system, its basic solution satisfies the original equation system. This basic solution equates each non-basic variable to zero, and it equates each basic variable to the right-hand-side value of the equation for which it is basic.

Pivoting lies at the core of an introductory account of the simplex method. In Chapter 4, it will be seen that:

- The simplex method executes Gauss-Jordan elimination and then *keeps on pivoting* in search of an optimal solution to the linear program.
- The terminology introduced here is used to describe the simplex method. These terms include pivot, basic variable, basic system, basic solution, basis, and basic solution.
- Geometry will help us to visualize the simplex method and to relate it to fundamental ideas in linear algebra.

In a starred section, it was observed, that "lower" pivots and back-substitution are preferable to the "full" pivots of Gauss-Jordan elimination. Lower pivots are faster. They reduce the rate of fill-in and the accumulation of round-off error. When lower pivots are used in conjunction with the simplex method, the notation becomes rather involved, and the subject matter shifts the tenor of the discussion from linear algebra to numerical analysis, which we eschew.

17. Homework and Discussion Problems

1. To solve the following system of linear equations, implement Gauss-Jordan elimination on a spreadsheet. Turn your spreadsheet in, and indicate the functions that you have used in your computation.

$$\begin{aligned} 1\text{A} - 1\text{B} + 2\text{C} &= 10 \\ -2\text{A} + 4\text{B} - 2\text{C} &= 0 \\ 0.5\text{A} - 1\text{B} - 1\text{C} &= 6 \end{aligned}$$

2. Consider the following system of three equations in three unknowns.

$$
\begin{aligned}
2A + 3B - 1C &= 12 \\
-2A + 2B - 9C &= 3 \\
4A + 5B \quad &= 21
\end{aligned}
$$

   (a) Use Gauss-Jordan elimination to find a solution to this equation system.

   (b) Plot those solutions to this equation system in which each variable is nonnegative. Complete this sentence: The solutions that have been plotted form a ________________.

   (c) What would have happened if one of the right-hand-side values had been different from what it is? Why?

3. Use a spreadsheet to find *all* solutions to the system of linear equations that appears below. (*Hint*: construct a dictionary.)

$$
\begin{aligned}
2x_1 + 4x_2 - 1x_3 + 8x_4 + 10x_5 &= 4 \\
1x_1 + 2x_2 + 1x_3 + 1x_4 + 2x_5 &= 1 \\
2x_3 - 4x_4 - 4x_5 &= -4/3 \\
-1x_1 + 1x_2 - 1x_3 + 1x_4 - 1x_5 &= 0
\end{aligned}
$$

4. Redo the spreadsheet computation in Tables 1-4 using lower pivots in place of (full) pivots. Turn in your spreadsheet. On it, indicate the functions that you used.

5. Consider system (1) with $p = -4/3$. Alter any single coefficient of x_1 in equation (1.1) or (1.2) or (1.3) and then re-execute the pivots that produced system (4). *Remark:* No grunt-work is needed if you use spreadsheets.

   (a) What happens?

   (b) Can you continue in a way that produces a basic solution? If so, do so.

6. The matrix A given by (27) consists of the coefficients of the decision variables in system (1). For this matrix A:

(a) Use Gauss-Jordan elimination to show that A^3 is a linear combination of A^1 and A^2. *Remark:* This can be done without grunt-work if you apply the pivot function to the homogeneous equation $A^T y = 0$.

(b) Determine whether or not A^4 is a linear combination of A^1, A^2 and A^3.

(c) Which subsets of $\{A^1, A^2, A^3, A^4\}$ are a basis for the row space of A? Why?

7. Tables 1-4 showed how to execute Gauss-Jordan elimination on a spreadsheet for the special case in which the datum p equals −4/3. Re-do this spreadsheet for the general case in which the datum p can be any number. *Hint*: Append to Table 3.1 a column whose heading (in row 1) is p and whose coefficients in rows 2, 3, 4 and 5 are 0, 0, 1, and 0, respectively.

8. *(a basis)* This problem concerns the four vectors that are listed below. Solve parts (a), (b) and (c) *without* doing any numerical computation.

$$\begin{bmatrix} 2 \\ 1 \\ 0 \\ -1 \end{bmatrix}, \begin{bmatrix} 4 \\ 2 \\ 0 \\ 1 \end{bmatrix}, \begin{bmatrix} -1 \\ 1 \\ 2 \\ -1 \end{bmatrix}, \begin{bmatrix} 8 \\ 1 \\ -4 \\ 1 \end{bmatrix}.$$

(a) Show that the left-most three of these vectors are linearly independent.

(b) Show that the left-most three of these vectors span the other one.

(c) Show that the left-most three of these vectors are a basis for the vector space that consists of all linear combinations of these four vectors.

9. **(Opposite columns)** In the equation Ax=b, columns j and k are said to be **opposite** if $A_j = -A_k$. Suppose columns 5 and 12 are opposite.

(a) After one Gaussian operation, columns 5 and 12 ___________.

(b) After any number of Gaussian operations, columns 5 and 12 ___________.

(c) If a pivot makes x_5 basic for some equation, then x_{12} ___________.

10. (*Homogeneous systems*) True or false?

    (a) When Gauss-Jordan elimination is applied to a homogeneous system, it can produce an inconsistent equation.

    (b) Every (homogeneous) system $Ax=0$ has at least one *non-trivial* solution, that is, one solution that has $x \neq 0$

    (c) Application of Gauss-Jordan elimination to a homogeneous system constructs a non-trivial solution if one exists.

    (d) Every homogeneous system of four equations in five variables has at least one non-trivial solution.

11. Let A be an $m \times n$ matrix with $m < n$.

    (a) Show that the columns of A are linearly dependent.

    (b) Prove or disprove: There exists a nonzero vector x such that $Ax=0$.

12. True or false? Each subset V of $\Re^n$ that is a vector space has a basis. *Hint:* take care.

13. This problem concerns the matrix equation $Ax=b$. Describe the conditions on A and b under which this equation has:

    (a) No solutions.

    (b) Multiple solutions.

    (c) Exactly one solution.

14. Prove that a non-empty set $\{v_1, v_2, \ldots, v_K\}$ of n-vectors is linearly independent if and only if none of these vectors is a linear combination of the others.

15. Prove that a set V of n-vectors that includes the origin is a vector space if and only if V contains the vector $[(1-\alpha)u + \alpha v]$ for every pair u and v of elements of V and for every real number α.

16. A set W of n-vectors is called an **affine space** if W is not empty and if W contains the vector $[(1-\alpha)u + \alpha v]$ for every pair u and v of elements of V and for every real number α.

(a) If an affine space W contains the origin, is it a vector space?

(b) For the case n=2, describe three types of affine space, and guess the "dimension" of each.

17. Designate as X the set consisting of each vector x that satisfies the matrix equation Ax=b. Suppose X is not empty. Is X a vector space? Is X an affine space? Support your answers.

18. Verify that equations (19) and (20) are correct. *Hint:* Equation (18) might help.

19. (*Small pivot elements*) You are to solve following system twice, each time by Gauss-Jordan elimination. Throughout each computation, you are to approximate each coefficient by three significant digits; this would round the number 0.01236 to 0.0124, for instance.

$$0.001\text{A} + 1\text{B} = 10$$
$$1\text{A} - 1\text{B} = 0$$

(a) For the first execution, begin with a pivot on the coefficient of A in the topmost equation.

(b) For the second execution, begin with a pivot on the coefficient of B in the topmost equation.

(c) Compare your solutions. What happens? Why?

Remark: The final two problems (below) refer to the starred section on efficient computation.

20. *(Work for lower pivots and back-substitution)* Imagine that a system of m equations in n unknowns is solved by lower pivots and back-substitution and that no trite or inconsistent equations have been encountered.

(a) Show that the number of multiplications and divisions required by back-substitution equals $(1 + 2 + \cdots + m) = (m)(m + 1)/2$.

(b) For each $j < m$, show that the j-th lower pivot requires $(m + 1 - j)(n)$ multiplications and divisions.

(c) How many multiplications and divisions are needed to execute Gauss-Jordan elimination with lower pivots and back-substitution? *Hint:* summing part (b) gives $(2 + 3 + \cdots + m)(n) = (n)(m)(m + 1)/2 - n$.

21. *(Sparseness)* In the detached-coefficient tableau that follows, each non-zero number is represented by an asterisk (*). Specify a sequence of lower pivots that implements the myopic rule, with c_k equal to the number of non-zero coefficients of x_k in rows on which pivots have not yet occurred. How many Gaussian operations does this implementation require? How many multiplications and divisions does it require, assuming that you omit multiplication by zero?

Equation	x_1	x_2	x_3	x_4	x_5	RHS
(1)	*	*	*	*	*	*
(2)	*	*				*
(3)		*	*			*
(4)			*	*	*	*
(5)				*	*	*

Part II–The Basics

This section introduces you to the simplex method and prepares you to make intelligent use of the computer codes that implement it.

Chapter 4. The Simplex Method, Part 1

In Chapter 3, you saw that Gauss-Jordan elimination pivots until it finds a basic solution to an equation system. In Chapter 4, you will see that the simplex method keeps on pivoting – it aims to improve the basic solution's objective value with each pivot, and it stops when no further improvement is possible.

Chapter 5. Analyzing Linear Programs

In this chapter, you will learn how to formulate linear programs for solution by Solver and by Premium Solver for Education. You will also learn how to interpret the output that these software packages provide. A linear program is seen to be the ideal environment in which to relate three important economic concepts – shadow price, "relative" opportunity cost, and marginal benefit. This chapter includes a "Perturbation Theorem" that can help you to grapple with the fact that a linear program is a model, an approximation.

Chapter 6. The Simplex Method, Part 2

This chapter plays a "mop up" role. If care is not taken, the simplex method *can* pivot forever. In Chapter 6, you will see how to keep that from occurring. The simplex method, as presented in Chapter 4, is initiated with a feasible solution. In Chapter 6, you will see how to adapt the simplex method to determine whether a linear program has a feasible solution and, if so, to find one.

Chapter 4: The Simplex Method, Part 1

1. Preview

The simplex method is the principal tool for computing solutions to linear programs. Computer codes that execute the simplex method are widely available, and they run on nearly every computer. You *can* solve linear programs without knowing how the simplex method works. Why should you learn it? Three reasons are listed below:

- Understanding the simplex method helps you make good use of the output that computer codes provide.

- The "feasible pivot" that lies at the heart of the simplex method is central to constrained optimization, much as Gauss-Jordan elimination is fundamental to linear algebra. In later chapters, feasible pivots will be adapted to solve optimization problems that are far from linear.

- The simplex method has a lovely economic interpretation. It will be seen that each basis is accompanied by a set of "shadow prices" whose

E. V. Denardo, *Linear Programming and Generalizations*, International Series in Operations Research & Management Science 149,
DOI 10.1007/978-1-4419-6491-5_4, © Springer Science+Business Media, LLC 2011

values determine the benefit of altering the basic solution by engaging in any activity that is currently excluded from the basis.

The simplex method also has a surprise to offer. It actually solves a pair of optimization problems, the one under attack and its "dual." That fact may seem esoteric, but it will be used in Chapter 14 to formulate competitive situations for solution by linear programming and its generalizations.

2. Graphical Solution

The simplex method will be introduced in the context of a linear program that is simple enough to solve visually. This example is

Problem A. Maximize $\{2x + 3y\}$ subject to the constraints

$$\begin{aligned} x \quad\quad &\leq 6, \\ x + y &\leq 7, \\ 2y &\leq 9, \\ -x + 3y &\leq 9, \\ x \quad\quad &\geq 0, \\ y &\geq 0. \end{aligned}$$

Before the simplex method is introduced, Problem A is used to review some terminology that was introduced in Chapter 1.

Feasible solutions

A *feasible solution* to a linear program is an assignment of values to its decision variables that satisfy all of its constraints. Problem A has many feasible solutions, one of which is the pair $(x, y) = (1, 0)$ in which $x = 1$ and $y = 0$. Because Problem A has only two decision variables, its feasible solutions can be depicted on the plane. Figure 4.1 does so. In it, each constraint in Problem A is represented as a line on which that constraint holds as an equation, accompanied by an arrow pointing into the half-space that satisfies it strictly. For instance, the pairs (x, y) that satisfy the constraint $-x + 3y \leq 9$ as an equation form the line through (0, 3) and (6, 5), and an arrow points from that line into the region that satisfies the constraint as a strict inequality.

Figure 4.1. The feasible solutions to Problem A.

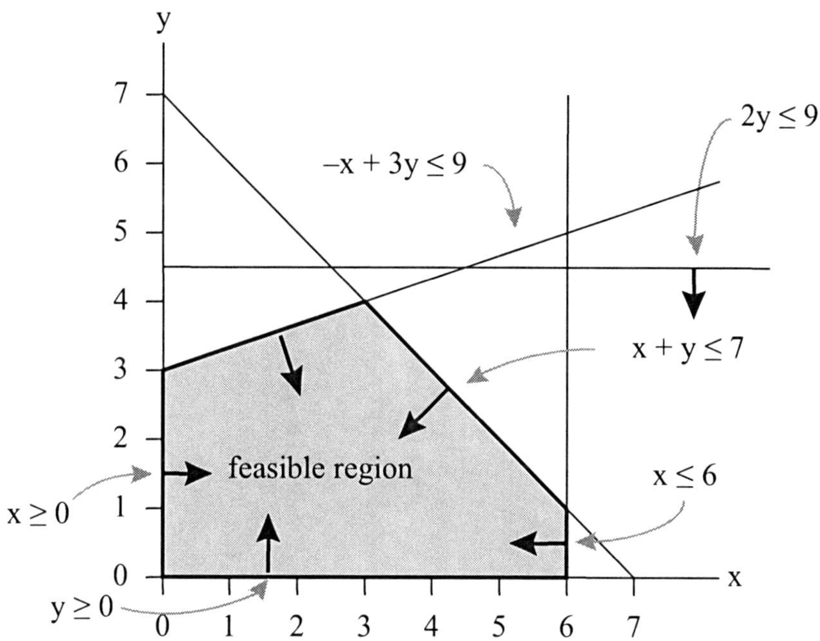

In Problem A and in general, the *feasible region* is the set of values of the decision variables that satisfy all of the constraints of the linear program. In Figure 4.1, the feasible region is shaded. Let us recall from Chapter 3 that the feasible region of a linear program is a *convex set* because it contains the *interval* (line segment) between each pair of points in it. A constraint in a linear program is said to be **redundant** if its removal does not change the feasible region. Figure 4.1 makes it clear that the constraint $2y \le 9$ is redundant.

Iso-profit lines

Figure 4.1 omits any information about the objective function. Each feasible solution assigns a value to the objective in the natural way; for instance, feasible solution (5, 1) has objective value $2x+3y=(2)(5)+(3)(1)=13$.

An *iso-profit line* is a line on which profit is constant. Figure 4.2 displays the feasible region for Problem A and four iso-profit lines. Its objective, $2x+3y$, equals 6 on the iso-profit line that contains the points (3, 0) and (0, 2). Similarly, the iso-profit line on which $2x+3y=12$ contains the points (6, 0) and (0, 4). In this case and in general, the iso-profit lines of a linear program are parallel to each other. Notice in Figure 4.2 that the point (3, 4) has a profit of 18 and that no other feasible solution has a profit as large as 18. Thus, x = 3 and y = 4 is the unique optimal solution to Problem A, and 18 is its optimal value.

Figure 4.2. Feasible region for Problem A, with iso-profit lines and objective vector (2, 3).

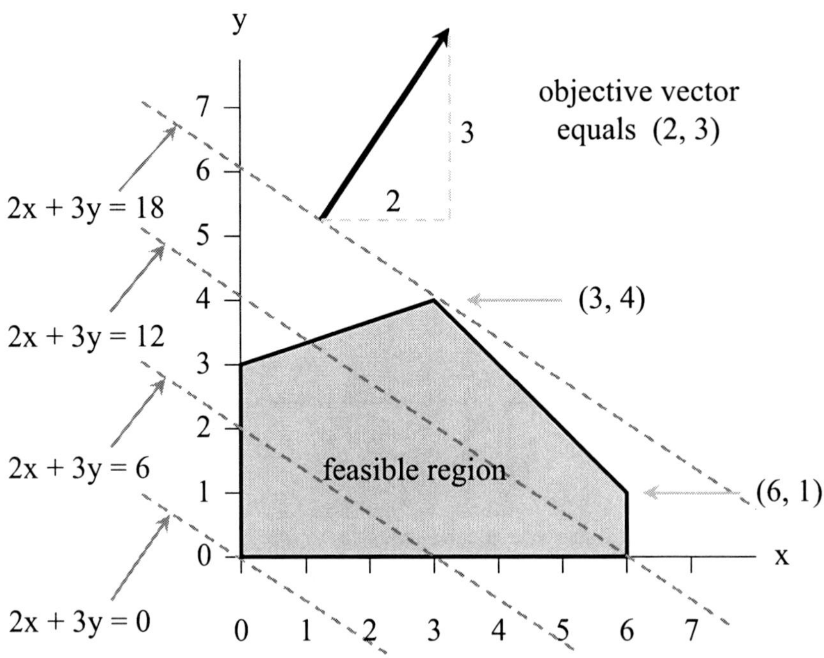

Each feasible solution to a linear program assigns a value to its objective function. An *optimal solution* to a linear program is a feasible solution whose objective value is largest in the case of a maximization problem, smallest in the case of a minimization problem. The *optimal value* of a linear program is the objective value of an optimal solution to it. It's clear from Figure 4.2 that (3, 4) is an optimal solution to Problem A and that 18 is its optimal value.

The objective vector

There is a second way in which to identify the optimal solution or solutions to a linear program. The object of Problem A is to maximize the expression (2x + 3y). The coefficients of x and y in this expression form the *objective vector* (2, 3). A vector connotes motion. We think of the vector (2, 3) as moving 2 units toward the right of the page and 3 units toward the top. In Figure 4.2, the objective vector shown touching the iso-profit line $2x + 3y = 18$.

The objective vector can have its tail "rooted" anywhere in the plane. In Figure 4.2 and in general, the objective vector is perpendicular to the iso-profit lines. It's the *direction* in which the objective vector points that matters. In a maximization problem, we seek a feasible solution that lies farthest in the direction of the objective vector. Similarly, in a minimization problem, we seek a feasible solution that lies farthest in the direction that is opposite to the objective vector. It is emphasized:

> The objective vector points "uphill" – in the direction of increase of the objective.

In Figure 4.2, for instance, the optimal solution is (3, 4) because, among feasible solutions, it lies farthest in the direction of the objective vector.

Extreme points

It is no surprise that the feasible region in Figure 4.2 is a convex set. In Chapter 3, it was observed that the feasible region of every linear program is a convex set. It is recalled that an element of a convex set is an **extreme point** of that set if it is not a convex combination of two other points in that set. The feasible region in Figure 4.2 has five extreme points (corners). The optimal solution lies at the extreme point (3, 4). The other four extreme points are (0, 0), (6, 0), (6, 1) and (0, 3).

Edges

The mathematical definition of an "edge" is a bit involved. But it is clear, visually, that the feasible region in Figure 4.2 has five **edges**. Each of these edges is a line segment that connects two extreme points. The line segment connecting extreme points (0, 0) and (6, 0) is an edge, for instance. Not every line segment that connects two extreme points is an edge. The line segment connecting extreme points (0, 0) and (3, 4) is not an edge (because it intersects the "interior" of the feasible region).

Optimality of an extreme point

In Figure 4.2, suppose the objective vector pointed in some other direction. Would an extreme point still be optimal? Yes, it would, but it could be a different extreme point. Suppose, for instance, that the objective vector is (3, 3). In this case, the objective vector has rotated clockwise, and extreme points (3, 4) and (6, 1) are both optimal, as is each point in the edge connecting them. If the objective vector is (4, 3), the objective vector has rotated farther clockwise, and the unique optimal solution is the extreme point (6, 1).

Adjacent extreme points

Two extreme points are said to be **adjacent** if the interval between them is an edge. In Figure 4.2, extreme points (0, 0) and (0, 3) are adjacent. Extreme points (0, 0) and (3, 4) are not adjacent.

Simplex pivots

"Degeneracy" is discussed later in this chapter. If a simplex pivot is degenerate, the extreme point does not change. If a simplex pivot is nondegenerate, it occurs to an adjacent extreme point, and each such pivot improves the objective value. The simplex method stops pivoting when it discovers that the current extreme point has the best objective value.

When the simplex method is applied to Problem A, the first pivot will occur from extreme point (0, 0) to extreme point (0, 3), and the second pivot will occur to extreme point (3, 4), which will be identified as optimal.

Bounded feasible region

A linear program is said to have a *bounded* feasible region if at least one feasible solution exists and if there exists a positive number K such that no feasible solution assigns any variable a value below –K or above +K. The feasible region in Figure 4.2 is bounded; no feasible solution has $|x| > 6$ or $|y| > 6$. A feasible region is said to be *unbounded* if it is not bounded.

Bounded linear programs

A linear program is said to be *feasible and bounded* if it has at least one feasible solution and if its objective cannot be improved without limit. Problem A is feasible and bounded. It would not be bounded if the constraints $x + y \le 7$ and $x \le 6$ were removed. It is easy to convince oneself, visually, of the following:

> If a linear program whose variables are constrained to be nonnegative is feasible and bounded, at least one of its extreme points is optimal.

A linear program can be feasible and bounded even if its feasible region is unbounded. An example is: Minimize {x}, subject to $x \geq 0$.

Unbounded linear programs

A maximization problem is said to be *unbounded* if it is feasible and if no upper bound exists on the objective value of its feasible solutions. Similarly, a minimization problem is *unbounded* if it is feasible and if no lower bound exists on the objective value of its feasible solutions. Unbounded linear programs are unlikely to occur in practice because they describe situations in which one can do infinitely well. They do occur from inaccurate formulations of bounded linear programs.

3. A Format that Facilitates Pivoting

The simplex method consists of a deft sequence of pivots. Pivots occur on systems of *equality* constraints. To prepare Problem A for pivoting, it is first placed in the format called **Form 1**, namely, as a liner program having these properties:

- The object is to maximize or minimize the quantity z.
- Each decision variable other than z is constrained to be nonnegative.
- All of the other constraints are linear equations.

Form 1 introduces z as the quantity that we wish to make largest in a maximization problem, smallest in a minimization problem. Form 1 requires each decision variable other than z to be nonnegative, and it gets rid of the inequality constraints, except for those on the decision variables.

A canonical form?

The simplex method will be used to solve every linear program that has been cast in Form 1. Can *every* linear program be cast in Form 1? Yes. To verify that this is so, observe that:

- Form 1 encompasses maximization problems and minimization problems.
- An equation can be included that equates z to the value of the objective.

- Each inequality constraint can be converted into an equation by insertion of a nonnegative (slack or surplus) variable.
- Each variable that is unconstrained in sign can be replaced by the difference of two nonnegative variables.

A **canonical form** for linear programs is any format into which every linear program can be cast. Form 1 is canonical form. Since Form 1 is canonical, describing the simplex method for Form 1 shows how to solve every linear program. It goes without saying, perhaps, that it would be foolish to describe the simplex method for linear programs that have not been cast in a canonical form.

Recasting Problem A

Let us cast Problem A in Form 1. The quantity z that is to be maximized is established by appending to Problem A the "counting" constraint,

$$2x + 3y = z,$$

which equates z to the value of the objective function. Problem A has four "$\leq$" constraints, other than those on its decision variables. Each of these inequality constraints is converted into an equation by inserting a **slack variable** on its left-hand side. This re-writes Problem A as

Problem A'. Maximize {z}, subject to the constraints

$$\begin{aligned}
&(1.0) \qquad 2x + 3y \qquad\qquad\qquad\qquad - z = 0,\\
&(1.1) \qquad 1x \qquad + s_1 \qquad\qquad\qquad = 6,\\
&(1.2) \qquad 1x + 1y + \qquad s_2 \qquad\qquad = 7,\\
&(1.3) \qquad\qquad 2y + \qquad\qquad s_3 \qquad = 9,\\
&(1.4) \qquad -1x + 3y + \qquad\qquad\quad s_4 = 9,\\
&\qquad\qquad x \geq 0, \quad y \geq 0, \quad s_i \geq 0 \ \text{ for } i = 1, 2, 3, 4.
\end{aligned}$$

Problem A' is written in Form 1. It has seven decision variables and five equality constraints. Each decision variable other than z is constrained to be nonnegative. The variable z has been shifted to the left-hand side of equation (1.0) because we want all of the decision variables to be on the left-hand sides of the constraints.

To see where the "slack variables" get their name, consider the constraint $x + y \le 7$. In the constraint $x + y + s_2 = 7$, the variable s_2 is positive if $x + y < 7$ and s_2 is zero if $x + y = 7$. Evidently, s_2 "takes up the slack" in the constraint $x + y \le 7$.

The variable –z

In Form 1, the variable z plays a special role because it measures the objective. We elect to think of –z as a decision variable. In Problem A', the variable –z is basic for equation (1.0) because –z has a coefficient of +1 in equation (1.0) and has coefficients of 0 in all other equations. During the entire course of the simplex method, no pivot will ever occur on any coefficient in the equation for which –z is basic. Consequently, –z will stay basic for this equation.

Reduced cost

The equation for which –z is basic plays a guiding role in the simplex method, and its coefficients have been given names. The coefficient of each variable in this equation is known as that variable's **reduced cost**. In equation (1.0), the reduced cost of x equals 2, the reduced cost of y equals 3, and the reduced cost of each slack variable equals 0. The term "reduced cost" is firmly established in the literature, and we will use it. But it will soon be clear that "marginal profit" would have been more descriptive.

The feasible region for Problem A'

Problem A' has seven decision variables. It might seem that the feasible region for Problem A' can only be "visualized" in seven-dimensional space. Figure 4.3 shows that a 2-dimensional picture will do. In Figure 4.3, each line in Figure 4.1 has been labeled with the variable in Problem A' that equals zero on it. For instance, the line on which the inequality $x + y \le 7$ holds as an equation is relabeled $s_2 = 0$ because s_2 is the slack variable for the constraint $x + y + s_2 = 7$.

Figure 4.3. The feasible region for Problem A'.

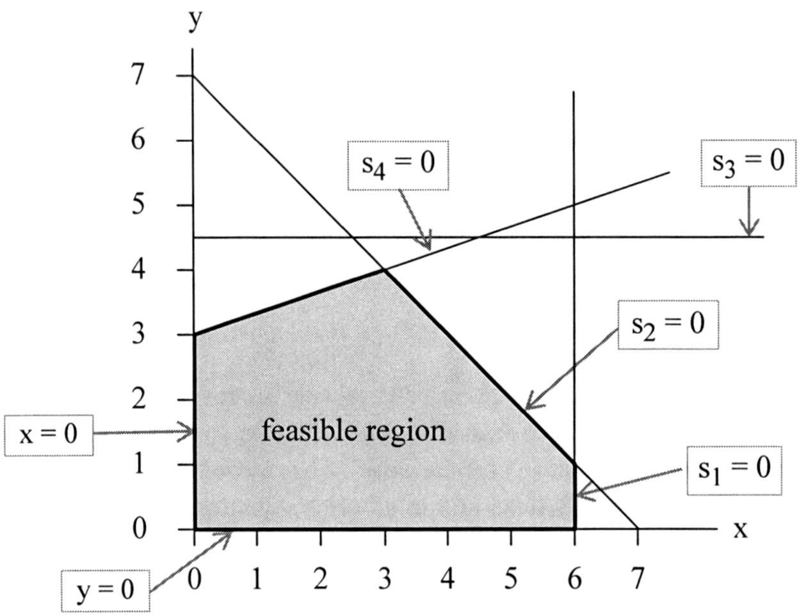

Bases and extreme points

Figure 4.3 also enables us to identify the extreme points with basic solutions to system (1). Note that each extreme point in Figure 4.3 lies at the intersection of two lines. For instance, the extreme point (0, 3) is the intersection of the lines $x = 0$ and $s_4 = 0$. The extreme point (0, 3) will soon be associated with the basis that excludes the variables that x and s_4.

System (1) has five equations and seven variables. The variables $-z$ and s_1 through s_4 form a basis for system (1). This basis consists of five variables, one per equation. A fundamental result in linear algebra (see Proposition 10.2 on page 334 for a proof) is that every basis for a system of linear equations has the same number of variables. Thus, each basis for system (1) contains exactly five variables, one per equation. In other words, each basis *excludes* two of the seven decision variables. Each basis for system (1) has a basic solution, and that basic solution equates its two nonbasic variables to zero. This identifies each extreme point in Figure 4.3 with a basis. Extreme point (0, 3) corresponds to the basis that excludes x and s_4 because (0, 3) is the intersection of the lines $x = 0$ and $s_4 = 0$. Similarly, extreme point (3, 4) corresponds to the basis that excludes s_2 and s_4 because (3, 4) is the intersection of the lines $s_2 = 0$ and $s_4 = 0$.

4. First View of the Simplex Method

Problem A' will now be used to introduce the simplex method, and Figure 4.3 will be used to track its progress. System (1) is basic because each of its equations has a basic variable. The basis for system (1) consists of –z and the slack variables. This basis excludes x and y. Its basic solution equates to zero its nonbasic variables (which are x and y) and is

$$x = 0, \quad y = 0, \quad -z = 0, \quad s_1 = 6, \quad s_2 = 7, \quad s_3 = 9, \quad s_4 = 9.$$

A feasible basis

A basis for Form 1 is now said to be **feasible** if its basic solution is feasible, that is, if the values of the basic variables are nonnegative, with the possible exception of –z. Evidently, the basis $\{-z, s_1, s_2, s_3, s_4\}$ is feasible.

Phases I and II

For Problem A, a feasible basis sprang immediately into view. That is *not typical.* Casting a linear program in Form 1 does not automatically produce a basis, let alone a feasible basis. Normally, a feasible basis must be wrung out of the linear program by a procedure that is known as **Phase I** of the simplex method. Using Problem A to introduce the simplex method begins with "Phase II" of the simplex method. Phase I has been deferred to Chapter 6 because it turns out to be a minor adaptation of Phase II.

Phase II of the simplex method begins with a feasible basis and with –z basic for one of its equations. Phase II executes a series of pivots. None of these pivots occurs on any coefficient in the equation for which –z is basic. Each of these pivots:

- keeps –z basic;
- changes the basis, but keeps the basic solution feasible;
- improves the basic solution's objective value or, barring an improvement, keeps it from worsening.

Phase II stops pivoting when it discerns that the basic solution's objective value cannot be improved. How this occurs will soon be explained.

A simplex tableau

A basic system for Form 1 is said to be a **simplex tableau** if –z is basic for the top-most equation and if the right-hand-side values of the other equations are nonnegative. This guarantees that the basic solution is feasible (equates all variables other than –z to nonnegative values) and that it equates z to the basic solution's objective value. A simplex tableau is also called a **basic feasible tableau**; these terms are synonyms.

The dictionary

We wish to pivot from simplex tableau to simplex tableau, improving – or at least not worsening – the objective with each pivot. It is easy to see which pivots do the trick if system (1) is cast in a format that has been dubbed a **dictionary**.[1] System (1) is placed in this format by executing these two steps.

- Shift the non-basic variables x and y to the right-hand sides of the constraints.
- Multiply equation (1.0) by –1, so that z (and not –z) appears on its left-hand side.

Writing system (1) in the format of a dictionary produces system (2), below.

$$\text{(2.0)} \qquad z = 0 + 2x + 3y$$

$$\text{(2.1)} \qquad s_1 = 6 - 1x + 0y$$

$$\text{(2.2)} \qquad s_2 = 7 - 1x + 1y,$$

$$\text{(2.3)} \qquad s_3 = 9 - 0x - 2y$$

$$\text{(2.4)} \qquad s_4 = 9 + 1x - 3y$$

[1] The term "dictionary" is widely attributed to Vašek Chvátal, who popularized it in his lovely book, *Linear Programming*, published in 1983 by W. H. Freedman and Co., New York. In that book, Chvátal attributes the term to J. E. Strum's, *Introduction to Linear Programming*, published in 1972 by Holden-Day, San Francisco.

In system (2), the variable z (rather than –z) is basic for the topmost equation, and the slack variables are basic for the remaining equations. The basic solution to system (2) equates each non-basic variable to zero and, consequently, equates each basic variable to the number on the right-hand-side value of the equation for which it is basic.

Perturbing a basic solution

The dictionary indicates what happens if the basic solution is perturbed by setting one or more of the nonbasic variables positive and adjusting the values of the basic variables so as to preserve a solution to the equation system. Equation (2.0) shows that the objective value is increased by setting x positive and by setting y positive.

Reduced cost and marginal profit

The coefficients of x and y in equation (2.0) equal their reduced costs, namely, their coefficients in equation (1.0). To see why this occurs, note that the reduced costs have been multiplied by –1 twice, once when equation (1.0) was multiplied by –1 and again when the nonbasic variables were transferred to its right-hand side.

In a simplex tableau for a maximization problem, the **marginal profit** of each nonbasic variable equals the change that occurs in its objective value when the basic solution is perturbed by setting that variable equal to 1 and keeping all other nonbasic variables equal to zero. The dictionary in system (2) makes the marginal profits easy to see. Its basic solution equates the nonbasic variables x and y to 0. Equation (2.0) shows that the marginal profit of x equals 2 and that the marginal profit of y equals 3. The marginal profit of each nonbasic variable is its so-called "reduced cost." It is emphasized:

> In each simplex tableau for a maximization problem, the "reduced cost" of each nonbasic variable equals the *marginal profit* for perturbing the tableau's basic solution by equating that variable to 1, keeping the other nonbasic variables equal to zero, and adjusting the values of the basic variables so as to satisfy the LP's equations.

Similarly, in a minimization problem, the "reduced cost" of each nonbasic variable equals the *marginal cost* of perturbing the basic solution by setting that nonbasic variable equal to 1 and adjusting the values of the basic variables accordingly.

As mentioned earlier, we cleave to tradition and call the coefficient of each variable in the equation for which –z is basic its reduced cost. Please interpret the "reduced cost" of each nonbasic variable as marginal profit in a maximization problem and as marginal cost in a minimization problem.

A pivot

Our goal is to pivot in a way that improves the basic solution's objective value. Each pivot on a simplex tableau causes one variable that had been nonbasic to become basic and causes one basic variable to become nonbasic. Equation (2.0) shows that the objective function improves if the basic solution is perturbed by setting x positive or by setting y positive. We could pivot in a way that makes x basic or in a way that makes y basic.

Perturbing system (2) by keeping $x = 0$ and setting $y > 0$ produces:

(3.0) $z = 0 + 3y;$

(3.1) $s_1 = 6$ so s_1 is positive for all values of y;

(3.2) $s_2 = 7 - 1y,$ so s_2 decreases to zero when $y = 7/1 = 7;$

(3.3) $s_3 = 9 - 2y,$ so s_3 decreases to zero when $y = 9/2 = 4.5;$

(3.4) $s_4 = 9 - 3y,$ so s_4 decreases to zero when $y = 9/3 = 3.$

Evidently, the largest value of y that keeps the perturbed solution feasible is $y = 3$. If y exceeds 3, the perturbed solution has $s_4 < 0$.

Graphical interpretation

Figure 4.3 is now used to interpret the ratios in system (3). The initial basis excludes x and y, and so the initial basic solution lies at the intersection of the lines $x = 0$ and $y = 0$, which is the point (0, 0). The perturbation in system (3) keeps $x = 0$ and allows y to become positive, thereby moving upward on the line $x = 0$. Each "ratio" in system (3) is a value of y for which (0, y) intersects a constraint. No ratio is computed for constraint (3.1) because the lines (0, y) and $s_1 = 0$ do not intersect. The smallest ratio is the largest value of y for which the perturbed solution stays feasible.

Feasible pivots

Rather than proceeding directly with the simplex method, we pause to describe a class of pivots that keeps the basic solution feasible. Specifically, starting with a basic feasible solution for Form 1, we select *any nonbasic variable* and call it the **entering variable.** In system (1), we take y as the entering variable. The goal is to pivot on a coefficient of y that keeps the basic solution feasible and keeps –z basic for the top-most equation. In a basic tableau for Form 1, which coefficient of the entering variable shall we pivot upon? Well:

- No coefficient of the equation for which –z is basic is pivoted upon in order to keep –z basic for the equation. For this reason, no "ratio" is ever computed for this equation.
- No coefficient that is negative is pivoted upon.
- Excluding the equation for which –z is basic, each equation whose coefficient of the entering variable is positive has a **ratio** that equals this equation's right-hand side value divided by its coefficient of the entering variable.
- The pivot occurs on the coefficient of the entering variable in an equation whose ratio is smallest.

System (1) is now used to illustrate feasible pivots. In this system, let y be the entering variable. No ratio is computed for equation (1.0) because –z stays basic for that equation. No ratio is computed for equation (1.1) because the coefficient of y in this equation is not positive. Ratios are computed for equations (1.2), (1.3) and (1.4), and these ratios equal 7, 4.5 and 3, respectively. The pivot occurs on the coefficient of y in equation (1.4) because that equation's ratio is smallest. Note that this pivot results in a basic tableau for which y becomes basic and the variable s_4 that had been basic for equation (1.4) becomes nonbasic. Equation (3.4) with $s_4 = 0$ shows that $y = 3$, hence that this pivot keeps the basic solution feasible. In this case and in general:

> In a feasible tableau for Form 1, pivoting on the coefficient of the entering variable in a row whose ratio is smallest amongst those rows whose coefficients of the entering variable are positive keeps the basic solution feasible.

A pivot is said to be **feasible** if it occurs on the coefficient of the entering variable in the "pivot row," where the **pivot row** has a *positive* coefficient of the entering variable and, among all rows having positive coefficients of the entering variable, the pivot row has the smallest **ratio** of RHS value to coefficient of the entering variable. The variable that had been basic for the pivot row is called the **leaving variable**. Thus, each feasible pivot causes the "entering variable" to join the basis and causes the "leaving variable" to depart.

With x (and not y) as the entering variable in system (1), ratios would be computed form equations (1.1) and (1.2), these ratios would equal $6/1 = 6$ and $7/1 = 7$, respectively, and a feasible pivot would occur on the coefficient of x in equation (1.1). This pivot causes s_1 to leave the basis, resulting in a basic tableau whose basic solution has $x = 6$ and remains feasible. By the way, the coefficient of x in equation (1.4) equals -1, which is negative, and a pivot on this coefficient would produce a basic solution having $x = 9/(-1) = -9$, which would not be feasible.

A simplex pivot

In a maximization problem, a **simplex pivot** is a feasible pivot for which the reduced cost (marginal profit) of the entering variable is positive. Compare equations (1.0) and (3.0) to see that the entering variable for a simplex pivot can be x or y. As noted previously, setting either of these variables positive improves the objective.

Is the simplex pivot unambiguous? No, it is not. More than one nonbasic variable can have marginal profit that is positive. Also, two or more rows can tie for the smallest ratio.

Rule #1

When illustrating the simplex method, *some* of the ambiguity in choice of pivot element is removed by employing **Rule #1**, which takes the entering variable as a nonbasic variable whose reduced cost is most positive in the case of a maximization problem, most negative in the case of a minimization problem. Rule #1 is not unambiguous. More than one nonbasic variable can have the most positive (negative) reduced cost in a maximization (minimization) problem, and two or more rows can tie for the smallest ratio.

The first simplex pivot

Table 4.1 shows how to execute a simplex pivot on a spreadsheet. In this table, the variable y has been selected as the entering variable (it has the largest reduced cost, and we are invoking Rule #1). The cell containing the label y has been shaded. The "IF" statements in column J of Table 4.1 compute ratios for the equations whose coefficients of y are positive. The smallest of these ratios equals 3 (which is no surprise), and the cell in which it appears is also shaded. The pivot element lies at the intersection of the shaded column and row, and it too is shaded.

To execute this pivot, select the block B12:I16, type the function **=pivot(C7, B3:I7)** and then hit **Ctrl+Shift+Enter** to remind Excel that this is an array function (because it sets values in an array of cells, rather than in a single cell).

Table 4.1. The first simplex pivot.

	A	B	C	D	E	F	G	H	I	J
1										
2	equation	x	y	s1	s2	s3	s4	-z	RHS	ratio
3	(2.0)	2	3					1	0	none
4	(2.1)	1		1					6	---
5	(2.2)	1	1		1				7	7
6	(2.3)		2			1			9	4.5
7	(2.4)	-1	3				1		9	3
8										
9							= IF(C7 > 0, I7/C7, "---")			
10										
11	equation	x	y	s1	s2	s3	s4	-z	RHS	ratio
12	(3.0)	3	0	-0	-0	-0	-1	1	-9	
13	(3.1)	1	-0	1	-0	-0	-0	-0	6	
14	(3.2)	1.333	0	-0	1	-0	-0.33	-0	4	
15	(3.3)	0.667	0	-0	-0	1	-0.67	-0	3	
16	(3.4)	-0.33	1	0	0	0	0.333	0	3	
17										
18				= PIVOT(C7, B3:I7)						

The pivot in Table 4.1 causes y to enter the basis and s_4 to depart. The basic solution that results from this pivot remains feasible because it equates each basic variable other than –z to a nonnegative value.

The change in objective value

This pivot improves z by 9, which equals the product of the reduced cost (marginal profit) of y and the ratio for its pivot row. This reflects a property that holds in general and is highlighted below:

> In each feasible pivot, the change in the basic solution's objective value equals the product of the reduced cost of the entering variable and the ratio for its pivot row.

This observation is important enough to be recorded as the equation,

(4)

$$\begin{bmatrix} \text{change in the basic solution's} \\ \text{objective value} \end{bmatrix} = \begin{bmatrix} \text{reduced cost of the} \\ \text{entering variable} \end{bmatrix} \times \begin{bmatrix} \text{ratio for its} \\ \text{pivot row} \end{bmatrix}.$$

In Problem A, each pivot will improve the basic solution's objective value. That does not always occur, however. The RHS value of the pivot row can equal 0. If it does equal 0, equation (4) shows that no change occurs in the basic solution's objective value. That situation is known as "degeneracy," and it is discussed in the next section.

The second simplex pivot

Let us resume the simplex method. For the tableau in rows 12-16 of Table 4.1, x is the only nonbasic variable whose marginal profit is positive; its reduced cost equals 3. So x will be the entering variable for the next simplex pivot. The spreadsheet in Table 4.2 identifies that 3 is the smallest ratio and displays the tableau that results from a pivot on the coefficient of x in this row. Equation (4) shows that this pivot will improve the basic solution's objective value by $9 = 3 \times 3$. This pivot causes x to become basic and causes s_2 (which had been basic for the pivot row) to become nonbasic. Rows 21-25 of Table 4.2 exhibit the result of this pivot.

The basic solution to the tableau in rows 21-25 of Table 4.2 has x = 3, y = 4 and z = 18. The nonbasic variables in this tableau are s_2 and s_4. In Figure 4.3, this basic solution lies at the intersection of the lines $s_2 = 0$ and $s_4 = 0$. Visually, it is optimal.

Table 4.2. The second simplex pivot.

	A	B	C	D	E	F	G	H	I	J
1										
11	equation	x	y	s1	s2	s3	s4	-z	RHS	ratio
12	(3.0)	3	0	-0	-0	-0	-1	1	-9	none
13	(3.1)	1	-0	1	-0	-0	-0	-0	6	6
14	(3.2)	1.333	0	-0	1	-0	-0.33	-0	4	3
15	(3.3)	0.667	0	-0	-0	1	-0.67	-0	3	4.5
16	(3.4)	-0.33	1	0	0	0	0.333	0	3	---
17										
18							= IF(B16 > 0, I16/B16, "---")			
19										
20	equation	x	y	s1	s2	s3	s4	-z	RHS	
21	(5.0)	0	0	0	-2.25	0	-0.25	1	-18	
22	(5.1)	0	-0	1	-0.75	0	0.25	0	3	
23	(5.2)	1	0	-0	0.75	-0	-0.25	-0	3	
24	(5.3)	0	0	0	-0.5	1	-0.5	0	1	
25	(5.4)	0	1	0	0.25	0	0.25	0	4	
26										
27				= PIVOT(B14, B12:I16)						

An optimality condition

To confirm, algebraically, that this basic solution is optimal, we write the equation system depicted in rows 20-25 in the format of a dictionary, that is, with the nonbasic variables on the right-hand side and with z (rather than –z) on the left-hand side of the topmost equation.

(5.0) $$z = 18 - 2.25s_2 - 0.25s_4,$$

(5.1) $$s_1 = 3 + 0.75s_2 - 0.25s_4,$$

(5.2) $$x = 3 - 0.75s_2 + 0.25s_4,$$

(5.3) $$s_3 = 1 + 0.50s_2 + 0.50s_4,$$

(5.4) $$y = 4 - 0.25s_2 - 0.25s_4.$$

In system (5), the variables s_2 and s_4 are nonbasic. The basic solution to system (5) is the unique solution to system (5) in which the nonbasic variables s_2 and s_4 are equated to zero. This basic solution has $z = 18$. Since the coefficients of s_2 and s_4 in equation (5.0) are negative, any solution that sets either s_2 and s_4 to a positive value has $z < 18$. In brief, the basic solution to system (5) is the unique optimal solution to Problem A'.

Test for optimality. The basic solution to a basic feasible system for Form 1 is optimal if the reduced costs of the nonbasic variables are:

- nonpositive in the case of a maximization problem;
- nonnegative in the case of a minimization problem.

Recap

Our introduction to Phase II of the simplex method is nearly complete. For a linear program that is written in Form 1, we have seen how to:

- Execute feasible pivots on a spreadsheet.
- Execute simplex pivots on a spreadsheet.
- Identify the optimal solution.

From the dictionary, we have seen that:

- The reduced cost of each nonbasic variable equals the change that occurs in the objective value if the basic solution is perturbed by setting that nonbasic variable equal to 1.
- If an equation has a ratio, this ratio equals the value of the entering variable for which the perturbed solution reduces the equation's basic variable to zero.
- The smallest of these ratios equals the largest value of the entering variable that keeps the perturbed solution feasible.

It would be hard to overstate the usefulness of the dictionary.

5. Degeneracy

In a feasible pivot, the RHS value of the pivot row must be nonnegative. A feasible pivot is said to be **nondegenerate** if the right-hand-side value of the pivot row is positive. Similarly, a feasible pivot is said to be **degenerate** if the RHS value of the pivot row equals 0.

Nondegenerate pivots

Equation (4) holds for every pivot that occurs on a basic tableau. If a pivot is nondegenerate:

- The RHS value of the pivot row is positive.
- The coefficient of the entering variable in the pivot row must be positive, so the ratio for the pivot row must be positive.
- Hence, equation (4) shows that each nondegenerate simplex pivot improves the basic solution's objective value.

It is emphasized:

> **Nondegenerate pivots:** If a simplex pivot is *nondegenerate*, the basis changes and objective value of the basic solution improves.

Degenerate pivots

Let us now interpret equation (4) for the case of a feasible pivot that is degenerate. In this case:

- The RHS value of the pivot row equals 0.
- This pivot (like any other) multiplies the pivot row by a constant, and it replaces the other rows by themselves less constants times the pivot row. Since the pivot is degenerate, the RHS value of the pivot row equals 0, so the pivot changes no RHS values.
- The variables that had been basic for rows other than the pivot row remain basic for those rows; their values in the basic solution remain as they were because the RHS values do not change.
- The variable that departs from the basis had equaled zero, and the variable that enters the basis will equal zero.

In brief:

> **Degenerate pivots:** If a feasible pivot is *degenerate*, the basis changes, but no change occurs in the basic solution or in its objective.

Cycling

Each nondegenerate simplex pivot improves the basic solution's objective value. Each degenerate simplex pivot preserves the basic solution's objective value. Hence, each nondegenerate simplex pivot results in a basis whose objective value improves on any seen previously. There are only finitely many bases because each basis is a subset of the variables and there are finitely many subsets. Thus, the simplex method can execute only finitely many nondegenerate simplex pivots before it terminates.

On the other hand, each degenerate pivot changes the basis without changing the basic solution. The simplex method is said to **cycle** if a sequence of simplex pivots leads to a basis visited previously. If a cycle occurs, it must consist exclusively of degenerate pivots.

The simplex method can cycle! In Chapter 6, an example will be exhibited in which Rule #1 *does* cycle. In that chapter, the ambiguity in Rule #1 will be resolved in a way that precludes cycling, thereby assuring finite termination.

In discussions of the simplex method, it is convenient to apply the terms "degenerate" and "nondegenerate" to basic solutions as well as to pivots. A basic solution is said to be **nondegenerate** if it equates every decision variable, with the possible exception of $-z$, to a nonzero value. Similarly, a basic solution is said to be **degenerate** if it equates to zero at least one basic variable, other than $-z$.

6. Detecting an Unbounded Linear Program

Let us recall that a linear program is unbounded if it is feasible and if the objective value of its feasible solutions can be improved without limit. What happens if Phase II of the simplex method is applied to an unbounded linear program? Phase II cannot find an optimal solution because none exists. To explore this issue, we introduce

Program B. Maximize $\{0x + 3y\}$, subject to the constraints

$$-x + y \leq 2,$$

$$x \geq 0, \quad y \geq 0.$$

Please sketch the feasible region of Problem B. Note that its constraints are satisfied by each pair (x, y) having $y \geq 2$ and $x = y - 2$; moreover, that each such pair has objective value of $0x + 3y = 3y$, which becomes arbitrarily large as y increases. To see what happens when the simplex method is applied to Problem B, we first place it in Form 1, as

Program B'. Maximize {z} subject to the constraints

$$(6.0) \qquad 0\,x + 3\,y \qquad - z = 0,$$

$$(6.1) \qquad -x + \quad y + s_1 \qquad = 2,$$

$$x \geq 0, \quad y \geq 0, \quad s_1 \geq 0.$$

Table 4.3 shows what happens when the simplex method is applied to Problem B'. The first simplex pivot occurs on the coefficient of y in equation (6.1), producing a basic feasible tableau whose basis excludes x and s_1 and is

$$x = s_1 = 0, \quad y = 2, \quad -z = -6.$$

Table 4.3. Application of the simplex method to Problem B.

	A	B	C	D	E	F	G
1							
2	equation	x	y	s1	-z	RHS	ratio
3	(6.0)	0	3	0	1	0	none
4	(6.1)	-1	1	1		2	2
5							
6	equation	x	y	s1	-z	RHS	ratio
7	(7.0)	3	0	-3	1	-6	none
8	(7.1)	-1	1	1	0	2	---

Writing rows 7 and 8 in the format of the dictionary produces

(7.0) $$z = 6 + 3x - 3s_1,$$

(7.1) $$y = 2 + 1x - 1s_1.$$

Perturbing the basic solution to system (7) by making x positive improves the objective and increases y. No basic variable decreases. No equation has a ratio. And the objective improves without limit as x is increased. In brief:

> **Test for unboundedness.** A linear program in Form 1 is unbounded if an entering variable for a simplex pivot has nonpositive coefficients in each equation other than the one for which $-z$ is basic.

A maximization problem is unbounded if the marginal profit (reduced cost) of a nonbasic variable is positive and if perturbing the basic solution by setting that variable positive causes no basic variable to decrease. The perturbed solution remains feasible no matter how large that nonbasic variable becomes, and its objective value becomes arbitrarily large.

7. Shadow Prices

A "shadow price" measures the marginal value of a change in a RHS (right-hand-side) value. Computer codes that implement the simplex method report the shadow prices for the basis with which the simplex method terminates. These shadow prices can be just as important as the optimal solution. In Chapter 5, we will see why this is so.

Shadow prices are present not just for the final basis, but at every step along the way. They guide the simplex method. In Chapter 11, we will see how they do that.

The Full Rank proviso

It will be demonstrated in Proposition 10.2 (on page 334) that every basis for the column space of a matrix has the same number of columns. Thus, every basic tableau for a linear program has the same number (possibly zero) of trite rows. A linear program is said to satisfy the **Full Rank proviso** if any basic tableau for its Form 1 representation has a basic variable for each row. Proposition 10.2 implies that the Full Rank proviso is satisfied if and only if *every* basic tableau has one basic variable for each row.

System (1) has a basic variable for each row, so Problem A satisfies the Full Rank proviso. If a linear program satisfies the Full Rank proviso, its equations must be consistent, and no basic tableau has a trite row.

A definition

For linear programs that satisfy the Full Rank proviso, each basis prescribes a set of shadow prices, one per constraint. Their definition is highlighted below.

> Each basis assigns to each constraint of a linear program a **shadow price** whose numerical value equals the change that occurs in the basic solution's objective value per unit change in that constraint's RHS value in the original linear program.

Evidently, each shadow price is a *rate of change* of the objective value with respect to the constraint's right-hand-side (RHS) value. (In math-speak, each shadow price is a partial derivative.)

Necessarily, the **unit of measure** of a constraint's shadow price equals the unit of measure of the objective divided by the unit of measure of that constraint. As an example, suppose that the objective is measured in dollars per week (\$/week) and that a particular constraint's right-hand side value is measured in hours per week (hours/week); this constraint's shadow price is measured in dollars per hour (\$/hour) because

$$(\$/\text{week}) \div (\text{hours/week}) = (\$/\text{week}) \times (\text{weeks/hour}) = (\$/\text{hour}).$$

An illustration of shadow prices

Problem A' is now used to illustrate shadow prices. It satisfies the Full Rank proviso because system (1) has one basic variable per equation. When applied to Problem A', the simplex method encountered three bases, each of which has its own set of shadow prices.

For the final basis, whose basic solution is in rows 22-25 of Table 4.2, the shadow price for the 2nd constraint will now be computed. That constraint's

RHS value in the original linear program value equals 7. Let us ask ourselves: What would happen to this basic solution if the RHS value of the 2nd constraint were changed from 7 to $7+\delta$? Table 4.4, below, will help us to answer this question. Table 4.4 differs from the initial tableau (rows 2-7 of Table 4.1) in that the dashed line records the locations of the "=" signs and in that the variable δ appears on the right-hand-side of each equation with a coefficient of 1 in the 2nd constraint and with coefficients of 0 in the other constraints. Effectively, the RHS value of the 2nd constraint has been changed from 7 to $7+\delta$.

Table 4.4. Initial tableau for Problem A' with perturbed RHS.

x	y	s_1	s_2	s_3	s_4	−z	RHS	δ
2	3	0	0	0	0	1	0	0
1	0	1	0	0	0	0	6	0
1	1	0	1	0	0	0	7	1
0	2	0	0	1	0	0	9	0
−1	3	0	0	0	1	0	9	0

The variables s_2 and δ have *identical* columns of coefficients in Table 4.4. Recall from Chapter 3 that identical columns stay identical after any sequence of Gaussian operations. Thus, performing on Table 4.4 the exact sequence of Gaussian operations that transformed rows 3-7 of Table 4.1 into rows 21-25 of Table 4.2 produces Table 4.5, in which the column of coefficients for δ duplicates that of s_2.

Table 4.5. The current tableau after the same two pivots.

x	y	s_1	s_2	s_3	s_4	−z	RHS	δ
0	0	0	−9/4	0	−1/4	1	−18	−9/4
0	0	1	−3/4	0	1/4	0	3	−3/4
1	0	0	3/4	0	−1/4	0	3	3/4
0	0	0	−1/2	1	−1/2	0	1	−1/2
0	1	0	1/4	0	1/4	0	4	1/4

Casting the basic solution to Table 4.5 in the format of a dictionary produces system (8), below. Equation (8.0) shows that the rate of change of the objective value with respect to the RHS value of the 2nd constraint equals 9/4. Thus, the shadow price of the 2nd constraint equals 9/4 or 2.25.

$$z = 18 + (9/4)\delta \tag{8.0}$$

$$s_1 = 3 - (3/4)\delta \tag{8.1}$$

$$x = 3 + (3/4)\delta \tag{8.2}$$

$$s_3 = 1 - (1/2)\delta \tag{8.3}$$

$$y = 4 + (1/4)\delta \tag{8.4}$$

The "range" of a shadow price

System (8) prescribes the values of the basic variables in terms of the change δ in the right-hand side of the 2nd constraint of Problem A. The **range** of a shadow price is the interval in its RHS value for which the basic solution remains feasible. It's clear from equations (8.1) through (8.4) that the basic variables stay nonnegative for the values of δ that satisfy the inequalities

$$s_1 = 3 - (3/4)\delta \geq 0,$$

$$x = 3 + (3/4)\delta \geq 0,$$

$$s_3 = 1 - (1/2)\delta \geq 0,$$

$$y = 4 + (1/4)\delta \geq 0.$$

These inequalities are easily seen to hold for δ in the interval

$$-4 \leq \delta \leq 2.$$

The largest value of δ for which the perturbed basic solution remains feasible is called the **allowable increase**. The negative of the smallest value of δ for which the perturbed basic solution remains feasible is called the **allowable decrease**. In this case, the allowable increase equals 2 and the allowable decrease equals 4.

A break-even price

Evidently, if the RHS value of the 2nd constraint can be increased at a per-unit cost p below 2.25 (which equals 9/4), it is profitable to increase it by as many as 2 units, perhaps more. Similarly, if the RHS value of the 2nd constraint can be decreased at a per-unit revenue p above 2.25, it is profitable to decrease it by as many as 4 units, perhaps more.

In this example and in general, each constraint's shadow price is a *break-even* price that applies to increases in a constraint's RHS value up to the "allowable increase" and to decreases down to the "allowable decrease."

Economic insight

It's often the case that the RHS values of a linear program represent levels of resources that can be adjusted upward or downward. When this occurs, the shadow prices give the break-even value of small changes in resource levels – they suggest where it is profitable to invest, and where it is profitable do divest.

Why the term, *shadow price*?

The term, shadow price, reflects the fact that these break-even prices are endogenous (determined within the model), rather than by external market forces.

Shadow prices for "≤" constraints

In Table 4.5, the reduced cost of the slack variable s_2 for the 2nd constraint equals –9/4, and the shadow price of the 2nd constraint equals 9/4. This is not a coincidence. It is a consequence of the fact that identical columns stay identical. In brief:

> In any basic tableau, the shadow price of each "≤" constraint equals (–1) times the reduced cost of that constraint's slack variable.

In Table 4.5, for instance, the shadow prices for the four constraints are 0, 9/4, 0 and 1/4, respectively.

Shadow prices for "≥" constraints

Except for a factor of (–1), the same property holds for each "≥" constraint.

> In any basic tableau, the shadow price of each "≥" constraint equals the reduced cost of that constraint's surplus variable.

This property also holds because identical columns stay identical.

Shadow prices for nonbinding constraints

An inequality constraint in a linear program is said to be **binding** when it holds as an equation and to be **nonbinding** when it holds as a strict inequality. Suppose the ith constraint in the original linear program is an inequality, and suppose that the current basis equates this constraint's slack or surplus variable to a positive value. Being positive, this variable is basic, so its reduced cost (top-row coefficient) equals zero. In brief:

> If a basic solution causes an inequality constraint to be nonbinding, that constraint's shadow price must equal 0.

This makes perfect economic sense. If a resource is not fully utilized, a small change in the amount of that resource has 0 as its marginal benefit.

Graphical illustration

For a graphical interpretation of shadow prices and their ranges, we return to Problem A. Figure 4.4 graphs its feasible region for various values of the RHS of its 2nd constraint, i.e., with that constraint as $x + y \le 7 + \delta$ and with δ between −4 and + 2. Figure 4.4 includes the objective vector, which is (2, 3). The optimal solution to Problem A is the feasible solution that lies farthest in the direction of its objective vector. For δ between −4 and +2, this optimal solution lies at the intersection of the lines

$$-x + 3y = 9 \quad \text{and} \quad x + y = 7 + \delta. \tag{9}$$

Solving these two equations for x and y gives

$$x = 3 + (3/4)\delta \quad \text{and} \quad y = 4 + (1/4)\delta, \tag{10}$$

and substituting these values of x and y in the objective function gives

$$z = 2x + 3y = 16 + (6/4)\delta + 12 + (3/4)\delta = 18 + (9/4)\delta. \tag{11}$$

This reconfirms that the shadow price of the 2nd constraint equals 9/4.

Note in Figure 4.4 that as δ ranges between −4 and + 2, the optimal solution shifts along the heavily-outlined interval in Figure 4.4. When δ equals +2, the constraint $2y \le 9$ holds as an equation. When δ exceeds +2, the perturbed solution violates the constraint $2y \le 9$. This reconfirms +2 as the allowable increase. A similar argument shows why 4 is the allowable decrease.

Figure 4.4. Perturbing the constraint $x + y \le 7$.

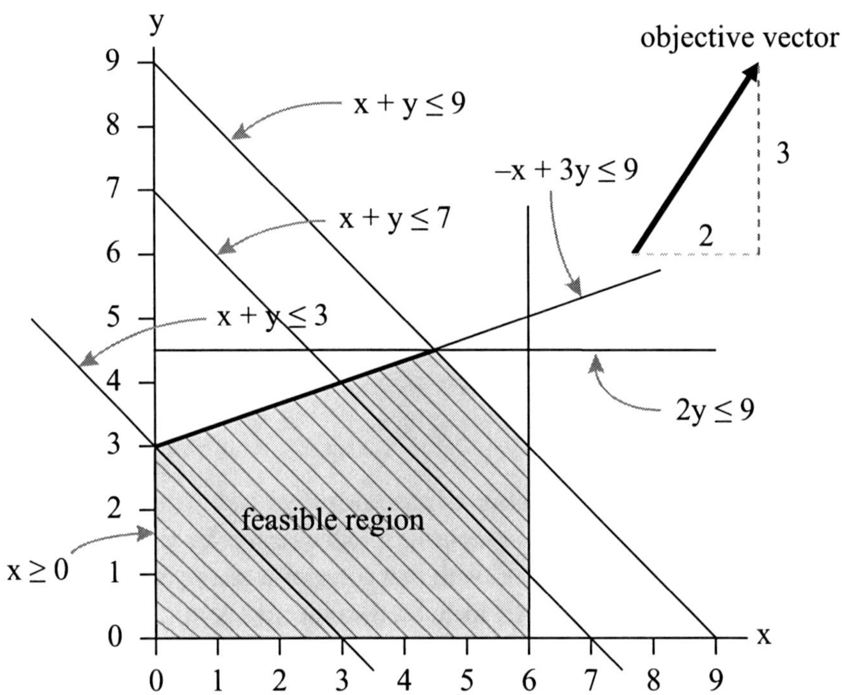

Perturbing multiple RHS values

For Problem A, consider the effect of adding δ_2 units to the RHS of the 2nd constraint and adding δ_4 units to the RHS of the 4th constraint. Let us ask ourselves: What effect would this have on the basic solution for the basis in Table 4.5? Inserting δ_4 on the RHS of Table 4.4 with a coefficient of +1 in the 4th constraint and repeating the above argument (the variables s_4 and δ_4 have identical columns of coefficients) indicates that the basic solution becomes

(12.0) $$z = 18 + (9/4)\delta_2 + (1/4)\delta_4,$$

(12.1) $$s_1 = 3 - (3/4)\delta_2 + (1/4)\delta_4,$$

(12.2) $$x = 3 + (3/4)\delta_2 - (1/4)\delta_4,$$

(12.3) $$s_3 = 1 - (1/2)\delta_2 - (1/2)\delta_4,$$

(12.4) $$y = 4 + (1/4)\delta_2 + (1/4)\delta_4.$$

Evidently, the objective value changes by $(9/4)\delta_2 + (1/4)\delta_4$. In this case and in general, the shadow prices apply to simultaneous changes in two or more RHS values.

These shadow prices continue to be break-even prices as long as the values of the basic variables s_1, x, s_3 and y remain nonnegative. In particular, the RHS of equation (12.1) is nonnegative for all values of δ_2 and δ_4 that satisfy the inequality

$$3 - (3/4)\delta_2 + (1/4)\delta_4 \geq 0.$$

In Chapter 3, it was noted that the set of ordered pairs (δ_2, δ_4) that satisfy a particular linear inequality, such as the above, is a convex set. It was also observed that the intersection of convex sets is convex. In particular, the set of pairs (δ_2, δ_4) for which the basic solution remains feasible (nonnegative) is convex. In brief:

> **Perturbed RHS values:** Each basis's shadow prices apply to simultaneous changes in two or more RHS values, and the set of RHS values for which the basis remains feasible is convex.

Note also that perturbing the RHS values of the original tableau affects *only* the RHS values of the current tableau. It has no effect on the coefficients of the decision variables in any of the equations. In particular, these perturbations have no effect on the reduced costs (top-row coefficients). If the reduced costs satisfy the optimality conditions before the perturbation occurs, they continue to satisfy it after perturbation occurs. It is emphasized:

> **Optimal basis:** Consider a basis that is optimal. If one or more RHS value is perturbed, its basic solution changes, but it remains optimal as long as it remains feasible.

In Chapter 5, we will see that the shadow prices are central to a key idea in economics, namely, the "opportunity cost" of doing something new. In Chapter 12, the shadow prices will emerge as the decision variables in a "dual" linear program.

Computer output, multipliers and the proviso

Every computer code that implements the simplex method finds and reports a basic solution that is optimal. Most of these codes also report a shadow price for each constraint, along with an allowable increase and an allowable decrease for each RHS value.

If the Full Rank proviso is violated, not all of the constraints can have shadow prices. These computer codes report them anyhow! What these codes are actually reporting are values of the basis's "multipliers" (short for Lagrange multipliers). In Chapter 11, it will be shown that these "multipliers" coincide with the shadow prices when they exist and, even if the shadow prices do not exist, the multipliers account correctly for the marginal benefit of perturbing the RHS values in any way that keeps the linear program feasible.

8. Review

Linear programming has its own specialized vocabulary. Learning the vocabulary eases access to the subject. In this review, the specialized terminology that was introduced in this chapter appears in italics. A crucial idea in this chapter is the *feasible pivot*. Before proceeding, make certain that you understand what feasible pivots are and that you can execute them on a spreadsheet.

Recap of the simplex method

Listed below are the most important of the properties of the simplex method.

- The simplex method pivots from one *basic feasible tableau* to another.

- Geometrically, each basic feasible tableau identifies an *extreme point* of the *feasible region*.

- In each basic feasible tableau, the *reduced cost* of each nonbasic variable equals the amount by which the basic solution's objective value changes if that nonbasic variable is set equal to 1 and if the values of the basic variables are adjusted to preserve a solution to the equation system.

- The *entering variable* in a simplex pivot can be any nonbasic variable whose reduced cost is positive in the case of a maximization problem, negative in the case of a minimization problem.

- Each simplex pivot occurs on a positive coefficient of the entering variable, and that coefficient has the smallest *ratio* (of RHS value to coefficient).

- If the RHS value of the *pivot row* is positive, the pivot is *nondegenerate*. Each nondegenerate simplex pivot improves the basic solution's objective value.

- If the RHS value of the pivot row is zero, the pivot is *degenerate*. Each degenerate pivot changes the basis, but causes no change in the basic solution or in its objective value.

- The simplex method identifies an optimal solution when it encounters a basic feasible tableau for which the reduced cost of each nonbasic variable is nonpositive in a maximization problem, nonnegative in a minimization problem.

- The simplex method identifies an unbounded linear program if the entering variable for a simplex pivot has nonpositive coefficients in every row other than the one for which $-z$ is basic.

- A linear program satisfies the *Full Rank proviso* if any basis has as many basic variables as there are constraints in the linear program's *Form 1* representation.

- If the Full Rank proviso is satisfied, each basic feasible tableau has these properties:

- The *shadow price* of each constraint equals the rate of change of the basic solution's objective value with respect to the constraint's RHS value.
- These shadow prices apply to simultaneous changes in multiple RHS values.
- If only a single RHS value is changed, the shadow price applies to increases as large as the *allowable increase* and to decreases as large as the *allowable decrease.*

What has been omitted?

This chapter is designed to enable you to make intelligent use of computer codes that implement the simplex method. Facets of the simplex method that are not needed for that purpose have been deferred to later chapters. In later chapters, we will see that:

- *Phase I* of the simplex method determines whether or not the linear program has a feasible solution and, if so, constructs a basic feasible tableau with which to initiate Phase II.
- Rule #1 can cause the simplex method to *cycle*, and the ambiguity in Rule #1 can be resolved in a way that precludes cycling, thereby guaranteeing finite termination.
- In some applications, decision variables that are unconstrained in sign are natural. They can be accommodated directly, without forcing the linear program into the format of Form 1.
- If the Full Rank proviso is violated, each basis still has "multipliers" that correctly account for the marginal value of any perturbation of the RHS values that keeps the linear program feasible.

Not a word has appeared in this chapter about the speed of the simplex method. For an algorithm to be useful, it must be fast. The simplex method is blazingly fast on nearly every practical problem. But examples have been discovered on which it is horrendously slow. Why that is so has remained a bit of a mystery for over a half century. Chapter 6 touches lightly on the speed of the simplex method.

9. Homework and Discussion Problems

1. On a spreadsheet, execute the simplex method on Problem A with x as the entering variable for the first pivot. Use Figure 4.3 to interpret its progress.

2. Rule #1 picks the most positive entering variable for a simplex pivot on a maximization problem. State a simplex pivot rule that makes the largest possible improvement in the basic solution's objective value. Use Problem A to illustrate this rule.

3. In system (1), execute a pivot on the coefficient of x in equation (1.2). What goes wrong?

4. *(graphical interpretation)* Each part of this problem refers to Figure 4.3.

   (a) The coefficient of y in equation (1.1) equals zero. How is this fact reflected in Figure 4.3? Does a similar interpretation apply to the coefficient of x in equation (1.3)?

   (b) With x as the entering variable, no ratio was computed for equation (1.4). If this ratio had been computed, it would have equaled $9/(-1) = -9$. Use Figure 4.3 to interpret this number.

   (c) True or false: Problem A' has a feasible basis whose nonbasic variables are x and s_4.

5. *(graphical interpretation)* It is clear from Figure 4.3 that system (1) has 5 bases that include –z and are feasible. Use Figure 4.3 to identify each basis that includes –z and is not feasible.

6. *(graphical interpretation)* True or false: for Problem A', every set that includes –z and all but two of the variables x, y and s_1 through s_4 is a basis.

7. Consider this linear program: Maximize {x}, subject to the constraints

$$-x + y \leq 1,$$

$$x + y \leq 4,$$

$$x - y \leq 2,$$

$$x \geq 0, \quad y \geq 0.$$

   (a) Solve this linear program by executing simplex pivots on a spreadsheet.

(b) Solve this linear program graphically, and use your graph to trace the progress of the simplex method.

8. *(no extreme points?)* Consider this linear program: Maximize (A – B) subject to the constraints

$$A - B \le 1,$$

$$-A + B \le 1.$$

(a) Plot this linear program's feasible region. Does it have any extreme points?

(b) Does this linear program have an optimal solution? If so, name one.

(c) Apply the simplex method to this linear program. What happens?

9. Consider this linear program: Maximize $\{x + 1.5y\}$, subject to the constraints

$$\begin{aligned} x \quad\quad &\le 4, \\ -x + y &\le 2, \\ 2x + 3y &\le 12, \\ x \ge 0, \quad y &\ge 0. \end{aligned}$$

(a) Solve this linear program by executing simplex pivots on a spreadsheet.

(b) Execute a feasible pivot that finds a second optimal solution to this linear program.

(c) Solve this linear program graphically, and use your graph to trace the progress of the simplex method.

(d) How many optimal solutions does this linear program have? What are they?

10. For the linear program that appears below, construct a basic feasible system, state its basis, and state its basic solution.

Maximize $\{2y - 3z\}$, subject to the constraints

$$\begin{aligned} x + y - z &= 16, \\ y + z &\leq 12, \\ 2y - z &\geq -10, \end{aligned}$$

$x \geq 0, \quad y \geq 0, \quad z \geq 0.$

(a) True or false: Problem A' has a feasible basis whose nonbasic variables are x and s_2.

(b) True or false: Problem A' has a feasible basis whose nonbasic variables are x and s_3.

11. *(an unbounded linear program)* Draw the feasible region for Problem B'. Apply the simplex method to Problem B', selecting y (and not x) as the entering variable for the first pivot. What happens? Interpret your result graphically.

12. *(degeneracy in 2-space)* This problem concerns the variant of Problem A in which the right-hand-side value of equation (1.4) equals 0, rather than 9.

(a) On a spreadsheet, execute the simplex method, with y as the entering variable for the first pivot.

(b) Draw the analog of Figure 4.3 for this linear program.

(c) List the bases and basic solutions that were encountered. Did a degenerate pivot occur?

(d) Does this linear program have a redundant constraint (defined above)?

13. *(degeneracy in 3-space)* This problem concerns the linear program: Maximize $\{x + 1.5y + z\}$ subject to the constraints $x + y \leq 1, \quad y + z \leq 1, \quad x \geq 0, \quad y \geq 0, \quad z \geq 0.$

(a) Use the simplex pivots with Rule #1 to solve this linear program on a spreadsheet. Did a degenerate pivot occur?

(b) Plot this linear program's feasible region. Explain why a degenerate pivot must occur.

(c) True or false: If a degenerate pivot occurs, the linear program must have a redundant constraint.

14. *(degeneracy in 3-space)* This problem concerns the linear program: Maximize $\{0.1x + 1.5y + 0.1z\}$ subject to the constraints $x + y \le 1$, $y + z \le 1$, $x \ge 0$, $y \ge 0$, $z \ge 0$.

    (a) Use the simplex pivots with Rule #1 to solve this linear program on a spreadsheet. Did a degenerate pivot occur?

    (b) True or false: The simplex method stops when it encounters an optimal solution.

15. True or false:

    (a) A nondegenerate pivot can result in a degenerate basic system.

    (b) A degenerate pivot can result in a nondegenerate basic system.

16. Consider a basic feasible tableau that is nondegenerate, so that its basic solution equates all variables to positive values, with the possible exception of $-z$. Complete the following sentence and justify it: A feasible pivot on this tableau will result in a degenerate tableau if and only if a tie occurs for________.

17. True or false: For a linear program in Form 1, feasible pivots are the only pivots that keep the basic solution feasible.

18. *(redundant constraints)* Suppose that you need to learn whether or not the ith constraint in a linear program is redundant.

    (a) Suppose the ith constraint is a "≤" inequality. How could you find out whether or not this constraint is redundant? *Hint:* use a linear program.

    (b) Suppose the ith constraint is an equation. *Hint:* use part (a), twice.

19. *(maximizing a decision variable)* Alter Program 1 so that its objective is to maximize y, but its constraints are unchanged. Adapt the simplex method to accomplish this directly, that is, *without* introducing an equation that defines z as the objective value. Execute your method on a spreadsheet.

20. *(bases and shadow prices)* This problem refers to Table 4.1.

    (a) For the basis solution in rows 2-7, find the shadow price for each constraint.

(b) For the basic solution in rows 11-16, find the shadow price for each constraint.

(c) Explain why some of these shadow prices equal zero.

21. Adapt Table 4.4 and Table 4.5 to compute the shadow price, the allowable increase, and the allowable decrease for the optimal basis of the RHS value of the constraint $-x + 3y \le 9$. Which previously nonbinding constraint becomes binding at the allowable increase? At the allowable decrease?

22. On the plane, plot the set S that consists of all pairs (δ_2, δ_4) for which the basic solution to system (12) remains feasible. For each point on the boundary of the set S that you have plotted, indicate which constraint(s) become binding.

23. Suppose every RHS value in Problem A is multiplied by the *same* positive constant, for instance, by 10.5. What happens to the optimal basis? To the optimal basic solution? To the optimal value? To the optimal tableau? Why?

24. This concerns the *minimization* problem whose Form 1 representation is given in the tableau that follows.

equation	A	B	C	s1	s2	-z	RHS
(1.1)	1	5	2			1	0
(1.2)	2	-5	1	1			-5
(1.3)	2	-1	5		1		4

(a) It this a basic tableau? Is its basis feasible?

(b) To make short work of Phase I, pivot on the coefficient of B in equation (1.2). Then continue Phase II to optimality.

25. True or false: When the simplex method is executed, a variable can:

(a) Leave the basis at a pivot and enter at the next pivot. *Hint:* If it entered, to which extreme point would it lead?

(b) Enter at a pivot and leave at the next pivot. *Hint:* Maximize $\{2y + x\}$, subject to the constraints $3y + x \le 3, \quad x \ge 0, \quad y \ge 0$.

26. The simplex method has been applied to a maximization problem in Form 1 (so that all variables other than $-z$ are constrained to be nonnegative). At some point in the computation, the tableau that is shown below has been encountered; in this tableau, u, v, w and x denote numbers.

A	B	C	D	E	-z	RHS
-2	-1	u			1	12
1	v		1			w
2	3	x		1		3

State the conditions on u, v, w and x such that:

(a) The basic solution to this tableau is the unique optimal solution.

(b) The basic solution to this tableau is optimal, but is not the unique optimal solution.

(c) The linear program is unbounded.

(d) The linear program has no feasible solution.

27. *(nonnegative column)* For a maximization problem in Form 1, the following tableau has been encountered. In it, * stands for an unspecified data element. Prove there exist no values of the unspecified data for which it is optimal to set A > 0. *Hint:* If a feasible solution exists with A > 0, show that it is profitable to decrease A to zero and increase the values of B, D and F in a particular way.

A	B	C	D	E	F	-z	RHS
-3		*		*		1	*
4	1	*		*			*
2		*	1	*			*
0		*		*	1		*

28. *(nonpositive row)* For a maximization or a minimization problem in Form 1, the following tableau has been encountered. In it, * stands for an unspecified data element.

A	B	C	D	E	F	-z	RHS
*		*		*		1	*
-4	1	-2		-2/3			11
*		*	1	*			*
*		*		*	1		*

(a) Prove that B is basic in every basic feasible tableau.

(b) Prove that deleting B and the equation for which it is basic can have no effect either on the feasibility of this linear program or on its optimal value.

Chapter 5: Analyzing Linear Programs

1. Preview

Dozens of different computer packages can be used to compute optimal solutions to linear programs. From this chapter, you will learn how to make effective use of these packages.

This chapter also addresses the fact that a linear program – like any mathematical model – is but an approximation to the situation that is under study. The information that accompanies the optimal solution to a linear program can help you to determine whether or not the approximation is a reasonable one.

E. V. Denardo, *Linear Programming and Generalizations*, International Series in Operations Research & Management Science 149,
DOI 10.1007/978-1-4419-6491-5_5,

Also established in this chapter is a close relationship between three economic concepts – the break–even (or shadow) price on each resource, the *relative* opportunity cost of engaging in each activity, and the marginal benefit of so doing. It will be seen that "relative opportunity cost" carries a somewhat different meaning than "opportunity cost," as that term is used in the economics literature.

Three sections of this chapter are starred because they can be read independently of each other. One of these starred sections provides a glimpse of duality.

2. All Terrain Vehicles

Much of the material in this chapter will be illustrated in the context of the optimization problem that appears below as

Problem A (All Terrain Vehicles)[1]**.** Three models of All Terrain Vehicle (ATV) are manufactured in a facility that consists of five shops. Table 5.1 names the vehicles and the shops. It specifies the capacity of each shop and the manufacturing time of each vehicle in each shop. It also specifies the con-

Table 5.1. The ATV Manufacturing Facility.

		Manufacturing times		
Shop	**Capacity**	**Standard**	**Fancy**	**Luxury**
Engine	120	3	2	1
Body	80	1	2	3
Standard Finishing	96	2		
Fancy Finishing	102		3	
Luxury Finishing	40			2
Contribution		840	1120	1200

Note on units of measure: In Table 5.1, capacity is measured in hours per week, manufacturing time in hours per vehicle, and contribution in dollars per vehicle.

[1] This example has a long history. An early precursor appears in the article by Robert Dorfman, "Mathematical or 'linear' programming: A nonmathematical exposition," *The American Economic Review*, V. 13, pp. 797-825, 1953.

tribution (profit) earned by manufacturing each vehicle. The plant manager wishes to learn the production rates (numbers of each vehicle to produce per week) that maximize the profit that can be earned by this facility.

Contribution

"Contribution," as used in this book, takes its meaning from accounting. When one contemplates taking an action, a **variable cost** is an expense that is incurred if the action is taken and only if the action is taken. When one contemplates an action, a **fixed cost** is a cost that has occurred or will occur whether or not the action is taken. Decisions should not be influenced by fixed costs.

When one is allocating this week's production capacity, the variable cost normally includes the material and energy that will be consumed during production, and the fixed cost includes depreciation of existing structures, property taxes, and other expenses that are unaffected by decisions about what to produce this week.

The **contribution** of an action equals the revenue that it creates less its variable cost. This usage abbreviates the accounting phrase, "contribution toward the recovery of fixed costs." Table 5.1 reports $840 as the contribution of each Standard model vehicle. This means that $840 equals the sales price of a Standard model vehicle less the variable cost of manufacturing it. When **profit** is used in this book, what is meant is contribution.

Maximizing contribution

The manager of the ATV plant seeks the mix of activities that maximizes the rate at which contribution is earned, measured in dollars per week. At a first glance, the Luxury model vehicle seems to be the most profitable. It has the largest contribution. Each type of vehicle consumes a total of 4 hours of capacity in the Engine and Body shops, where congestion is likely to occur. But we will see that no Luxury model vehicles should be manufactured, and we will come to understand why that is so.

The decision variables

Let us formulate the ATV problem for solution via linear programming. Its decision variables are the rates at which to produce the three types of vehicles, which are given the names:

S = the rate of production of Standard model vehicles (number per week),

F = the rate of production of Fancy model vehicles (number per week),

L = the rate of production of Luxury model vehicles (number per week).

Evidently, mnemonics (memory aids) are being used; the labels S, F and L abbreviate the production rates for Standard, Fancy and Luxury model vehicles.

Inequality constraints

The ATV problem places eight constraints on the values taken by the decision variables. Three of these constraints reflect the fact that the production quantities cannot be negative. These three are $S \geq 0$, $F \geq 0$, and $L \geq 0$. The remaining five constraints keep the capacity of each shop from being over-utilized. The top line of Table 5.1 shows that producing at rates S, F, and L vehicles per week consumes the capacity of the Engine shop at the rate of $3S + 2F + 1L$ hours per week, so the constraint

$$3S + 2F + 1L \leq 120$$

keeps the number of hours consumed in the Engine shop from exceeding its weekly capacity. The expression, $\{840S + 1120F + 1200L\}$, measures the rate at which profit is earned. The complete linear program is:

Program 1: Maximize $\{840S + 1120F + 1200L\}$, subject to the constraints

Engine:	$3S +$	$2F +$	$1L \leq$	$120,$
Body:	$1S +$	$2F +$	$3L \leq$	$80,$
Standard Finishing:	$2S$		$\leq$	$96,$
Fancy Finishing:		$3F$	$\leq$	$102,$
Luxury Finishing:			$2L \leq$	$40,$

$$S \geq 0, \quad F \geq 0, \quad L \geq 0.$$

Integer-valued variables?

As written, Program 1 allows the decision variables to take fractional values. This makes sense. The manager wishes to determine the profit-maximizing rate of production of each vehicle. For instance, setting S=4.25 amounts to producing Standard model vehicles at the rate of 4.25 per week.

If the production quantities had been required to be integer-valued, Program 1 would be an "integer program," rather than a linear program, and could be a great deal more difficult to solve. Integer programming is discussed in later chapters.

A spreadsheet

Table 5.2 prepares the ATV problem for solution on a spreadsheet. Note that:

- Information about Standard, Fancy and Luxury model vehicles appears in columns B, C and D, respectively. In particular:
  - Cells B2, C2 and D2 contain the labels of these decision variables.
  - Cells B9, C9 and D9 are reserved for the values of these decision variables, each of which has been set equal to 1, temporarily.
  - Cells B3, C3 and D3 contain their contributions.
  - Cells B4:D4 contain the manufacturing times of each vehicle in the Engine shop.
  - Rows 5 through 8 contain comparable data for the other four shops.
- Column E contains (familiar) sumproduct functions, and cells E4 through E8 record their values when each decision variable is set equal to 1.
- Column F contains "<=" signs. These are memory aids; they remind us that the quantities to their left must not exceed the RHS values to their right.
- Column G contains the capacities of the five shops.

It remains for Solver to select values in cells B9, C9 and D9 that maximize the value in cell E3 while enforcing the constraints of Program 1.

Table 5.2. A Spreadsheet for the ATV problem.

	A	B	C	D	E	F	G
1							
2	label of variable	S	F	L			RHS
3	contribution	840	1120	1200	3160		
4	Engine shop	3	2	1	6	<=	120
5	Body shop	1	2	3	6	<=	80
6	Standard Finishing shop	2			2	<=	96
7	Fancy Finishing shop		3		3	<=	102
8	Luxury Finishing shop			2	2	<=	40
9	value of variable	1	1	1	↑		
10							
11		= SUMPRODUCT(B8:D8 , B$9:D$9)					

A standard format

Table 5.2 presents the data for the ATV problem in a **standard format**, which consists of:

- One row for the labels of the decision variables (row 2 in this case).
- One row for the values of the decision variables (row 9 in this case).
- One row for the contribution of each decision variable (row 3).
- One row for each constraint (rows 4 to 8).
- One column for the coefficients of each decision variable, one column for the sumproduct functions that measure the consumption of each resource, one column for the RHS values, and one (optional) column that records the sense of each constraint.

This standard format is a handy way in which to prepare a linear program for solution on a spreadsheet. It will be used repeatedly in this book.

3. Using Solver

This is the first of two sections that describe slightly different ways to compute the optimal solution to the ATV problem. This section is focused on Solver, which comes with Excel. The next section is focused on Premium Solver for Education, which is on the disc that accompanies this book.

The Solver Parameters dialog box

Figure 5.1 displays a Solver dialog box, which has been filled out. This dialog box identifies E3 as the cell whose value we wish to maximize, it specifies cells B9:D9 as the changing cells, and it imposes constraints that keep the quantities in cells B9:D9 nonnegative and keep the quantities in cells E4 through E8 from exceeding the quantities in cells G4 through G8, respectively.

Figure 5.1. A Solver Dialog box for the ATV problem.

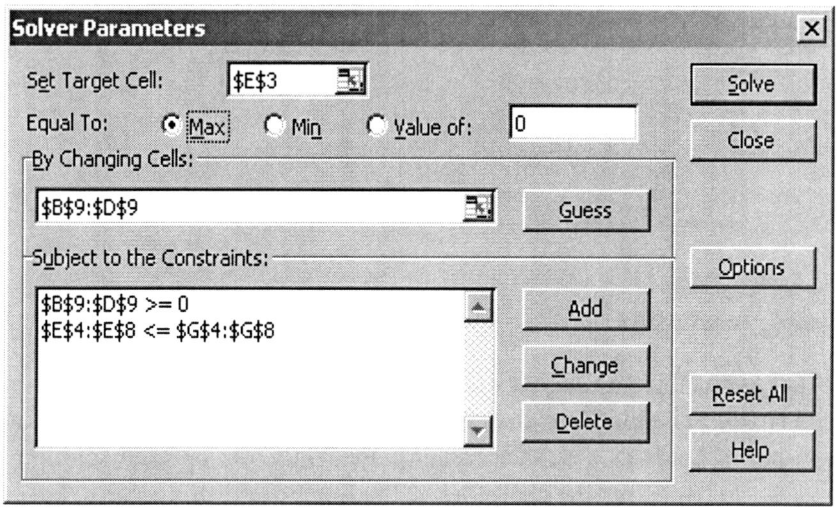

Chapter 2 tells how to fill out this dialog box. As was indicated in Chapter 2, the Solver dialog box for Excel 2010 differs slightly from the above, but is filled out in a similar way.

Getting an optimal solution

After you have reproduced Figure 5.1, remember to click on the Options button and, on the menu that appears, click on Assume Linear Model, then on OK. Then click on the Solve button. In a flash, Solver will report that it has found an optimal solution, which is S=20, F=30 and L=0. Solver will also report an optimal value of 50,400. Your spreadsheet will resemble Table 5.3, but no cells will (yet) be shaded.

Table 5.3. An optimal solution to the ATV problem.

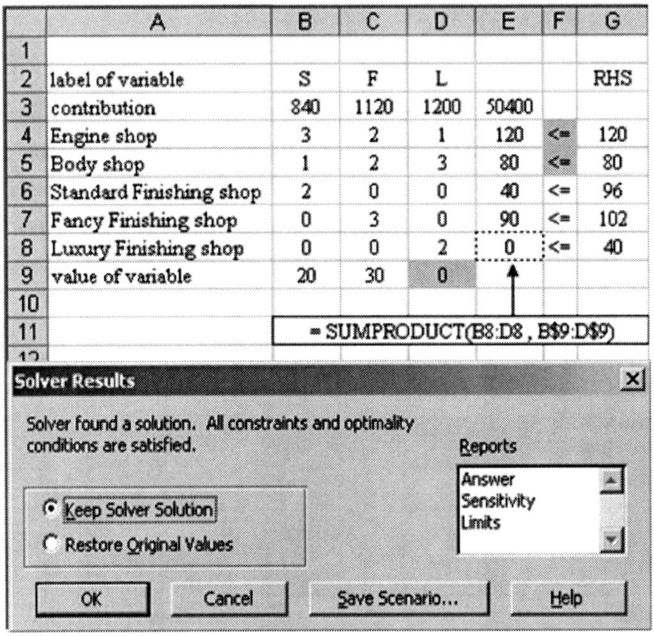

	A	B	C	D	E	F	G
1							
2	label of variable	S	F	L			RHS
3	contribution	840	1120	1200	50400		
4	Engine shop	3	2	1	120	<=	120
5	Body shop	1	2	3	80	<=	80
6	Standard Finishing shop	2	0	0	40	<=	96
7	Fancy Finishing shop	0	3	0	90	<=	102
8	Luxury Finishing shop	0	0	2	0	<=	40
9	value of variable	20	30	0			
10							
11		= SUMPRODUCT(B8:D8 , B$9:D$9)					

Binding constraints

Let us recall from Chapter 4 that an inequality constraint in a linear program is said to be **binding** when it holds as an equation and to be **nonbinding** when it holds as a strict inequality. Evidently, the optimal solution in Table 5.3 fully utilizes the capacities of the Engine and Body shops, but not of the finishing shops. This optimal solution has three binding constraints, which are:

- the nonnegativity constraint on L,
- the constraint on Engine shop's capacity,
- the constraint on Body shop's capacity.

In Table 5.3, the shaded cells identify these binding constraints. It will soon be argued that it is optimal to keep these constraints binding even if the model's data are somewhat inaccurate.

A sensitivity report

The Solver Results dialog box (see Table 5.3) has a window containing the word "sensitivity." Clicking on it creates a sheet containing a Sensitivity Report that is reproduced as Table 5.4.

Table 5.4. A Sensitivity Report for the ATV example.

Adjustable Cells

Cell	Name	Final Value	Reduced Cost	Objective Coefficient	Allowable Increase	Allowable Decrease
B9	value of variable S	20	0	840	200	280
C9	value of variable F	30	0	1120	560	100
D9	value of variable L	0	-200	1200	200	1E+30

Constraints

Cell	Name	Final Value	Shadow Price	Constraint R.H. Side	Allowable Increase	Allowable Decrease
E4	Engine shop	120	140	120	56	16
E5	Body shop	80	420	80	5 1/3	40
E6	Standard Finishing shop	40	0	96	1E+30	56
E7	Fancy Finishing shop	90	0	102	1E+30	12
E8	Luxury Finishing shop	0	0	40	1E+30	40

Each constraint in the ATV problem is an inequality, so the slack variables form a basis for its Form 1 representation. This guarantees that the Full Rank proviso is satisfied, hence that the shadow prices exist. Recall from Chapter 4 that the information in the Sensitivity Report has this interpretation:

- The shadow price for a constraint equals the rate of change of the basic solution's objective value with respect to the RHS value of that constraint. The basic solution remains feasible (and hence optimal) for increases in its RHS value up to the Allowable Increase and for decreases up to the Allowable Decrease.

- The shadow prices apply to simultaneous changes, albeit with smaller ranges.

- The optimal solution is unchanged if a single contribution is increased by an amount up to its Allowable Increase or is decreased by an amount up to its Allowable Decrease.

- The reduced cost of each basic variable equals zero, and the reduced cost of each nonbasic variable equals the amount by which the optimal value changes if that variable is set equal to 1 and the values of the basic variables are adjusted accordingly.

In particular:

- The capacity of the Engine shop has a break-even value of 140 $/hour, and this value applies to increases of up to 56 hs/wk and to decreases up to 16 hr/wk. Hence, an increase of up to 56 hours per week of Engine shop capacity is profitable if it can be obtained at a price below 140 dollars per hour. And a decrease in Engine shop capacity of up to 16 hours per week is profitable if it be put to an alternative use that is worth more than 140 dollars per hour.

- The optimal solution to the ATV problem is unchanged if the contribution of the Standard model vehicle is between $560 (because 560 = 840 – 280) and $1,040 (because 1040 = 840 + 200).

- Making one Luxury model vehicle reduces the rate of profit by 200 $/wk.

You forgot?

Sooner or later, nearly everyone who uses Solver will forget to check off Assume Linear Model before solving a linear program. If you forget, the "engine" that solves your linear program will not be the simplex method, but a more general algorithm. It computes the correct shadow prices but calls them **Lagrange multipliers**, it computes the correct reduced costs, but calls them **reduced gradients**, and it omits the Allowable Increases and Allowable Decreases because it presumes the problem is nonlinear.

4. Using the Premium Solver for Education

Premium Solver for Education has added features and fewer bugs than do the earlier versions of Solver. If you have a choice, use the Premium version. Chapter 2 tells how to install and activate it. After it is activated, Premium Solver will appear on the Add-Ins tab of the Excel? File menu.

To illustrate the use of the Premium Solver dialog box, we arrange for it to solve the ATV problem. The first step is to replicate the spreadsheet in Table 5.3. Next, click on the Add-Ins tab on the File menu. An icon labeled Premium Solver will appear at the left, just below the File tab. Click on it. A Solver Parameters dialog box will appear. To make it look like that in Figure 5.2, follow the procedure that is described in Chapter 2. After you succeed, click on Solve. In a flash, a solution will appear, along with the usual box that accords you the opportunity to obtain a sensitivity report.

Figure 5.2. A Premium Solver dialog box for the ATV problem.

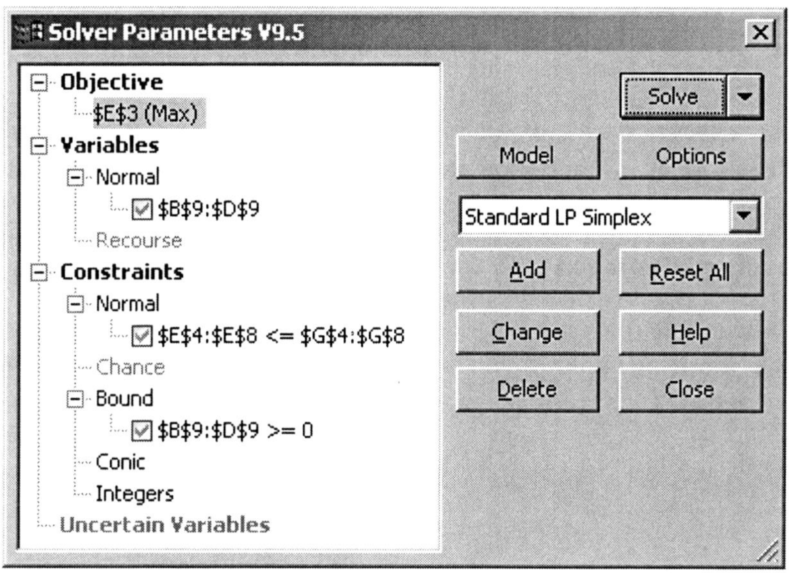

5. Differing Sign Conventions!

Solver and Premium Solver report reduced costs and shadow prices as they are defined in Chapter 4. Some computer packages use different conventions as to the *signs* (but not the magnitudes) of the reduced costs and the shadow prices. If you are using a different software package to solve linear programs, you will need to figure out what sign conventions it employs. An easy way to do that is described below.

A maximization problem

To see what sign conventions a particular computer package uses for maximization problems, you can ask it to solve a simple linear program, such as

Example 1. Maximize {2x + 4y}, subject to

$$3x + 3y \leq 6,$$

$$x \geq 0, \quad y \geq 0.$$

Clearly, the optimal solution to Example 1 is x=0 and y=2, and its optimal value equals 8. It is equally clear that:

- Increasing the RHS value of its constraint from 6 to 7 increases y by 1/3, so the shadow price of the constraint equals 4/3.
- Perturbing the optimal solution by setting x=1 reduces y by 1 and changes the optimal value by –2, so the reduced cost of y equals –2.

Whatever sign conventions your computer program reports for Example 1 will hold for all maximization problems. If it reverses the sign of the reduced cost in Example 1, it will do so for all maximization problems.

A simple minimization problem

Similarly, to find out what sign conventions a computer package employs when it is solving minimization problems, you can ask it to solve Example 2, below, and ask for a sensitivity report.

Example 2. Minimize {2x + 4y}, subject to

$$3x + 3y \geq 6,$$

$$x \geq 0, \quad y \geq 0.$$

The optimal solution to Example 2 is x=2 and y=0, and its optimal value equals 4. Evidently:

- Increasing the RHS value of its constraint from 6 to 7 increases x by 1/3, so the shadow price of the 1st constraint equals 2/3.
- Perturbing the optimal solution by setting y=1 reduces x by 1 and changes the optimal value by 2, so the reduced cost of y equals 2.

Whatever sign conventions a particular computer package reports for Example 2 will apply to all minimization problems.

6. A Linear Program as a Model

Like any mathematical model, a linear program is an inexact representation of reality. Linear programs approximate reality in three ways – by eliminating uncertainty, by aggregating activities, and by suppressing nonlinearities. The ATV problem is now used to illustrate these approximations.

Uncertain data

This model's data are **uncertain** because they cannot be measured precisely and because they can fluctuate in unpredictable ways. In the ATV model, it is presumed that, in an "representative" week, 120 machine hours are available in the Engine shop; 120 equals the number of machine hours during which the Engine shop is open for business *less* allowances for routine maintenance, machine breakdowns, shortages of vital parts, absences of key workers, power failures, and other unforeseen events. The actual number of machine hours available in a particular week could be larger or smaller than 120, depending on how things turned out that week.

Similarly, the contribution of $840 per Standard model approximates an uncertain quantity. The actual contribution could be larger or smaller than this figure, depending on the current prices of raw materials, defects that require abnormal amounts of rework, market conditions that affect the sales revenues, and changes in inventory carrying costs. Uncertainty in the data is one reason why models are approximate.

Aggregation

A second reason why models are approximate is **aggregation**, which refers to the lumping together of several activities in a single entity. The assembly times in the ATV model reflect aggregation. The Engine shop is modeled as a single entity, but it is actually a system consisting of people, tools and machines that can produce the engines and drive trains for the three vehicles at different rates. In our simplified view, it takes 3 hours of Engine shop time to make each Standard model vehicle. Aggregation is useful when it avoids detail that is unimportant.

Linearization

Linearization is the third way in which the ATV model is approximate. The capacity constraint for the Engine shop is $3S + 2F + 1L \le 120$. This constraint presumes linear interactions among the three types of vehicles that are produced there. The actual interactions are more complicated and are somewhat nonlinear. For example, this constraint accounts crudely for the set-up times that are needed to change from the production of one model to another.

The value of an approximation

The ATV example is aggregated and simplified. It has to be. Imagine how intractable. this model would become if it incorporated all of the complexities and details just mentioned. Yet, there is merit in starting with a simple and aggregated model. It will be relatively easy to build and debug. And if it is artfully built, its simplicity will cause the main insights to stand out starkly.

Robustness

It would be foolish to believe the results of a simplified model without first considering how its simplifications influenced its optimal solution. If the insights obtained from the model hold up in the real world, the model is said to be **robust**. To ascertain whether a model is robust, we can execute a **sensitivity analysis** – change those data that are suspect, rerun the model, and see whether the insights obtained from it are preserved.

In the case of a linear program, the Sensitivity Report provided by the simplex method helps us with a sensitivity analysis. The ATV example illustrates this point. Each Allowable Increase and Allowable Decrease in Table 5.4 is fairly large, which suggests that the optimal basis remains unchanged over fairly broad ranges of objective coefficients and RHS values.

The Perturbation Theorem

The Sensitivity Report describes how a change in a single datum affects the optimal solution to a linear program. What happens if we change several elements of data? A partial answer to this question lies in:

Proposition 5.1. *(the Perturbation Theorem).* Suppose that Premium Solver is used to find an optimal solution to a linear program via the simplex method and that its sensitivity report states that each Allowable Increase and each Allowable Decrease is positive. If the data of this linear program are perturbed by small amounts, then:

- No change occurs in the optimal basis.
- No change occurs in the set of binding constraints.
- The values taken by the basic variables may change.

Illustration

The ATV problem is used to illustrate the Perturbation Theorem. Table 5.4 shows that each Allowable Increase and Decrease is positive. Thus, the Perturbation Theorem shows that the binding constraints (identified by the shaded cells in Table 5.3) stay binding if any or all of the data in Table 5.1 are perturbed. An optimal solution to the ATV problem can be described in either of these ways:

- Make Standard model vehicles at the rate of 20 per week, make Fancy model vehicles at the rate of 30 per week and make no Luxury model vehicles.
- Keep the Engine and Body shops busy making Standard and Fancy model vehicles, and make no Luxury model vehicles.

If the model's data were exact (and that never occurs), both descriptions of the optimal solution would be correct. If the model's data are close, the latter is correct because it keeps the binding constraints binding.

Sketch of a proof

An air-tight proof of the Perturbation Theorem would entail an interpretation of the shadow prices that does not appear until Chapter 11, but a sketch of a proof can be provided here. The hypothesis of this theorem guarantees that:

- Each basis for the LP's equation system contains one variable per equation. (In other words, the Full Rank proviso is satisfied.)
- The linear program has only one optimal solution, and this optimal solution is nondegenerate. (It sets each basic variable that is constrained to be nonnegative to a positive value.)

Coupling these observations with the fact that the inverse of a matrix is a continuous function of its data would prove the theorem.

7. Relative Opportunity Cost

Linear programming is a natural environment to describe and illustrate three economic concepts, which are:

- The *break-even price* (or shadow price) of a resource.
- The *relative opportunity cost* of engaging in an activity.
- The *marginal benefit* of so doing.

It will soon be seen that three concepts are very closely linked.

Context

These three concepts are described in the context of a linear program that has been cast in Form 1 – with equality constraints and nonnegative decision variables. In such a linear program, each decision variable is now said to represent an **activity**. Each constraint (other than the nonnegativity constraints on the variables) measures the consumption of a **resource** and requires its consumption to equal its availability (RHS value). Each basis is now said to **engage** in those activities (decision variables) that the basis includes. The value assigned to each decision variable in a basic solution is now said to be the **level** of the activity that it represents.

Relative opportunity cost

Shadow prices and marginal benefit are familiar from Chapter 4, but relative opportunity cost was not discussed there. Like the other two terms, relative opportunity cost is defined in the context of a particular basis. The **relative opportunity cost** of each activity equals the reduction in contribution that occurs the levels of the activities in which the basis is engaged are altered so as to free up (make available) the resources needed to set the level of that activity equal to 1.

The ATV example

Shadow prices, relative opportunity costs and marginal benefit are defined for *every* basis, not merely for the optimal basis. To illustrate these concepts – and the relationship between them – we focus on the basis (and basic solution) that is optimal for the ATV problem. Table 5.4 reports its shadow prices. For convenient reference, these shadow prices are recorded in Table 5.5, with a label and unit of measure of each.

Table 5.5. Label and value of each constraint's shadow prices for the optimal solution to the ATV problem.

Constraint	label	Value
Engine shop capacity	E	140 $/hr
Body shop capacity	B	420 $/hr
Standard Finishing shop capacity	SF	0 $/hr
Fancy Finishing shop capacity	FF	0 $/hr
Luxury Finishing shop capacity	LF	0 $/hr

Let us recall that these shadow prices are break-even prices. For instance, a change of δ in the RHS of the Engine shop capacity constraint causes a change of 140 δ in the basic solution's objective value. These break-even prices apply to simultaneous changes in several RHS values.

The Luxury model vehicle

Table 5.5 presents the shadow prices for the optimal basis. With reference to that basis, the relative opportunity cost of making one Luxury model vehicle is now computed. Making one Luxury model vehicle requires 1 hour in the Engine shop, 3 hours in the Body shop, and 2 hours in the Luxury Finishing shop. The shadow prices apply to simultaneous changes in several RHS values, and the prices in Table 5.5 show that:

$$(1)\quad \begin{bmatrix}\text{relative opportunity cost of}\\ \text{one Luxury model vehicle}\end{bmatrix} = (1)\times(140) + (3)\times(420) + (2)\times(0),$$
$$= \$140 + \$1{,}260 + \$0 = \$1{,}400.$$

Evidently, contribution is reduced by $1,400 when the levels of the activities in which the basis is engaged are adjusted so as to free up the recourses needed to make one Luxury model vehicle.

The marginal profit of any activity equals its contribution *less* the relative opportunity cost of freeing up the resources needed to accomplish that activity. In particular,

$$(2)\quad \begin{bmatrix}\text{marginal profit of one}\\ \text{Luxury model vehicle}\end{bmatrix} = \begin{bmatrix}\text{its}\\ \text{contribution}\end{bmatrix} - \begin{bmatrix}\text{its relative}\\ \text{opportunity cost}\end{bmatrix},$$
$$= \$1{,}200 - \$1{,}400 = -\$200.$$

Equation (2) tells us nothing new because Table 5.4 reports –200 as the reduced cost of L, namely, as the change in profit if the basic solution is perturbed by setting L=1 and adjusting the values of the basic variables accordingly.

Why the optimal solution is what it is

Equation (2) tells us nothing new, but equation (1) does. It indicates *why* the Luxury model vehicle is unprofitable. Making one Luxury model vehicle requires 3 hours in the Body shop, which has a break-even price of $420/hour, and $(3) \times (420) = \$1{,}260$. This (alone) exceeds the contribution of the Luxury model vehicle. In this example and in general:

> To learn *why* the optimal solution to a linear program is what it is, use the shadow prices to parse the relative opportunity cost of each activity in which it does not engage.

The Luxury model vehicle would become profitable if its relative opportunity cost could be reduced below $1,200, and equation (1) shows that this would occur if the time it required in the Body shop could be reduced below 2 11/21 hours.

The Nifty model vehicle

To further illustrate the uses of relative opportunity cost, suppose that the manager of the ATV plant has been asked to manufacture a new model, the Nifty. She wonders whether it will turn a profit. Discussions with the engineering department lead her to conclude that making each Nifty model would require about 2.5 hours in the Engine shop, 1.5 hours in the Body shop, and 3 hours in the Fancy Finishing shop. From information provided by the marketing and operations departments, she estimates that making each Nifty would contribute approximately $900. To determine whether or not Nifties are profitable, she calculates their relative opportunity cost and marginal profit:

$$\begin{aligned}\begin{bmatrix}\text{relative opportunity cost of}\\ \text{one Nifty model vehicle}\end{bmatrix} &= (2.5)\times(140)+(1.5)\times(420)+(3)\times(0)\\ &= 350+630=\$980,\\ \begin{bmatrix}\text{marginal profit of}\\ \text{one Nifty model vehicle}\end{bmatrix} &= \$900-\$980=-\$80.\end{aligned}$$

Making one Nifty reduces profit by approximately \$80. Nifties are not profitable. Their relative opportunity cost shows that they would become slightly profitable. if their manufacturing time in the Body shop could be reduced by 0.2 hours.

The Standard model vehicle

Still in the context of the optimal plan for the ATV facility, let's compute the relative opportunity cost of making one Standard model vehicle. To do so, we must free up the resources that it requires, which are 3 hours in the Engine shop (at a cost of \$140 per hour), 1 hour in the Body shop (at a cost of \$420 per hour) and 2 hours in the Standard finishing shop (at a cost of \$0 per hour), so that:

$$(3) \quad \begin{bmatrix} \text{relative opportunity cost of} \\ \text{one Standard model vehicle} \end{bmatrix} = (3) \times (140) + (1) \times (420) + (2) \times (0),$$
$$= \$420 + \$420 + \$0 = \$840.$$

Note that \$840 *is* the contribution of one Standard model vehicle. Is it an accident that the relative opportunity cost of one Standard model vehicle equals its contribution? No. It makes perfect economic sense because:

- Contribution is maximized by producing at the three vehicles at the rates S=20, F=30 and L=0.
- If we remove the resources needed to make one Standard model vehicle, it must be optimal to produce at the rates S=19, F=30 and L=0.
- Doing that reduces contribution by exactly \$840.

Please pause to verify that the relative opportunity cost of one Fancy model vehicle equals \$1,120. This illustrates a point that holds in general and is highlighted below:

Consider any basis for a linear program. The contribution of each basic variable equals its relative opportunity cost.

The above has been justified on economic grounds. It also follows from the fact that the reduced cost of each basic variable equals zero.

Computing the shadow prices

The observation that is highlighted above shows how to compute the shadow prices for *any* basis. To illustrate, consider the basis for the ATV problem that includes the decision variables S, L, s_3, s_4 and s_5. The relative opportunity cost of each basic variable equals its contribution, so the shadow prices for this basis must satisfy:

$$
\begin{array}{lllllll}
S \text{ is basic} & \Rightarrow & 3E + 1B + 2SF & & & = & 840 \\
L \text{ is basic} & \Rightarrow & 1E + 3B & & + 2LF & = & 1200 \\
s_3 \text{ is basic} & \Rightarrow & 1SF & & & = & 0 \\
s_4 \text{ is basic} & \Rightarrow & & 1FF & & = & 0 \\
s_5 \text{ is basic} & \Rightarrow & & & 1LF & = & 0
\end{array}
$$

That's five linear equations in five unknowns. The lower three equations set SF = FF = LF = 0. This reduces the upper two equations to 3E + 1B = 840 and 1E + 3B = 1200; their solution is E = 165 and B = 345.

By the way, the shadow price that this basis assigns to a particular *constraint* could have been computed by adding δ to its RHS value and seeing what happens to the basic solution and finding the change in its objective value. For the ATV example, this would have required solution of six linear equations (not 5). And it would have given us only one of the shadow prices.

No shadow price?

The ATV problem satisfies the Full Rank proviso because each constraint in its Form-1 representation has a slack variable. This guarantees that the equation system has a solution *and* that it continues to have a solution if a RHS value is perturbed, hence that each basis assigns a shadow price to each constraint.

What about a linear program whose Form-1 representation violates the Full Rank proviso? Every basic feasible solution to this linear program has at least one trite row. The linear program becomes infeasible if at least one RHS value is perturbed. Not every row can have a shadow price. What then?

Multipliers

Solver is actually reporting values of the **multipliers**. If a constraint has a shadow price, its multiplier is that shadow price. Constraints that lack shadow prices have infinitely many multipliers, but *every set* of multipliers can be shown to have the property that is highlighted below:

The relative opportunity cost of each decision variable is determined by using the multpliers as though they were shadow prices.

Demonstration that this is so is deferred to Chapter 11. An interpretation is provided here.

An illustration

To illustrate the role of multipliers in a linear program that lacks shadow prices, we turn our attention to

Program 2. Maximize $\{4x_1 + 2x_2 + 5x_3\}$, subject to

$$
\begin{aligned}
y_1: \quad & 1x_1 + 2x_2 + 3x_3 = 6,\\
y_2: \quad & 2x_1 + 4x_2 + 6x_3 = 12,\\
y_3: \quad & 3x_1 + 2x_2 + 1x_3 = 6,\\
& x_1 \geq 0, \quad x_2 \geq 0, \quad x_3 \geq 0.
\end{aligned}
$$

"Multipliers" y_1, y_2 and y_3 have been assigned to the constraints of Program 2.

The 2nd constraint in Program 2 is twice the 1st constraint. If the RHS value of either of these constraints was perturbed, the linear program would become infeasible. Neither constraint can have a shadow price. Solver and Premium Solver report "shadow prices" anyhow.

Multipliers, by hand

What these software packages are actually reporting are values of the "multipliers." As noted above, the values of the multipliers y_1, y_2 and y_3 are such that the relative opportunity cost of each basic variable equals its contribution. For the basis that includes x_1 and x_3 and excludes x_2, the multipliers must satisfy:

$$x_1 \text{ is basic} \quad \Rightarrow \quad 1y_1 + 2y_2 + 3y_3 = 4,$$

$$x_3 \text{ is basic} \quad \Rightarrow \quad 3y_1 + 6y_2 + 1y_3 = 5.$$

That's 2 equations in 3 unknowns. It cannot have a unique solution. One of its solutions is $y_1 = 11/8$, $y_2 = 0$, and $y_3 = 7/8$. It can be shown (and is shown in Chapter 11) that that this solution *and all others* correctly account reduced cost of the nonbasic variable x_2:

$$\text{reduced cost of } x_2 = 2 - [2y_1 + 4y_2 + 2y_3] = 2 - 36/8 = -2.5.$$

The reduced cost of x_2 is negative, so an optimal solution to Program 2 is at hand.

Multipliers, with Premium Solver

Premium Solver has been used to solve Program 2. The sensitivity report in Table 5.6 records its optimal solution and multipliers, which are

$$x_1 = 3/2, \quad x_2 = 0, \quad x_3 = 3/2, \quad y_1 = 11/8, \quad y_2 = 0, \quad y_3 = 7/8.$$

This is the same basis and the same set of multipliers that are computed above.

Table 5.6. A sensitivity report for Program 2.

Decision variables

Cell	Name	Final Value	Reduced Cost	Obj Coef	Allow Incr	Allow Decr
B3	x1	1.5	0	4	1E+30	5.00
C3	x2	0	-2.5	2	2.5	1E+30
D3	x3	1.5	0	5	1E+30	5.00

Constraints

Cell	Name	Final Value	Shadow Price	Constr RHS	Allow Incr	Allow Decr
E5		6	1.375	6	0	0
E6		12	0	12	0	0
E7		6	0.875	6	12	4

Premium Solver reports 0 as the Allowable Increase and Decrease of the RHS values of the 1st and 2nd constraints. This is correct. These constraints are linearly dependent, and the LP becomes infeasible if the RHS value of either is perturbed.

Multipliers, with Solver

When Solver is applied to Program 2, it reports the same optimal solution as in Table 5.6. The version of Solver with which Excel 2010 is equipped fails to report correct values of the Allowable Increase and Allowable Decrease of the RHS values of linearly dependent constraints.

Sneak preview

As mentioned earlier, the current discussion of multipliers anticipates material in Chapter 11. That chapter includes a version of the simplex method that uses multipliers to determine which columns might profitably enter the basis. That version is known by two names, which are the "revised simplex method" and the "simplex method with multipliers."

8. Opportunity Cost

This section is focused on opportunity cost, as that term is used in economics. Its definition appears below:

> The **opportunity cost** of doing something is the benefit one can obtain from the best alternative use of the resources needed to do that thing.

Friedrich von Wieser

The Austrian economist, Friedrich von Wieser (1851-1926), is credited with coining the term, "opportunity cost." To illustrate his use of it, consider a businessperson who has purchased a quantity of iron and who contemplates using it for a particular purpose. That person should be concerned with the direct profit (contribution) obtained from that particular use less its opportunity cost, the latter being the profit from the best alternative use of the same quantity of iron. Barring ties, only one use of this quantity of iron will have a direct profit that exceeds its opportunity cost, and that use will be best.

Paul Samuelson

In Paul Samuelson's classic text on microeconomics, the concept of opportunity cost is illustrated using Robinson Crusoe. He is thinking of devoting the afternoon to picking raspberries. His best alternative use of the time and effort he would spend picking raspberries is in picking strawberries. Crusoe anticipates that spending the afternoon picking strawberries or raspberries will be equally pleasurable. The opportunity cost of picking raspberries is the value that Crusoe places on the strawberries he might have picked that afternoon.

In both illustrations, the motivating idea is to ascertain the marginal benefit of a decision – this being the direct benefit (contribution) less its opportunity cost. The definition works well in settings for which the "best alternative use" exists and is easily identified.

A difficulty

Consider a slightly more complicated situation, namely, the ATV problem. What is the opportunity cost of the resources needed to make one Luxury model vehicle? There is *no alternative use* of this bundle of resources. They cannot have a best alternative use. Their opportunity cost is not defined.

A puzzle

At the 2005 meeting of the American Economics Association, Paul J. Ferraro and Laura J. Taylor[2] posed the question that is paraphrased below:

> "You have been given a ticket to see an Eric Clapton concert tonight. This ticket has no resale value. Bob Dylan is performing tonight, and is your next-best alternative activity. Tickets to see Dylan cost \$40. If the concerts were on separate evenings, you would be willing to pay up to \$50 to see Dylan whether or not you see Clapton tonight. There are no other costs to seeing either performer. The opportunity cost of seeing Clapton is one of the following:
>
> a) \$0, b) \$10, c) \$40, d) \$50.
>
> Which is it?"

Fewer than 22% of the roughly 200 professional economists who responded to this question got the correct answer. That is pretty dismal performance.

[2] Paul J. Ferraro and Laura O. Taylor, "Do Economists Recognize an Opportunity Cost When They See One? A Dismal Performance from the Dismal Science," Contributions to Economic Analysis & Policy: Vol. 4, Issue 1, Article 7, 2005.

If every one of them had chosen amongst the four choices at random (a pure guess), a statistic this low would occur with a probability that is below 0.2.

Extra work

To place opportunity cost and relative opportunity cost on a "level playing field," consider an example in which the bundles of resources needed to engage in each activity has at least one alternative use. We wish to determine the *program* (set of activities) that is most profitable (optimal). How can we check whether or not a particular program is optimal? This question is answered twice.

- With relative opportunity costs:
  - Compute the shadow prices for this program of activities.
  - Then, for each activity that is excluded from the program, use these shadow prices to compute the relative opportunity cost of the activity.
  - The current program is optimal if no activity's contribution exceeds its relative opportunity cost.
- With opportunity costs:
  - For each activity, find the best alternative use of the bundle of resources it requires. (This requires solution of one optimization problem per activity.)
  - The current program is optimal if the contribution of each activity in which it is engaged is at least as large as that activity's opportunity cost and if the contribution of each activity in which it is not engaged is no greater than that activity's opportunity cost.

Using opportunity costs to determine whether or not a particular program is optimal can require the solution of one optimization problem *per activity*.

Recap

The *relative* opportunity cost focuses the decision maker on the marginal benefit of an action – this being the benefit (contribution) obtained from the action less the cost of freeing up the resources needed to make it possible. The classic definition of opportunity cost is also motivated by marginal analysis, but it is not always defined, it can be difficult to grasp, and it can require extensive side-computation.

The connection between shadow prices, relative opportunity costs and marginal benefit has been presented in the context of a resource allocation problem, but it holds in general. It is central to constrained optimization and to competitive equilibria. It has not yet been fully assimilated into mainstream economics, however.

The economic insight of George B. Dantzig

Virtually all of the ideas that have appeared so far in this book are due to George B. Dantzig. Not all of the terminology is due to him. In his classic text[3], Dantzig used the term **relative cost** instead of "reduced cost." *Relative cost* reflects the fact that it is relative to the current basis. In place of "shadow price," he used the terms **price** and **multiplier**, the latter as an abbreviation of **Lagrange multiplier**. Dantzig understood that the shadow prices exist if and only if the Full Rank proviso is satisfied. He fully understood the role of multipliers in marginal analysis and in the "revised" simplex method. Consider this:

- Prior to Dantzig's simplex method, no systematic method existed to determine an optimal allocation of resources.
- Dantzig's simplex method has remained the principal tool for finding an optimal allocation of resources ever since he devised it.
- Dantzig fully understood the relation of the simplex method to economic reasoning, including the fact that each basis has multipliers (prices) that, even when non-unique, correctly account for the marginal benefit of every feasible perturbation of the its basic solution.

A perplexing decision

In 1975, the Nobel Prize in Economics was awarded to Leonid V. Kantorovich and Tjalling C. Koopmans for their contributions to the "optimal allocation of resources." That is perplexing. George B. Dantzig had done the above work well before 1975. At the time the prize was awarded, he was with-

[3] George B. Dantzig, *Linear programming and extensions*, R-366-PR, The RAND Corporation, August, 1963 and Princeton University Press, Princeton, NJ, 1963.

out peer as concerns the optimal allocation of resources. Since that time, no one has approached his stature in this field.

9. A Glimpse of Duality*

This is the first of three starred sections that are starred because they are independent of each other. Here, the ATV problem is used to glimpse a topic that is known as duality. Let's begin by recalling the labels of the shadow prices – E for Engine shop, B for Body shop, SF for Standard Finishing shop, and so forth. Table 5.5 reports the values that the optimal solution assigns to the shadow prices, namely,

$$E = 140, \qquad B = 420, \qquad SF = 0, \qquad FF = 0, \qquad LF = 0.$$

These shadow prices are break-even prices, and their unit of measure is $/hour.

Bidding for the resources

The labels E, B, SF, FF and LF will be used to describe the decision variables in a second linear program. Think of yourself as an outsider who wishes to rent the ATV facility for one week. Imagine that you face the situation in:

Problem B *(renting the ATV facility).* The ATV company has agreed to rent its ATV facility to you for one week according to the following terms:

- You must offer them a price for each unit of capacity of each shop.
- You must set each price high enough that they have no economic motive to withhold any capacity of any shop from you.

What prices should you set, and how much must you spend to rent the facility for one week?

The ATV company can earn $50,400 by operating this facility for one week. You must set your prices high enough that they have no motive to withhold any capacity from you. Intuitively, it seems clear that you will need to pay at least $50,400 to rent their entire capacity. But must you spend more than $50,400? And what prices should you offer? To answer these questions, we will build a linear program.

Decision variables

The decision variables in this linear program are the prices that you will offer. By agreement, you must offer five prices, one per shop. These prices are labeled:

E = the price ($/hour) you offer for each unit of Engine shop capacity.

B = the price ($/hour) you offer for each unit of Body shop capacity.

SF = the price ($/hour) you offer for each unit of Standard Finishing shop capacity.

FF = the price ($/hour) you offer for each unit of Fancy Finishing shop capacity.

LF = the price ($/hour) you offer for each unit of Luxury Finishing shop capacity.

Renting the capacity

Let us compute the cost you will incur for renting the entire capacity of the ATV facility. The Engine shop has a capacity of 120 hours, and you must pay E dollars for each hour you rent. The cost to you of renting the entire capacity of the Engine shop is 120E. The cost of renting the entire capacity of the Body shop capacity is 80B. And so forth. The total cost that you will pay to rent every unit of every shop's capacity is given by

$$\{120\text{E} + 80\text{B} + 96\text{SF} + 102\text{FF} + 40\text{LF}\}.$$

You wish to minimize this expression, which is your rental bill, subject to constraints that keep the ATV company from withholding any capacity from you.

Leaving resources idle

The ATV company need not make full use of the capacity of any of its shops. That fact constrains the prices that you can offer. For instance, the capacity constraint on the Engine shop is the inequality,

$$3\text{S} + 2\text{F} + 1\text{L} \le 120.$$

Can you offer a price E that is negative? No. If you did, the ATV company would not rent you any of the capacity of its Engine shop. Instead, it would leave those resources idle. You must offer a price E that is nonnegative. The decision variable E must satisfy the constraint $E \geq 0$.

Each shop's capacity constraint is a "≤" inequality. For this reason, each of the prices that you offer must be nonnegative. In other words, the decision variables must satisfy the constraints,

$$E \geq 0, \qquad B \geq 0, \qquad SF \geq 0, \qquad FF \geq 0, \qquad LF \geq 0.$$

Producing vehicles

The ATV facility can be used to manufacture vehicles. Your prices must be high enough that manufacturing each type of vehicle becomes unprofitable. The price of the bundle of resources needed to produce each vehicle must be at least as large as its contribution.

Let us begin with the Standard model vehicle. Column B of Table 5.2 shows that the company would earn \$840 for each Standard model vehicle that it made, that making this vehicle would require 3 hours in the Engine shop, 1 hour in the Body shop, and 2 hours in the Standard Finishing shop. Thus, it becomes unprofitable to make any Standard model vehicles if the prices you offer satisfy

S: $$3E + 1B + 2SF \geq 840.$$

Similarly, the data in column C of Table 5.2 shows that it becomes unprofitable to make any Fancy model vehicles if you offer prices that satisfy

F: $$2E + 2B + 3FF \geq 1120.$$

In the same way, the data in column D of Table 5.2 shows the Luxury model vehicle becomes unprofitable if the prices you offer satisfy

L: $$1E + 3B + 2LF \geq 1200.$$

A price-setting linear program

The constraints and objective of a linear program that rents the ATV manufacturing facility have now been presented. Assembling them produces

Program 3. Minimize $\{120\text{E} + 80\text{B} + 96\text{SF} + 102\text{FF} + 40\text{LF}\}$, subject to

$$\begin{array}{llr} \text{S:} & 3\text{E} + 1\text{B} + 2\text{SF} & \geq 840, \\ \text{F:} & 2\text{E} + 2\text{B} + \quad 3\text{FF} & \geq 1120, \\ \text{L:} & 1\text{E} + 3\text{B} + \quad 2\text{LF} & \geq 1200, \end{array}$$

$$\text{E} \geq 0, \quad \text{B} \geq 0, \quad \text{SF} \geq 0, \quad \text{FF} \geq 0, \quad \text{LF} \geq 0.$$

Program 3 calculates the prices that minimize the cost of renting the facility for one week, subject to constraints that make it unprofitable for the ATV company to withhold any capacity from you, the renter. From Solver, we could learn that the optimal solution to this linear program is

$$\text{E} = 140, \quad \text{B} = 420, \quad \text{SF} = 0, \quad \text{FF} = 0, \quad \text{LF} = 0,$$

that its optimal value is 50,400 $/wk, and the shadow prices of its three constraints are

$$\text{S} = 20, \quad \text{F} = 30, \quad \text{L} = 0.$$

Program 1 and Program 3 have the same optimal value, and the shadow prices of each form an optimal solution to the other!

Duality

The properties exhibited by Programs 1 and 3 are no coincidence. They illustrate a general "duality" principle that is highlighted below:

> **Duality:** Each linear program is paired with another. If either linear program in a pair is feasible and bounded, then:
>
> - The other linear program is feasible and bounded, and both linear programs have the same optimal value.
> - The shadow prices (multipliers) for either linear program form an optimal solution to the other.

Is duality a curiosity? No! From a practical viewpoint, a number of competitive situations can be formulated not as a single linear program, but as a linear program and its dual. Also, in economic reasoning, there is often a duality (pairing) of production quantities and break-even prices. Finally, from a theoretical viewpoint, duality is a widely-used tool in the analysis of optimization problems.

A surprise?

Did duality surprise you? If so, you are in very good company. It surprised George B. Dantzig too. In retrospect, it's eminently reasonable. To see why, we note that the optimal solution to Program 1 and its shadow prices have these properties:

- Each constraint in Program 1 is a "≤" inequality and increasing its RHS value can only cause the optimal value to improve, so each shadow price is nonnegative.
- These shadow prices are such that no vehicle's contribution exceeds its relative opportunity cost, so they satisfy the "≥" constraints of Program 3.

Thus, these shadow prices form a feasible solution to Program 3. It remains to argue that multiplying each constraint's shadow price by its RHS value and summing them up totals 50,400. That's not difficult; see Problem 7.

10. Large Changes and Shadow Prices*

Each shadow price is accompanied by an "allowable increase" and an "allowable decrease" that determines the range of RHS values over which it applies. This section probes the question: What happens if a RHS value is perturbed by an amount that falls outside the range for which the shadow price applies?

From Table 5.4, we see that the shadow price on Engine shop capacity is 140 $/hr, with an Allowable Increase of 56 and an Allowable Decrease of 16. This price applies for capacity levels in the range between 104 hours (because $104=120-16$) and 176 hours (because $176=120+56$).

We can use Solver to find optimal solutions to Program 1 for Engine shop capacities that are just below 104 and just above 176, find the range and shadow price in each case, and repeat. Figure 5.3 plots the result. This figure shows that the slope decreases with quantity. This is no accident. A linear program exhibits decreasing marginal return on each capacity (right-hand side value). On reflection, the reason is clear. When there is only a tiny amount of a particular capacity, that capacity is as profitable as it can be. As this capacity increases, other resources become more fully utilized, slack constraints become tight, and one can make less and less profitable use of the added capacity.

Figure 5.3. Contribution versus Engine shop capacity.

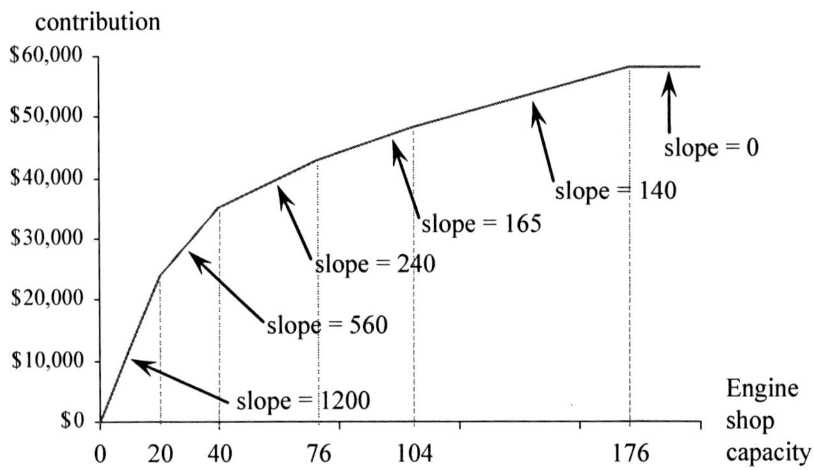

Figure 5.3 exhibits **decreasing marginal return** because the marginal benefit (slope) can only decrease as the quantity increases. This means that, in an important sense, the current shadow prices are the most favorable. Starting with a capacity of 120, small decreases in capacity cost $140 per unit. Larger decreases cost $165 per unit. Still larger increases cost even more. And so forth. Similarly, starting with a capacity of 120, small increases earn $140 per unit. Larger increases earn less. It is emphasized:

> The current shadow prices are the *most favorable*; larger increases will be less profitable, and larger decreases will be more costly.

11. Linear Programs and Solid Geometry*

The example in Chapter 4 had only two decision variables, so its geometry could be visualized on the plane. The ATV problem has three decision variables, which are S, F and L, so a geometric view of it requires three-dimensional (or solid) geometry. Solid geometry has been familiar since birth, even if it is omitted from typical high-school geometry courses.

Cartesian coordinates

In Chapter 4, Cartesian coordinates were used to identify each ordered pair (x, y) of real numbers with a point in the plane. The point (0, 0) on the page was called the origin, and the point (1, 2) was located 1 unit to the right of the origin and 2 units toward the top of the page.

In a similar way, Cartesian coordinates identify each ordered triplet (A, B, C) of real numbers with a point in three-dimensional space. The point (0, 0, 0) on the page is called the origin, and the point (1, 2, 3) is located 1 unit toward the right of the origin, 2 units toward the top of the page, and 3 units above the page, for instance.

In this way, Cartesian coordinates identify every feasible solution (S, F, L) to Program 1 with a point in three-dimensional space. The feasible region (set of feasible solutions) becomes the polyhedron in Figure 5.4. For instance, the triplet (40, 0, 0) lies 40 units to the right of the origin, and the triplet (0, 34, 0) lies 34 units toward the top of the page from the origin. And the triplet (0, 0, 20) lies 20 units in front of the origin.

Figure 5.4. Feasible region for Program 1, and its extreme points.

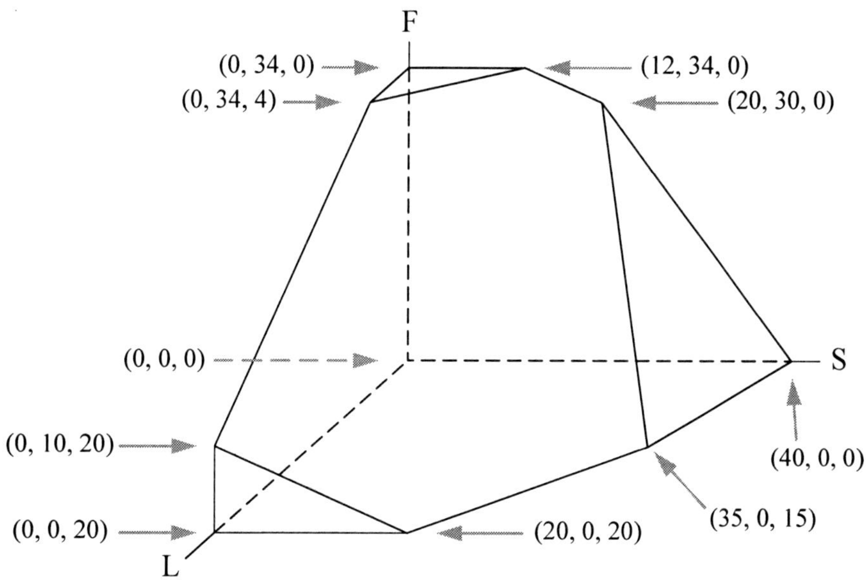

Vertices, edges and faces

The feasible region for the ATV problem is polyhedron that has 10 vertices (extreme points), 15 edges, and 7 faces. One constraint is binding on each face, two on each edge, and three at each vertex. Two vertices are adjacent if an edge connects them.

When the simplex method is applied to the ATV example, each pivot shifts it from a vertex to an adjacent vertex whose objective value is improved. It stops pivoting after reaching vertex (20, 30, 0).

Watching Solver pivot

It is possible to watch Solver pivot. To do so, open the Excel spreadsheet for Chapter 5, and click on the sheet entitled Table 5.3. Then click on Solver or on Premium Solver. Either way, click on its Options box, and then click on the Show Iteration Results window. When you run Solver or Premium Solver, you will see that the first pivot occurs to the triplet (S, F, L) given by (0, 0, 20), the second to (0, 10, 20), the third to (20, 0, 20), the fourth to (34, 0, 15) and the fifth to (20, 30, 0). Each of these triplets corresponds to an extreme pivot of the feasible region in Figure 5.4, each pivot occurs to an adjacent extreme point, and each pivot improves the basic solution's objective value.

Higher dimensions

Plane geometry deals with only two variables, solid geometry with three. Linear programs can easily have dozens of decision variables, or hundreds. Luckily, results that hold for plane geometry *and* for solid geometry tend to remain valid when there are many variables. That is why geometry is relevant to linear programs.

12. Review

This chapter has shown how to formulate a linear program for solution by Solver and by Premium Solver for Education. A focal point of this chapter has been the interpretation of the information accompanies the optimal solution. We have seen:

- How the shadow prices determine the relative opportunity costs of the basic and nonbasic variables.

- How the relative opportunity costs help us to understand why the optimal solution is what it is.
- How the Allowable Increases and Allowable Decreases help determine whether the optimal solution is robust.
- How the optimal solution responds to small changes in the model's data; the values of the basic variables may change, but the binding constraints stay binding.

Relative opportunity cost has been used to determine whether or not a plan of action can be improved. It has been argued that it is more difficult – and that it may be impossible – to use the classic definition of opportunity cost to determine whether or not a plan of action can be improved.

Material in later chapters has been glimpsed. In Chapter 11, shadow prices, relative opportunity cost, and marginal profit will be used in to guide the "revised" simplex method as it pivots. Chapter 12 is focused on duality and its uses. In Chapter 14, duality will be used to construct a simple (stylized) model production and consumption in an economy in general equilibrium. Nonlinear programs will be studied in Chapter 20, where it is seen that the natural generalization of decreasing marginal return produces nonlinear programs that are relatively easy to solve.

13. Homework and Discussion Problems

1. For the ATV problem, Table 5.4 (on page 161) reports an Allowable Decrease on the objective coefficient of L of ∞. That is no accident. Why?

2. For the ATV problem, Table 5.4 (on page 161) reports a shadow price of 140 for the Engine shop constraint and an Allowable Decrease of 16 in this constraint's right-hand-side value. Thus, renting 4 hours of Engine shop for one week decreases contribution by \$560 because $560 = 4 \times 140$. *Without* re-running the linear program, show how the optimal solution changes when the Engine shop capacity is decreased by 4. *Hint:* The Perturbation Theorem reduces this problem to solving two equations in two unknowns.

3. Eliminate from the ATV problem the Luxury model vehicles. The linear program that results has only two decision variables, so its feasible region is a portion of the plane.

   (a) Write down this linear program.

   (b) Display its feasible region graphically.

   (c) Display its iso-profit lines and objective vector graphically.

   (d) Show, graphically, that its optimal solution sets S=20 and F=30, so that its optimal value equals 50,400.

   (e) Set aside the capacity needed to make one Luxury model vehicle. Resolve the linear program graphically. Show that making one Luxury model vehicle decreases contribution by 200.

4. Suppose, in the ATV problem, that the contribution of the Standard model vehicle is $900 apiece for the first 12 made per week and $640 apiece for production above that level.

   (a) Revise Program 1 to account for this diseconomy of scale, and solve it.

   (b) You can figure out what the optimal solution would be without doing Part (a). How? Support your answer.

5. Consider a company that can use overtime labor at an hourly wage rate that is 50% in excess of regular time labor cost. Does this represent an economy of scale? A diseconomy of scale? Will a profit-maximizing linear program have an unintended option? If so, what will it be? Will this unintended option be selected by optimization, or will it be ruled out?

6. Consider a linear program that is feasible and bounded. Let us imagine that each right-hand-side value in this linear program was multiplied by 0.75. Complete the following sentences, and justify them. (*Hint:* To educate your guess, re-solve Program 1 with each right-hand-side value multiplied by 0.75.)

   (a) The optimal solution would be multiplied by ________, and the optimal value would be multiplied by ________.

(b) On the other hand, the shadow prices would be _______.

(c) There was nothing special about the factor 0.75 because ________.

7. The shadow prices are supposed to apply to small changes in right-hand side values. Compute the amount that the manager of the ATV shop could earn by renting the entire capacity of her shops at the shadow prices. Is this amount familiar? If so, why? *Hint:* It might help to review the preceding problem.

8. The sensitivity report seems to omit the shadow prices of the nonnegativity constraints. True or false:

   (a) In a maximization problem, the reduced cost of each nonnegative variable x equals the shadow price of the constraint $x \geq 0$.

   (b) In a minimization problem, the reduced cost of each nonnegative variable x equals the shadow price of the constraint $x \geq 0$.

9. In a linear program, a decision variable x is said to be **free** if neither the constraint $x \geq 0$ nor the constraint $x \leq 0$ is present in the linear program. A free variable is allowed to take values that are positive, negative or zero. In the optimal solution to a maximization problem, what can you say about the reduced cost of each free variable? Why?

10. This problem refers to the survey by Paul J. Ferraro and Laura J. Taylor that is cited in Section 8 of this chapter.

    (a) With the economists' definition of opportunity cost, which of the four answers is correct?

    (b) Suppose that exactly 200 professional economists answered their question and that 44 of them got the correct answer (that's 22%). Of the 44 answered correctly, 10 taught micro and knew it. The rest guessed. What is the probability of as few as 34 correct answers from the remaining 190 economists if each of them answered at random?

    (c) Suppose you had planned to attend the Dylan concert when someone offers you a free ticket to the Clapton concert. What can you say of the relative opportunity cost of seeing the Clapton concert?

11. This problem refers to the Robinson Crusoe example that is discussed in Section 8. Suppose Crusoe has a third alternative. In addition to spending the afternoon picking strawberries or strawberries, he could spend it lolling on the beach.

    (a) Suppose he would rather loll on the beach than pick strawberries. Carefully describe:

        (i) The opportunity cost of picking raspberries.

        (ii) The opportunity cost of lolling on the beach.

    (b) Suppose he planned to pick raspberries when the sun came out, at which time it occurred to him that he might enjoy an afternoon on the beach. Describe the relative opportunity cost of an afternoon on the beach.

12. Write down linear programs that have each of these properties:

    (a) It has no feasible solution.

    (b) It is feasible and unbounded.

    (c) Its feasible region is bounded, and it has multiple optima.

    (d) It is bounded, it has an unbounded feasible region, it has multiple optimal solutions, but only one of them occurs at an extreme point.

    (e) Its feasible region is unbounded, and it has multiple optima, none of which occur at an extreme point.

13. Perturbing the optimal solution to Program 1 by making one Luxury model vehicle decreases profit by $200. Complete the following sentence and justify it: Perturbing this optimal solution by making 10 Luxury model vehicles decreases profit by at least ________ because ____________________________.

14. In Table 5.6, the Allowable Increase and Allowable Decrease for rows 5 and 6 are zero because Program 2 becomes infeasible if the RHS value of either of the first two constraints is perturbed. These constraints do not have shadow prices. Table 5.6 reports that row 7 *does* have a shadow price, and it reports an Allowable Increase and Allowable Decrease for its RHS

value. Why does this row have a shadow price? What accounts for the range on this shadow price?

15. Consider a constraint in a linear program that has no shadow price. Does this constraint have a multiplier? What can be said about the Allowable Increase and the Allowable Decrease of that constraint's RHS value?

16. With the Engine shop capacity fixed at 120 hours per week, use Solver to compute the optimal value of Program 1 for all values of the Body shop capacity. Plot the analog of Figure 5.3. Do you observe decreasing marginal return?

17. With Figure 5.4 in view, have Solver or Premium Solver use the simplex method to solve Program 1, but use the Options tab to have it show the results of each iteration. Record the sequence of basic solutions that it followed. Did it pivot from extreme point to extreme point? Did each pivot occur along an edge?

18. *(A farmer)* A 1,200 acre farm includes a well that has a capacity of 2,000 acre-feet of water per year. (One acre-foot is one acre covered to a depth of one foot.) This farm can be used to raise wheat, alfalfa, and beef. Wheat can be sold at $550 a ton and beef at $1,300 a ton. Alfalfa can be bought or sold at the market price of $220 per ton. Each ton of wheat that the farmer produces requires one acre of land, $50 of labor, and 1.5 acre-feet of water. Each ton of alfalfa that she produces requires 1/3 acre of land, $40 of labor and 0.6 acre-feet of water. Each ton of beef she produces requires 0.8 acres of land, $50 of labor, 2 acre-feet of water, and 2.5 tons of alfalfa. She can neither buy nor sell water. She wishes to operate her farm in a way that maximizes its annual profit. Below are the data in a spreadsheet formulation, the solution that Solver has found, and a Sensitivity Report.

	A	B	C	D	E	F	G	H	I
1									
2		label of variable	W	AR	B	AS			
3		value of variable	0	0	1000	-2500			RHS
4		acreage	1	1/3	0.8	0	800	<=	1200
5		water	1.5	0.6	2	0	2000	<=	2000
6		balance	0	-1	2.5	1	0	=	0
7		contribution	500	-40	1250	220	700000		
8							↑		
9			= SUMPRODUCT(C$3:F$3 , C7:F7)						

Adjustable Cells

Cell	Name	Final Value	Red Cost	Obj Coef	All. Incr.	All. Decr.
C3	value of variable W	0	-25	500	25	1E+30
D3	value of variable AR	0	-30	-40	30	1E+30
E3	value of variable B	1000	0	1250	1E+30	33 1/3
F3	value of variable AS	-2500	0	220	13 1/3	1E+30

Constraints

Cell	Name	Final Value	Shadow Price	Const. rhs	All. Incr.	Allowable Decr.
G4	acreage	800	0	1200	1E+30	400
G5	water	2000	350	2000	1000	2000
G6	balance	0	220	0	1E+30	1E+30

(a) Write down the linear program. Define each variable. Give each variable's unit of measure. Explain the objective function and each constraint. Explain why the constraint AS ≥ 0 is absent. What is the unit of measure of the objective? What is the unit of measure of each constraint?

(b) State the optimal solution in a way that can executed when the data are inexact.

(c) As an objective function coefficient or right-hand side value varies within its allowable range, how does she manage the farm? That is, in which activities does she engage, and which resources does she use to capacity?

(d) What would have to happen to the price of wheat in order for her to change her production mix?

(e) What would have to happen to the price of alfalfa for her to change her production mix?

Note: Parts (f) through (i) refer to the original problem and are independent of each other.

(f) The government has offered to let her deposit some acreage in the "land bank." She would be paid to produce nothing on those acres. Is she interested? Why?

(g) The farmer is considering soybeans as a new crop. The market price for soybeans is $800 per ton. Each ton of soybeans requires 2 acres of land, 1.8 acre feet of water and $60 of labor. Without re-running the linear program, determine whether or not soybeans are a profitable crop.

(h) A neighbor has a 400 acre farm with a well whose capacity is 500 acre-feet per year. The neighbor wants to retire to the city and to rent his entire farm for $120,000 per year. Should she rent it? If so, what should she do with it?

(i) The variable AS is unconstrained in sign. Rewrite the linear program with AS replaced by (ASOLD – ABOUGHT), where ASOLD and ABOUGHT are nonnegative decision variables. Solve the revised linear program. Did any changes occur? Does one formulation give more accurate results than the other? If so, how and why?

19. *(pollution control)* A company makes two products in a single plant. It runs this plant for 100 hours each week. Each unit of product A that that the company produces consumes 2 hours of plant capacity, earns the company a contribution of $1,000 and causes, as an undesirable side effect, the emission of 4 ounces of particulates. Each unit of product B that the company produces consumes 1 hour of capacity, earns the company a contribution of $2,000 and causes, as undesirable side effects, the emission of 3 ounces of particulates and of 1 ounce of chemicals. The EPA (Environmental Protection Agency) requires the company to limit particulate emission to at most 240 ounces per week and chemical emission to at most 60 ounces per week.

(a) Formulate this problem for solution by linear programming. In this linear program, what is the unit of measure of each decision variable? Of the objective function? Of each shadow price?

(b) Solve this linear program on a spreadsheet. Describe its optimal solution in a way that can be implemented when its data are inexact.

(c) What is the value to the company of the EPA's relaxing the constraint on particulate emission by one ounce per week? What is the value to the company of the EPA's relaxing the constraint on Chemical emissions by one ounce per week?

(d) *(an emissions trade-off)* By how much should the company be willing to reduce its weekly emission of chemicals if the EPA would allow it to emit one additional ounce of particulates each week?

(e) *(an emissions tax)* The EPA is considering the control of emissions through taxation. Suppose that the government imposes weekly tax rates of P dollars per ounce of particulate emissions and C dollars per ounce of chemical emission. Find tax rates, P and C, that keep the company's pollutants at or below the current levels and minimize the company's tax bill. *Hint:* With the constraints on emissions deleted, the feasible region becomes a triangle, so the tax rates must be sufficiently large to make the extreme point(s) that cause excess pollution to become undesirable.

(f) By how much does the taxation scheme in part (e) reduce profit?

Chapter 6: The Simplex Method, Part 2

1. Preview

This chapter completes this book's introductory account of the simplex method. In this chapter, you will see:

- How Phase I of the simplex method determines whether a linear program has a feasible solution and, if so, how it constructs a basic feasible tableau with which to initiate Phase II.
- That the simplex method can "cycle" (fail to terminate finitely) and that it can be kept from doing so.
- That the simplex method readily accommodates variables that are not required to be nonnegative.

Also discussed here is the speed of the simplex method. Decades after its discovery and in spite of the best efforts of scores of brilliant researchers, the simplex method remains the method of choice for solving large linear programs.

E. V. Denardo, *Linear Programming and Generalizations*, International Series in Operations Research & Management Science 149,
DOI 10.1007/978-1-4419-6491-5_6, © Springer Science+Business Media, LLC 2011

2. Phase I

Phase II of the simplex method is initialized with a linear program for which a basic feasible tableau has been found. Thus, two tasks remain for **Phase I:**

- Determine whether or not a linear program has a feasible solution.
- If it has a feasible solution, find a basic feasible tableau with which to initiate Phase II.

These tasks can be accomplished in several different ways. And several different versions of Phase I appear in the literature. No matter how it is organized, Phase I is a bit complicated. The version of Phase I that is presented here appends *one* artificial variable α to the original linear program. The coefficients of α are selected so that a single pivot creates a basic feasible tableau, except that the basis will include α and the basic solution will equate α to a positive value. The simplex method will then be used to drive α toward 0. If α can be reduced to 0, it is removed from the basis, and a basic feasible tableau for the original linear program results. If α cannot be reduced to zero, the linear program has no feasible solution.

This method is described below as a six-step procedure. Each step is illustrated in the context of

Problem A. Maximize $\{4p + 1q + 2r\}$, subject to the constraints

(1.1) $$-1p + 1q + 2r \geq 6,$$

(1.2) $$1p - 3.5q - 3r = -10,$$

(1.3) $$-2p - 3q \leq 0,$$

(1.4) $$p \geq 0, \quad q \geq 0, \quad r \geq 0.$$

Step 1 of Phase I

The 1st step of Phase I is to cast the linear program in Form 1, preserving its sense of optimization. Executing this step on Problem A rewrites it as

Program 1. Maximize {z}, subject to the constraints

(2.0)	$4p + 1q + 2r \qquad - z = 0,$
(2.1)	$-1p + 1q + 2r - s_1 \qquad = 6,$
(2.2)	$1p - 3.5q - 3r \qquad = -10,$
(2.3)	$-2p - 3q \qquad + s_3 = 0,$
(2.4)	$p \geq 0, \quad q \geq 0, \quad r \geq 0, \quad s_1 \geq 0, \quad s_3 \geq 0.$

In Program 1, equation (2.0) defines z as the value of the objective. The surplus variable s_1 converts inequality (1.1) to an equation, and the slack variable s_3 converts inequality (1.3) to an equation.

Step 2

The 2nd step of Phase I is to ignore the nonnegativity constraints on the decision variables and apply Gauss-Jordan elimination to the equations that remain, keeping –z basic for the top-most equation. This step either constructs a basic system or an inconsistent equation. If it finds an inconsistent equation, no solution exists to the equation system, so no feasible solution can exist to the linear program, which has additional constraints. Presented in Table 6.1 is the result of executing Step 2 on Program 1.

Rows 2-6 of Table 6.1 mirror system (2). Rows 4 and 5 lack basic variables. A choice exists as to the elements on which to pivot. Table 6.1 exhibits the result of pivoting on the coefficient of s_1 in row 4 and then on the coefficient of p in row 11. These pivots produce the basic tableau in rows 14-18. This tableau's basic solution sets $p=-10$ and $s_3=-20$. If this solution were feasible, Phase I would be complete. It is not feasible, so Phase I continues.

Step 3

The 3rd step is to insert on the left-hand side of the equation system an **artificial variable** α with a coefficient of –1 in each equation whose RHS value has the wrong sign and with a coefficient of 0 in each of the remaining equations. Displayed in rows 20-24 of Table 6.2 is the result of executing Step 3 on Program 1.

Table 6.1. Pivoting to create a basic system for Program 1.

	A	B	C	D	E	F	G
1							
2	p	q	r	s1	s3	-z	RHS
3	4	1	2			1	0
4	-1	1	2	-1			6
5	1	-3.5	-3				-10
6	-2	-3			1		0
7							
8	p	q	r	s1	s3	-z	RHS
9	4	1	2	-0	-0	1	0
10	1	-1	-2	1	-0	-0	-6
11	1	-3.5	-3	-0	-0	-0	-10
12	-2	-3	-0	-0	1	-0	0
13							
14	p	q	r	s1	s3	-z	RHS
15	0	15	14	0	0	1	40
16	0	2.5	1	1	0	0	4
17	1	-3.5	-3	-0	-0	-0	-10
18	0	-10	-6	-0	1	-0	-20

Table 6.2. Creating a basic feasible tableau, except that $\alpha > 0$.

	A	B	C	D	E	F	G	H
1								
20	p	q	r	s1	s3	-z	α	RHS
21	0	15	14	0	0	1		40
22	0	2.5	1	1	0	0		4
23	1	-3.5	-3	-0	-0	-0	-1	-10
24	0	-10	-6	-0	1	-0	-1	-20
25								
26	p	q	r	s1	s3	-z	α	RHS
27	0	15	14	0	0	1	-0	40
28	0	2.5	1	1	0	0	-0	4
29	1	6.5	3	0	-1	0	0	10
30	-0	10	6	0	-1	0	1	20

To see what Step 3 accomplishes, we write the equations represented by rows 20-24 of Table 6.2 in dictionary format, with the nonbasic variables on the right.

$$(3.0) \qquad z = -40 + 15q + 14r,$$

$$(3.1) \qquad s_1 = 4 - 2.5q - 1r,$$

$$(3.2) \qquad p = -10 + 3.5q + 3r + 1\alpha,$$

$$(3.3) \qquad s_3 = -20 + 11q + 6r + 1\alpha.$$

In system (3), setting $q=0$, $r=0$ and $\alpha \geq 20$ equates the variables s_1, p and s_3 to nonnegative values. Moreover, a pivot on the coefficient of α in the equation for which s_3 is basic will remove s_3 from the basis and will produce a basic solution in which $\alpha=20$. This motivates the next step.

Step 4

Step 4 is to select equation whose RHS value is most negative and pivot upon the coefficient of α in that equation. When applied to the tableau in rows 20-24 of Table 6.2, this pivot occurs on the coefficient of α in row 24. This pivot produces the basic tableau in rows 26-30. That tableau's basic solution sets $s_1=4$, $p=10$ and $\alpha=20$, exactly as predicted from system (3)

Step 4 has produced a **Phase I simplex tableau**, namely, a basic tableau in which the artificial variable α is basic and whose basic solution equates all basic variables (with the possible exception of $-z$) to nonnegative values.

Step 5

What remains is to drive α down toward zero, while keeping the basic variables (other than $-z$) nonnegative. This will be accomplished by a slight adaptation of the simplex method. To see how to pivot, we write the equations represented by rows 26-30 in dictionary format, as:

$$(4.0) \qquad z = -40 + 15q + 14r,$$

$$(4.1) \qquad s_1 = 4 - 2.5q - 1r,$$

$$(4.2) \qquad p = 10 - 7.5q - 3r + 1s_3,$$

$$(4.3) \qquad \alpha = 20 - 10q - 6r + 1s_3.$$

The goal is to decrease the value of α, which is the basic variable for equation (4.3). The nonbasic variables q and r have negative coefficients in equation (4.3), so setting either of them positive decreases α.

In a Phase I simplex tableau, the entering variable can be any nonbasic variable that has a *positive* coefficient in the equation for which α is basic. (The positive coefficients became negative when they were switched to the right-hand side.) The usual ratios keep the basic solution feasible. As in Phase II, no ratio is computed for the row for which –z is basic, and no ratio is computed for any row whose coefficient of the entering variable is not positive. The pivot occurs on a row whose ratio is smallest. But if the row for which α is basic has the smallest ratio, pivot on that row because it removes α from the basis. In brief:

In a **Phase I simplex pivot** for Form 1, the entering variable and pivot element are found as follows:

- The entering variable can be any nonbasic variable that has a positive coefficient in the row for which α is basic.
- The pivot row is selected by the usual ratios, which keep the basic solution feasible and keep –z basic.
- But if the row for which α is basic ties for the smallest ratio, pivot on that row.

To reduce the ambiguity in this pivot rule, let's select as the entering variable a nonbasic variable that has the *most positive* coefficient in the row for which α is basic. Table 6.3 displays the Phase I simplex pivots that result.

Rows 26-30 of Table 6.3 indicate that for the first of these pivots, q is the entering variable (its coefficient in row 30 is most positive), and row 29 has the smallest ratio, so q enters the basis and p departs. Rows 35-38 result from that pivot.

Rows 34-38 of Table 6.3 indicate that for the second pivot, r is the entering variable (its coefficient in row 38 is most positive), and α is the departing variable because the row for which α is basic ties for the smallest ratio.

Rows 40-44 display the basic tableau that results from that pivot. The variable α has become nonbasic. The numbers in cells H42-H44 are non-

Table 6.3. Illustration of Step 5.

	A	B	C	D	E	F	G	H	I
1									
26	p	q	r	s1	s3	-z	α	RHS	ratio
27	0	15	14	0	0	1	-0	40	---
28	0	2.5	1	1	0	0	-0	4	1.6
29	1	6.5	3	0	-1	0	0	10	1.538
30	-0	10	6	0	-1	0	1	20	2
31									
32				= IF(B30 > 0 , H30/B30 , "---")					
33									
34	p	q	r	s1	s3	-z	α	RHS	ratio
35	-2.308	0	7.077	0	2.308	1	-0	16.92	
36	-0.385	0	-0.154	1	0.385	0	-0	0.154	---
37	0.154	1	0.462	0	-0.154	0	0	1.538	3.333
38	-1.538	0	1.385	0	0.538	0	1	4.615	3.333
39									
40	p	q	r	s1	s3	-z	α	RHS	
41	5.556	0	0	0	-0.444	1	-5.111	-6.667	
42	-0.556	0	0	1	0.444	0	0.111	0.667	
43	0.667	1	0	0	-0.333	0	-0.333	-4E-16	
44	-1.111	0	1	0	0.389	0	0.722	3.333	

negative, so this tableau's basic solution equates the basic variables q, r and s_1 to nonnegative values. Deleting α and its column of coefficients deleted produces a basic feasible tableau for Program 1.

Step 6 of Phase I

The 6th and final step is to delete α and its column of coefficients. This step produces a basic feasible tableau with which to begin Phase II. Applying this step to rows 40-44 of Table 6.3 casts Program 1 as the linear program:

Program 1. Maximize {z}, subject to the constraints

$$(5.0)\qquad 5.556p \qquad - 0.444s_3 - z = -6.667,$$

$$(5.1)\qquad -0.556p \qquad + 1s_1 + 0.444s_3 = 0.667,$$

$$(5.2)\qquad 0.667p + 1q \qquad - 0.333s_3 = 0.000,$$

$$(5.3)\qquad -1.111p \qquad + 1r \qquad + 0.389s_3 = 3.333,$$

$$(5.4)\qquad p \geq 0, \quad q \geq 0, \quad r \geq 0, \quad s_1 \geq 0, \quad s_3 \geq 0.$$

Phase II of the simplex method commences by selecting p as the entering variable and executing a (degenerate) pivot on the coefficient of p in equation (5.2).

No entering variable?

One possibility has not yet been accounted for. Phase I pivots can result in a basic tableau whose basic solution sets $\alpha > 0$ but in which no nonbasic variable has a positive coefficient in the row for which α is basic. If this occurs, no entering variable for a Phase I simplex pivot can be selected. What then?

To illustrate this situation, imagine that we encounter rows 34-38 of Table 6.3, except that the coefficients of r and s_3 in row 38 are –1.616 and –0.333. Row 38 now represents the equation

$$\text{(6)} \qquad \alpha = 4.615 + 1.538p + 1.616r + 0.333s_3.$$

The basic solution to this equation system has $\alpha = 4.615$, and the variables p, r and s_3 are constrained to be nonnegative, so equation (6) demonstrates that no feasible solution can have $\alpha < 4.615$. The artificial variable α cannot be reduced below 4.615, so the linear program is infeasible.

Recap – infeasible LP

To recap Phase I, we first consider the case of a linear program that is infeasible. When an infeasible linear program is placed in Form 1 and Phase I is executed, one of these two things must occur:

- Gauss-Jordan elimination produces an inconsistent equation.
- Gauss-Jordan elimination produces a basis whose basic solution is infeasible. An artificial variable α is inserted, and the simplex method determines that the value of α cannot be reduced to zero.

Recap – feasible LP

Now consider a linear program that is feasible. Neither of the above conditions can occur. If Gauss-Jordan elimination (Step 2) produces a feasible basis, Phase II is initiated immediately. If not, an artificial variable α is inserted, and a pivot produces a basic solution that is feasible, except that it creates α to a positive value. The simplex method reduces the value of α to 0 and eliminates α from the basis, thereby exhibiting a feasible basis with which to initiate Phase II.

Commentary

Let's suppose that certain variables are likely to be part of an optimal basis. Phase I can be organized for a **fast start** by pivoting into the initial basis as many as possible of these variables.

A disconcerting feature of Phase I is that the objective value z is ignored while feasibility is sought. Included in Chapter 13 is a *one-phase* scheme (that is known by the awkward name "the parametric self-dual method") that uses one artificial variable, α, and seeks feasibility and optimality simultaneously.

3. Cycling

Does the simplex method terminate after finitely many pivots? The answer is a qualified "yes." If no care is taken in the choice of the entering variable and the pivot row, the simplex method *can* keep on pivoting forever. If care is taken, the simplex method is guaranteed to be finite. This section describes the difficulty that can arise and shows how to avoid it.

The difficulty

This difficulty is identified in this subsection. A linear program has finitely many decision variables. It can only have finitely many bases because each basis is a subset of its decision variables, and there are only finitely many such subsets. Let us recall from Chapter 4 that:

- Each nondegenerate simplex pivot changes the basis, changes the basic solution, and improves its objective value.
- Each degenerate pivot changes the basis, but does not change any RHS values, hence causes no change in the basic solution or in the basic solution's objective value.

As a consequence, each *nodegenerate* simplex pivot results in a basis whose basic solution improves on all basic solutions seen previously. That is good. No nondegenerate pivot can result in a basis that had been visited previously. Also, since there are finitely many bases, so only finitely many nondegenerate simplex pivots can occur prior to termination.

On the other hand, a sequence of degenerate pivots (none of which changes the basic solution) can **cycle** by leading to a basis that had been visited previously. That is not good! If the simplex method cycles once and if it employs a consistent rule for selecting the entering variable and the pivot row, it will cycle again and again.

Ambiguity in the pivot element

Whether or not the simplex method cycles depends on how the ambiguity in its pivot rule is resolved. The entering variable can be *any* variable whose reduced cost is positive in a maximization problem, negative in a minimization problem. The pivot row can be *any* row whose ratio is smallest.

Rule A

To specify a particular pivot rule, we must resolve these ambiguities. For a linear program that is written in Form 1, each decision variable is assigned a *column* of coefficients, and these columns are listed from left to right. Let us dub as **Rule A** the version of the simplex method that chooses as follows:

- In a maximization (minimization) problem, the entering variable is a nonbasic variable whose reduced cost is most positive (negative). Ties, if any, are broken by picking the variable that is listed farthest to the left.
- The pivot row has the lowest ratio. Ties, if any, are broken by picking the row whose basic variable is listed farthest to the left.

The tableau in Table 6.4 will be used to illustrate Rule A. In that tableau, lower-numbered decision variables are listed to the left. For the first pivot, x_1 is the entering variable because it has the most positive reduced cost, and rows 4 and 5 tie for the smallest ratio. The basic variable for row 4 is x_5, and the basic variable for row 5 is x_6. Row 4 is the pivot row because its basic variable x_5 is listed to the left of x_6. No ties occur for the second pivot. A tie does occur for the third pivot, which occurs on row 16 because x_1 is listed to the left of x_2. Evidently, the first three pivots are degenerate. They change the basis but do not change the basic solution.

Table 6.4. Illustration of Rule A.

	A	B	C	D	E	F	G	H	I	J	K
1											
2		x1	x2	x3	x4	x5	x6	x7	-z	RHS	ratio
3		0.75	-10	0.2	-6	0	0	0	1	0	none
4		0.25	-4	-0.4	9	1	0	0	0	0	0
5		0.5	-6	-0.2	3	0	1	0	0	0	0
6		0	0	1	0	0	0	1	0	1	---
7											
8		x1	x2	x3	x4	x5	x6	x7	-z	RHS	ratio
9		0	2	1.4	-33	-3	0	0	1	0	none
10		1	-16	-1.6	36	4	0	0	0	0	---
11		0	2	0.6	-15	-2	1	0	0	0	0
12		0	0	1	0	0	0	1	0	1	---
13											
14		x1	x2	x3	x4	x5	x6	x7	-z	RHS	ratio
15		0	0	0.8	-18	-1	-1	0	1	0	none
16		1	0	3.2	-84	-12	8	0	0	0	0
17		0	1	0.3	-7.5	-1	0.5	0	0	0	0
18		0	0	1	0	0	0	1	0	1	1

A cycle

Rule A *can* cycle. In fact, when Rule A is applied to the linear program in Table 6.4, it does cycle. After six degenerate pivots, the tableau in rows 3-6 reappears.

An anti-cycling rule

Abraham Charnes was the first to publish a rule that precludes cycling. The key to his paper, published in 1952[1], was to pivot as though the RHS values were *perturbed* in a way that breaks ties. Starting with a basic feasible tableau (either in Phase I or in Phase II), imagine that the RHS value of the 1st non-trite constraint is increased by a very small positive number ε, that the RHS value of the 2nd non-trite constraint is increased by ε^2, and so forth. Standard results in linear algebra make it possible to demonstrate that, for all sufficiently small positive values of ε, there can be no tie for the smallest ratio. Consequently, each basic feasible solution to the perturbed problem equates each basic variable (with the possible exception of –z) to a positive value. In the perturbed problem, each simplex pivot is nondegenerate. This guarantees that the simplex method cannot cycle. Termination must occur after finitely many pivots.

[1] Charnes, A [1952]., "Optimality and Degeneracy in Linear Programming," *Econometrica*, V. 20, No. 2, pp 160-170.

The **perturbation** argument that Charnes pioneered has had a great many uses in optimization theory. From a computational viewpoint, perturbation is unwieldy, however. Integrating it into a well-designed computer code for the simplex method requires extra computation that slows down the algorithm.

A simple cycle-breaker

In 1977, Robert Bland published a simple and efficient anti-cycling rule[2]. Let's call it **Rule B**; it resolves the ambiguity in the simplex pivot in this way:

- The entering variable is a nonbasic variable whose reduced cost is positive in a maximization problem (negative in a minimization problem). Ties are broken by choosing the variable that is listed farthest to the left.
- The pivot row has the smallest ratio. Ties are broken by picking the row whose basic variable is listed farthest to the left.

When Rule B is applied to a maximization problem, the entering variable has a positive reduced cost, but it needn't have the largest reduced cost. Among the variables whose reduced costs are positive, the entering variable is listed farthest to the left.

Rule B is often called **Bland's rule**, in his honor. Proving that Bland's rule precludes cycles is a bit involved. By contrast, incorporating it in an efficient computer code is easy, and it adds only slightly to the computational burden. Bland's rule can be invoked after encountering a large number of consecutive degenerate pivots.

The early days

Initially, it was not clear whether the simplex method could cycle if no special care was taken to break ties for the entering variable and the pivot row. George Dantzig asked Alan Hoffman to figure this out. In 1951, Hoffman found an example in which Rule A cycles. The data in Hoffman's example entail the elementary trigonometric functions ($\sin \varphi$, $\cos^2 \varphi$ and so forth). In Hoffman's memoirs[3], he reports:

---

[2] Robert G. Bland, "New finite pivot rules for the simplex method, *Mathematics of Operations Research*, V. 2, pp. 103-107, 1977.

[3] Page 171 of *Selected Papers of Alan Hoffman with Commentary*, edited by Charles Micchelli, World Scientific, Rver Edge, NJ.

> "On Mondays, Wednesdays and Fridays I thought it could (cycle). On Tuesdays, Thursdays and Saturdays I thought it couldn't. Finally, I found an example which showed it could … I was never able to … explain what was in my mind when I conceived the example."

The example in Table 6.4 is simpler than Hoffman's; it was published by E. M. L. Beale in 1955.

Charnes was the first to publish an anti-cycling rule, but he may not have been the first to devise one. In his 1963 text, George Dantzig[4] wrote,

> "Long before Hoffman discovered his example, simple devices were proposed to avoid degeneracy. The main problem was to devise a way to avoid degeneracy with as little extra work as possible. The first proposal along these lines was presented by the author in the fall of 1950 …. Later, A. Orden, P. Wolfe and the author published (in 1954) a proof of this method based on the concept of lexicographic ordering …."

Perturbation and lexicographic ordering are two sides of the same coin; they lead to the same computational procedure, and it is a bit unwieldy.

Following Charnes's publication in 1952 of his perturbation method, a heated controversy developed as to whether he or Dantzig was primarily responsible for the development of solution methods for linear programs. Researchers found themselves drawn to one side or the other of that question. A quarter century elapsed before Robert Bland published his anti-cycling rule. It is Bland's rule that achieves the goal articulated by Dantzig – avoid cycles with little extra work.

4. Free Variables

In a linear program, a decision variable is said to be **free** if it is not constrained in sign. A free variable can take any value – positive, negative or zero. Free variables do occur in applications. To place a linear program that has one or more free variables in Form 1, we must replace each free variable by the difference of two nonnegative variables. That is no longer necessary. Modern

---

[4] Page 231 of *Linear Programming and Extensions* by George B. Dantzig, published by Princeton University Press, 1963.

codes of the simplex method accommodate free variables. How they do so is the subject of this section.

Form 2

Form 2 generalizes Form 1 by allowing any subset of the decision variables to be free, that is, unconstrained in sign. In the presence of free variables, the simplex method must pivot a bit differently. To see how, we consider:

Problem B. Max $(-0.5a - 1.25b - 5.00c + 3d + 10e + 25f)$, subject to

$$
\begin{aligned}
0.8a - 1.30b \quad\quad\quad - 1d \quad\quad\quad &= 12.0,\\
1b - 1c \quad - 1e \quad\quad &= 0.6,\\
1c \quad\quad - 1f &= 1.2,\\
1b &\le 2.5,\\
1c &\le 9.6,\\
0.5a + 0.8b + 4c &\le 45,\\
0.9a + 1.5b &\le 27,\\
a \ge 0, \quad b \ge 0, \quad c \ge 0.&
\end{aligned}
$$

In Problem B, the decision variables d, e and f are free; they can take any values.

Free variables do arise in applications. One of their uses is to model the quantity of a commodity that can be bought or sold at the same (market) price. In Problem B, the decision variables d, e and f can represent the *net* sales quantities of commodities whose market prices are \$3, \$10 and \$25 per unit, respectively.

Getting started

Problem B is placed in Form 2 by inserting slack variables in the bottom four constraints and introducing an equation that defines z as the objective value. The tableau in rows 2-10 of Table 6.5 results from this step. In this tableau, rows 4, 5 and 6 lack basic variables.

The tableau in rows 12-20 of Table 6.5 results from pivoting on the coefficients of d, e and f in rows 4, 5 and 6, respectively. This tableau is basic. Its basic solution is feasible because d, e and f are allowed to take any values.

Table 6.5. Phase I for Problem B.

	A	B	C	D	E	F	G	H	I	J	K	L	M	N
1														
2		a	b	c	d	e	f	s4	s5	s6	s7	-z	RHS	
3		-0.5	-1.25	-5	3	10	25					1		
4		0.8	-1.3		-1								12	
5			1	-1		-1							0.6	
6				1			-1						1.2	
7			1					1					2.5	
8				1					1				9.6	
9		0.5	0.8	4						1			45	
10		0.9	1.5								1		27	
11														
12		a	b	c	d	e	f	s4	s5	s6	s7	-z	RHS	ratio
13		1.9	4.85	10								1	71	none
14		-0.8	1.3		1								-12	none
15			-1	1		1							-0.6	none
16				-1			1						-1.2	none
17			1					1					2.5	---
18				1					1				9.6	9.6
19		0.5	0.8	4						1			45	11.25
20		0.9	1.5								1		27	---

Keeping free variables basic

Once a free variable becomes basic, the RHS value of the equation for which it is basic can have any value, positive, negative or zero. To keep a free variable basic, compute no ratio for the row for which it is basic. To keep d, e and f basic for the equations represented by rows 14, 15 and 16 of Table 6.5, we've placed "none" in cells N14, N15 and N16. In this example and in general:

> After a free variable becomes basic, compute *no ratio* for the equation for which it is basic. This keeps the free variable basic, allowing it to have any sign in the basic solution that results from each simplex pivot.

Rows 13-20 of Table 6.5 are a basic feasible tableau with which to initiate Phase II, and its first pivot occurs on the coefficient of c in row 18. Pivoting continues until the optimality condition or the unboundedness condition occurs.

Nonbasic free variables

Problem B fails to illustrate one situation that can arise: In a basic feasible tableau for a linear program that has been written in Form 2, a free variable can be nonbasic. Let us suppose we encounter a basic feasible tableau in which the free variable x_j is not basic and in which the reduced cost of x_j is not zero. What then?

- If the reduced cost of x_j is positive, select x_j as the entering variable, and pivot as before.
- If the reduced cost of x_j is negative, aim to bring x_j into the basis at a negative level by computing ratios for rows whose coefficients of x_j are *negative* and selecting the row whose ratio is closest to zero (least negative).
- In either case, compute no ratio for any row whose basic variable is free.
- If no row has a ratio, the linear program is unbounded.

Needless work

Accommodating free variables is easy. To conclude this section, let's see why it is a good idea to do so. To cast Problem B in Form 1, we would need to introduce one new column per free variable. The coefficients in these columns would be *opposite* to the coefficients of d, e and f. Columns that start opposite stay opposite. Even so, updating opposite columns requires extra work per pivot. Furthermore, if a pivot reduces a previously-free variable to zero, the next pivot is quite likely to introduce its opposite column. That's an extra pivot. Finally, forcing a linear program into Form 1 can cause the ranges of the shadow prices to become artificially low, which makes the optimal basis seem less robust than it is.

5. Speed

In a Form 1 representation of a linear program, let m denote the number of equations (other than the one defining z as the objective value), and let n denote the number of decision variables (other than z).

Typical behavior

The best codes of the simplex method quickly solve practical linear programs having m and n in the thousands or tens of thousands. No one really understands why the simplex method is as fast as it is. On carefully-constructed examples (one of which appears as Problem 5), the simplex method is exceedingly slow. Any attempt to argue that the simplex method is fast "on average" must randomize in a way that bad examples occur with miniscule probability.

In Chapter 12 of his text, Robert J. Vanderbei[5] provided a heuristic rationale as to why the parametric self-dual method (that is described in Chapter 13) should require approximately $(m+n)/2$ pivots, and he reported the number of pivots required to solve each member of a standard family of test problems that is known as the NETLIB suite[6]. He made a least-squares fit of the number of pivots to the function $\alpha(m+n)^{\beta}$, and he found that the best fit is to the function

$$0.488(m+n)^{1.0515}, \tag{7}$$

moreover, that the quality of the fit is quite good. Expression (7) is strikingly close to $(m+n)/2$.

Atypical behavior

The simplex method does not solve *all* problems quickly. In their 1972 paper, Klee and Minty[7] showed how to construct examples having m equations and 2 m decision variables for which Rule A requires $2^m - 1$ pivots. (Problem 5 presents their example for the case $m=3$.) Even at the (blazing) speed of one million pivots per second, it would take roughly as long as the universe has existed for Rule A to solve a Klee-Minty example with $m=100$.

A conundrum

The gap between typical performance of roughly $(m+n)/2$ pivots and atypical performance of 2^{m-1} pivots has been a thorn in the side of every person who wishes to measure the efficiency of a computational procedure by its worst-case performance. Over the decades, several brilliant works have been written on this issue. The interested reader is referred to a paper by Daniel

---

[5] Vanderbei, Robert J., *Linear Programming: Foundations and Extensions,* Kluwer Academic Publishers, Boston, Mass., 1997.

[6] Gay, D. "Electronic mail distribution of linear programming test problems," *Mathematical Programming Society COAL Newsletter*, V. 13, pp 10-12, 1985.

[7] V. Klee and G. J. Minty, "How good is the simplex algorithm?" In O. Shisha, editor, *Inequalities III*, pp. 159-175, Academic Press, New York, NY, 1972.

Spielman and Shang-Huia Tang that has won both the Gödel and the Fulkerson prize[8].

The ellipsoid method

In 1979, Leonid G. Kachian[9] created a sensation with the publication of his paper on the **ellipsoid method.** It is a divide-and-conquer scheme for finding the solutions to the inequalities that characterize optimal solutions to a linear program and its dual. An upper bound on the number of computer operations required by the ellipsoid method (this counts the square root as a single operation) is a fixed constant times n^4 L, where L is the number of bits needed to record all of the nonzero numbers in A, b and c, along with their locations.

From a theoretical viewpoint, Kachian's work was a revelation. It showed that linear programs *can* be solved with a method whose worst-case work bound is a polynomial in the size of the problem. From a computational viewpoint, the ellipsoid method was disappointing, however. It is not used because it solves practical linear programs far more slowly than does the simplex method.

Interior-point methods

In 1984, Narendra Karmarkar created an even greater sensation with the publication of his paper on **interior-point methods.**[10] These methods move through the interior of the feasible region, avoiding the extreme points entirely. One of the methods in his paper has the same worst-case work bound as the ellipsoid method, and Karmarkar claimed running times that were many times faster than the simplex method on representative linear programs.

A controversy erupted. Karmarkar's running times proved to be difficult to duplicate, and they seemed to be for an "affine scaling" method that was not polynomial.

---

[8] Spielman, D and S.-H. Teng, "Smoothed analysis of algorithms: Why the simplex method usually takes polynomial time," Journal of the ACM, V. 51, pp. 385-463 (2004).

[9] L. G. Kachian, "A polynomial algorithm in linear programming," *Soviet Mathematics Doklady*, V. 20, pp 191-194 (1979).

[10] N. Karmarkar, "A new polynomial-time algorithm for linear programming," *Proceedings of the 16th annual symposium on Theory of Computing,* ACM New York, pp 302-311 (1979).

AT&T weighs in

In an earlier era, when AT&T had been a highly-regulated monopoly, it had licensed its patents free of charge. By 1984, when Karmarkar published his work, this had changed. AT&T had become eager to earn royalties from its patents. AT&T sought and obtained several United States patents that were based on Karmarkar's work. This was surprising because:

- Patents are routinely awarded for processes, rarely for algorithms.
- Interior-point methods were hardly novel; beautiful work on these methods had been by done in the 1960's by Fiacco and McCormick[11], for instance.
- The "affine scaling" method in Karmarkar's paper had been published in 1967 by Dikin[12].
- Karmarkar's fastest running times seemed to have been for Dikin's method.
- Karmakar's claims of faster running times than the simplex method could not be substantiated, and AT&T would not release the test problems on which these claims were based!

The AT&T patents on Karmarkar's method have not been challenged in a United States court, however. The validity of these patents might now be moot, as the interior-point methods that Karmarkar proposed have since been eclipsed by other approaches.

A business unit

Aiming to capitalize on its patents for interior-point methods, AT&T formed a business unit named Advanced Decision Support Systems. The *sole* function of this business unit was to produce and sell a product named KORBX (short for nothing) that consisted of a code that implemented interior

---

[11] Fiacco, A. V. and G. McCormick, *Nonlinear programming: sequential unconstrained minimization techniques,"* John Wiley & Sons, New York, 1968, reprinted as *Classics in applied mathematics volume 4*, SIAM, Philadelphia, Pa., 1990.

[12] Dikin, I. I.. "Iterative solution of problems of linear and quadratic programming," *Soviet Math. Doklady*, V. 8, pp. 674-675, 1967.

point methods on a parallel computer made by Alliant Corporation of Acton, Massachusetts.

This implementation made it difficult (if not impossible) to ascertain whether KORBX ran faster than the simplex method. This implementation also made it difficult for AT&T to keep pace with the rapid improvement in computer speed.

As a business unit, Advanced Decision Support Systems existed for about seven years. It was on a par, organizationally, with AT&T's manufacturing arm, which had been Western Electric and which would be spun off as Lucent Technologies. At its peak, in 1990, Advanced Decision Support Systems had roughly 200 full-time employees. It sold precisely two KORBX systems, one to the United States Military Airlift Command, the other to Delta Airlines. As a business venture, Advanced Decision Support Systems was unprofitable and, in the eyes of many observers, predictably so.

Seminal work

Karmarkar's 1984 paper sparked an enormous literature, however. Hundreds of brilliant papers were written by scores of talented researchers. Any attempt to cite a few of these papers overlooks the contributions of others as well as the many ways in which researchers interacted. That said, the candidates for the fastest interior-point methods may to be the "path-following" algorithm introduced by J. Renegar[13] and the self-dual homogeneous method of Y. Ye, M. Todd and S. Muzino[14]. While this research was underway, the simplex method was vastly improved by incorporation of modern sparse-matrix techniques.

What's best?

For extremely large linear programs, the best of the interior-point method might run a bit faster than the simplex method. The simplex method enjoys an important advantage, nonetheless. In Chapter 13, we will see how to solve an integer program by solving a sequence of linear programs. The simplex

---

[13] Renegar, J., "A polynomial-time algorithm, based on Newton's method for linear programming," *Mathematical Programming*, V. 40, pp 59-93, 1988

[14] Ye, Yinyu, Michael J. Todd and Shinji Mizuno, "On $O(\sqrt{n}\,L)$ iteration homogeneous and self-dual linear programming algorithm," *Mathematics of Operations Research*, V. 19, pp. 53-67, 1994.

method is far better suited to this purpose because it finds an optimal solution that is an extreme point; interior-point methods find an extreme point only if the optimum solution is unique.

Currently, the main use of interior-point methods is to solve classes of nonlinear programs for which the simplex method is ill-suited. For computing optimal solutions to linear programs, large or small, Dantzig's simplex method remains the method of choice.

6. Review

The key to the version of Phase I that is presented here is to introduce a single artificial variable and then attempt to pivot it out of the basis. The same device will be used in Chapter 15 to compute solutions to the "bi-matrix game."

The simplex method can cycle, and cycles can be avoided. Bland's method for avoiding cycles is especially easy to implement. Even so, the perturbation method of Charnes (equivalently, the lexicographic method of Dantzig) has proved to be a useful analytic tool in a number of settings.

Decision variables that are not constrained in sign are easy to accommodate within the simplex method. Once a free variable is made basic, it is kept basic by computing no ratio for the equation for which it is basic.

No one fully understands why the simplex method is as fast as it is on practical problems. Any effort to prove that the simplex method is fast on average (in expectation) must assign miniscule probabilities to "bad examples."

Modern interior-point methods may run a bit faster than the simplex method on enormous problems, but the simplex method remains the method of choice, especially when integer-valued solutions are sought.

7. Homework and Discussion Problems

1. *(Phase I)* In Step 2 of Phase I, would any harm be done by giving the artificial variable α a coefficient of -1 in every equation other than the one for which $-z$ is basic?

2. *(Phase I)* For the tableau in rows 35-39 of Table 6.3, rows 37 and 38 tie for the smallest ratio. Execute a pivot on the coefficient of r in row 37. Does this result in a basis that includes α and whose basic solution sets α=0? If so, indicate how to remove α from the basis and construct a basic feasible tableau with which to initiate Phase II.

3. *(Phase I)* In Phase 2, an entering variable can fail to have a pivot row, in which case the linear program is unbounded. This cannot occur in Phase I. Why?

4. *(Phases I and II)* Consider this linear program: Maximize {2x + 6y}, subject to the constraints

$$
\begin{aligned}
2x - 5y \qquad &\le -3,\\
4x - 2y + 2z &\le -2,\\
1x + 2y \qquad &\le \ 4,\\
x \ge 0, \quad y \ge 0, \quad & z \ge 0.
\end{aligned}
$$

   (a) On a spreadsheet, execute Phase I of the simplex method.

   (b) If Phase I constructs a feasible solution to the linear program, execute Phase II on the same spreadsheet.

5. The spreadsheet that appears below is a **Klee-Minty example** in which the number m of constraints equals 3 and the number n of decision variables (other than –z) equals 2 m. The goal is maximize z.

	A	B	C	D	E	F	G	H
1								
2	x1	x2	x3	s1	s2	s3	-z	RHS
3	4	2	1				1	0
4	1			1				5
5	4	1			1			25
6	8	4	1			1		125

   (a) For this example, execute the simplex method with Rule A. (You will need seven pivots.)

   (b) For each extreme point encountered in part (a), record the triplet (x_1, x_2, x_3,).

(c) Plot the triplets you recorded in part (b). Identify the region of which they are the extreme points. Does it resemble a deformation of the unit cube? Could you have gotten from the initial extreme point to the final extreme point with a one simplex pivot?

(d) What do you suppose the comparable example is for the case m=2? Have you solved it?

(e) Write down but do not solve the comparable example for m=4.

6. Apply the simplex method with Rule A to the maximization problem in Table 6.4, but stop when a cycle occurs.

7. Apply the simplex method with Rule B to the maximization problem in Table 6.4. Did it cycle? Identify the first pivot at which Rule B selects a different pivot element than does Rule A.

8. In Rule B, ties are broken by picking the variable that is farthest to the left. Would it work equally well to pick the variable that is farthest to the right?

9. The idea that motivates Charnes's perturbation scheme is to resolve the ambiguity in the variable that will leave the basis by perturbing the RHS values by miniscule amounts, but in a nonlinear way. The tableau that appears below reproduces rows 2-6 of Table 6.4, with the dashed line representing the "=" signs and with the quantity ε^j added to the jth constraint, for j=1, 2, 3.

x1	x2	x3	x4	x5	x6	x7	-z	RHS	ε	ε^2	ε^3
0.75	-10	0.2	-6				1	0			
0.25	-4	-0.4	9	1				0	1		
0.5	-6	-0.2	3		1			0		1	
		1				1		1			1

(a) Execute Charnes's pivot rule (for maximization) on this tableau, selecting the nonbasic variable whose reduced cost is most positive as the entering variable.

(b) Identify the first pivot at which Charnes's rule selects a different pivot element than does Rule A.

(c) Complete and justify the sentence: If a tie were to occur for the smallest ratio when Charnes's pivot rule is used, two rows would need to have coefficients of ε^1, ε^2, and ε^3 that are __________, and that cannot occur because elementary row operations keep _______ rows _______.

(d) There is a sense in which Charnes's rule is lexicographic. Can you spot it? If so, what is it?

10. Cycling can occur in Phase I. Cycling in Phase I can be precluded by Rule B or by Charnes's perturbation scheme. At what of the six steps of Phase I would Charnes perturb the RHS values? Which RHS values would he perturb?

11. Consider a linear program that is written in Form 1 and is feasible and bounded. By citing (but not re-proving) results in this chapter, demonstrate that this linear program has a basic feasible solution that is optimal.

12. *(free variables)* This problem concerns the maximization problem that is described by rows 12-20 of Table 6.5, in which d, e and f are free.

    (a) On a spreadsheet, execute the simplex method with Rule A, but computing no ratios for the rows whose basic variables are free.

    (b) Did any of the free variables switch sign? If so, what would have occurred if this problem had been forced into Form 1 prior to using Rule A? *Remark*: Part (b) requires no computation.

13. *(free variables)* The tactic by which free variables are handled in Section 4 of this chapter is to make them basic and keep them basic. Here's an alternative:

    (i) After making a free variable basic, set aside this variable, and set aside the equation for which it just became basic. (This reduces by one the number of rows and the number of columns.)

    (ii) At the end, determine the values taken by the free variables from the values found for the other variables.

    Does this work? If it does work, why does it work? And how would you determine the "values taken by the free variables."

14. *(extreme points and free variables)* A feasible solution to a linear program is an *extreme point* of the feasible region if that feasible solution is not a convex combination of two other feasible solutions. Consider a linear program that is written in Form 2. Suppose this linear program is feasible and bounded. Is it possible that no extreme point is an optimal solution? *Hint*: can a feasible region have no extreme points?

Part III–Selected Applications

Part III surveys optimization problems that involve one decision-maker.

Chapter 7. A Survey of Optimization Problems

This chapter is built upon 10 examples. When taken together, these examples suggest the range of uses of linear programs and their generalizations. These examples include linear programs, integer programs, and nonlinear programs. They illustrate the role of optimization in operations management and in economic analysis. Uncertainty plays a key role in several of them. Also discussed in this chapter are the ways in which Solver and Premium Solver can be used to solve problems that are not linear.

Chapter 8. Path-Length Problems and Dynamic Programming

This chapter is focused on the problem of finding the shortest or longest path from one node to another in a directed network. Several methods for doing so are presented. Linear programming is one of these methods. Path-length problems are the ideal setting in which to introduce "dynamic programming," which is a collection of ideas that facilitate the analysis of decision problems that unfold over time.

Chapter 9. Flows in Networks

Described in this chapter are "network flow" models and the uses to which they can be put. If the "fixed" flows such a model are integer-valued, the simplex method is shown to find an integer-valued optimal solution.

Chapter 7: A Survey of Optimization Problems

1. Preview

The variety of optimization problems that can be formulated for solution by linear programming and its generalizations is staggering. The "survey" in this chapter is selective. It must be. Each problem that appears here illustrates one or more of these themes:

- Exhibit the capabilities of the Premium Solver software package.
- Relate optimization to economic reasoning.

E. V. Denardo, *Linear Programming and Generalizations*, International Series in Operations Research & Management Science 149,
DOI 10.1007/978-1-4419-6491-5_7, © Springer Science+Business Media, LLC 2011

- Relate optimization to operations management.
- Relate optimization to situations in which uncertainty plays a central role.

Only a few of the optimization problems in this chapter are linear programs. That's because of the need to make room for optimization problems that include integer-valued variables and nonlinearities. Linear programs are strongly represented in three other chapters – in Chapter 8 (dynamic programming), in Chapter 9 (network flow) and in Chapter 14 (game theory).

Three sections of this chapter are starred. The starred sections delve into probabilistic modeling. These starred sections present all of the "elementary" probability that they employ, but readers who are new to that subject may find those sections to be challenging.

To a considerable extent, each section is independent of the others. They can be read selectively. An exception occurs in the starred sections. The "normal loss function" is introduced in the first starred section, and it is used in all three. Another exception consists of Sections 10-12. They form a coherent account of basic ideas in operations management and might best be read as a unit.

2. Production and Distribution

The initial example is a rudimentary version of a problem that is faced in the petroleum industry.

Problem 7.A A vertically-integrated petroleum products company produces crude oil in three major fields, which are labeled U, V and W, and ships it to four refineries, which are labeled 1 through 4. The top nine rows of Table 7.1 contain the relevant data. Cells H5, H6 and H7 of this table specify the production capacities of fields U, V and W, respectively. Cells I5, I6 and I7 contain the production costs for these fields. Cells D9 through G9 contain the demand for crude oil one week hence at the refineries 1 through 4. These demands must be met by production during the current week. Each entry in the array D5:G7 is the cost of shipping from the field in its row to the refinery in its column. Capacities and demands are measured in thousands of barrels per week. Production and shipping costs are measured in dollars per barrel. The company wants to minimize the cost of satisfying these demands. How shall it do this?

Table 7.1. Spreadsheet formulation of Problem 7.A.

	A	B	C	D	E	F	G	H	I	J
1										
2					Refinery					
3				1	2	3	4			
4					shipping cost			capacity	prod. cost	
5		Field	U	5.4	9.2	7.4	10.2	250	6.2	
6			V	8.2	7.1	2.1	8.1	400	9.4	
7			W	5.8	5.8	8.3	5.7	350	11.3	
8										
9				200	300	250	150	demand		
10										
11					shipping quantity			prod. q.		
12		Field	U	200	50	0	0	250		
13			V	0	150	250	0	400		
14			W	0	100	0	150	250		
15				200	300	250	150			
16									= SUM(D14:G14)	
17					prod.	ship	total			
18				cost	8135	4565	12700		= SUM(G12:G14)	
19										
20							= E18 + F18			
21						= SUMPRODUCT(D5:G7 , D12:G14)				
22					= SUMPRODUCT(H12:H14 , I5:I7)					

A tailored spreadsheet

In earlier chapters, a "standardized" spreadsheet was used to build a linear program. Each decision variable was represented by a column, and each constraint was depicted as a row. For Problem 7.A, the decision variables are the shipping quantities, and it is natural to organize them in the same pattern as the shipping costs.

The "tailored" spreadsheet in Table 7.1 presents the shipping quantities in the array D12:G14. The sum across a row of this array is the quantity produced in the corresponding field, and the sum down a column of this array is the quantity shipped to the corresponding refinery.

A linear program

The functions in cells E18 and F18 compute the shipping and production costs. Solver has been asked to minimize the quantity in cell G18, which is the total cost. Its changing cells are the shipping quantities in cells D12:G14. Its constraints are H12:H14 ≤ H5:H7 (production quantities cannot exceed production capacities), D15:H15 = D9:H9 (demands must be met) and D12:G14 ≥ 0 (shipping quantities must be nonnegative). Table 7.1 reports the optimal solution to this linear program.

The petroleum industry

In Chapter 1, it had been observed that a paper on the use of a linear program to find a blend of aviation fuels had excited great interest in the petroleum industry. Problem 7.A suggests why. Linear and nonlinear programs offered the promise of integrating the production, refining, distribution, and marketing of petroleum products in ways that maximize after-tax profit.

A coincidence?

Table 7.1 reports an optimal solution that is integer-valued. This is not an accident. Problem 7.A happens to be a type of "network flow" problem for which every basic solution is integer-valued.

3. A Glimpse of Network Flow

Figure 7.1 depicts the constraints of Problem 7.A as a network flow model. Each "flow" occurs on a "directed arc" (line segment with an arrow). The amount flowing into each node (circle) must equal the amount flowing out of that node. All "flows" are nonnegative. Some flows are into a node, some flows are out of a node, and some flows are from one node to another. The flows can have bounds, and they can be fixed.

Figure 7.1 has 7 nodes, one for each field and one for each refinery. The node for field U accounts for the production in that field and for its shipment to the four refineries. The node for refinery 1 accounts for the demand at its refinery and the ways in which this demand can be satisfied. The flow into node U cannot exceed 250, which is the capacity of field 1, and the flow out of node 1 must equal 200, which is the demand at refinery 1.

Figure 7.1. A network flow interpretation of Problem 7.A.

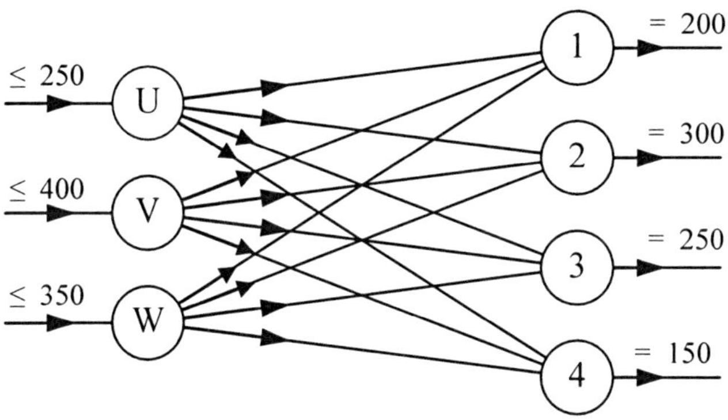

The Integrality Theorem

A network flow model is said to have **integer-valued data** if each of its bounds and each of its fixed flows is integer-valued. In Figure 7.1, each fixed flow and each bound is integer-valued. This network flow model does have integer-valued data. This model's costs are not integer-valued, but that does not matter. An important property of network flow models is highlighted below.

> **The Integrality Theorem:** Consider a network flow model that has integer-valued data. Each of its basic solutions is integer-valued.

The Integrality Theorem is proved in Chapter 9.

The simplex method for network flow

Let us consider what happens when the simplex method is applied to a network flow model that has integer-valued data. The simplex method pivots from one basic solution to another. Each basic solution that it encounters is integer-valued. The simplex method stops with a basic solution that is optimal, and it too is integer-valued. For this class of optimization problems, the simplex method is guaranteed to produce an optimal solution that is integer-valued.

The Integrality Theorem is of little consequence in Problem 7.A. Petroleum is no longer shipped in barrels. Even if it were, little harm would be done by rounding off any fractions to the nearest integer.

In other contexts, such as airline scheduling, it is vital that the decision variables be integer-valued. If the network flow model of an airline scheduling problem has iteger-valued data, the simplex method produces a basic solution that is optimal and is integer-valued.

4. An Activity Analysis

An **activity analysis** is described in terms of *goods* and *technologies*. Each technology transforms one bundle of goods into another. The *inputs* to a technology are the goods it consumes, and the *outputs* of a technology are the goods it produces. Each technology can be operated at a range of nonnegative *levels*. The decision variables in an activity analysis include the level at which to operate each technology. If a model of an activity analysis has constant returns to scale, it leads directly to a linear program. To illustrate this type of model, consider

Problem 7.B. *(Olde England).* In an early era, developing nations shifted their economies from agriculture toward manufacturing. Old England had three principal technologies, which were the production of food, yarn and clothes. It traded the inputs and outputs of these technologies with other countries. In particular, it exported the excess (if any) of yarn production over internal demand.

The Premier asked you to determine the production mix that would maximize the net value of exports for the coming year. Your first step was to accumulate the "net output" data that appear in cells B4:D10 of Table 7.2. Columns B records the net output for food production; evidently, producing each unit of food requires that Olde England import £0.50 worth of goods (e.g., fertilizer), consume 0.2 units of food (e.g., fodder to feed to animals), consume 0.5 units of labor, and use 0.9 units of land. Column C records the net outputs for yarn production; producing each unit of yarn requires that Olde England import £1.25 worth of goods, consume 1 unit of labor, and use 1.5 units of land. Column D records the net outputs for clothes production; producing each unit of clothes requires the nation to import £5.00 worth of goods, consume 1 unit of yarn, and consume 4 units of labor.

Cells J5:J7 record the levels of internal consumption of food, yarn and clothes, respectively; in the coming year, Olde England will consume 11.5

million units of food, 0.6 million units of yarn and 1.2 million units of clothes. Cells J9:J12 record the nation's capacities, which are 65 million units of labor and 27 million units of land, as well as the capability to produce yarn at the rate of 10.2 million units per year and clothes at the rate of 11 million units per year.

Row 4 records the world market prices of £3 per unit for food, £10 per unit for yarn and £16 per unit for clothes. The amounts that Olde England imports or exports will have negligible effect on these prices.

Table 7.2. An activity analysis for Olde England.

	A	B	C	D	E	F	G	H	I	J
1										
2	label	FP	YP	CP	FE	YE	CE			
3	value	13	10.2	11	-1.1	-1.4	9.5			
4	contribution	-0.5	-1.25	-5	3	10	16	60.5		balance
5	food	0.8			-1			11.5	=	11.5
6	yarn		1	-1		-1		0.6	=	0.6
7	clothes			1			-1	1.5	=	1.5
8										capacity
9	labor	0.5	1	4				60.7	<=	65
10	land	0.9	1.5					27	<=	27
11	yarn cap.		1					10.2	<=	10.2
12	clothes cap.			1				11	<=	11

Decision variables

This activity analysis has two types of decision variables. The symbols FP, YP and CP stand for the quantity of food, yarn and clothes to produce in the coming year. The symbols FE, YE and CE stand for the *net* exports of food, yarn and clothes during the coming year. The unit of measure of each of these quantities is millions of units per year. The production quantities FP, YP and CP must be nonnegative, of course. The net export quantities FE, YE and CE can have any sign; setting FE = −1.5 accounts for importing 1.5 million units of food next year, for instance.

A linear program

The linear program whose data appear in Table 7.2 maximizes the net value of exports. Column H contains the usual sumproduct functions. Cell H4 measures the contribution (value of net exports). Rows 5-7 account for the uses of food, yarn and clothes. Rows 9-10 account for the uses of land and labor. Rows 11 accounts for the loom capacity, and row 12 accounts for the clothes-making capacity.

The decision variables in cells B3:D3 are required to be nonnegative, but the decision variables in cells E3:G3 are not. Solver has been asked to maximize the value of net exports (the number in cell H4) subject to constraints H5:H7 = J5:J7 and H9:H12 ≤ J9:J12. Row 3 of Table 7.2 reports the optimal values of its decision variables.

Evidently, the net trade balance is maximized by making full use of the land, full use of the capacity to weave yarn, and full use of the capacity to produce clothes. Clothes are exported. The nation produces most, but not all, of the food and yarn it requires.

Some "what if" questions

Activity analyses like this one make it easy to respond to a variety of "what if" questions. Here are a few: What would occur if Olde England decided that it ought to be self-sufficient as concerns food? Would it pay to increase the capacity to produce yarn? What would occur if the market price of clothes decreased by 20%?

A bit of the history

The phrase "activity analysis" was first used by Tjalling Koopmans; its initial appearance is in the title[1] of the proceedings of a famous conference that he organized shortly after George Dantzig developed the simplex method. Well before that time (indeed, well before any digital computers existed) Wassily Leontief (1905-1999) built large input-output models of the American economy and used them to answer "what if" questions. Leontief received the Nobel Prize in 1973 "for the development of the input-output method and for its application to important economic problems." As Leontief had observed, an activity analysis is the natural way in which to describe the production side of a model of an economy that is in general equilibrium. One such model appears in Chapter 14.

[1] *Activity analysis of production and allocation: Proceedings of a conference*, Tjalling C. Koopmans, ed., John Wiley & Sons, 1951.

5. Efficient Portfolios

The net return (profit) that will be earned on an investment is uncertain. Table 7.3 specifies the net return that will be earned by the end of a six-month period per unit invested in each of three assets. These returns depend on the state that will occur at that time. Cells C4 through C6 specify the probability distribution over these states. Cells D4 through D6 specify the net return R1 per unit invested in asset 1 if each state occurs. Cells E4 through E6 and F4 through F6 specify similar information about assets 2 and 3. Evidently, if state a occurs, these assets have return rates of −20%, 40% and −30%, respectively. The returns on these assets are dependent; if you know the value taken by the return on one of the assets, you know the state and, consequently, the returns on the other assets.

Table 7.3. Rates of return on three assets.

	A	B	C	D	E	F
1						
2				return on asset		
3		state	prob.	R1	R2	R3
4		a	0.2	-0.2	0.4	-0.3
5		b	0.5	0.0	0.2	-0.1
6		c	0.3	0.3	-0.5	0.5
7						
8				0.05	0.03	0.04
9						↑
10		= SUMPRODUCT($C4:$C6 , F4:F6)				

The functions in row 8 of this table compute the mean (expected) rate of return on each asset; these are 5%, 3%, and 4%, respectively.

A portfolio

A **portfolio** is a set of levels of investment, each in a particular asset. The net return (profit) R on a portfolio is uncertain; it depends on the state that will occur. The portfolio that invests the fractions 0.6, 0.3 and 0.1 in assets 1, 2, and 3, respectively, is evaluated in Table 7.4. The functions in cells G4 through G6 specify the value taken by R under outcomes a through c. Cell G4 reports that, if outcome a occurs, the rate of return on this portfolio will be $-0.03 = (0.6)^*(-0.2) + (0.3)^*(0.4) + (0.1)^*(-0.3)$, for example.

The function in cell G11 computes the expectation E(R) of the return on this portfolio. The functions in cells H4 through H6 compute the difference R – E(R) between the return R and its expectation if states a, b and c occur. The function in cell H11 computes Var(R) because the variance equals the expectation of the squared difference between the outcome and the mean.

Table 7.4. The return on a particular portfolio.

	A	B	C	D	E	F	G	H	I
1									
2					return on asset		total		
3		state	prob.	R1	R2	R3	R	R - E(R)	
4		a	0.2	-0.2	0.4	-0.3	-0.03	-0.073	
5		b	0.5	0.0	0.2	-0.1	0.05	0.007	
6		c	0.3	0.3	-0.5	0.5	0.08	0.037	
7									
8		= SUMPRODUCT(D6:F6 , D$11:F$11)						= G6 - G$11	
9									
10				f1	f2	f3	E(R)	Var(R)	StDev(R)
11				0.6	0.3	0.1	0.043	0.0015	0.038743
12									
13			= SUMPRODUCT(C4:C6 , G4:G6)						
14			= SUMPRODUCT(C4:C6 , H4:H6 , H4:H6)						
15								= SQRT(H12)	

Efficiency

Individuals and companies often take E(R) as a measure of desirability (higher expectation being preferred) and Var(R) as a measure of risk (lower variance being preferred). With these preferences, a portfolio is said to be **efficient** if it achieves the smallest variance in profit over all portfolios whose expected profit is at least as large as its expected profit. If a portfolio is not efficient, some other portfolio has less risk and has mean return that is at least as large. To illustrate the construction of an efficient portfolio, consider

Problem 7.C, part (a). For the data in Table 7.3, find the minimum-variance portfolio whose expected rate return rate is at least 3%.

It is not difficult to show (we omit this) that Var(R) is a convex quadratic function of the fractions invested in the various assets. For that reason, minimizing Var(R) subject to a constraint that keeps E(R) from falling below a prescribed bound is a garden-variety (easily solved) nonlinear program.

The spreadsheet in Table 7.5 exhibits the portfolio that minimizes Var(R) subject to $E(R) \geq 0.03$. The data and functions in Table 7.4 are reproduced in Table 7.5. In addition, cell C9 contains the lower bound on the return rate, which equals 0.03, and cell B9 contains a function that computes the sum $(f1 + f2 + f3)$ of the fractions invested in the three assets.

The GRG nonlinear code has been used to minimize the variance in the return (cell H9) with the fractions invested in the three assets (cells D9:F9) as the changing cells, subject to constraints that keep the fractions nonnegative, keep their total equal to 1, and keep the mean return (cell G9) at least as large as the number in cell C9. This portfolio invests roughly 47% in asset 2 and roughly 53% in asset 3. It achieves a mean return rate of 3.53%. The standard deviation in its rate of return is roughly 0.005. Evidently, if an investor seeks a higher mean rate of return than 3.53%, she or he must accept more risk (a higher variance, equivalently, a higher standard deviation).

Table 7.5 An efficient portfolio

	A	B	C	D	E	F	G	H	I
1									
2					return on asset		total		
3		state	prob.	R1	R2	R3	R	R - E(R)	
4		a	0.2	-0.2	0.4	-0.3	0.02653	-0.0088	
5		b	0.5	0.0	0.2	-0.1	0.03994	0.00461	
6		c	0.3	0.3	-0.5	0.5	0.03353	-0.0018	
7									
8			min	f1	f2	f3	E(R)	Var(R)	StDev(R)
9		1	0.03	0	0.46647	0.53353	0.03534	2.7E-05	0.005205
10		↑							
11			= SUM(D9:F9)						

The efficient frontier

The set of all pairs [E(R), Var(R)] for efficient portfolios is called the **efficient frontier**. If a rational decision maker accepts E(R) as the measure of desirability and Var(R) as the measure of risk, that person chooses a portfolio on the efficient frontier. If a portfolio is not on the efficient frontier, some other portfolio is preferable.

Problem 7.C, part (b). For the data in Table 7.3, find the portfolios that are on the efficient frontier.

No asset returns more than 5%, so placing a value greater than 0.05 in cell C9 guarantees infeasibility. To find a family of portfolios that are on the efficient frontier, one can repeat the calculation whose result is exhibited in Table 7.5 with the number in cell C9 equal to a variety of values between 0.03 and 0.05. There is a technical difficulty, however.

Using Solver repeatedly

Suppose we solve the NLP with 0.034 in cell C9, then change that number to 0.038, and then solve again. The new solution *replaces* the only one. This difficulty has been anticipated. Row 9 contains all of the information we might want to keep from a particular run. Before making the 2nd run, "Copy" row 9 onto the Clipboard and then use the Paste Special command to put only its "Values" in row 14. After changing the entry in cell C9 and re-optimizing, use the Paste Special command to put the new "Values" in row 15. And so forth. Reported in Table 7.6 is the result of a calculation done with values of C9 between 0.03 and 0.05 in increments of 0.004.

Table 7.6. Portfolios on the efficient frontier.

	A	C	D	E	F	G	H	I
1								
13		min	f1	f2	f3	E(R)	Var(R)	StDev(R)
14		0.03	0.0000	0.4665	0.5335	0.0353	0.0000	0.0052
15		0.034	0.0000	0.4665	0.5335	0.0353	0.0000	0.0052
16		0.03534	0.0000	0.4662	0.5338	0.0353	0.0000	0.0052
17		0.038	0.2212	0.4213	0.3575	0.0380	0.0001	0.0121
18		0.042	0.5537	0.3538	0.0925	0.0420	0.0005	0.0225
19		0.0434	0.6698	0.3302	0.0000	0.0434	0.0007	0.0261
20		0.046	0.8000	0.2000	0.0000	0.0460	0.0058	0.0764
21		0.05	1.0000	0.0000	0.0000	0.0500	0.0325	0.1803

Piecewise linearity

These portfolios exhibit *piecewise linearity*. As the mean rate of return increases from 3.53% to 4.34%, the portfolio varies linearly. When the mean rate of return reaches 4.34%, the fraction invested in asset 3 decreases to 0. As the rate of return increases from 4.34% to 5%, the portfolio again varies linearly, with f3 = 0 in this interval. Evidently, as the mean return rate increases, the optimal portfolio "pivots" from one extreme point to another. This is the sort of behavior that one expects in the optimal solution to a *linear* program. One is led to wonder whether this nonlinear program is mimicking a linear program.

Using Premium Solver repeatedly

The calculation whose results are reported in Table 7.6 is a bit unwieldy. To change the value in cell C9, one needs to close Solver, insert the new number, and then reopen Solver. That's because Solver is "modal." When Premium Solver is run off the Tools menu, it too is modal, and it is equally unwieldy.

When Premium Solver is operated off the ribbon, it is "modeless," and it can easily be used to solve an optimization problem repeatedly with a variety of values of a parameter. How to accomplish this is described with reference to Figure 7.2. The left-hand side of this figure displays the pull-down menu that appears when you click on Premium Solver on the ribbon. If you then click on the drop-down entry entitled Model, the dialog box to the right of Figure 7.2. appears.

Figure 7.2. Premium Solver on the ribbon.

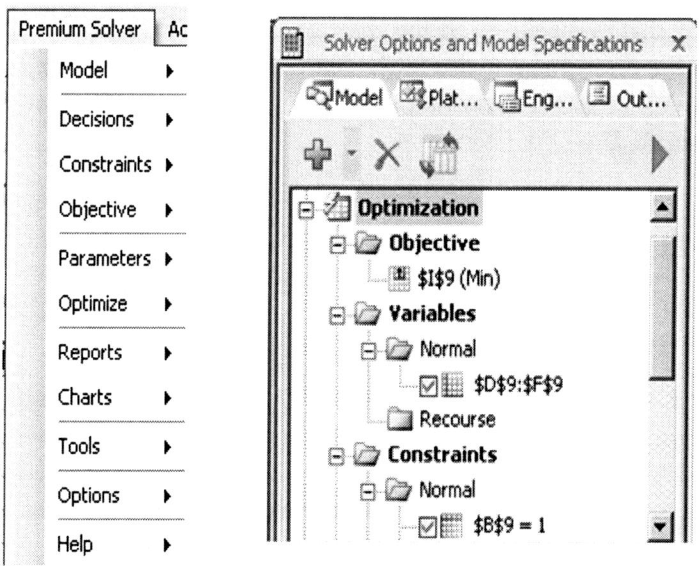

Suppose you wish to solve the portfolio optimization with cell C9 (the lower bound on the mean return) equal to the 11 equally-spaced values 0.03, 0.032, 0.034, …, 0.05. To do so, follow this protocol:

- Select cell C9 of the spreadsheet exhibited in Table 7.5.
- Click on the Premium Solver on the ribbon. The drop-down menu to the left of Figure 7.2. will appear. On it, click on Parameters and then click on Optimization. In the dialog box that appears, enter 0.03 as the Lower value and 0.05 as the Upper value. This causes the function = **PsiOptParam(0.03,0.05)** to appear in cell C9.
- Next, click again on Premium Solver on the ribbon. On the drop-down menu that appears, click on Model. The dialog box to the right of Figure 7.2 will appear. In the menu at its top, click on "Plat…" A dialog box will appear. In the "Optimizations to Run" window, enter **11**
- Next, return to the Model tab on the dialog box to the right of Figure 7.2 Then click on the row containing the variables, cells D9:F9 in this case. Make sure that the Monitor Value of these cells is set equal to True. (If it is set equal to False, switch it.)
- Finally, either click on the green triangle to the right of the dialog box that is displayed to the right of Figure 7.2 or click on Optimize in the drop-down menu to the left of Figure 7.2 Either action causes Premium Solver to solve the 11 optimization problems that you have specified.

You can then scroll through the solutions to these optimization problems by clicking on the window in a ribbon that currently reads "Opt. 11." You can also create a chart by clicking Charts on the drop-down menu.

The ribbon

The ribbon can also be used to specify an optimization problem. The drop-down menu at the left of Figure 7.2 lets you specify the model's decision cells, constraints and objective. Using the ribbon can be easier because it allows you to alter your spreadsheet without closing Premium Solver.

Measures of risk

By longstanding tradition, variance is used as the measure of risk. As noted in Chapter 1, the variance puts heavy weight on observations that are far from the mean. With $\mu = E(R)$, it might make better sense to accept $MAD(R) = E|R - \mu|$ as the measure of risk.

These two measures of risk share a defect; Var(R) and MAD(R) place large penalties on outcomes that are far *better* than the mean. It might make still better sense to minimize the expectation of the amount by which the mean exceeds the outcome, i.e., to accept $E[(\mu - R)^+]$ as the measure of risk.

With either $E|R - \mu|$ or $E[(\mu - R)^+]$ as the measure of risk, an efficient portfolio can be found by solving a linear program, rather than a nonlinear program. The optimal portfolio will continue to be a piecewise linear function of E(R), but the Allowable Increase and Allowable Decrease will determine the points at which the basis changes.

Getting the data

If the assets in a portfolio are common stocks of widely-traded companies, data from which to build a model like that in Table 7.3 can be obtained from the historical record. For each of, say, twenty six-month periods, record the "real" rate of return on each stock, this being the excess of its return over the "risk-free" return, i.e., that of a six-month treasury bill for the same period. Place the real rates of return for each period in a row. Assume that each row represents a state that occurs with probability 1/20.

This approach relies on the "efficient markets hypothesis," which states that all of the publicly-available information about the future of a company is contained in the current price of its stock. This hypothesis discounts the possibility of "bubbles." It does not predict the violent swings in market prices that occur from time to time. It is widely used, nonetheless.

A bit of the history

The ideas and results in this section were developed by Harry Markowitz while he was a Ph. D. student at the University of Chicago. He published a landmark paper in 1952, and he shared in the 1990 Nobel Prize in Economics, which was awarded for "pioneering work in the theory of financial economics."

6. Modeling Decreasing Marginal Cost

As was noted in Chapter 1, when a linear program is used to model *increasing* marginal cost, unintended options are introduced, but they are ruled out by optimization. The opposite occurs when one attempts to use a linear program to model decreasing marginal cost; unintended options are introduced, and they are selected by optimization. Decreasing marginal cost – equivalently, increasing marginal profit – cannot be handled by linear programs. A method that does handle these situations will be developed in the context of

Problem 7.D. This problem appends to the ATV problem in Chapter 5 the possibility of leasing tools that improve efficiency and thereby lower manufacturing costs. Tools α and β facilitate more efficient manufacture of Fancy and Luxury model vehicles, respectively. Leasing tool α costs \$1,800 per week, and this tool reduces the cost of manufacturing each Fancy model vehicle by \$120. Similarly, leasing tool β costs \$3,000 per week, and that tool reduces the cost of producing each Luxury model vehicle by \$300. The goal remains unchanged; it is to operate the ATV plant in a way that maximizes contribution. What production rates accomplish this?

Binary variables

Problem 7.D can be formulated as an optimization problem that differs from a linear program in that two of its decision variables are required to take the value 0 or 1. A decision variable whose values are restricted to 0 and 1 is said to be **binary**.

An integer program

Throughout this text – and throughout much of the literature – the term **integer program** is used to describe an optimization problem that would be a linear program if requirement that some or all of its decision variables be integer-valued were deleted. An integer program can have no quadratic terms, for instance. It might be more precise to describe this type of optimization problem as an "integer linear program," but that usage never took root. Two different methods for solving integer programs are discussed in Chapter 14. Both of these methods solve a sequence – often surprisingly short – of linear programs.

Break-even values

Our goal is to formulate Problem 7.D for solution as an integer program, rather than as a more complicated object. Let us begin by computing a *break-even value* for each tool. The equation \$120 F=\$1,800 gives the value of F at which we are indifferent between leasing tool α and not leasing it. Evidently, leasing this tool is worthwhile when F > 15 = \$1,800/\$120. Similarly, the break-even equation \$300 L = \$3,000 indicates that leasing tool β is worthwhile if L > 10.

Binary variables will be used to model the leasing of these tools. Equating the binary variable a to 1 corresponds to leasing tool α. Equating the binary variable b to 1 corresponds to leasing tool β. Our goal is to formulate Problem 7.D as an optimization problem that differs from a linear program *only* in that the variables a and b are binary.

Accounting for the contribution

Leasing tool α increases the contribution of each Fancy model vehicle by \$140, from \$1,120 to \$1,340, but it incurs a fixed cost of \$1,800. The contribution of the Fancy model vehicle can be accounted for by using the binary variable a in the linear expression and constraints that appear below:

$$1120\ \mathrm{F1} + 1240\ \mathrm{F2} - 1800a,$$

$$a \in \{0,1\}, \qquad \mathrm{F2} \leq 40a,$$

$$\mathrm{F1} \geq 0, \qquad \mathrm{F2} \geq 0, \qquad \mathrm{F} = \mathrm{F1} + \mathrm{F2}.$$

The linear expression measures the contribution earned from Fancy model vehicles. If a = 0, the constraint F2 ≤ 40a keeps F2 = 0, so F = F1 and the linear expression reduces to 1120 F, which is the contribution earned without the tool. If a = 1, the linear expression is maximized by setting F1 = 0 and F2 = F, which reduces it to 1240 F − 1800. As noted above, this is preferable to 1120 F if F exceeds 15.

The binary variable b accounts in a similar way for leasing the tool that reduces the cost of producing Luxury model vehicles.

A spreadsheet

The spreadsheet in Table 7.7 prepared this optimization problem for solution by Solver. Rows 5-9 account for the capacities of the five shops. Rows 10 and 11 model the constraints F2 ≤ 40a and L2 ≤ 40b. Rows 12 and 13 model the constraints F = F1 + F2 and L = L1 + L2.

Table 7.7. A spreadsheet for Problem 7.D.

	A	B	C	D	E	F	G	H	I	J	K	L	M
1													
2		S	F	L	F1	F2	L1	L2	a	b			RHS
3		20	30	0	0	30	0	0	1	0			
4		840	0	0	1120	1240	1200	1500	-1800	-3000	52200		
5		3	2	1							120	<=	120
6		1	2	3							80	<=	80
7		2									40	<=	96
8			3								90	<=	104
9				2							0	<=	40
10						1			-40		-10	<=	0
11								1		-40	0	<=	0
12			1		-1	-1					0	=	0
13				1			-1	-1			0	=	0

Reported in Table 7.7 is the optimal solution to Problem 7.D. This optimal solution has been found by maximizing the value in cell K4 with B3:J3 as changing cells, subject to constraints B3:H3 ≥ 0, I3:J3 binary, K5:K11 ≤ M5:M11, and K12:K13 = M12:M13. Evidently, it is profitable to lease tool α but not tool β. And it remains optimal to produce no Luxury model vehicles.

Constraining variables to be binary

To solve Problem 7.D with Solver or with Premium Solver, we need to require that the decision variables in cells I3 and J3 be binary. An easy way to do that is to call upon the "Add Constraints" dialog box in Figure 7.3. In the left-hand window of Figure 7.3, enter I3:J3. Then click on the center window and scroll down to "bin" and then release. After you do so, "bin" will appear in the center window and the word "binary" will appear in the right window. It will not work to select "=" in the center window and enter the word "binary" in the right-hand window, incidentally.

Figure 7.3. Specifying binary variables.

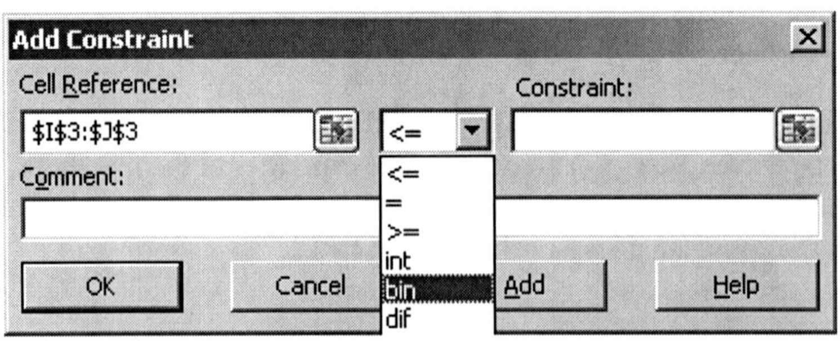

Solving integer programs

After you formulate your integer program, but before you click on the Solver button:

- with Solver in Excel 2003, click on "Assume Linear Model;"
- with Solver in Excel 2010, select "Simplex LP;"
- with Premium Solver, select "Standard LP/Quadratic."

If you follow these rules, a method akin to those in Chapter 14 will be used, with good results. If you do not follow these rules, a more sophisticated method will be used. That method seeks a "local optimum," which may not be a global optimum.

No shadow prices?

If you present Solver or Premium Solver with an optimization problem that includes any integer-valued variables, it does *not* report shadow prices. Let us see why that is so.

First, consider the case in which all of the decision variables must be integer-valued. In this case, shadow prices cannot exist because perturbing a RHS value by a small amount causes the optimization problem to become infeasible.

Next, consider the case in which only some of the decision variables must be integer-valued. In this case, perturbing a RHS value may preserve feasibility, but it may cause an abrupt change in the objective value. When that occurs, the shadow price cannot exist.

Finally, suppose a constraint did have a shadow price. It applies to a small change in a RHS value, but it gives no information about the effect of larger changes. If a constraint's shadow price equals 2, for instance, increasing that constraints RHS value by δ increases the objective by 2δ if δ is close enough to 0. But the objective could increase by more than 2δ if δ were larger.

A nonlinear integer program

The term **nonlinear integer program** is used to describe an optimization problem that would be a nonlinear program if we omitted the requirement that some or all of its decision variables be integer-valued. The GRG code tackles such problems, but it seeks a *local* optimum, which may or may not be a global optimum.

Problem 7.D illustrates this phenomenon. It is not hard to show that the feasible solution S=35, F=0 and L=15 is a local maximum. Perturbing this solution by setting L=1 decreases the objective value by $50, for instance. If the GRG code encounters this feasible solution, it will stop; it has found a local maximum that is not a global maximum.

7. The Traveling Salesperson

The data in the "traveling salesperson problem" are the number of cities that the salesperson is to visit and the travel times from city to city. A **tour** occurs if the salesperson starts at one of these cities and visits each of the other cities *exactly once* prior to returning to the city at which he or she began. The length of the tour is the sum of the times it takes to travel from each city to the next. The **traveling salesperson problem** is that of finding a tour whose length is smallest. The traveling salesperson problem may sound a bit contrived, but it arises in a variety of contexts, including

Problem 7.E *(scheduling jobs).* Five different jobs must be done on a single machine. The needed to perform each job is independent of the job that preceded it, but the time needed to reset the machine to perform each job does vary with the job that preceded it. Rows 3 to 9 of Table 7.8 specifies the times needed to reset the machine to accomplish each of the five jobs. "Job 0" marks the start, and "job 6" marks the finish. Each reset time is given in minutes. This table shows, for instance, that doing job 1 first entails a 3-minute setup and that doing job 4 immediately after job 1 entails a 17-minute reset time. Reset times of 100 minutes represent job sequences that are not allowed. The goal is to perform all five jobs in the shortest possible time, equivalently, to minimize the sum of the times needed to set up the machine to perform the five jobs.

Table 7.8. Data and solution of Problem 7.E.

	A	B	C	D	E	F	G	H	I
1									
2						-- to --			
3				1	2	3	4	5	6
4			0	3	5	11	13	7	100
5			1	100	11	21	17	15	0
6		-- from --	2	9	100	17	25	15	0
7			3	17	19	100	21	7	0
8			4	11	13	27	100	11	0
9			5	21	15	13	29	100	0
10									
11	seq. =		0	2	1	4	5	3	6
12				5	9	17	11	13	0
13				↑					55
14		= OFFSET($C4, C11, D11)							↑
15							= SUM(D12:I12)		

The offset function

The reset times in Table 7.8 form a two-dimensional array. Excel's "offset" function identifies a particular element in such an array. If the Excel function **=OFFSET(X, Y, Z)** is entered in a cell, that cell records the number in the cell that is Y rows below and Z rows to the right of cell X. For instance, entering the function **=OFFSET(C4, 1, 3)** in cell K2 would cause the number 21 to appear in cell K2; this occurs because 21 is the number that's 1 row below and 3 columns to the right of cell C4.

A job sequence and its reset times

Row 11 of Table 7.8 records a particular sequence in which the jobs are performed, namely, job 2, then job 1, then job 4, and so forth. The "offset" functions in row 12 record the times needed to prepare the machine to perform each of these jobs. Note that the offset function in Cell D12 gives the setup time needed to do job 2 first. Also, the offset function in cell E12 records the reset time needed to do job 1 second given that job 2 is done first. And so forth.

The Evolutionary Solver*

This subsection describes a solution method that uses the Standard Evolutionary Solver, which exists only in Premium Solver. If you do not have access to Premium Solver, please skip to the next subsection.

Table 7.8 records result of applying the Standard Evolutionary Solver to Problem 7.e. The quantity in cell I15 was minimized with D11:H11 as changing cells, subject to constraints that the numbers in cells D11:H11 be integers between 1 and 5 and that these integers be different from each other. The requirement that these integers be different from each other was imposed by selecting "dif" in the middle window of the Add Constraints dialog box. The Evolutionary Solver found the solution in Table 7.8. It did *not* find it quickly, and that's for the case of 5 jobs.

The assignment problem

The traveling salesperson problem has been widely studied, and several different methods of solution have been found to work well even when the number n of cities is fairly large. One of these methods is based on the "assignment problem." A network flow model is called an **assignment problem** if it has 2 m nodes and m^2 directed arcs with these properties:

- The network has m "supply" nodes, with a fixed flow of 1 into each supply node.
- The network has m "demand" nodes, with a fixed flow of 1 out of each demand node.
- It has a directed arc pointing from each supply node to each demand node. The flows on these arcs are nonnegative.

Each fixed flow equals 1, so the assignment problem has integer-valued data. The Integrality Theorem guarantees that that each basic solution to the assignment problem is integer-valued.

An assignment problem with side constraints

In Table 7.9, Problem 7.E is viewed as an assignment problem with "side constraints." Rows 2-10 of this spreadsheet are identical to rows 2-10 of Table 7.8. These rows have been hidden to save space. The rows that are displayed in Table 7.9 have these properties:

- Each cell in the array D12:I17 contains the shipping quantity from the "supply node" in its row to the "demand node" in its column.
- The SUMPRODUCT function in cell B20 computes the cost of the shipment.

Solver had been asked to find the least-cost assignment. This assignment ships one unit out of each supply node and one unit into each demand node. The solution to this assignment problem is *not* reported in Table 7.9. With x(i, j) as the flow from source node i to demand node j, the least-cost assignment sets

$$1 = x(0,\ 2) = x(2,\ 1) = x(1,\ 4) = x(4,\ 6)\ ,$$
$$1 = x(3,\ 5) = x(5,\ 3)\ ,$$

and has 51 minutes as its objective value.

Table 7.9. Viewing Problem 7.E as an assignment problem with side constraints.

	A	B	C	D	E	F	G	H	I	J	K	L
1												
11				1	2	3	4	5	6			
12			0	0	1	0	0	0	0	1		
13			1	0	0	0	1	0	0	1		
14			2	1	0	0	0	0	0	1		
15			3	0	0	0	0	0	1	1		
16			4	0	0	0	0	1	0	1		
17			5	0	0	1	0	0	0	1		
18				1	1	1	1	1	1			
19												
20		55		= SUM(E12 : E17)				= SUM(D17 : I17)				1
21												
22		= SUMPRODUCT(D12:I17 , D4:I9)							= H15 + F17			

Subtours

This optimal solution identifies the job sequences 0-2-1-4-6 and 3-5-3. Neither of these is a tour. These job sequences correspond to *subtours* because neither of them includes all of the jobs (cities in case of a traveling salesperson problem).

A subtour elimination constraint

To eliminate the subtour 3-5-3, it suffices to append to the assignment problem the constraint $x(3,\ 5) + x(5,\ 3) \leq 1$. The function in cell L20 and the constraint $L20 \leq 1$ enforce this constraint. There is no guarantee that the resulting linear program will have an integer-valued optimal solution, and there is no guarantee that it will not have some other subtour.

An optimal solution

Table 7.9 reports the optimal solution to the assignment problem supplemented by this constraint. This optimal solution *is* integer-valued, and it corresponds to the tour (job sequence) 0-2-1-4-5-3-6. This job sequence requires 55 minutes of reset time, and no job sequence requires less.

We could have imposed the constraint that eliminates the other subtour. That constraint is $x(0, 2) + x(2, 1) + x(1, 4) + x(4, 6) \leq 3$.

The general situation

In larger problems, it can be necessary to solve the constrained assignment problem repeatedly, each time with more subtour elimination constraints. It can be necessary to require particular decision variables to be binary. There is no guarantee that this approach converges quickly to an optimal solution to the traveling salesperson problem, but it often does.

8. College Admissions*

This section discusses a subject with which every college student is familiar. This section is starred because readers who have not studied elementary probability may find it to be challenging.

Problem 7.F. You are the Dean of Admissions at a liberal arts college that has a strong academic tradition and has several vibrant sports programs. You seek a freshman class of 510 persons. An agreement has been reached with the head coach of each of several sports. These agreements allow each coach to admit a limited number of academically-qualified applicants who that coach seeks to recruit for his or her team. The coaches have selected a total of 280 such persons. From past data, you estimate that each of these 280 people will join the entering class with probability of 0.75, independent of the others. Your college has no dearth of qualified applicants. From past experience, you estimate that each qualified person you accept who has not been selected (and courted) by a coach will join the entering class with probability of 0.6. Your provost is willing to risk one chance in 20 of having an entering class that is larger than the target of 510. How many offers should you make to non-athletes? What is the expectation of the number of students who will join the freshman class?

The binomial distribution

The "binomial" distribution is the natural model for situations of this type. If n students are offered admission and if each of them joins the class with probability p, independently of the others, the number N who join the class has the *binomial distribution* with parameters n and p. The mean and variance of this binomial distribution are easily seen to be $E(N) = n\,p$ and $Var(N) = n\,p\,(1-p)$.

In particular, the number A of athletes who will join the class has the binomial distribution with parameters $n = 280$ and $p = 0.75$. Thus, the mean and variance of A are given by

$$E(A) = 280 \times 0.75 = 210\,, \qquad Var(A) = 280 \times 0.75 \times 0.25 = 52.5\,.$$

The decision you face as Dean of Admissions is to determine the number n of offers of admission to make to applicants who are not being recruited for athletic teams. If you offer admission to n such people, the number N of them who will join the freshman class also has the binomial distribution, with

$$E(N) = n \times 0.6\,, \qquad Var(N) = n \times 0.6 \times 0.4\,.$$

The random variables A and N are mutually independent because students decide to come to your college independently of each other. The total number, $A + N$, of students in the entering class would be binomial if each person who is admitted joins with the same probability. That is not the case, however. The total number, $A + N$, of persons in the entering class does not have the binomial distribution.

A normal approximation

If a binomial distribution with parameters n and p has an expected number $n\,p$ of "successes" and an expected number $n(1-p)$ of "failures" that are equal to 7 or more, it is well-approximated by a random variable that has the normal distribution with the same mean and variance. The quality of the approximation improves as the numbers $n\,p$ and $n(1-p)$ grow larger. The binomially-distributed random variable A and N have values of $n\,p$ and $n(1-p)$ that are far larger than 7, for which reason A and N are very well approximated by random variables whose distributions are normal.

Adding normal random variables

The sum $N_1 + N_2$ of independent normal random variables N_1 and N_2 is a random variable whose distribution is normal. Thus, the number of people who will join the freshman class is very well approximated by a random variable C whose distribution is normal with mean and variance given by

$$E(C) = n \times 0.6 + 280 \times 0.75 \ ,$$

$$Var(C) = n \times 0.6 \times 0.4. + 280 \times 0.75 \times 0.25 \ .$$

A spreadsheet

The spreadsheet in Table 7.10 evaluates the yield from the pool of athletes and non-athletes. Cell C4 contains the number of offers to make to non-athletes. This number could have been required to be integer-valued, but doing so would make little difference. The functions in cells F3 and G3 compute the mean and variance of the yield from the athletes. The functions in cells F4 and G4 compute the mean and variance of the yield from the others. The functions in cells C8, C9 and C10 compute the mean, variance and standard deviation of the class size C. The function in cell C12 computes the probability that C does not exceed the target of 510.

Table 7.10. The yield from admissions.

	A	B	C	D	E	F	G
1							
2		cohort	# offers	p	1 - p	mean	variance
3		athletes	280	0.75	0.25	210	52.5
4		others	464.88	0.6	0.4	278.93	111.57
5						↑	↑
6						= C4*D4	= F4*E4
7							
8		E(C) =	488.93	←	= F3 + F4		
9		Var(C) =	164.07	←	= G3 + G4		
10		StDev(C) =	12.81	←	= SQRT(C7)		
11		target =	510				
12		P(C ≤ target) =	0.950000	←	= NORMDIST(C11, C8, C10, 1)		
13		probability =	0.95				
14		E(C - 510)+ =	0.268	←	= nl(C9, C6, C8)		
15		E(class size) =	488.66	←	= C6 - C12		

Solver has been asked to find the number in cell C4 such that C12=C13. Evidently, you should offer admission to approximately 465 non-athletes.

How's that?

A binomially-distributed random variable N assigns values only to *integers*. A normally-distributed random variable X assigns probabilities to *intervals*; the probability that X takes any particular value equals 0. How can an integer-valued random variable N be approximated by a normally distributed random variable X that has the same mean and variance? The approximation occurs when X is *rounded off* to the nearest integer. For a given integer t, the probability that N=t is approximated by the probability that X falls in the interval between $t-0.5$ and $t+0.5$.

Fine tuning

For example, the probability that the class size does not exceed 510 is well approximated by the probability that the normally distributed random variable C does not exceed 510.5. A slightly more precise answer to the problem you face as the Director of Admissions can be found by making these changes to the spreadsheet in Table 7.10:

- Require that the decision variable in cell C4 be integer-valued.
- Arrange for Solver or Premium Solver to place the largest number in cell C4 for which $P(C \geq 510.5)$ does not exceed 0.05.

If you make these changes, you will find that they result in a near-imperceptible change in the number of non-athletes to whom admission is to be offered.

What's in cell C14?

The function in cell C14 requires explanation. The **positive part** $(x)^+$ of the number x is defined by $(x)^+=\max\{0, x\}$. Interpret $(x)^+$ as the larger of x and 0. When D denotes a random variable and q is a number $(D-q)^+$ is the random variable whose value equals the amount, if any, by which D exceeds q.

For a random variable D whose distribution is normal, the quantity $E[(D-q)^+]$ is known as the **normal loss function** and is rather easy to compute. Calculus buffs are welcome to work out the formula, but that is not necessary. One of the functions in OP_Tools is =**NL(q, μ, σ)** and this function returns the value of $E[(D-q)^+]$ where D is a normally distributed random variable whose mean and standard deviation equal μ and σ, respectively.

In the College Admissions problem, the random variable C denotes the class size, and $(C-510)^+$ equals the amount, if any, by which the class size exceeds the target of 510 students. This random variable does have the normal distribution. The function in cell C14 of Table 7.10 computes the expectation of the excess, if any, of C over 510. This number equals 0.268. Thus, in the event that C does exceed 510, the expectation of the amount by which C exceeds 510 equals $0.268/(0.05)=5.36$.

9. Design of an Electric Plant*

This section is starred because readers who have not had a course in "elementary" probability may find it to be challenging. In many of the United States, electric utilities are allowed to produce the power required by their customers, and they are allowed to purchase power from other utilities. Problem 7.G, below, concerns a utility that is in such a state.[1]

Problem 7.G *(a power plant)* You are the chief engineer for a utility company. Your utility must satisfy the entire demand for electricity in the district it serves. The rate D at which electricity is demanded by customers in your district is uncertain (random), and it varies with the time of day and with the season. It is convenient to measure this demand rate, D, in units of electricity per year, rather than units per second or per hour. The *load curve* specifies for each value of t the fraction F(t) of the year during which D does not exceed t. This load curve is known. The distribution of D is approximately normal with a mean of 1250 thousand units per year and a standard deviation of 200 thousand units per year. Your utility has no way to store electricity. It can produce electricity efficiently with "base load" plant or less efficiently with "peak load" plant. It can also purchase electricity from neighboring utilities that have spare capacity. The "transfer" price at which this occurs has been set – tentatively – at 6.20 dollars per unit of electricity. Of this transfer price, only the fuel cost is paid to the utility providing the power; the rest accrues to the state. The transfer price is intended to be

[1] Connecticut is not such a state, and its utility rates in 2009 are exceeded only by Hawaii's.

high enough to motivate each utility to satisfy at least 98% of its annual power requirement from its own production. The relevant costs are recorded in Table 7.11. Annualized capital costs are incurred whether or not the plant is being used to generate electricity. Fuel costs are incurred only for fuel that is consumed.

Table 7.11. Capital and fuel costs per unit of electricity.

source of power	base load plant	peak load plant	Transfer
annualized capital cost ($/yr)	2.00	1.30	0.00
fuel cost ($/unit)	1.10	2.10	6.20

Your goal is to design the plant that minimizes the expected annualized cost of supplying power to your customers. What is that cost? How much of each type of plant should your utility possess? Will your utility produce at least 98% of the power that its customers consume?

The plant

Base load plant is cheaper to operate (see Table 7.11), so you will not use any peak-load plant unless your base-load plant is operating at capacity. For the same reason, you will not purchase any electricity from other utilities unless your base-load and peak-load capacities are fully utilized. This leads to the introduction of two decision variables:

q_1 = the capacity of the base-load plant.
q_2 = the total capacity of the base-load and peak-load plant.

The variables q_1 and q_2 are measured in units of electricity per year. From Table 7.11, we see that base-load and peak-load plant have annualized capital costs of 2.00 dollars per unit of capacity and 1.30 dollars per unit of capacity, respectively. The annualized cost C of the plant is given by

$$C = 2.00\, q_1 + 1.30\, (q_2 - q_1)\,,$$

and the unit of measure of C is \$/year.

The electricity

To C must be added the expected cost G of the generating or purchasing the electricity that your utility's customers consume over the course of the year.

The random variable $(D-q_2)^+$ equals the annualized rate at which electricity is purchased from other utilities, this being the excess of D over the total capacity of the base-load and peak-load plant. This electricity costs \$6.20 per unit, so its expected annual cost equals

$$(1) \qquad 6.20\, E[(D - q_2)^+]\,.$$

Similarly, the random variable $(D - q_1)^+ - (D - q_2)^+$ equals the annualized rate at which electricity is satisfied by peak-load plant, this being the excess of D over the capacity q_1 of the base-load plant less the rate of purchase from other utilities. Peak-load electricity costs \$2.10 per unit. The expectation of the difference of two random variables equals the difference of their expectations, even when they are dependent. For these reasons, the expected annual cost of the fuel burned in peak-load plant equals

$$(2) \qquad 2.10\, E[(D - q_1)^+] - 2.10\, E[(D - q_2)^+].$$

Finally, $D - (D - q_1)^+$ equals the annualized rate at which electricity is satisfied by base-load plant, this being D less the excess, if any, of D over the capacity of the base-load plant. This electricity costs \$1.10 per unit. Again, the expectation of the difference equals the difference of the expectations. The expected annualized cost of the fuel burned in base-load plant equals

$$(3) \qquad 1.10\, E[D] - 1.10\, E[(D - q_1)^+].$$

The expectation G of the cost of the electricity itself equals the sum of expressions (1) and (2) and (3). Since D has the normal distribution, each of these expressions can be found from the normal loss function.

A spreadsheet

The spreadsheet in Table 7.12 calculates the annualized capital cost C, the annual generating cost G, and the total cost of the plant whose values of q_1 and q_2 are in cells C7 and D7, respectively. The functions in cells C10 and D10 compute the annualized investment in base-load and peak-load plant. The functions in cells C11, D11 and E11 use expressions (1), (2) and (3) to compute the generating costs of electricity obtained from base-load plant, peak-load plant, and other utilities, respectively.

Table 7.12. Annualized cost of electrical plant.

	A	B	C	D	E	F	G	H
1								
2			base	peak	grid			
3		plant	2.00	1.30	0.00			
4		fuel	1.10	2.10	6.20			
5								
6			q1	q2		μ =	1250	
7			1152.86	1339.84		σ =	200	
8								
9			base	peak	grid			
10		capital	2305.72	243.07	0	C =	$2,548.79	
11		generation	1223.65	199.08	265.28	G =	$1,688.01	
12							$4,236.80	
13								
14							= G10 + G11	
15					= E4 * nl(D7,G6,G7)			
16				= D4 * (nl(C7,G6,G7) - nl(D7,G6,G7))				
17			= C4 * (G6 - nl(C7,G6,G7))					

The GRG solver

The goal of this optimization problem is to minimize the quantity in cell G12. The decision variables are in cells C7:D7. The constraints are C7:D7 $\geq$ 0 and C7 $\leq$ D7. If you attack this problem with the GRG Solver, you will learn that it has more than one local minimum.

The Standard Evolutionary Solver

The solution that is displayed in Table 7.12 was found with the Standard Evolutionary Solver, and it was found quickly. If you explore this solution, you will see that the design problem exhibits a "flat bottom." Eliminating peak-load plant capacity increases the annualized cost by less than 1%, for instance.

It is left for you, the reader, to explore these questions: Is the transfer price large enough to motivate the utility to produce at least 98% of the power its customers require? If not, what is the smallest price that would motivate it to do so?

10. A Base Stock Model

Many retail stores face the problem of providing appropriate levels of inventory in the face of uncertain demand. These stores face a classic tradeoff:

Large levels of inventory require a large cash investment. Low levels of inventory risk stock-outs and their attendant costs.

A simple "base stock" model illustrates this trade-off. Let us suppose that an item is restocked each evening after the store closes. Let us suppose that the demands the store experiences for this item on different days are uncertain, but are independent and identically distributed. The decision variable in this model is the *order up to* quantity q, which equals the amount of inventory that is to be made available when the store opens each morning. This model is illustrated by

Problem 7.H *(a base stock problem).* You must set the stock levels of 100 different items. The demand for each item on each day has the Poisson distribution. The demands on different days are independent of each other. From historical data, you have accurate estimates of the mean demand for each item. If a customer's demand cannot be satisfied, he or she buys the item from some other store. Management has decreed that you should run out of each item infrequently, not more than 2% of the time, but that you should not carry excessive inventory. What is your stocking policy?

Let the random variable D denote the demand for a particular item on a particular day. Your order-up-to quantity for this item is the smallest integer q such that $P(D \le q)$ is at least 0.98. Row 3 of the spreadsheet in Table 7.13 displays the optimal order quantity for items whose expected demand E(D) equals 10, 20, 40, 80, 160 and 320.

Table 7.13. Base stock with a 2% stockout rate.

	A	B	C	D	E	F	G	H
1								
2		E(D)	10	20	40	80	160	320
3		q	17	30	53	99	186	357
4		P(D ≤ q)	0.986	0.987	0.980	0.983	0.980	0.981
5		safety stock	7	10	13	19	26	37
6		s.stk/E(D)	0.7	0.5	0.325	0.2375	0.1625	0.11563
7		s.stk/√E(D)	2.214	2.236	2.055	2.124	2.055	2.068
8								
9			= C5/SQRT(C2)					

Safety stock

In order to provide a high level of service to your customers, you begin each period with a larger quantity q on hand than the mean demand E(D)

that will occur until you are able to restock. The excess of the order-up-to quantity q over the mean demand E(D) is known as the **safety stock.** Row 5 of the spreadsheet in Table 7.13 specifies the safety stock for various levels of expected demand. Row 6 shows that the safety stock grows less rapidly than the expected demand. Row 7 shows that the safety stock is roughly proportional to the *square root* of the expected demand.

An economy of scale

For the base stock model, the safety stock is not proportional to the mean demand. If the mean demand doubles, the safety stock grows by the factor of approximately $\sqrt{2}$, not by a factor of 2. This economy of scale is common to nearly every inventory model. The safety stock needed to provide a given level of service grows as the square root of the mean demand.

11. Economic Order Quantity

In many situations, the act of placing an order entails a cost K that is independent of the size of the order. This cost K might include the expense of the paperwork needed to write the order and the cost of dealing with the merchandise when it is received. If this cost K is large, ordering frequently cannot be optimal. The trade-off between ordering to much and too little is probed in the context of

Problem 7.I *(cash management).* Mr. T does not use credit or debit cards spends, and he spends cash at a constant rate, with a total of $2,000 each month. He obtains the cash he needs by withdrawing it from an account that pays "simple interest" at the rate of 5% per year. His paycheck is automatically deposited in the same account, and he is careful never to let the balance in that account go negative. Each withdrawal requires him to spend 45 minutes traveling to the bank and waiting in line, and he values his free time at $20/hour. How frequently should he visit the bank, and how much should he withdraw at each visit?

Opportunity cost

It is optimal for Mr. T to arrive at the bank with no cash in his pocket. When Mr. T does visit the bank, he withdraws some number q of dollars. Because he spends cash at a uniform rate, the average amount of cash that he has in his possession is q/2, and the opportunity cost of not having that amount of cash in an account that pays 5% per year equals (0.05)(q/2).

Annualized cost

Over the course of a 12 month year, Mr. T withdraws a total of $24,000 from his account. He withdraws q dollars at each visit to the bank, so the number of visits he makes to the bank per year equals 24,000/q, and the cost to him of each visit equals $15. Thus, Mr. T's aggregate annualized cost C(q) of withdrawing q dollars at each visit is given by

$$(4) \qquad C(q) = \frac{(24,000)\,(15)}{q} + \frac{(q)\,(0.05)}{2}\,.$$

As q increases, the number of visits to the bank decreases, but the opportunity cost of the cash that is not earning interest increases. A trade-off exists.

Inventory control

Problem 7.I becomes a classic problem in inventory control if the symbols A, K and H are introduced, where

A = the annual demand for an item,
H = the opportunity cost of keeping one item in inventory for one year,
K = the cost of placing each order.

Here, the demand for a product is assumed to occur at a time-invariant rate, with total demand of A units per year. The numbers A, K and H are assumed to be positive. It is optimal to place an order only when the inventory is reduced to 0. The annualized cost C(q) of ordering q units each time the inventory decreases to 0 is given by the analogue of expression (4), which is

$$(5) \qquad C(q) = \frac{A\,K}{q} + \frac{q\,H}{2}\,.$$

The EOQ

Finding the optimal order quantity q^* is an exercise in calculus. Differentiating C(q) with respect to q gives

(6) $$\frac{d}{dq}C(q) = \frac{-A\,K}{q^2} + \frac{H}{2}.$$

When q is small, the derivative is negative. As q increases, the derivative increases. As q become very large, the derivative approaches the positive number, H/2. The optimal order quantity q^* is the unique value of q for which the derivative equals 0. Equating to 0 the RHS of equation (6) produces

(7) $$q^* = \sqrt{\frac{2A\,K}{H}}.$$

The number q^* given by (7) has been known for nearly a century as the **economic order quantity** (or **EOQ** for short).

Bank withdrawals

When particularized to Mr. T's cash management problem, the amount q^* to withdraw at each visit is given by

$$q^* = \sqrt{\frac{24{,}000 \times 15}{0.05}} = \$3{,}794,$$

and the number A/q^* of visits to the bank over the course of the year equals 6.32. Evidently, Mr. T is not troubled about having a large amount of cash in his pocket.

An economy of scale

It is easy to verify, by plugging the formula for q^* that is given by equation (7) into the expression for C(q), that

(8) $$C(q^*) = \sqrt{2AKH}.$$

If the annual demand doubles, equations (7) and (8) show that the economic order quantity q^* and the annualized cost $C(q^*)$ increase by the factor of $\sqrt{2}$, rather than by the factor of 2. This is the same sort of economy of scale that was exhibited by the base stock model.

A flat bottom

Algebraic manipulation of the expressions for C(q) and for C(q*) produces the equation

$$\text{(9)} \qquad \frac{C(q)}{C(q^*)} = \frac{1}{2}\left(\frac{q^*}{q} + \frac{q}{q^*}\right).$$

This ratio is easily seen to be a convex function of q. It is minimized by setting $q=q^*$, of course. For $q = q^*\sqrt{2}$ and for $q = q^*/\sqrt{2}$, the ratio in (9) equals $(3/4)\sqrt{2}$, and $(3/4)\sqrt{2} \cong 1.06$, which exceeds the minimum by only 6%. It is emphasized:

> *Flat bottom:* In the EOQ model, the annualized cost C(q) exceeds C(q*) by not more than 6% as q varies by a factor of 2, between 0.707 q* and 1.414 q*.

This "flat bottom" can be good news. An EOQ model can result from simplifying a situation that is somewhat more complex. Its flat bottom connotes that the simplification may have little impact on annualized cost.

A bit of the history

The EOQ model was developed in 1913 by F. W. Harris of the Westinghouse Corporation. It was widely studied two decades later by R. W. Wilson, and it is also known as the "Wilson lot size model."

The cash management problem (Problem 7.I) is an instance of the EOQ model. This instance is often referred to as the **Baumol/Tobin model**. William Baumol published a paper with this interpretation of the EOQ model in 1952. Independently, in 1956, James Tobin published a similar paper. Baumol and Tobin do have a joint paper on this model. In 1989, they pointed out that Maurice Allais had published this result in 1947.

12. EOQ with Uncertain Demand*

In the EOQ model, demand is assumed to occur at a fixed rate. In this section, that assumption is relaxed. The rate of demand of an item is now assumed to be uncertain, but with a probability distribution that is stable over the course of the year.

Stationary independent increments

The demand for a product is has **stationary independent increments** if the demands that occur in non-overlapping intervals of the same time length have the same distribution and are mutually independent.

Let us consider a product whose demand has stationary independent increments. Interpret D(t) as the demand for this product that occurs during a period of time that is t units in length. The means and the variances of independent random variables add, and it is not difficult to show that the expectation of D(t) and the variance of D(t) grow linearly with the length t of the period.

A replenishment interval

In the model that is under development, it is assumed that replenishment does not occur immediately – that it takes a fixed number k of days to fill each order. The demand D(k) that occurs during the replenishment interval is uncertain, but its mean and variance are assumed to be known. Let them be denoted as μ and σ^2, respectively:

$$\mu = \mathrm{E}[\mathrm{D}(\mathrm{k})], \qquad \sigma^2 = \mathrm{Var}[\mathrm{D}(\mathrm{k})].$$

In this model, the symbol A denotes the expectation of the total demand that occurs during a 365 day year. Because demand has stationary independent increments,

$$\mathrm{A} = \mu \times (365/\mathrm{k}).$$

Backorders

The demand D(k) that occurs during the replenishment interval can exceed the supply at the start of that period. When it does, a stock-out occurs. In the model that is under development, it is assumed that demands that cannot be met from inventory are **backordered**, that is, filled when the merchandise becomes available.

Costs

This model has three different types of cost:

- Each unit of demand that is backordered incurs a penalty b, which can include the loss of good will due to requiring the customer to wait for his or her order to be filled.
- Each unit that is held in inventory accrues an opportunity cost at the rate of H per year.
- Each order that is placed incurs a fixed ordering cost K that is independent of the size of the order.

Customers' demands must be satisfied, for which reason the per-unit purchase cost is independent of the ordering policy, hence can be omitted from the model.

The decision variables

For the model that has just been specified, it is reasonable – and it can be shown to be optimal – to employ an ordering policy that is determined by numbers r and q, where

- The number r is the **reorder point**. An order is placed at each moment at which the inventory position decreases to r.
- The number q is the **reorder quantity**. Each moment at which the inventory position is reduced to r, an order is placed for q units.

In this context, the quantity $r - E[D(k)] = r - \mu$ is the *safety stock.*

In general, the expectation of the amount by which inventory is depleted between orders is called the **cycle stock**. In this model, q is the cycle stock. On average, all of the safety stock and half of the cycle stock will be on hand. The average inventory position is $(r - \mu + q/2)$, and the annualized inventory carrying cost is given by

$$(r - \mu + q/2)H. \tag{10}$$

The ordering cost

The number of orders placed per year is uncertain, but the average number of orders placed per year equals the ratio A/q of the expected annual demand A to the size q of the order. Each order incurs cost K, and the expected annualized ordering cost is given by the (familiar) expression

$$KA/q. \tag{11}$$

The backorder cost

The number of units backordered at the moment before the order is filled equals the excess $[D(k) - r]^+$ of the demand during the k-day period over the stock level r at the moment the order is placed. Each unit that is backordered incurs a cost that equals b, and the expected number of orders placed per year equals A/q. Hence, the expectation of the annualized cost of backorders is given by

$$(A/q)(b)E\{[D(k) - r]^+\}. \tag{12}$$

The optimization problem is to select values of q and r that minimize the sum of the expressions (10), (11) and (12).

A cash management problem

The EOQ model with uncertain demand is illustrated in the context of

Problem 7.J *(more cash management).* Rachael is away at college. She and her mom have established a joint account whose sole use is to pay for Rachael's miscellaneous expenses. Rachael charges these expenses on a debit card. The bank debits withdrawals from this account immediately, and the bank credits deposits to this account 16 days after they are made. This account pays no interest, and it charges a penalty of $3 per dollar of overdraft. Rachael's miscellaneous expenses have stationary independent increments, and the amount of miscellaneous expense that she incurs during each 16-day period is approximately normal with a mean of $160 and a standard deviation of $32. Rachael and her mom practice inventory control. When the balance in Rachael's account is reduced to r dollars, she phones home to request that a deposit of q dollars be credited to this account. Her mother attends to this immediately. The transfer takes her mom 30 minutes, and she values her time at $30 per hour. Rachael's mom transfers this money from a conservative investment account that returns 5% simple interest per year. What values of r and q do Rachael and her mom choose?

A spreadsheet

The spreadsheet in Table 7.14 presents their cash management problem for solution as a nonlinear program. In this spreadsheet, cell H3 contains the value of q, and cell I3 contains the value of r. The functions in cells D6, D7 and D8 evaluate expressions (10), (11) and (12), respectively.

Table 7.14 reports the optimal solution that was found by Solver's GRG code. It had been asked to minimize the number in cell D9 (total annualized expense) with changing cells H3 and I3. As mentioned earlier, the GRG code works best when it is initialized with reasonable values of the decision variables in the changing cells. This run of Premium Solver was initialized with the EOQ (roughly 1500) in cell H3 and with 160 (the mean demand during the replenishment interval) in cell I3. Solver reports a optimal order quantity $q^* = 1501$ and a reorder point $r^* = 239$, which provides a safety stock of $79 = 239 - 160$.

Table 7.14. Rachael's cash management problem.

	A	B	C	D	E	F	G	H	I	J
1										
2		A	K	H	b	μ	σ	q	r	
3		3650	15	0.05	3	160	32	1501	238.9	
4										
5		type of cost		$						
6		inventory		41.48	←	= D3 * (I3 - F3 + H3/2)				
7		ordering		36.47	←	= B3 * C3/H3				
8		backorder		0.522	←	= (B3/H3) * E3 * nl(I3, F3, G3)				
9		total		78.47	←	= SUM(D6:D8)				

Rachael's account is replenished about twice a year. The reorder point is almost 2.5 standard deviations above the mean demand during the replenishment interval because $(r - \mu)/\sigma = (239 - 160)/32 = 2.49$. This safety factor guarantees that Rachael and he mom rarely fall prey to overdraft fees.

For a variant of Rachael's cash management problem in which her mom replenishes her account at fixed intervals, with some uncertainty in the replenishment amount, see Problem 15 at the end of this chapter.

13. Review

Optimal solutions to the ten constrained optimization problems in this chapter have been found by the simplex method and its generalizations. Only two of these problems are linear programs. Of the others, some have objectives and constraints that are nonlinear, and some have decision variables that must be integer-valued.

If an optimization problem has some integer-valued variables, strive for a formulation that becomes a linear program when the integrality requirements are omitted. That allows you to use the "Standard LP Simplex" code, which is faster and more likely to find a global optimum than is the "Standard GRG Solver." To require decision variables to be binary or to be integer-valued, use the "bin" or "int" feature of the Add Constraints dialog box.

The GRG Solver works best if you can initialize it with values of the decision variables that are reasonably close to the optimum. Tips on getting good results with it can be found in Section 11 of Chapter 20. Look there if you are having trouble.

It might have seemed, at first glance, that the simplex method and its generalizations apply solely to optimization problems that are deterministic. That is not so. Uncertainty plays a central role in several of the examples in this chapter, and that is true of other chapters as well.

14. Homework and Discussion Problems

1. *(Olde England)* Write down the linear program whose spreadsheet formulation is presented as Table 7.2.

2. *(Olde England)* At a cabinet meeting, The Minister of the Interior expressed her conviction that Olde England should be self-sufficient as concerns food production. What would this imply?

3. *(efficient portfolios)* Redo part (a) of Problem 7.C with MAD as the measure of risk. Compare your results with those in Table 7.5.

4. *(efficient portfolios)* Redo part (b) of Problem 7.C with MAD as the measure of risk. Compare your results with those in Table 7.6.

5. *(decreasing marginal cost)* In Problem 7.D, the price of \$3,000 proved to be high enough that leasing the tool that increases the contribution of the Luxury model vehicle from \$1200 to \$1500 was unprofitable. Is there a price at which it becomes profitable? If so, what is that price? If not, why not?

6. *(decreasing marginal cost)* Does the integer program in Table 7.7 introduce unintended options that are ruled out by optimization? If so, what are they?

7. *(decreasing marginal cost)* In Problem 7.D, the constraints $F2 \le 40a$ and $a \in \{0, 1\}$ place an upper bound of 40 on the decision variable F2. Is this justifiable? If so, why?

8. *(decreasing marginal cost)* In Problem 7.D, consider a formulation in which the constraint $F2 \le 40a$ is replaced with the constraint $F1 \le 15(1-a)$.

   (a) Does this work? If so, why?

   (b) Is there a situation in which this type of formulation is preferable to the type used in Table 7.7. What is it?

9. *(college admissions)* In Problem 7.F, you, as Dean of Admissions, place some number w of non-athletes whom you did not admit on the wait list. If the yield from the athletes and the regular admits is below 510, you offer admission to persons on the wait list one by one in an attempt to fill your quota – to achieve a freshman class of 510 students. Each person who is offered a place on the wait list will join the class with probability of 0.32 if he or she is later offered admission. You are willing to run one chance in 20 of ending up with fewer than 510 freshmen.

   (a) How many people should you place on the wait list?

   (b) With what probability does your admissions policy produce a freshman class that contains precisely 510 persons?

   (c) What is the expected number of vacant positions in next year's Freshman class.

10. *(college admissions)* In Problem 7.F, rework the spreadsheet to account for the suggestions in the subsection entitled "Fine Tuning." How may offers of admission will be made?

11. *(a power plant)* In Problem 7.G, by how much does expected annual cost increase if peak-load plant is eliminated? *Hint:* You might re-optimize with C7 as the only decision variable and with the function =**C7** in cell C8.

12. *(a power plant)* In Problem 7.G, is the transfer price of 6.20 $/unit large enough to motivate the utility to satisfy at least 98% of the power its customers demand with its own production capacity"? If not, how much larger does the transfer price need to be? *Hint:* You might wish to optimize with a variety of values in cell E4.

13. *(a power plant)* In Problem 7.G, suppose that base-load plant emits 1 unit of carbon per unit of electricity produced and that peak-load plant emits 2 units of carbon per unit of electricity produced. How large a tax on carbon is needed to motivate the utility to produce no electricity with peak-load plant? What impact would this tax have on the utility's expected annual cost?

14. *(Rachael's cash management problem)* For the data in Table 7.14:

    (a) What is the safety stock? With what probability does Rachael incurs an overdraft prior to replenishment of the account?

    (b) As the reorder point q is varied, does the annualized cost continue to display a "flat bottom?" akin to that of the EOQ model?

    (c) Suppose that both the mean *and* the standard deviation of D(16) were doubled, from 160 and 32 to 320 and 64. Does the optimal solution display an economy of scale akin to that of the EOQ model?

15. *(Rachael, yet again)*. This problem has the same data as in Problem 7.J. Rachael's mom has found it inconvenient to supply cash at uncertain times. She would prefer to supply uncertain amounts of cash at pre-determined times. Rachael and her mom have revised the structure of the cash management policy. Every t days, Rachael requests the amount needed to raise her current bank balance to x dollars.

    (a) Rachael's miscellaneous expense has stationary independent increments, and the demand D(16) is normal with mean of 160 dollars and standard deviation of 32 dollars. As a consequence, the demand D(z) during z days is normal with mean equal to αz and standard deviation equal to $\beta\sqrt{z}$. What are α and β?

(b) A deposit is credited to Rachael's account 16 days after it is made. What can you say about the balance in her account the moment *after* the check is credited to it?

(c) What can you say about the balance in her account at the moment *before* the deposit is credited to it?

(d) On a spreadsheet, compute the values of t and x that minimize the expected annualized cost of maintaining this account.

(e) What is the probability distribution of the amount of cash that Rachael's mom transfers to her account?

(f) What would happen to the expected annualized cost of this account if Rachael's mom made a deposit every six months.

16. In a winter month, an oil refinery has contracted to supply 550,000 barrels of gasoline, 700,000 barrels of heating oil and 240,000 barrels of jet fuel. It can purchase light crude at a cost of $60 per barrel and heavy crude at a cost of $45 per barrel. Each barrel of light crude it refines produces 0.35 barrels of gasoline, 0.35 barrels of heating oil and 0,15 barrels of jet fuel. Each barrel of heavy crude it refines produces 0.25 barrels of gasoline, 0.4 barrels of heating oil and 0.15 barrels of jet fuel.

(a) Formulate and solve a linear program that satisfies the refinery's contracts at least cost.

(b) Does this refinery meet its demands for gasoline, heating oil and aviation fuel exactly? If not, why not?

17. *(a staffing problem)* Police officers work for 8 consecutive hours. They are paid a bonus of 25% above their normal pay for work between 10 pm and 6 am. The demand for police officers varies with the time of day, as indicated below:

period	minimum
2 am to 6 am	12
6 am to 10 am	20
10 am to 2 pm	18
2 pm to 6 pm	24
6 pm to 10 pm	29
10 pm to 2 am	18

The goal is to minimize the payroll expense while satisfying or exceeding the minimal staffing requirement in each period.

(a) Formulate this optimization problem for solution by Solver or by Premium Solver. Solve it.

(b) It is not necessary in part (a) to require that the decision variables be integer-valued. Explain why. *Hint:* it is relevant that 6 is an even number.

18. *(a traveling salesperson).* The spreadsheet that appears below specifies the driving times in minutes between seven state capitals. Suppose that you are currently at one of these capitals and that you wish to drive to the other six and return where you started, spending as little time on the road as possible.

(a) Formulate and solve an assignment problem akin to the one in the chapter. Its optimal solution will include some number k of subtours.

(b) Append to your assignment problem k subtour elimination constraints. Solve it again. Did you get a tour? If so, explain why that tour is optimal.

	A	B	C	D	E	F	G	H	I
1									
2									
3			Alb	Aug	Bur	Con	Har	Pro	Spr
4		Albany NY	5000	325	188	200	116	167	91
5		Augusta ME	325	5000	314	169	261	232	253
6		Burlington VT	188	314	5000	153	233	273	206
7		Concord NH	200	169	153	5000	159	239	145
8		Hartford CT	116	261	233	159	5000	103	30
9		Providence RI	167	232	273	239	103	5000	94
10		Springfield MA	91	253	206	145	30	94	5000

19. *(departure gates)* As the schedule setter for an airline, you must schedule exactly one early-morning departure from Pittsburgh to each of four cities. Due to competition, the contribution earned by each flight depends on its departure time, as indicated below. For instance, the most profitable departure time for O'Hare is at 7:30 am. Your airline has permission to schedule these four departures at any time between 7 am and 8 am, but you have only two departure gates, and you cannot schedule more than two departures in any half-hour interval.

Time	Newark	O'Hare	Logan	National
7:00 am	8.2	7.0	5.6	9.5
7:30 am	7.8	8.2	4.4	8.8
8:00 am	6.9	7.8	3.1	7.0

Contribution per flight, in thousands of dollars

(a) Formulate the problem of maximizing contribution as an integer program.

(b) Solve the integer program you formulated in part (a).

(c) Another airline wishes to rent one departure gate for the 7:00 am time. What is the smallest rent that would be profitable for you to charge?

20. *(redistricting a state)* A small state is comprised of ten counties. In the most recent reapportionment of the House of Representatives, this state has been allocated three seats (Congressional Districts). By longstanding agreement between the parties, no county can be split between two Congressional Districts. Each Congressional District must represent between 520 and 630 thousand persons. The Governor wishes to assign each county to a Congressional District in a way that maximizes the number of districts in which registered Democrats are at least 52% of the population. Rows 3 and 4 of the spreadsheet that appears below list the population of each district and the number of registered Democrats in it, both in thousands.

	A	B	C	D	E	F	G	H	I	J	K	L
1												
2		county	1	2	3	4	5	6	7	8	9	10
3		population	226	218	76	75	234	255	168	155	47	274
4		Democrats	126	129	44	25	122	96	128	114	31	112
5												
6		district 1	1	0	0	0	0	0	0	1	0	1
7		district 2	0	0	1	1	1	0	1	0	0	0
8		district 3	0	1	0	0	0	1	0	0	1	0

(a) Assigning each county to a district can be accomplished as follows: In cell C8, enter the function **=1 – C6 – C7** and drag that function across row 8 as far as cell L8. Require that the decision variables in cells C6:L7 be binary, and require that the numbers in cells C8:L8 be nonnegative. Why does this work?

(b) Denote as T(i) the total population of district i, and denote as D(i) the number of registered Democrats in district i. Use Solver or Premium Solver to compute T(i) and D(i) for each district, to enforce the constraints $520 \le T(i) \le 630$ for each i, to enforce the constraints

$$0.52\ T(i) \le D(i) + 630[1 - f(i)] \quad \text{and } f(i) \text{ binary}$$

for each i, and to maximize the number of districts in which at least 52% of the population is registered Democrats. Discuss its optimal solution.

(c) Is the optimization problem that you devised an integer linear program, or is it an integer nonlinear program?

21. *(A perfume counter)* Your company's perfume counter in a chi-chi department store is restocked weekly. You sell three varieties of perfume in this store. Storage space is scarce. You have room for 120 bottles. The demand weekly demand for each type of perfume you offer is approximately normal, and the demands for different types of perfume are mutually independent. The table that appears below specifies the mean and standard deviation of each demand, the profit (contribution) from each sale you make, and the loss of good will from each sale you are unable to make. Your goal is to maximize the excess of the expected sales revenue over the expected loss of good will. How many bottles of each type of perfume

type of perfume	A	B	C
expected demand	50	45	30
standard deviation of demand	15	12	10
Contribution	$30	$43	$50
loss of good will	$20	$30	$40

should you stock?

22. *(A perfume counter, continued)* The chi-chi department store in the preceding problem is open from Monday through Saturday each week. The demand that occurs for a particular type of perfume is not dependent on

the day of the week, and demands on different days are independent of each other. Your supplier had been resupplying each Thursday evening, after the store closes. For an extra fee of $350, your supplier has offered to resupply a second time each week, after the close of business on Monday. How many bottles of each type of perfume should you stock with resupply on a twice-a-week basis? Is it worthwhile to do so?

Chapter 8: Path Length Problems and Dynamic Programming

1. Preview

This is the first of a pair of chapters that deal with optimization problems on "directed networks." This chapter is focused on path-length problems, the next on network-flow problems. Path-length problems are ubiquitous. A variety of path-length problems will be posted in this chapter. They will be solved by linear programming and, where appropriate, by other methods.

The phrases "linear programming" and "dynamic programming" are similar, but they have radically different meanings. Dynamic programming is an ensemble of concepts that can be used to analyze decision problems that unfold over time. Dynamic programming plays a vital role in fields as diverse as macroeconomics, operations management, and control theory. Path-length problems are the ideal environment in which to introduce the subject.

E. V. Denardo, *Linear Programming and Generalizations*, International Series in Operations Research & Management Science 149,
DOI 10.1007/978-1-4419-6491-5_8, © Springer Science+Business Media, LLC 2011

2. Terminology

The network optimization problems in this chapter and the next employ terminology that is introduced in this section. Most of these definitions are easy to remember because they are suggested by normal English usage.

A "directed network" consists of "nodes" and "directed arcs." Figure 8.1 depicts a directed network that has 5 nodes and 7 directed arcs. Each node is represented as a circle with an identifying label inside, and each directed arc is represented as a line segment that connects two nodes, with an arrow pointing from one node to the other.

Figure 8.1. A directed network.

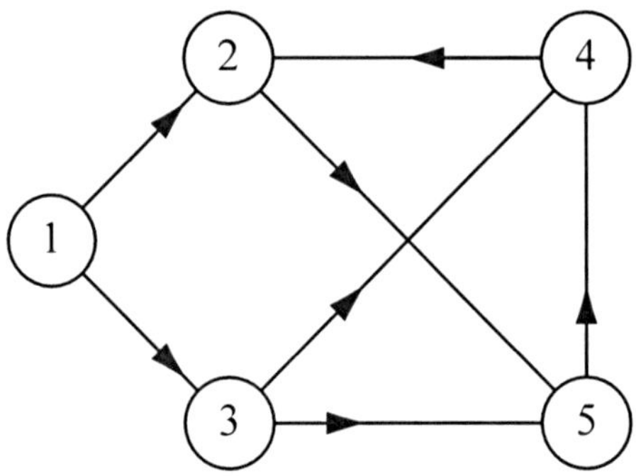

In general, a **directed network** consists of a finite set N and a finite set A each of whose members is an ordered pair of elements of N. Each member of N is called a **node**, and each member of A is called a **directed arc**. The directed network in Figure 8.1 has

$$N = \{1, 2, 3, 4, 5\},$$

$$A = \{(1, 2), (1, 3), (2, 5), (3, 4), (3, 5), (4, 2), (5, 4)\}.$$

None of the optimization problems discussed in this book entail "undirected" networks, whose arcs lack arrows. For that reason, "directed network" is sometimes abbreviated to **network**, and "directed arc" is sometimes abbreviated to **arc**.

Paths

Directed arc (i, j) is said to have node i as its **tail** and node j as its **head**. A **path** is a sequence of n directed arcs with $n \geq 1$ and with the property that the head of each arc other than the nth is the tail of the next. A path is said to be **from** the tail of its initial arc **to** the head of its final arc. In Figure 8.1, the arc (2, 5) is a path from node 2 to node 5, and the arc sequence {(2, 5), (5, 4)} is a path from node 2 to node 4.

To interpret a path, imagine that when you are at node i, you can *walk* across any arc whose tail is node i. Walking across arc (i, j) places you at node j, at which point you can walk across any arc whose tail is node j. In this context, any *sequence* of arcs that can be walked across is a path.

Cycles

A path from a node to itself is called a **cycle**. In Figure 8.1, the path {(2, 5), (5, 4), (4, 2)} from node 2 to itself is a cycle, for instance. A path from node j to itself is said to be a **simple cycle** if node j is visited exactly twice and if no other node is visited more than once.

A directed network is said to be **cyclic** if it contains at least one cycle. A directed network that contains no cycles is said to be **acyclic**. The network in Figure 8.1 is cyclic. This network would become acyclic if arcs (5, 4) and (4, 2) were removed or reversed.

Trees

A set T of directed arcs is said to be a **tree from** node i if T contains exactly one path from node i to each node $j \neq i$. The network in Figure 8.1 has several trees from node 1 to the others; one of these trees is the set T = {(1, 2), (1, 3), (2, 5), (3, 4)}. Similarly, a set T of directed arcs is said to be a **tree to** node j if T contains exactly one path from each node i other than j to node j. The network in Figure 8.1 has several trees to node 4, including T = {(1, 2), (2, 5), (5, 4), (3, 4)}. No tree can contain a cycle.

Arc lengths

Let us consider a directed network in which each arc (i, j) has a datum c(i, j) that is dubbed the **length** of arc (i, j). Figure 8.2 depicts a network that has 8 nodes and 16 directed arcs. The length of each arc is adjacent to it. Four of these so-called "lengths" are negative. In particular, c(5, 7) = – 5.4.

Figure 8.2. A directed network whose arcs have lengths.

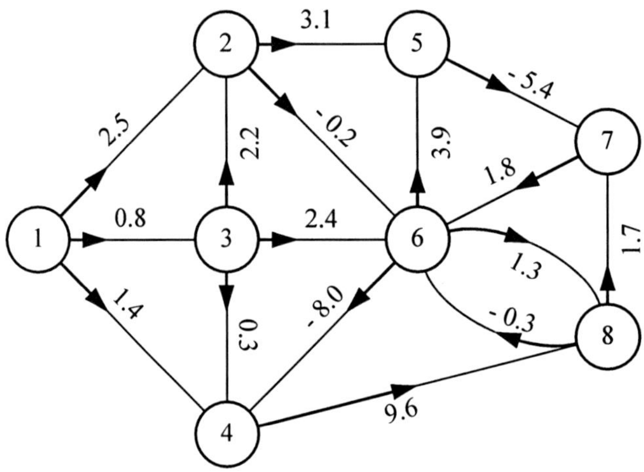

Path lengths

The **length** of each path is normally taken to be the sum of the lengths of its arcs. In Figure 8.2, for instance, the path {(1, 2), (2, 6)} has length $2.3 = 2.5 + (-0.2)$. Also, the path {{6, 8), (8, 6), (6, 8)} has length $2.3 = 1.3 - 0.3 + 1.3$.

Path-length problems

A directed network can have many paths from node i to node j. The *shortest path* problem is that of finding a path from a given node i to a given node j whose length is smallest. The *longest path* problem is that of finding a path from given node i to a given node j whose length is longest. Path-length problems are important in themselves, and they arise as components of other optimization problems. Solution methods for path-length problems are introduced in the context of

Problem 8.A. For the directed network depicted in Figure 8.2, find the shortest path from node 1 to node 8.

Problem 8.A can be solved by trial-and-error. The shortest path from node 1 to node 8 follows the node sequence (1, 2, 5, 7, 6, 8) and has length of $3.3 = 2.5 + 3.1 - 5.4 + 1.8 + 1.3$.

3. Elements of Dynamic Programming

Having solved Problem 8.A by trial and error, we will now use it to introduce a potent group of ideas that are known, collectively, as "dynamic programming." These ideas will lead us to a variety of ways of solving Problem 8.A, and they have a myriad of other uses.

States

In dynamic programming, a **state** is a summary of what has transpired so far that contains enough detail about the past to enable rational decisions about what to do now and in the future. Implicit in the idea of a state are:

- *A sense of time.* This sense of time may be an artifice. For our shortest-route problem, think of each transition from node to node as taking an amount of time that is indeterminate and immaterial, but positive.
- *A measure of performance*: In Problem 8.A, it is rational to choose the shortest path to node 8.
- *A notion of parsimony*: A summary that includes less information about the past is preferable.

For our shortest-path problem, the only piece of information that needs to be included the state is the node i that we are currently at. How we got to node i doesn't matter; we seek the shortest path from node i to node 8.

Embedding

Dynamic programming begins by taking what may seem to be a large step backwards. Rather than attacking the problem directly, it is **embedded** in a family of related problems, one per state. For our shortest-path problem, we elect to find the shortest path from *each* node i to node 8. To this end, we denote f(i) by

$$f(i) = \text{the length of the shortest route from node i to node 8.}$$

(A choice as to embedding has just been made; it would work equally well to find, for each node j, the length F(j) of the shortest path from node 1 to node j.)

Linking

The optimization problem with which we began has now been replaced with a family of optimization problems, one per state. Members of this family are closely related in a way that will make them easy to solve. For the shortest-path problem at hand, each arc (i, j) in Figure 8.2 establishes the relationship,

$$(1) \qquad f(i) \leq c(i, j) + f(j) \qquad \text{for each arc } (i, j),$$

because $c(i, j) + f(j)$ is the length of some path from node i to node 8, and f(i) is the length of the shortest such path. Moreover, with $f(8) = 0$,

$$(2) \qquad f(i) = \min_j \{c(i, j) + f(j)\} \qquad \text{for } i = 1, 2, \ldots, 7.$$

Equation (2) holds because the shortest path from node i to node 8 has as its first arc (i, j) for some node j and its remaining arcs form the shortest path from node j to node 8. Expression (2) **links** the optimization problems for the various starting states.

In the jargon of dynamic programming, equation (2) is called an **optimality equation** because it links the solutions to a family of optimization problems, one per state. In our example and in many others, the easy way to compute f(i) for a particular state i is to use the optimality equation to compute f(j) for every state j.

4. Shortest Paths via Linear Programming

Imagine, for the moment, that correct numerical values have been assigned to f(2) through f(8). The value of f(1) that satisfies (2) is the *largest* value of f(1) that satisfies the inequalities in (1) for the arcs that have node 1 as their tail.

To compute f(1), our original goal, it would suffice to maximize f(1) subject to the constraints in system (1). This would give the correct f-value for each node on the shortest path from node 1 to node 8, but it might given incorrect f-values for the others. A linear program that gives the correct f-value for every node is

Program 8.1. Maximize {f(1) + f(2) + … + f(7)} subject to f(8) = 0 and the constraints in system (1)

A standard-format representation of Program 8.1 appears as the spreadsheet in Table 8.1. In particular, the function in cell E5 and the constraint E5 ≥ 0. implement the constraint $c(1, 2) + f(2) - f(1) \geq 0$.

Table 8.1. The optimal solution to Program 8.1: Solver has maximized the value in cell E21 with F22:L22 as changing cells and with constraints E5:E20 >= 0.

	A	B	C	D	E	F	G	H	I	J	K	L	M
1													
2					= C5 + SUMPRODUCT(F5:M5 , F$22:M$22)								
3									node i				
4		arc	length			1	2	3	4	5	6	7	8
5		(1, 2)	2.5		0	-1	1						
6		(1, 3)	0.8		0.5	-1		1					
7		(1, 4)	1.4		7.7	-1			1				
8		(2, 5)	3.1		0		-1			1			
9		(2. 6)	-0.2		0.3		-1				1		
10		(3, 2)	2.2		0		1	-1					
11		(3, 4)	0.3		6.9			-1	1				
12		(3, 6)	2.4		0.7			-1			1		
13		(4, 8)	9.6		0				-1				1
14		(5, 7)	-5.4		0					-1		1	
15		(6, 4)	-8		0.3				1		-1		
16		(6, 5)	3.9		0.3					1	-1		
17		(6, 8)	1.3		0						-1		1
18		(7, 6)	1.8		0						1	-1	
19		(8, 6)	-0.3		1						1		-1
20		(8, 7)	1.7		4.8							1	-1
21					18.8								
22					f(i) =	3.3	0.8	3.0	9.6	-2.3	1.3	3.1	0
23													
24			= SUM(F22 : L22)										

Recorded in Table 8.1 is the optimal solution that Solver has found. The seven arcs whose constraints are binding have been shaded. These arcs form a *tree* of shortest paths to node 8. This tree is displayed in Figure 8.3, as is the length f(i) of the shortest path from each node i to node 8.

Figure 8.3. Shortest paths to node 8, with the length of each.

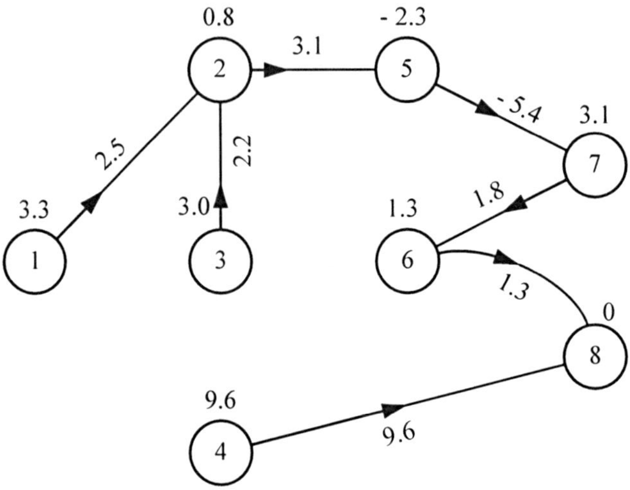

5. The Principle of Optimality

Nearly all of the elements of dynamic programming have been introduced, but the most elusive element has not. It is known as the "principle of optimality." It will be presented in the context of Problem 8.A.

A preliminary definition is needed. In the lingo of dynamic programming, a **policy** is any rule that picks an admissible decision for each state. The states in our formulation of Problem 8.A are the nodes 1 through 7, and a policy is any rule that assigns to each node i (with $i \neq 8$) an arc whose tail is node i. One such policy is depicted in Figure 8.3. This policy assigns arc (1, 2) to node 1, arc (2, 5) to node 2, and so forth.

To **use** a particular policy is to begin at whatever state one is placed and, for each state one encounters, to choose the decision (or action) that this policy specifies for that state. A policy is said to be optimal for state i if no other policy is preferable, given state i as the starting state. A policy is said to be **optimal** if it is optimal, simultaneously, for every starting state. The policy depicted in Figure 8.3 *is* optimal; its use provides the shortest path from *each* node i to node 8.

Version #1

The principle of optimality exists in several versions, three of which are presented in this section. The first of these is the

> **Principle of optimality (1st version).** There exists a policy that is optimal, simultaneously, for every starting state.

Figure 8.3 illustrates this version of the principle of optimality – it exhibits a policy whose use prescribes a shortest path from each node i to node 8.

Evidently, dynamic programming describes a family of optimization problems in which there is *no tradeoff* between starting states; in order to do best for one starting state, it is not necessary to do less than the best for another.

Version #2

Before discussing a different version of the principle of optimality, we pause to write a path as a sequence of nodes (rather than as a sequence of arcs) like so: The node sequence $(i_0, i_1, \ldots, i_n)$ is a **path** from node i_0 to node i_n if (i_{i-1}, i_i) is a directed arc for $i=0, 1, \ldots, n-1$. For any integers p and q that satisfy $0 \le p < q \le n$, this path is said to have path $(i_p, i_{p+1}, \ldots, i_q)$ as a **subpath**. In Figure 8.3, path (2, 5, 7, 6) has subpath (5, 7, 6), for instance. A version of the principle of optimality that is keyed to path-length problems appears below as the

> **Principle of optimality (2nd version).** Consider an optimal path from some node to some other node. Each subpath $(i_p, \ldots, i_q)$ of this path is an optimal path from node i_p to node i_q.

We have made use of the 2nd version! Please pause to convince yourself that equation (2) does so.

Version #3

The 3rd version of the principle of optimality rests on the notion of a **cycle of events** – observe a state, select a decision, wait for transition to occur to a new state, observe that state, and repeat. This version is the

> **Principle of optimality (3rd version).** An optimal policy has the property that whatever the initial state is and no matter what decision is selected initially, the remaining decisions in the optimal policy are optimal for the state that results from the first transition.

Problem 8.A illustrates the 3rd version as well. This version states, for instance, that if one begins at node 3 and chooses any arc whose tail is node 3, the remaining arcs in the optimal policy are optimal for the node to which transition occurs. The 3rd version a verbal counterpart of the optimality equation.

The 1st version of the principle of optimality can be stated as a mathematical theorem. The 2nd version is particular to path-length problems. The 3rd version is the traditional one, and it is due to Richard Bellman.

Recap

Dynamic programming is an ensemble of related thought processes, which are to:

- Identify the states, each of which is a summary of what's happened to date that suffices to make rational decisions now and in the future.
- Embed the problem of interest in a family of related problems, one per starting state.
- Link these problems through an optimality equation.
- Solve the optimality equation and thereby obtain an optimal policy, e.g., a decision procedure that performs as well as possible, simultaneously, for every starting state.
- Use the principle of optimality to verbalize the optimality equation and the type of policy it identifies.

A linear program has been used to find an optimal policy. This illustrates a link between linear and dynamic programming. Do there exist dynamic programming problems whose optimal policies cannot be found by solving linear programs? Yes, there do, but they are rare.

A bit of the history

At The RAND Corporation in the fall of 1950, Richard E. Bellman (1920-1984) was asked to investigate the mathematics of multi-stage decision processes. He quickly observed common features in an enormous variety of optimization problems and coined the language of dynamic programming. Bellman used **functional equation** in place of *optimality equation*; his term is snazzier, but more mysterious.

Bellman used the methods he had devised to solve hundreds of seemingly-different problems a variety of fields – including control theory, economics, mathematics, operations research, medicine, and physics. His many papers and his many books[1] spanned a myriad of applications, launched a thousand research careers, and helped awaken the academic community to the importance of problem-based (i.e., applied) mathematics.

On page 159, of his autobiography[2], Bellman reports that he dubbed his approach *dynamic programming* to mask its ties to mathematical research, a subject he reports to have been anathema to Charles E. Wilson, who as Secretary of Defense from 1953 to 1957 was the person to whom The RAND Corporation reported.

Cycles and their lengths

The directed network in Figure 8.2 is cyclic, which is to say that at least one of its paths is a cycle. The node sequence (5, 7, 6, 5) describes a cycle whose length equals $0.3 = -5.4 + 1.8 + 3.9$. This network has several cycles, but it has no cycle whose length is negative.

In fact, if this network did have a cycle whose length were negative, the shortest-path problem would be ill-defined: There would be no shortest path from node 1 to node 8 because a path from node 1 to node 8 could repeat this (negative) cycle any number of times en route. By the way, if the network in Figure 8.2 did have a negative cycle, Program 8.1 would be infeasible.

The longest-path problem

What about the longest path from node 1 to node 8? That problem is ill-defined because a path from node 1 to 8 can repeat the cycle (5, 7, 6, 5) an arbitrarily large number of times.

You might wonder, as have many others, whether it might be easy to find the longest path from one node to another that contains no cycle. It isn't. That is equivalent to the "traveling salesman problem," which is to say that it is NP-complete. (No polynomial algorithm is known to solve it, and if you did find an algorithm that solves it for all data sets, you would have proved that

---

[1] Richard Bellman's books include the classic, *Dynamic Programming*, Princeton University Press, 1957, reprinted by Dover Publications, 2003.

[2] Richard Bellman, *Eye of the Hurricane: an Autobiography*, World Scientific Publishing Co, Singapore, 1984.

P=NP,) This is one case – amongst many – in which one of a pair of closely-related problems is easy to solve, and the other is not.

6. Shortest Paths via Reaching

Linear programming is one way to solve a shortest-path problem. Linear programming works when the network has no cycle whose length is negative. A method that we call "reaching" is presented in this section. Reaching works when the arc lengths are nonnegative. Reaching is faster. It will be introduced in the context of

Problem 8.B. For the network in Figure 8.4, find the tree of shortest paths from node 1 to all others.

Figure 8.4. A network whose arc lengths are nonnegative.

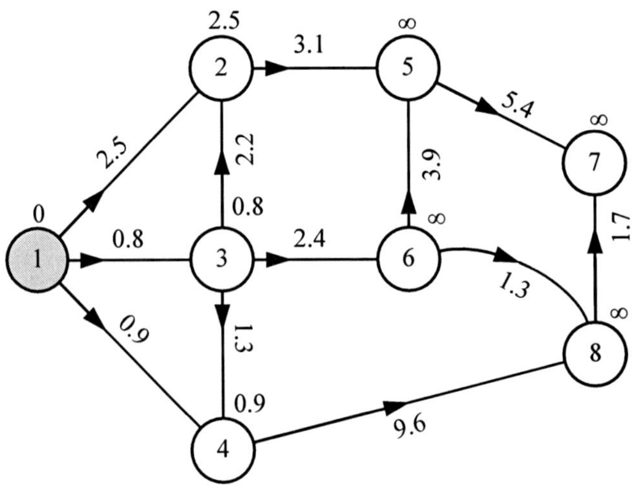

Reaching

All arc lengths in Figure 8.4 are nonnegative. Figure 8.4 hints at the algorithm that is about to be introduced. This algorithm is initialized with $v(1) = 0$ and with $v(j) = +\infty$ for each $j \neq i$. Initially, each node is unshaded. The general step is to select an unshaded node i whose label is *smallest* (node 1 initially) and execute the

Reaching step: Shade node i. Then, for each arc (i, j) whose tail is node i, update v(j) by setting

$$v(j) \leftarrow \min\{v(j),\ v(i) + c(i,\ j)\}. \tag{3}$$

Figure 8.4 describes the result of the first application of the Reaching step. Node 1 has been shaded, and the labels of nodes 2, 3 and 4 have been reduced to 2.5, 0.8 and 0.9, respectively. Evidently, there is a path from node 1 to node 3 whose length equals 0.8. The fact that arc lengths are nonnegative guarantees that all other paths from node 1 to node 3 have lengths of 0.9 or more. As a consequence, node 3 has v(3) = f(3) = 0.8. The second iteration of the reaching step will shade node 3 and will execute (3) for the arcs (3, 2), (3, 4) and (3, 6). This will not change v(2) or v(4), but it will reduce v(6) from $+\infty$ to 3.2.

The update in (3) "reaches" out from node i to update the labels for some unshaded nodes. After any number of executions of the Reaching step:

- If node j is shaded, its label v(j) equals the length of the shortest path from node 1 to node j.
- If node j is not shaded, its label v(j) equals the length of the shortest path from node 1 to node j whose final arc (i, j) has i shaded.

The fact that arc lengths are nonnegative suffices for an easy inductive proof of the properties that are highlighted above.

Recording the minimizer

As soon as a label v(j) becomes finite, it equals the length of some path from node 1 to node j. To build a shortest-path tree, augment the Reaching step to record at node j the arc (i, j) that reduced v(j) most recently.

E. W. Dijkstra

The algorithm that has just been sketched bears the name of its inventor. It is known as **Dijkstra's method**, after the justly-famous Dutch computer scientist, E. W. Dijkstra (1930-2002). Dijkstra is best known, perhaps, for his recommendation that the GOTO statement be abolished from all higher-level programming languages, i.e., from everything except machine code.

For large sparse networks, the most time-consuming part of Dijkstra's method is the selection of the unshaded node whose label is lowest. This can be accelerated by adroit use of a data structure that is known as a *heap*.

Reaching with buckets

If all arc lengths are positive, it is not necessary to pick the unshaded node whose label is smallest. Note in Figure 8.4 that:

- Each arc whose head and tail are unshaded has length of 1.3 or more.
- No unshaded node whose label is within 1.3 of the smallest can have its label reduced. In particular, since $v(4) = 0.9 \leq 0.8 + 1.3$, it must be that $v(4) = f(4)$.

Denote as m the length of the shortest arc whose head and tail are unshaded. (In Figure 8.4, m equals 1.3.) As just noted, each unshaded node j whose label v(j) is within m of the smallest has $v(j) = f(j)$. The unshaded nodes can be placed in a system of **buckets**, each of whose width m, where the pth bucket contains each unshaded node j having label v(j) that satisfies $pm \leq v(j) < (p+1)m$. The reaching step in (3) can be executed for each node i in the lowest-numbered nonempty bucket. After a bucket is emptied, it can be re-used, and a system of $1 + M/m$ buckets suffices, where M is the length of the longest arc whose head is unshaded.

Recap

Dijkstra's method works when the arc lengths are nonnegative. It works whether or not the network is cyclic. For large sparse networks, the time-consuming part of Dijkstra's method is determination of the unshaded node whose label is smallest. When the reaching step is executed for node i, the shortest path to node i is known; this can be used to *prune* the network of arcs that will not be needed to compute shortest paths to the rest of the nodes. If arc lengths are positive, reaching can be speeded up by the use of buckets. The uses of reaching with pruning and buckets are explored in a series of papers written with Bennett L. Fox[3].

---

[3] See, for instance, E. V. Denardo and B. L. Fox, "Shortest-route methods: 1. Reaching, pruning and buckets," *Operations Research*, 27, pp. 161-186, 1979.

7. Shortest Paths by Backwards Optimization

Reaching works if the arc lengths are nonnegative. If the network is acyclic, an even simpler method is available. That method is introduced in the context of

Problem 8.C. Find the tree of shortest paths to node 9 for the network that is depicted in Figure 8.5.

Figure 8.5. An acyclic network.

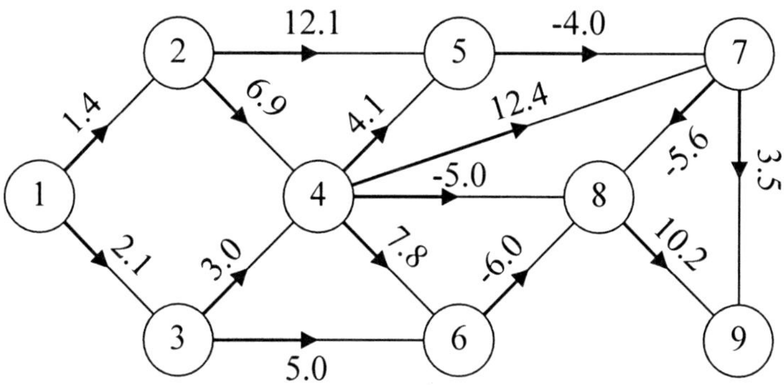

The network in Figure 8.5 has 9 nodes and 15 directed arcs. Each arc (i, j) in this network has $i < j$, which guarantees that the network is acyclic. Each arc (i, j) in this network also has a length c(i, j). Some arc lengths are negative, e.g., $c(5, 7) = -4.0$.

The optimality equation

The states for Problem 8.C are the integers 1 through 8, and f(i) denotes the length of the shortest path from node i to node 9. With $f(9) = 0$, the optimality equation takes the form,

$$f(i) = \min_j \{c(i, j) + f(j)\} \quad \text{for } i = 1, 2, \ldots, 8, \tag{4}$$

where it is understood that the minimum is to be taken over all j such that (i, j) is an arc.

Since the head of each arc has a higher number than the tail, this optimality equation is easy to solve by a method that is known as **backwards optimi-**

zation. This method solves (4) *backwards*, that is, in decreasing i. Backwards optimization is easily executed by hand. Doing so gives f(8) = 10.2, then

$$f(7) = \min \begin{cases} c(7,\ 8) + f(8) \\ c(7,\ 9) + f(9) \end{cases} = \min \begin{cases} -5.6 + 10.2 \\ \ \ 3.5 + 0 \end{cases} = 3.5,$$

then f(6) = − 6.0 + f(8) = 4.2, and so forth.

Spreadsheet computation

Backwards optimization can also be executed on a spreadsheet using Excel's "**offset function.**" This function identifies a cell that is *offset* from a specified cell by a given number of rows and by a given number of columns. To see how this works, we note that cell E14 of Table 8.2 contains the function

=D14 + OFFSET(H3, C14, 0)

Cell C14 contains the integer 7, so this function takes the sum of the number in cell D14 and the number that is in the cell that is 7 rows below cell H3 and 0 columns to the right of cell H3, thereby computing c(5, 7) + f(7).

Table 8.2. Backwards optimization on a spreadsheet.

	A	B	C	D	E	F	G	H	I
1									
2		arc (i, j)							
3		i	j	c(i, j)	c(i, j) + f(j)		j	f(j)	
4		1	2	1.4	11.9		1	8.7	
5		1	3	2.1	8.7		2	10.5	
6		2	4	6.9	10.5		3	6.6	
7		2	5	12.1	11.6		4	3.6	
8		3	4	3.0	6.6		5	-0.5	
9		3	6	5.0	9.2		6	4.2	
10		4	5	4.1	3.6		7	3.5	
11		4	6	7.8	12.0		8	10.2	
12		4	7	12.4	15.9		9	0	
13		4	8	-5.0	5.2				
14		5	7	-4.0	-0.5				
15		6	8	-6.0	4.2		= MIN(E10 : E13)		
16		7	8	-5.6	4.6				
17		7	9	3.5	3.5				
18		8	9	10.2	10.2				
19									
20					= D14 + OFFSET(H3, C14, 0)				

Dragging the function in cell E14 up and down the column computes c(i, j) + f(j) for each arc (i, j). The *min* functions in column H equate f(i) to the RHS of equation (4). The arcs that attain these minima have been shaded. The shades arcs form a tree of shortest paths to node 9.

8. The Critical Path Method

It is perverse to seek the longest path through an acyclic network? No, as will become evident from

Problem 8.D (project management). Lara is undertaking a project that entails five different tasks, which are labeled A through E. Each task requires a period of time, and certain of these tasks cannot begin until others have been completed. For each task, Table 8.3 specifies the time it requires (in weeks) and its list of predecessors. This table indicates, for instance, that task D requires 5 weeks, and work on task D cannot commence until tasks A and C have been completed. Lara wants to complete this project as quickly as possible. She wonders how long it will take and which tasks cannot be delayed without extending the project's completion time.

Table 8.3. Lara's project.

task	A	B	C	D	E
completion time	9	6	4	5	8
predecessors	--	--	B	A, C	B

A seemingly-natural approach to this type of problem is to build a network whose arcs depict tasks and whose nodes depict the start and/or end of tasks. If a task has more than one predecessor, extra arcs of length 0 will be required, and the network will become a bit unwieldy.

A different type of network

A simpler approach is to identify the tasks with the nodes and the preference relations with the arcs. In Figure 8.6, node S represents the start of the project and node F represents its completion. Each task is represented as an ellipse (node) with its length inside it, and each precedence relationship is represented as a directed arc. For instance, arcs (A, D) and (C, D) exist because task D cannot begin until tasks A and C have been completed.

Figure 8.6. A network representation of Problem 8.D.

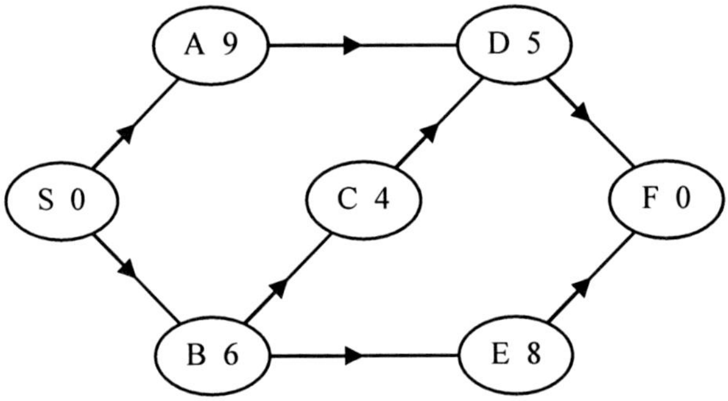

The network in Figure 8.6 is acyclic. That's no surprise. If this network had a cycle, the project could never be completed.

In Figure 8.6, the lengths are associated with the nodes, rather than with the arcs. A **path** is now a sequence of tasks each of which is a predecessor of the next. Task sequences (S, A, D) and (B, E) and (C) are paths. The **length** of a path equals the sum of the lengths of the tasks that it contains. The longest path that includes the start and finish tasks is (S, B, C, D, F) and its length is $15=6+4+5$. The project takes 15 weeks to complete.

Any path whose length is longest is said to be a **critical path**. Each task in a critical path is called **critical**. The critical tasks cannot be delayed without increasing the project completion time. For the data in Figure 8.6, tasks B, C and D are critical.

The critical path method

Problem 8.D illustrates the **critical path method**, whose components are:

- Construct a precedence diagram akin to that in Figure 8.6.
- Find the shortest time needed to complete the project, and identify the tasks that cannot be delayed without increasing the completion time.

The critical path method (also known as **CPM**) is commonly used in large-scale construction projects and in the management of research and development projects. In both contexts, building the diagram (the analogue of Figure 8.6) helps the manager to get things into focus.

Figure 8.6 has only seven nodes and is easy to solve by "eyeball." If a project had a large number of tasks, a systematic solution procedure would be called for. The network representation of a project management problem must be acyclic, so backwards optimization can be adapted to find the shortest completion time and the critical tasks. The arc lengths are positive, so reaching can also be adapted to this purpose.

Earliest completion times

Let us consider, briefly, how one might compute the earliest time at which each task can be completed. For each task x, designate

$$t(x) = \text{the earliest time at which task x can be completed.}$$

These earliest completion times satisfy an optimality equation, and t(x) can be computed from this equation as soon as the task completion times for its predecessors have been determined.

The data in Figure 8.6 are used to illustrate this recursion. With $t(S)=0$, this recursion gives $t(A)=t(S)+9=9$, then $t(B)=t(S)+6=6$, then $t(C)=t(b)+4=10$, then

$$t(D) = \max\begin{cases} t(A) + 5 \\ t(C) + 5 \end{cases} = \max\begin{cases} 9 + 5 \\ 10 + 5 \end{cases} = 15,$$

and so forth. The method that has just been sketched is just like backwards optimization, except that it begins at the start of the network. This method is sometimes called **forwards optimization**.

Reaching can also be used to compute the earliest completion times, and they can even be found by solving a linear program. Using a linear program to solve a problem as simple as this seems a bit like overkill, however.

Crashing and a linear program

On the other hand, linear programming has a role to play when the problem is made a little more complicated. **Crashing** accounts for the situation in which some or all of the task times can be shortened at added expense. To illustrate, we alter Problem 8.D by supposing that:

- Each task's duration can be shortened by as much as 25% by the use of overtime labor at a cost of \$1,000 per week of shortening.
- The economic benefit of shortening the project completion time is \$2,500 per week.

Without using any overtime labor, the project can be completed in 15 weeks. To allocate overtime labor efficiently, we designate

w(x) = the number of weeks by which the time needed to perform task x is reduced.

Task times cannot be reduced by more than 25% of their durations, so

$$w(A) \le 9/4, \quad w(B) \le 6/4, \quad \ldots, \quad w(E) \le 8/4.$$

The aggregate number w of weeks by which task times are reduced is given by

$$w = w(A) + w(b) + \ldots + w(E).$$

Each of the eight arcs in Figure 8.6 gives rise to an inequality on a latest completion time. Two of these inequalities are

$$t(A) \ge 9 - w(A), \qquad t(D) \ge t(A) + 5 - w(D).$$

The linear program selects nonnegative values of the decision variables that minimize {2,500t(F) + 1,000w}, subject to constraints that are illustrated above.

Critique

One defect of CPM is that it fails to deal with *nonconcurrence* constraints. When building a house, for instance, one can sand the floors before or after one paints the interior walls, but not while one paints the interior walls. Nonconcurrence constraints can be handled by shifting to an integer programming formulation.

A second weakness of CPM lies in its assumption that the tasks and their duration times are fixed and known. One is actually working with estimates. As time unfolds, unforeseen events occur. When they do, one can revise the network, re-estimate the task times, re-compute its critical path, and re-determine which tasks need close monitoring. It is sometimes practical to model the uncertainty in the task duration times. A technique that is known as **PERT** (short for Program Evaluation and Review Technique) is a blend of simulation and critical-path computation. PERT is sketched below.

Step 1: For each task, estimate these three elements of data:

- Its most optimistic (smallest possible) duration, A.
- Its most pessimistic (largest possible) duration, B.
- Its most likely duration, M.

Model the duration time of this task by a random variable X whose distribution is triangular with the above parameters.

Step 2: Simulate the project a large number of times. For each simulation, record the project completion time, which tasks are critical, and the difference between each task's earliest and latest start times.

PERT allows one to discern which of the tasks are most likely to be critical.

A bit of the history

A pioneering application of CPM and PERT was to the development of the world's first nuclear-powered submarine, the Nautilus. This project entailed considerable technical innovation. It was completed *on schedule and within budget* in 1955 under the direction of the legendary Admiral Hyman Rickover (1900-1986).

On the 1950's, when PERT was developed, computer simulation was exceedingly arduous. That has changed. In the current era, simulations are easy to execute on a spreadsheet. Today, PERT and CPM are routinely used in the management of large-scale development projects.

9. Review

This chapter has exposed you to the terminology that is used to describe directed networks and to a few representative path-length problems. The shortest-path problem is well-defined when the network has no cycle whose length is negative. It can be solved by linear programming. If all arc lengths are nonnegative, it can also be solved by reaching. If the network is acyclic, it can also be solved by backwards optimization. All three of these methods produce a tree of shortest paths.

Path-length problems have been used to introduce the components of a thought process that is known as dynamic programming. These components include:

- *State* – enough information about what has happened so far.
- *Embedding* – placing a problem of interest in a family of related problems, one per state.
- *Linking* – relating the solutions of the problems through an *optimality equation*,
- *Solving* – finding a *policy* that is optimal for each state.

The phrase "dynamic programming" describes a perspective on modeling, and "linear programming" describes the analysis of a particular model. These two subjects have much in common, nonetheless. Each has a substantial body of theory, and each provides insight into a variety of field as diverse as economics and physics.

Let us close with mention of the fact that dynamic programming is especially well-suited to the analysis of **Markov decision models**; these models describe situations in which decisions must be taken in the face of uncertainty. These models are important. They are not explored in this book, but the ideas that have just been reviewed may help to provide access to them.

10. Homework and Discussion Problems

1. The network in Figure 8.2 has a tree of shortest paths from node 1 to all others.

(a) Write down the optimality equation whose solution gives the lengths of the paths in this tree.

(b) Find this shortest-path tree by any method (including trial and error) and draw the analogue of Figure 8.3.

(c) Check that the path lengths in your tree satisfy the optimality equation you wrote in part (a).

2. Use the reaching method to compute the tree of shortest paths to node 7 of the network in Figure 8.4.

3. Consider any directed network.

   (a) If this network is acyclic, show that its nodes can be relabeled so that each arc (i, j) has $i<j$. *Hint:* Add an arc at the end of a path repeatedly, and see what happens.

   (b) If this network is cyclic, show that its nodes cannot be relabeled so that each arc (i, j) has $i<j$. *Hint:* if each arc (i, j) has $i<j$, can there be a cycle?

4. For the network in Figure 8.4, there does not exist a tree of paths from node 2 to all others. Adapt reaching to find the shortest path from node 2 to each node that can be reached from node 2.

5. For the network in Figure 8.5, there exists a tree of shortest paths from node 1 to all others.

   (a) Write down the optimality equation that is satisfied by the lengths of the paths in this tree.

   (b) Adapt backwards optimization to find the lengths of the paths in this tree.

6. Reaching was proposed for a network whose arc lengths are nonnegative. Does Reaching work on acyclic networks whose arc lengths can have any signs? In particular, for the network in Figure 8.5, does reaching find the tree of shortest paths from node 1 to the others. Support your answer.

7. *(commuting by bicycle)* Vince cycles from home (node 1 of Figure 8.5) to the office (node 9 of that figure). He wishes to choose a route that

minimizes the steepest grade he must climb. Each arc represents a road segment, and the number adjacent each arc is the steepest grade he will encounter when traveling that road segment in the indicated direction.

(a) Embed Vince's problem in a family of problems, one per state.

(b) Write an optimality equation for the set of problems you identified in part (a).

(c) Solve that optimality equation. What route do you recommend? What is its maximum grade?

(d) Which versions of the principle of optimality are valid for this problem?

8. Represent the project management problem whose data are in Table 8.3 as a longest-path problem in which each of the five activities is an arc. You will need more than 5 arcs. Why?

9. **(crashing)** Sketched in Section 8.8 are some of the constraints of the formulation of a project management problem with crashing as a linear program.

(a) Write out the entire linear program.

(b) Solve it on a spreadsheet.

(c) With crashing, what is the optimal project completion time? Which tasks are critical?

(d) Why is it not economical to shorten the completion time below the value you reported in part (c)?

10. *(Dawn Grinder)* Dawn Grinder has 10 hours to prepare for her exam in linear algebra. The exam covers four topics, which are labeled A through D. Dawn wants to maximize her score on this test. She estimates that devoting j hours to topic x will improve her score by b(j, x) points.

(a) What are the states in a dynamic programming formulation of Dawn's problem? *Hint:* Nothing is lost by studying the subjects in alphabetic order.

(b) Write the optimality equation for the formulation you proposed in part (a).

(c) Dawn has estimated that the benefit of each hour spent on subjects A or B and C or D are as indicated below. How shall she allocate her time? How many points can she gain? How many points will she lose if she takes an hour off? *Remark:* For these data, the solution should be pretty clear.

hours	0	1	2	3	4
A or B	0	2	2	3	4
C or D	0	1	1	2	3

11. *(No Wonder bakers)* No Wonder Bakers currently has 120 bakers in its employ. Corporate policy allows bakers to be hired at the start of each month, but never allows firing. Training each new baker takes 1 month and requires a trained baker to spend half of that month supervising the trainee, rather than making bread for the company. Eight percent of the bakers and eight percent of the trainees quit at the end of each month. The demand D(j) for trained bakers in each of the next seven months, is, respectively, 100, 105, 130, 110, 140, 120, and 100. In particular, D(1) = 100. Demand must be met by production in the current month. Trained bakers and trainees receive the same monthly wage rate. The company wishes to satisfy demand at minimum payroll cost. Denote as x(j) a number of the number of trained bakers that is large enough to satisfy the demand during months j through 7, and denote as t(j) the number of trainees that are hired at the start of month j.

    (a) How do x(j) and t(j) relate to D(j) and x(j + 1)?

    (b) Does the company wish to minimize x(j)?

    (c) Write an optimality equation whose solution minimizes the payroll cost during months 1 through 7. Solve this equation, rounding t(j) up to the nearest integer as is needed.

12. *(bus stops)* A city street consists of 80 blocks of equal length. Over the course of the day, exactly S(j) people start uptown bus trips at block j, and exactly E(j) people end uptown bus trips at block. Necessarily,

$$S(1) + \cdots + S(k) \geq E(1) + \cdots + E(k) \qquad \text{for } k = 1, \ldots, 80,$$

and this inequality holds as an equation when k=80. The director of public services wishes to locate 12 bus stops on this street so as to minimize the total number of blocks that people have to walk to and from bus stops.

(a) Suppose that bus stops are located at blocks p and q>p, but not in between. Interpret

$$W(p, q) = \sum_{j=p}^{q} [S(j) + E(j)] \min \begin{cases} j - p \\ q - j \end{cases}.$$

(b) Suppose that the first stop is located at block p and that the last stop is located at block q. Interpret

$$B(p) = \sum_{j=1}^{p} S(j)(p - j), \qquad T(q) = \sum_{j=q}^{80} E(j)(j - q).$$

(c) Suppose bus stops are located at blocks $p_1 < p_2 < \cdots < p_{12}$. Express the total number of blocks walked by bus users in terms of the functions specified in parts (a) and (b). Justify your answer.

(d) Can you relate the bus-stop location problem to a dynamic program each of whose states is a pair (k, q) in which k bus stops are located in blocks 1 through q, with the highest-numbered stop at block q? Justify your answer. *Hint:* look at

$$\begin{aligned} f(1, q) &= B(q), \\ f(k, q) &= \min_{p<q} \{f(k-1, p) + W(p, q)\}, \\ F(80) &= \min_{q \le 150} \{f(12, q) + T(q)\} \end{aligned}$$

13. *(bus stops, continued)* For the data in the preceding problem, the director of public services wishes to optimize with a different objective. She now wishes to locate bus stops (not necessarily 12 in number) so as to minimize the total travel time of the population of bus users. People walk at the rate of one block per minute. Buses travel at the rate of 5 blocks per minute, but they also take 1.5 minutes to decelerate to a stop, allow passengers to get off and on, and reaccelerate.

(a) Suppose that stops are located at blocks p and $q > p$, but not in between. Give a formula for the number $K(p, q)$ of people who are on the bus while it travels from the stop at block p to the stop at block q.

(b) With $W(p, q)$ as given in part (a) of the preceding problem, interpret

$$B(p, q) = W(p, q) + K(p, q)\ [1.5 + (q - p)/5].$$

(c) Can you relate this bus-stop location problem to a dynamic program each of whose states is a singleton q? *Hints:* Is it the case that

$$f(p) + B(p, q) \geq f(q)?$$

If so, how can you account for the people who walk uptown to the first stop and uptown from the final stop? and in which k bus stops are located in blocks 1 through q, with the highest-numbered stop at block q?

Chapter 9: Flows in Networks

1. Preview

In this chapter, you will see how the simplex method simplifies when it is applied to a class of optimization problems that are known as "network flow models." You will also see that if a network flow model has "integer-valued data," the simplex method finds an optimal solution that is integer-valued.

Also included in this chapter is a different method for solving a particular network flow model that is known as the "assignment problem." That method is known as the "Hungarian method," and it is very fast.

2. Terminology

Figure 9.1 depicts a directed network that has 5 nodes and 7 directed arcs. As was the case in the previous chapter, each node is represented as a circle with an identifying label inside, and each directed arc is represented

E. V. Denardo, *Linear Programming and Generalizations*, International Series in Operations Research & Management Science 149,
DOI 10.1007/978-1-4419-6491-5_9, © Springer Science+Business Media, LLC 2011

Figure 9.1. A directed network.

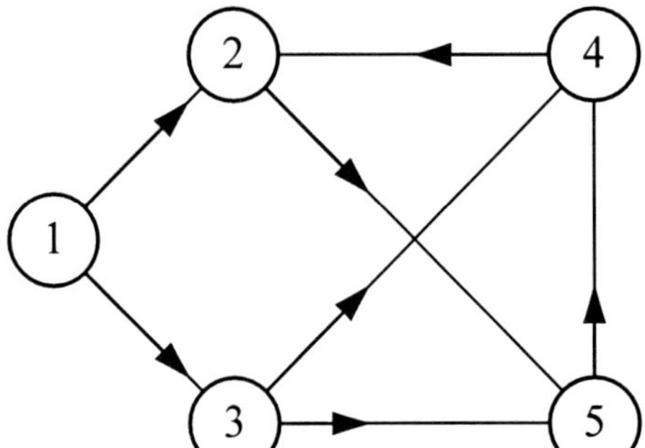

as a line segment that connects two nodes, with an arrow pointing from one node to the other.

As in the preceding chapter, directed arc (i, j) is said to have node i as its *tail* and has node j as its *head.* Again, a *path* is a sequence of n directed arcs with n ≥ 1 and with the property that the head of each arc other than the nth is the tail of the next. A path is said to be *from* the tail of its initial arc *to* the head of its final arc. A path from a node j to itself is called a *cycle.* Again, directed network is sometimes abbreviated to *network* and directed arc is sometimes abbreviated to *arc.*

Chains

Roughly speaking, a "chain" is a path of distinct arcs that can be traversed if we allow ourselves to walk across each arc in either direction, with or against its arrow. In this context, $(i, j)^F$ is interpreted as arc (i, j) when it is traversed in the **forward** direction, from node i to node j. Similarly, $(i, j)^R$ is interpreted as arc (i, j) when it is traversed in the **reverse** direction, from node j to node i.

Arcs $(i, j)^F$ and $(i, j)^R$ are said to be **oriented.** Arc $(i, j)^F$ has node i as its **tail** and node j as its **head.** Arc $(i, j)^R$ has node j as its **tail** and node i as its **head.** In this context, a **chain** is a sequence of n distinct oriented arcs, with n ≥ 1, whose orientations are such that the head of each arc but the nth is the tail of the next arc. A chain is said to be **from** the tail of its initial arc **to** the head of its final arc. In Figure 9.1, $(3, 5)^R$ and $\{(5, 4)^F, (3, 4)^R\}$ are chains from node 5 to node 3.

Loops

A chain from a node to itself is called a **loop.** In Figure 9.1, the chain $\{(5, 4)^F, (3, 4)^R, (3, 5)^F\}$ is a loop, namely, a chain from node 5 to itself. A loop from node i to itself is said to be a **simple loop** if node i is visited exactly twice and if no other node is visited more than once.

Spanning trees

A subset T of the arcs in a network is said to be a **spanning tree** if T includes no loop and if T includes a chain from each node i in the network to each node j ≠ i. A spanning tree cannot contain two different chains from node i to node j; if it did, it would contain a loop.

The network in Figure 9.1 has several spanning trees, one of which is the set T of directed arcs in Figure 9.2. These arcs do contain exactly one chain from every node to every other node. Their chain from node 4 to node 1 is $\{(4, 2)^F, (1, 2)^R\}$, for instance.

Figure 9.2. A spanning tree T for the network in Figure 9.1.

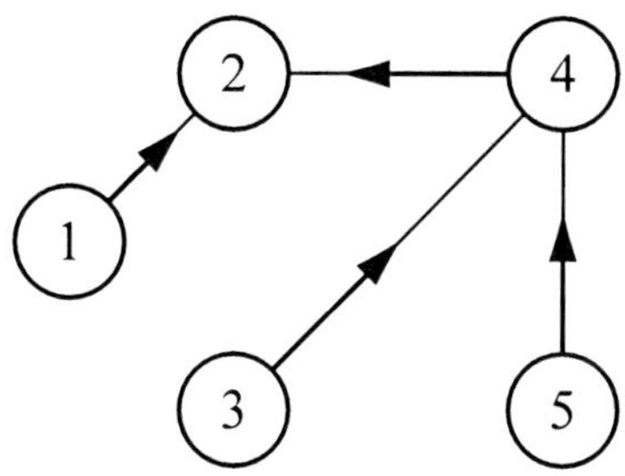

When S is a set, |S| denotes the number of elements in S. The spanning tree in Figure 9.2 illustrates properties that hold in general. Let us consider

Proposition 9.1. Let (N, A) be any directed network. A subset T of its arcs is a spanning tree if and only if T satisfies any two of the following:

(a) T contains no loops.

(b) T contains a chain from each node i in the network to each node j ≠ i.

(c) $|T| = |N| - 1$.

Proof. The result is trite when $|N| = 1$, and it is easily verified by induction on the number $n = |N|$ of nodes in N. ■

Evidently, a spanning tree must contain one fewer arc than the number of nodes in the network. Spanning trees play an important role in network flow. Later in this chapter, the spanning trees will be shown to be the bases for the "transportation problem," for instance.

3. The Network Flow Model

The "network flow model" is a linear program whose decision variables are flows on the arcs in a directed network. In particular:

- The amount that flows on each arc must be nonnegative.
- The flow on each arc occurs from its tail to its head.
- Each arc can have a positive lower bound on the amount of its flow.
- Each arc can have a finite upper bound on the amount of its flow.
- Each node can have a fixed flow, and a node's fixed flow may be into or out of that node.
- Flow is conserved at each node; the sum of the flows into each node equals the sum of the flows out of that node.

It will prove convenient to allow this network flow model to have an "unseen" node. Why that is so is suggested by Figure 7.1, which, for easy reference, is reproduced here as Figure 9.3.

The requirement that "flow in" equals "flow out" applies to each of the seven nodes in this network. Nodes 1 through 4 have fixed outward flows. The sum of these fixed outward flows is 900. Because flow is conserved at each node, the sum of the flows into nodes U, V and W must also equal 900.

The "unseen" node

The flows into nodes U, V and W are decision variables. On what arcs do these flows occur? Each arc must have a head and a tail. Implicitly, the network in Figure 9.3 has an **unseen node**. Let us label it node α. The flows into node U, V and W occur on directed arcs (α, U), (α, V), and (α, W). Also, the fixed flows out of nodes 1 through 4 occur on arcs $(1, \alpha)$ through $(4, \alpha)$.

Figure 9.3. A network flow model.

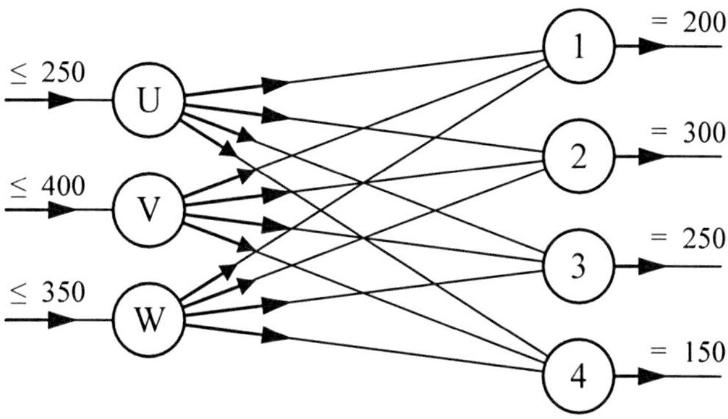

Flow is conserved at the unseen node, and this occurs *automatically*. The flow conservation constraints for the seven nodes in Figure 9.3 guarantee that the total flow into node α equals 900 and that the total flow out of node α equals 900.

The model

The example in Figure 9.3 prepares for a precise description of the network flow model. Its backbone is a directed network that consists of:

- A finite set N whose members arc called *nodes* and, possibly, one "unseen" node α that is not included in N.

- A finite set A of *arcs*. Each arc (i, j) in A has i and j in $N \cup \{\alpha\}$, but no arc has $i = j = \alpha$.

The **network flow model** is a linear program that is cast in the format of

Program 9.1. Minimize $\left\{\sum_{(i,j)\in A} c_{ij}x_{ij}\right\}$, subject to

(1) $$L_{ij} \le x_{ij} \le U_{ij} \quad \text{for each arc } (i, j) \in A,$$

(2) $$\textstyle\sum_j x_{ji} = D_i + \sum_j x_{ij} \quad \text{for each node } i \in N.$$

Program 9.1 has three elements of data per arc; each arc (i, j) has a *unit cost* c_{ij}, a *lower bound* L_{ij} and an *upper bound* U_{ij}. Each lower bound is nonnegative. Each arc's upper bound must be at least as large as its lower bound, and each upper bound can be as large as $+\infty$. Program 9.1 also has one element of data per node in N; the number D_i for node i is called node i's *net outward flow*. The number D_i can be positive, negative, or zero. If D_i is positive, it is a fixed flow out of node i. If D_i is negative, then $-D_i$ it is a fixed flow into node i.

Program 9.1 minimizes the cost of the flow subject to constraints that keep each arc's flow between its lower and upper bound and require the total flow into each node in N to equal the total flow out of that node. The network may include one node α that is not in N. The flow-conservation constraint for node α can be (and is) omitted from (2) because this constraint is implied by the others.

An example

The network flow model is illustrated in Figure 9.4. Adjacent to each arc in this figure is its unit cost (e.g., $c_{42} = -1.3$). The stubby grey arrows in Figure 9.4 identify those fixed flows that do not equal zero. In particular D_1 equals −12, which is represented as a fixed flow of 12 units into node 1. This network flow problem has a fixed flow of 12 into node 1, and it has fixed flows of 8 and 4 out of nodes 4 and 5, respectively.

Figure 9.4. A network flow problem.

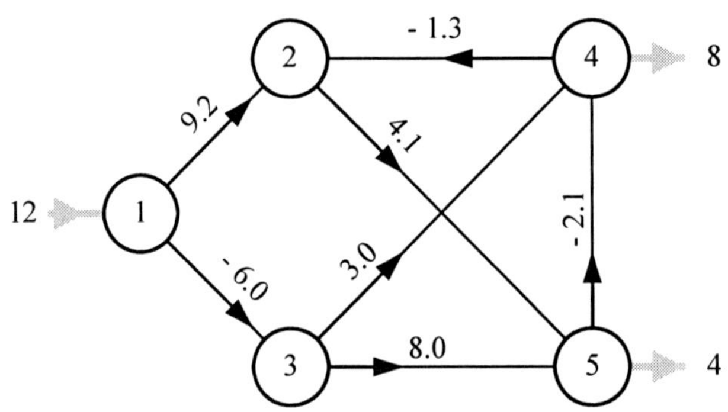

Problem 9.A. For the directed network in Figure 9.4, each arc has 0 as its lower bound and has + ∞ as its upper bound. Find a least-cost flow.

The decision variables in a network flow problem are the flows on the arcs. For the network in Figure 9.4, x_{ij} denotes the flow on arc (i, j). Program 9.2, below, minimizes the cost of the flow, subject to constraints that keep the flows nonnegative and that conserve flow at each node. Each node's flow conservation constraint is written in the format, *flow in equals flow out.*

Program 9.2. Min $\{9.2\ x_{12} - 6\ x_{13} + 4.1\ x_{25} + 3\ x_{34} + 8\ x_{35} - 1.3\ x_{42} - 2.1\ x_{54}\}$, s. to

(3.1) $12 = x_{12} + x_{13}$ (node 1),

(3.2) $x_{12} + x_{42} = x_{25}$ (node 2),

(3.3) $x_{13} = x_{34} + x_{35}$ (node 3),

(3.4) $x_{34} + x_{54} = 8 + x_{42}$ (node 4),

(3.5) $x_{25} + x_{35} = 4 + x_{54}$ (node 5),

(3.6) $x_{ij} \geq 0$ for each $(i, j) \in A$.

A spreadsheet

Writing equations (3.1) through (3.6) with the decision variables on the left-hand side and the data on the right-hand side produces rows 5 through 9 of the spreadsheet in Table 9.1. In this spreadsheet, the labels of the arcs appear in row 2, the flows appear in row 3, and the unit costs are in row 4.

Table 9.1. A spreadsheet for Problem 9.A.

	A	B	C	D	E	F	G	H	I	J	K	L
1												
2		label	x12	x13	x25	x34	x35	x42	x54			RHS
3		value	0	12	4	12	0	4	0			
4		unit cost	9.2	-6	4.1	3	8	-1.3	-2.1	-24.8		
5		node 1	-1	-1	0	0	0	0	0	-12	=	-12
6		node 2	1	0	-1	0	0	1	0	0	=	0
7		node 3	0	1	0	-1	-1	0	0	0	=	0
8		node 4	0	0	0	1	0	-1	1	8	=	8
9		node 5	0	0	1	0	1	0	-1	4	=	4
10												
11												
12							= SUMPRODUCT(C$3:I$3, C9:I9)					

Solver has minimized the quantity in cell J4 of Table 9.1, with C3:I3 as its changing cells and subject to the constraints J5:J9 = L5:L9 and C3:I3 ≥ 0. By doing so, Solver has minimized the cost of the flow that satisfies the flow-conservation constraints and keeps the flows nonnegative. Table 9.1 reports an optimal solution that sets

(4) $$x_{13} = 12, \quad x_{25} = 4, \quad x_{34} = 12, \quad x_{42} = 4,$$

and that equates the remaining flows to zero. That the values of these flows are integers is no accident, as will soon be evident.

4. The Integrality Theorem

A network-flow model is said to have **integer-valued data** if the following conditions are satisfied:

- Each fixed flow is integer-valued.
- Each lower bound is integer-valued.
- Each upper bound either is integer-valued or is infinite.

Problem 9.A does have integer-valued data; its fixed flows are integers, its lower bounds are 0, and its upper bounds are infinite. A property of network flow problems with integer-valued data appears below as

Proposition 9.2 *(the Integrality Theorem).* Consider a network flow model that has integer-valued data. Each of its basic solutions is integer-valued.

Remark: This result is dubbed the **Integrality Theorem.** The proof earns a star because of its length. The gist of this proof appears as the next subsection, which is not starred.

Proof*. In a basic solution to an equation system, the nonbasic variables equal zero, and the values of the basic variables are unique. Each decision variable whose value is not zero must be basic.

To study basic solutions for Program 9.1, we write it in Form 1. If L_{ij} is positive, the Form-1 representation of (1) includes the equation $L_{ij} = x_{ij} - y_{ij}$ and the inequality $y_{ij} \geq 0$. If U_{ij} is finite, the Form-1 representation includes the equation $U_{ij} = x_{ij} + z_{ij}$ and the inequality $z_{ij} \geq 0$.

Consider any basic solution to (2) and to the Form-1 representation of (1). Denote as B the set containing each arc $(i, j) \in A$ for which the flow x_{ij} is not integer-valued. Aiming for a contradiction, suppose that B is not empty.

Claim #1: Consider any arc $(i, j) \in B$. The decision variable x_{ij} is basic. If L_{ij} is positive, then y_{ij} is basic. If U_{ij} is finite, then z_{ij} is basic.

Proof of Claim: Since x_{ij} is not zero, it must be basic. Suppose L_{ij} is is positive. Since x_{ij} is not integer-valued and since L_{ij} is integer-valued, the decision variable $y_{ij} = x_{ij} - L_{ij}$ is not zero, hence is basic. Similarly, if U_{ij} is a finite integer, z_{ij} cannot equal zero, hence must be basic. This proves the claim.

By hypothesis, B is nonempty, so B contains at least one arc, (i, j). A chain will be "grown" whose first oriented arc is $(i, j)^F$ and each of whose arcs is in B. By hypothesis, the net fixed flow into node j is integer-valued. Since x_{ij} is not integer-valued, the flow-conservation constraint for node j guarantees that at one other arc touches node j and has a flow that is not integer-valued. For some k, that arc is either (k, j) or (j, k), and it is in B. In the former case, the next oriented arc in the chain is taken as $(k, j)^R$. In the latter case, the next oriented arc in the chain is taken as $(j, k)^F$. In either case, j is replaced by k, and the chain-growing step is repeated until a node in the chain is revisited. That must occur because the number of nodes is finite.

When a node repeats, a simple loop T has been found. Perturb x, y, and z as follows. Add a positive number K to the flow on each forward arc in this simple loop and subtract K from each reverse arc in this simple loop. If an arc in this loop has a positive lower bound L_{ij}, decrease or increase by K the value of the basic variable y_{ij} so as to preserve a solution to $L_{ij} = x_{ij} - y_{ij}$. Do the same for the arcs in this loop that have finite upper bounds. The perturbed solution satisfies (2) and the Form-1 representation of (1). Claim #1 shows that only the values of basic variables have been perturbed. This contradicts the fact that the basic solution is unique. Hence, B must be empty. ∎

The gist

To provide a simple illustration of Proposition 9.2, consider the case in which the lower bounds are zero and the upper bounds are infinite. In this case, (1) reduces to the requirement that each flow be nonnegative. If a basic solution to (2) is not integer-valued, the integrality condition guarantees that there exists a simple loop whose flows are not integer-valued, hence must be basic. Figure 9.5 illustrates such a loop.

Perturbing the solution x to (2) by "shipping" K around this loop results in a new solution to (2). This cannot occur because the basic solution to (2) is unique.

Figure 9.5. A loop whose arcs are basic.

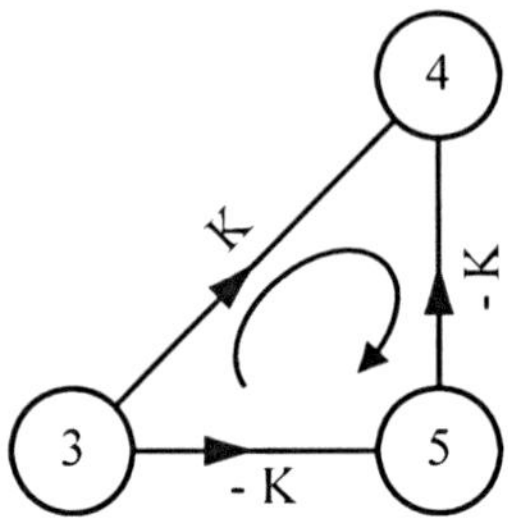

An important result

The importance of the Integrality Theorem would be hard to overstate. A great many practical network flow problems *do* have integer-valued data. When the simplex method is used to solve such a problem, it pivots from basis to basis, and each basis assigns an integer value to each flow, x_{ij}. This occurs *automatically,* without the need to require decision variables to be integer-valued. If these flows represent items that exist in integer quantities (such as airplanes or ships), the simplex method finds an optimal solution that is integer-valued.

Simplex versus interior-point methods

Large network flow problems that have integer-valued data tend to have many alternative optima. For those problems, the simplex method enjoys an advantage over interior-point methods. Each simplex pivot produces a basic solution, which is guaranteed to be integer-valued. In particular, the optimal solution with which the simplex method terminates is integer-valued. By contrast, interior-point methods converge to the "center" of the set of optimal solutions, which is not integer-valued when there are multiple optima.

5. The Transportation Problem

The simplex method simplifies markedly when it is tailored to network flow problems. This will be illustrated for a particular type of network flow problem that is known as the "transportation problem." The transportation problem has these properties:

- There are m **supply** nodes, which are numbered 1 through m, and the positive datum S_i is the number of units available for shipment out of supply node i.

- There are n **demand** nodes, which are numbered 1 through n, and the positive datum D_j is the number of units that must be shipped to demand node j.

- Shipment can occur from each supply node to each demand node. For each pair (i, j), the cost of shipping each unit from supply node i to demand node j equals c_{ij}.

The decision variables in the transportation problem are the quantity x_{ij} to ship from each supply node i to each demand node j. The **transportation problem** is the linear program,

Program 9.3. Minimize $\left\{\sum_{i=1}^{m} \sum_{j=1}^{n} c_{ij}\, x_{ij}\right\}$, subject to the constraints

(5.i) $$\sum_{j=1}^{n} x_{ij} \le S_i \qquad \text{for } i = 1, 2, \ldots, m,$$

(6.j) $$\sum_{i=1}^{m} x_{ij} = D_j \qquad \text{for } j = 1, 2, \ldots, n,$$

(7) $$x_{ij} \ge 0 \qquad \text{for each } i \text{ and } j.$$

Equation (5.i) requires that not more than S_i units are shipped out of supply node i. Equation (6.j) requires that exactly D_j units are shipped into demand node j. The constraints in (7) keep the shipping quantities nonnegative.

By summing (5.i) over i, we see that the sum of the shipping quantities cannot exceed the sum of the supplies. By summing (6, j) over j, we see that the sum of the shipping quantities must equal the sum of the demands. As a consequence, Program 9.3 cannot be feasible unless its data satisfy

(8) $$\sum_{i=1}^{m} S_i \ge \sum_{j=1}^{n} D_j.$$

In brief, aggregate demand cannot be satisfied if it exceeds aggregate supply.

A "dummy" demand node

Testing for (8) is easy. For the remainder of this discussion, it is assumed that (8) holds. In fact, a seemingly-stronger assumption is invoked. It is that expression (8) holds as an *equation*. Thus, for the remainder of this section, it is assumed that the aggregate supply equals the aggregate demand. This entails no loss of generality; it can be obtained by including a **dummy** demand node, say node m, whose demand equals the excess of the aggregate supply over the aggregate demand and with shipping cost $c_{im} = 0$ for each (supply) node i.

An equality-constrained transportation problem

When (8) holds as an equation, every solution to (5) and (6) satisfies each inequality as an equation. Assuming that aggregate supply equals aggregate demand lets us switch our attention from Program 9.3 to

Program 9.3^E. Minimize $\left\{\sum_{i=1}^{m}\sum_{j=1}^{n} c_{ij}\,x_{ij}\right\}$, subject to the constraints

(9) $$\sum_{j=1}^{n} x_{ij} = S_i \qquad \text{for } i = 1, 2, \ldots, m,$$

(10) $$\sum_{i=1}^{m} x_{ij} = D_j \qquad \text{for } j = 1, 2, \ldots, n,$$

$$x_{ij} \geq 0 \qquad \text{for each i and j.}$$

The remainder of this section is focused on Program 9.3^E (the "E" being short for equality-constrained), and it is assumed that (8) holds as an equation.

An example

Figure 9.6 presents the data for a transportation problem that has m = 3 (three source nodes) and n = 5 (five demand nodes). The supplies S_1 through S_3 are at the right of the rows. The demands D_1 through D_5 are at the bottom of the columns. The number in the upper left-hand corner of each cell is the shipping cost from the supply in its row to the demand in its column. By reading across the first row, we see that $c_{11} = 4$, $c_{12} = 7$, $c_{13} = 3$, and so forth.

The total of the supplies equals 10,000, and the sum of the demands equals 10,000, so aggregate supply does equal the aggregate demand. Demand node 5 has $D_5 = 1{,}000$, and each shipping cost in its column equals 0. Evidently, demand node 5 is a dummy node that "absorbs" at zero cost the excess supply of 1,000 units.

Problem 9.B. For the transportation problem whose data are presented in Figure 9.6, find a least-cost flow.

The simplex method will soon be executed directly on diagrams like that in Figure 9.6. There is no need for simplex tableaus.

Figure 9.6. Data for a Transportation Problem.

4	7	3	5	0	2500 = S_1
10	9	3	6	0	4000 = S_2
3	6	4	4	0	3500 = S_3
2000 = D_1	3000 = D_2	2500 = D_3	1500 = D_4	1000 = D_5	

Initializing Phase II

Figure 9.6 has 15 "cells." Shipping quantities will be placed in certain of these cells. For these shipping quantities to represent a feasible solution:

- Each shipping quantity must be nonnegative.
- Their sum across each row must equal that row's supply.
- Their sum down each column must equal that column's supply.

The above conditions guarantee a feasible solution. In addition, we want these shipping quantities to represent a basis. That's easily accomplished by the procedure that's been dubbed the

Northwest Corner rule:

- Start with i = 1 and j = 1 and proceed as follows.
- Record as x_{ij} the smaller of the unsatisfied supply in row i and the unsatisfied demand in column j.
  - If this exhausts the supply in row i but not the demand in column j, increase i by 1 and repeat.

- If this exhausts the demand in column j but not the supply in row i, increase j by 1 and repeat.

- If this exhausts both the supply in row i and the demand in column j, either increase i by 1 and repeat *or* increase j by 1 and repeat.

Figure 9.7 displays the result of applying this rule to the example in Figure 9.6. The first step sets x_{11} = 2,000. This reduces S_1 to 500, and it reduces D_1 to 0, so j is increased to 2. The second step sets x_{12} = 500, which exhausts the supply at node 1, so i is increased to 2. And so forth.

Figure 9.7. Initializing Phase II.

4 2000	7 500	3	5	0	~~500~~ ~~2500~~
10	9 2500	3 1500	6	0	~~1500~~ ~~4000~~
3	6	4 1000	4 1500	0 1000	~~1000~~ ~~2500~~ ~~3500~~
~~2000~~	~~3000~~ ~~2500~~	~~2500~~ ~~1000~~	~~1500~~	~~1000~~	

The entries in Figure 9.7 do form a feasible solution: These entries are nonnegative. Their sum across each row equals that row's supply. And the sum down each column equals that column's demand.

A spanning tree

The directed network in Figure 9.8 records this feasible solution. Its arcs correspond to the cells in Figure 9.7 to which flows have been assigned, and the value of each flow appears beside its arc. The fixed flows into supply nodes 1-3 and out of demand nodes 1-5 are recorded next to stubby arrows into and out of their nodes.

The seven arcs in Figure 9.8 form a *spanning tree*; these arcs contain no loop, and they contain a chain from every node to every other. It will soon be shown that the flows in Figure 9.8 are a basis, moreover, that the bases for the transportation problem correspond to the spanning trees. That is the content of

Figure 9.8. The spanning tree and basic solution constructed by the Northwest Corner rule.

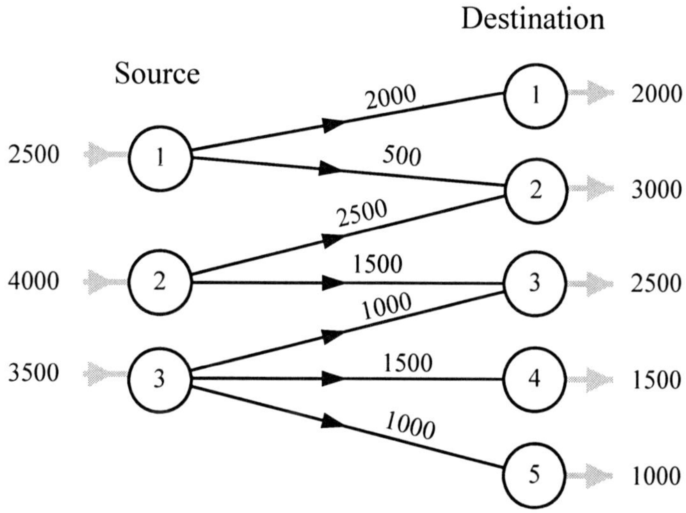

Proposition 9.3. Suppose (8) holds as an equation. In Program 9.3^E, consider any subset S of the decision variables. This set S is a basis if and only if the set B of directed arcs to which it corresponds is a spanning tree.

Remark: It's important to understand that the bases correspond to the spanning trees. It's less important to know why. The proof draws on information in Chapter 10 and is starred.

Proof*. Each set S of decision variables for Program 9.3^E corresponds to a set B of arcs. The proof is organized as a series of three claims.

Claim #1: The rank of equation system (9)-(10) is at most $m + n - 1$.

Proof. Multiply each constraint in (9) by +1, multiply each constraint in (10) by −1, take the sum, cancel terms, and obtain "0 = 0," which shows that the $m + n$ constraints in (9)-(10) are linearly dependent. Thus, the rank of (9)-(10) cannot exceed $m + n - 1$.

Claim #2: A spanning tree exists, and the rank of (9)-(10) equals $m + n - 1$. If B is a spanning tree, then S is a basis.

Proof: That a spanning tree T exists is obvious. Proposition 9.1 shows that $|T| = n + m - 1$. The set S of decision variables that correspond to the arcs in

T is easily seen to be linearly independent, and Proposition 9.1 shows that $|S| = |T| = m + n - 1$. Thus, from Claim #1, the rank of (9)-(10) equals $m + n - 1$, and this set S is a basis. The same holds for any spanning tree B.

Claim #3: Suppose that S is a basis. Then B is a spanning tree.

Proof. Let S be a basis. Claim #2 shows that $|S| = m + n - 1$. Aiming for a contradiction, suppose B contains a loop. Adding a positive number ε to the forward flows in this loop and subtracting ε from the reverse arcs in this loop would produce a new flow in which only the basic variables were altered. That is not possible, so B can contain no loop. And, since $|B| = m + n - 1$, Proposition 9.1 shows that B is a spanning tree. ■

A correspondence between bases and spanning trees has been established for the transportation problem with equality constraints. Similar results hold for all network flow problems whose arcs have 0 and $+\infty$ as the lower and upper bounds on their flows.

Multipliers

In a Form-1 representation of a linear program, the simplex method executes a sequence of pivots. None of these pivots occurs on a coefficient in the top row, for which the variable $-z$ is kept basic. The tableau that results from any sequence of pivots is shown in Proposition 11.1 to have this property; the top row of the current tableau equals the top row of the initial tableau less a linear combination of the other rows of the initial tableau. The amount by which each row is multiplied in this linear combination is called the **multiplier** for that row.

In the application of the simplex method to Program 9.3^E, let us denote as u_i the multiplier for the ith constraint in (9), and let us denote as v_j the multiplier for the lth constraint in (10).

Proposition 9.4. Suppose (8) holds as an equation. When the simplex method is applied to Program 9.3^E, each tableau that it encounters has multipliers that satisfy

(11.ij) $$\bar{c}_{ij} = c_{ij} - u_i - v_j \qquad \text{for each i and j,}$$

(12.ij) $$c_{ij} = u_i + v_j \qquad \text{if } x_{ij} \text{ is basic}$$

Proof*: The variable x_{ij} has coefficients of 0 in all but one constraint in (9) and in all but one constraint in (10). It has coefficients of +1 in the constraints whose multipliers are u_i and v_j, which justifies (11). Also, if x_{ij} is basic, its top-row coefficient (reduced cost) $\bar{c}_{ij}$ equals zero, which justifies (12). ■

The multipliers for Program 9.3^E are not unique. To see why, consider any solution to (11). If we add to each multiplier u_i a constant K and subtract from each multiplier v_j the same constant K, we obtain another solution to (11). As a consequence, we can begin by picking one multiplier and equating it to any value we wish, and then use (11) to compute the values of the other multipliers.

Computing the multipliers

Figure 9.7 presents a basic solution. Let us compute a set of multipliers. We can begin by equating any single multiplier to any value we wish. Let's start by setting $u_1 = 0$. In Figure 9.9, the multiplier for each source node appears to the left of its row, and the multiplier for each demand node appears above its column.

Figure 9.9. Multipliers for the initial basis.

	$v_1 = 4$	$v_2 = 7$	$v_3 = 1$	$v_4 = 1$	$v_5 = -3$	
$u_1 = 0$	4 2000	7 500	3	5	0	2500
$u_2 = 2$	10	9 2500	3 1500	6	0	4000
$u_3 = 3$	3	6	4 1000	4 1500	0 1000	3500
	2000	3000	2500	1500	1000	

Having set $u_1 = 0$, the fact that x_{11} is basic lets us compute v_1 from (11) because

$$c_{11} = u_1 + v_1, \qquad \text{so} \qquad 4 = 0 + v_1,$$

which gives $v_1 = 4$. Similar arguments show that $v_2 = 7$, that $u_2 = 2$, and so forth.

An entering variable

The reduced costs are given in terms of the multipliers by equation (9), which is $\bar{c}_{ij} = c_{ij} - u_i - v_j$. The reduced cost of each basic variable equals 0. Figure 9.10 records the reduced cost $\bar{c}_{ij}$ of each nonbasic variable x_{ij} in that variable's cell.

Figure 9.10. An entering variable and its loop.

	$v_1 = 4$	$v_2 = 7$	$v_3 = 1$	$v_4 = 1$	$v_5 = -3$	
$u_1 = 0$	[4] ⊖ 2000	[7] ⊕ 500	[3] +2	[5] +4	[0] +3	2500
$u_2 = 2$	[10] +4	[9] ⊖ 2500	[3] ⊕ 1500	[6] +3	[0] +1	4000
$u_3 = 3$	[3] ⊕ -4	[6] -4	[4] ⊖ 1000	[4] 1500	[0] 1000	3500
	2000	3000	2500	1500	1000	

In a minimization problem, the entering variable can be any variable whose reduced cost is negative. A glance at Figure 9.10 shows us that the variables x_{31} and x_{32} have negative reduced costs. In particular,

$$\bar{c}_{31} = c_{31} - u_3 - v_1 = 3 - 3 - 4 = -4. \tag{13}$$

Evidently, perturbing the basic solution by setting $x_{31} = K$ changes cost by -4 K.

A loop

In Figure 9.10, the variable x_{31} has been selected as the entering variable, as is indicated by the "+" sign in its cell. Setting $x_{31} = K$ requires the values of the basic variables to be perturbed like so:

- To preserve the flow into demand node 1, we must decrease x_{11} by K.
- To preserve the flow out of source node 1, we must increase x_{12} by K.

- To preserve the flow into demand node 2, we must decrease x_{22} by K.
- To preserve the flow out of source node 2, we must increase x_{23} by K.
- Finally, to preserve the flow into demand node 3, we must decrease x_{33} by K.

In Figure 9.10, the cells whose flows increase are recorded with a "+" sign and the cells whose flows decrease are recorded with a "–" sign. The effect of this perturbation is to ship K units around the loop

$$\{(3, 1)^F, (1, 1)^R, (1, 2)^F, (2, 2)^R, (2, 3)^F, (3, 3)^R\},$$

and the shipping costs for the arcs in this loop indicate that shipping K units around this loop changes cost by

$$K(3 - 4 + 7 - 9 + 3 - 4) = -4K,$$

exactly as is predicted by equation (13).

A leaving variable

The largest value of K for which the perturbed solution stays feasible is 1,000, the smallest of the shipping quantities in the cells that are marked with "–" signs. Thus, the simplex pivot calls for the variable x_{31} to enter the basis and x_{33} to leave the basis. Figure 9.11 records the basic feasible solution that results from this pivot.

Figure 9.11. The second basis.

4 / 1000	7 / 1500	3	5	0	2500
10	9 / 1500	3 / 2500	6	0	4000
3 / 1000	6	4	4 / 1500	0 / 1000	3500
2000	3000	2500	1500	1000	

The second pivot

Recorded in Figure 9.12 are the multipliers for the second basis. Also recorded in that figure is the reduced cost of x_{25}, which equals −3. Selecting x_{25} as the entering variable for the next pivot creates the loop for which arcs have "+" and "–" signs. Note that a tie occurs for the departing variable. Setting the value K of the entering variable equal to 1,000 causes the values of x_{11} and x_{35} to equal 0.

Figure 9.12. The second pivot.

	$v_1 = 4$	$v_2 = 7$	$v_3 = 1$	$v_4 = 5$	$v_5 = 1$	
$u_1 = 0$	4 (−) 1000	7 (+) 1500	3	5	0	2500
$u_2 = 2$	10	9 (−) 1500	3 2500	6	0 (+) -3	4000
$u_3 = -1$	3 (+) 1000	6	4	4 1500	0 (−) 1000	3500
	2000	3000	2500	1500	1000	

Degeneracy

When such a tie occurs, it is necessary to remove *exactly one* of the variables that are tied, as this preserves a spanning tree (basis). The new basic solution will equate to 0 the variable (or variables) that tied but were not removed. Record their "0's" in the tableau. This will enable you to do subsequent pivots, which may be degenerate.

To firm up your understanding of the simplex method for transportation problems, begin with Figure 9.12 and execute a pivot with x_{25} as the entering variable and either x_{11} or x_{35} as the departing variable. Continue until you encounter a basic optimal solution.

The assignment problem

The special case of the transportation problem in which m = n and in which each supply node i has $S_i = 1$ and in which each demand node j has $D_j = 1$ is known as the **assignment problem**. The bases for the assignment problem correspond to spanning trees, and each spanning tree has 2m – 1 arcs (one fewer than the number of nodes). Thus, each basic feasible solution to the assignment problem equates m basic variables to the value 1 and equates the remaining m – 1 basic variables to the value 0. That's a *lot* of degeneracy. Is the assignment problem a curiosity? No, as will soon be evident.

Aircraft scheduling

Let us consider, briefly, the problem of assigning a fleet of identical aircraft to a schedule. Think of the termination of each flight in this schedule as a supply node with a supply of 1. Think of the beginning of each flight in this schedule as a demand node with a demand of 1. Interpret c_{ij} as the cost of assigning the aircraft that whose flight termination is i to flight j. If this assignment is impossible (e.g., if flight j takes of before flight i lands), take c_{ij} as a large positive number. If this assignment is possible but flight j does not begin at the airport where flight i ends, take c_{ij} as the cost of ferrying the plane to the desired airport. Finally, if flight j departs from the same airport at which flight i lands and departs after flight i lands, take $c_{ij} = 0$. What results is an assignment problem! The simplex method will determine whether the schedule can be satisfied and, if so, find a least-cost way to do so.

The general situation

Moderate-sized transportation problems are easy to solve by hand. And they introduce ideas that work for *all* network flow problems. These ideas are that:

1. In each basic solution, the flows that lie strictly between their lower and upper bounds can form no loop.
2. Multipliers for each basis are easily found by an analogue of (11).
3. The entering variable identifies a loop.
4. The largest amount K that can be shipped around this loop while preserving feasibility identifies a variable to remove from the basis.

The simplex method for network flow does not require simplex tableaus. Nor does it require diagrams akin to Figure 9.10. Efficient implementations use two or three "pointers" per node, instead. Given a basic solution and its multipliers, these pointers enable quick identification of: (i) the entering variable; (ii) the loop created by the entering variable; (iii) the leaving variable; and (iv) the change in flows and multipliers due to the pivot. Updating the pointers after a pivot occurs is not difficult, but it lies outside the scope of this book.

Speed

On large practical network flow problems, the general-purpose simplex code runs very quickly. Simplex codes that exploit the structure of network flow problems are faster still. On the other hand, the worst-case behavior of the simplex method for network flow is *not polynomial!* In 1973, Norman Zadeh[1] published a family of examples showing that the simplex method can require more than 2^r pivots when it is applied to transportation problems having r nodes.

Thus, network flow problems share a peculiarity with the general formulation – excellent performance by the simplex method on practical problems coupled with horrendous performance in worst case.

6. The Hungarian Method*

This section is starred because it can be skipped with no loss of continuity. This section is interesting, nonetheless, because it describes an instance (and there are not many) in which the simplex method can be beat. This method that is described in this section is due to Harold K. Kuhn.[2] He dubbed it the **Hungarian method** to acknowledge its ties to work done prior to the advent of linear programming by the Hungarian mathematicians, J. Egerváry and D. Konig. When this method is used to solve the assignment problem, it:

---

[1] Norman Zadeh, "A bad network problem for the simplex method and other minimum cost flow algorithms, "*Mathematical Programming*, V. 5, p 255-266, 1973.

[2] H. K. Kuhn, "The Hungarian method for the assignment problem," *Naval Research Logistics Quarterly*, V 2, pp. 83-97, 1955.

- Runs as quickly as does the simplex method on typical problems.
- Has good worst-case behavior.
- Produces integer-valued solutions to problems whose data are integer-valued.

The Hungarian method will be introduced in the context of the transportation problem in Figure 9.6. Let us begin with a simple observation:

> Subtracting a constant K from each cost in a row or column of the transportation problem has no effect on the relative desirability of various shipping plans.

To illustrate, suppose that the number 3 is subtracted from each cost in column 1 of Figure 9.6. Exactly 2,000 units must be shipped to demand node 1, so this subtracts 6,000 from the cost of *every* shipping plan. It has no effect on the relative desirability of different shipping plans.

One might hope to subtract constants from the rows and columns of the cost matrix so that:

- Each shipping cost is nonnegative.
- The current plan ships only on arcs whose costs are 0.

This might seem to be something new. But it isn't new. Equations (11) and (12) indicate that this is *exactly* what the multipliers for the optimal basis accomplish.

Revised shipping costs

As noted above, the relative desirability of different shipping plans is preserved if constants are subtracted from the rows and columns of the shipping cost matrix. It's easy to subtract constants so that the shipping costs become nonnegative and have least one zero in each row and in each column.

For the example in Figure 9.6, this can be accomplished by subtracting 3 from each cost in column 1, subtracting 6 from each cost in column 2, subtracting 3 from each cost in column 3, and subtracting 4 from each cost in column 4. Table 9.2 displays the data that results.

Table 9.2. Data for a transportation problem equivalent to that in Figure 9.6.

		demand nodes						
		4	5	6	7	8		
supply nodes	1	1	1	0	1	0	2500	supplies
	2	7	3	0	2	0	4000	
	3	0	0	1	0	0	3500	
		2000	3000	2500	1500	1000		
		demands						

Relabeled demand nodes

The demand nodes had originally been labeled 1 through 5. In Table 9.2, the demand nodes have been relabeled nodes 4 through 8. A node's number now identifies it uniquely; e.g., node 4 is the left-most of the demand nodes. With this (revised) notation, c_{24} denotes the shipping cost from node 2 to node 4 (from the 2nd supply node to the left-most demand node). Similarly, x_{24} denotes the quantity shipped from node 2 to node 4.

A partial shipping plan

A **partial shipping plan** ships only on arcs whose costs are zero. A *myopic* rule for establishing a partial shipping plan is as follows: Repeatedly, identify an arc whose cost equals zero and ship as much as possible on that arc. For the data in Table 9.2, one implementation of this myopic rule sets

(14) $$x_{16} = 2500, \quad x_{28} = 1000, \quad x_{34} = 2000, \quad x_{35} = 1500.$$

This partial shipping is recorded in the cells of Table 9.3 that have "chicken wire" background. The other cells in the boxed-in array contain shipping costs. Recorded to the right of this array is the residual supply, which is 3,000 units at node 2. Recorded below this array are the residual demands, which are 1,500 units at node 5 and 1,500 units at node 7.

The reachable network

Four of the nodes in Table 9.3 are labeled "R." These are the nodes in the "reachable network" whose arcs appear as solid lines in Figure 9.13. The **reachable network** contains those arcs on which the residual supply can be shipped at zero cost. This network includes arcs (2, 6) and (2, 8) because the

Table 9.3. A partial shipping plan.

		demand nodes					
		4	5	6 R	7	8 R	
supply nodes	1 R	1	1	2500	1	0	
	2 R	7	3	0	2	1000	3000 (residual supplies)
	3	2000	1500	1	0	0	
			1500		1500		
		residual demands					

residual supply at node 2 can be sent on arcs (2, 6) and (2, 8) at zero cost. This network also includes arc (1, 6) because flow that reaches node 6 can be forwarded to node 1 by *decreasing* x_{16} from its current value of 2,500.

Figure 9.13. The reachable network.

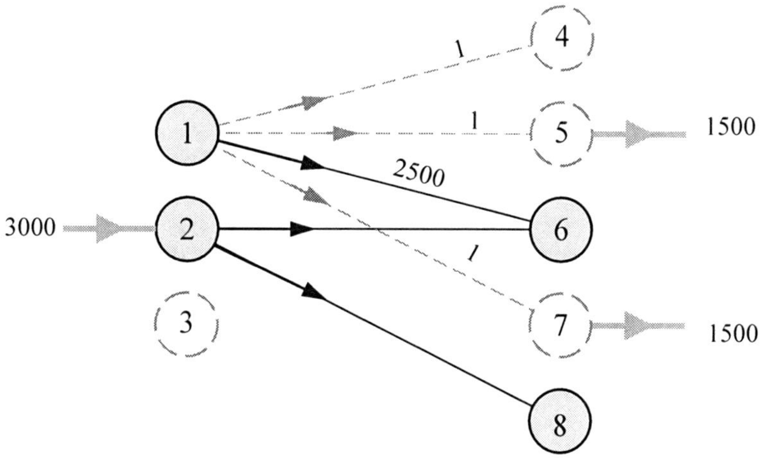

The set R of nodes in the reachable network is given by

$$R = \{1, 2, 6, 8\}. \tag{15}$$

These nodes are shaded in Fig. 9.13 and they are labeled "R" in Table 9.3. Note that each arc (i, j) with $i \in R$ and $j \notin R$ has a shipping cost that is positive; if its cost were 0, node j would have been included in R. Figure 9.13 displays as dashed lines the three arcs (i, j) with $i \in R$ and $j \notin R$ for which costs are smallest. These three arcs are (1, 4), (1, 5), and (1, 7).

Revising the costs

Figure 9.13 suggests how to revise the shipping costs in a way that keeps all shipping costs nonnegative, keeps the cost of the partial shipment equal to zero, and allows the set R of reachable nodes to be enlarged. Denote as Δ the smallest of the shipping costs on arcs (i, j) with $i \in R$ and $j \notin R$. For the shipping costs in Table 9.3,

$$(16) \qquad \Delta = \min\{c_{ij} : i \in R,\ j \notin R\} = 1 = c_{14} = c_{15} = c_{17}.$$

This number Δ is positive, as must be.

Some shipping costs will be increased by Δ, some costs will be reduced by Δ, and the rest will not be unchanged. The shipping costs are now **revised** by:

- Reducing by Δ each shipping cost c_{ij} with $i \in R$ and $j \notin R$;
- Increasing by Δ each shipping cost c_{ij} with $i \notin R$ and $j \in R$.

In this example and in general, the revised costs have the properties that are listed below.

(a) If x_{ij} is positive, its cost c_{ij} remains equal to zero. (That occurs because R contains either both i and j or neither.)

(b) No cost becomes negative. (That occurs because each arc (i, j) for which cost c_{ij} will be reduced has $c_{ij} \geq \Delta$.)

(c) The cost of each arc in the reachable network remains equal to zero. (Each such arc has its head and its tail in R, so its cost is not revised.)

(d) Each arc (i, j) that attains the minimum in (16) has cost c_{ij} decreased to zero. Each such arc has i in R.

Point (a) guarantees that the cost of the partial shipping plan remains equal to zero. Point (b) guarantees that the shipping costs remain nonnegative. Point (c) guarantees that the arcs that had been in the reachable network remain in that network. Point (d) allows at least one arc to be added to the reachable network. In brief, the revised costs stay nonnegative, the cost of the partial shipping plan stays equal to zero, and set of reachable nodes can be enlarged.

Incremental shipment

This revision reduces to zero the shipping costs on arcs (1, 4), (1, 5) and (1, 7). As Figure. 9.13 attests, it becomes possible to ship some of the residual supply from node 2 to residual demand nodes 5 and 7. Shipping 1500 units from node 2 to node 5 on the chain $\{(2, 6)^F, (1, 6)^R, (1, 5)^F\}$ satisfies the residual demand at node 5, and it reduces x_{26} from 2,500 to 1,000. Shipping an additional 1,000 units from node 2 to node 7 on the on the chain $\{(2, 6)^F, (1, 6)^R, (1, 7)^F\}$ reduces the residual demand at node 7 from 1,500 to 500 units, and it reduces x_{26} from 1,000 to 0.

Table 9.4 reports the current shipment costs and the partial shipment plan that results from these shipments. A residual supply of 500 units remains at node 2, and a residual demand of 500 units remains at node 7.

Table 9.4. Revised shipping costs and partial shipping plan.

		demand nodes					
		4	5	6 R	7	8 R	residual supplies
supply nodes	1	0	1500	0	1000	0	
	2 R	6	2	2500	1	1000	500
	3	2000	1500	2	0	1	
residual demands					500		

For the partial shipment plan in Table 9.4, it is evident that

$$R = \{2, 6, 8\} \qquad \text{and} \qquad \Delta = 1 = c_{27},$$

moreover, that the next revision of the shipping costs will reduce c_{27} to 0, thereby allowing the remaining 500 units to be shipped directly from node 2 to node 7. Evidently, an optimal shipping plan sets

$$x_{15} = 1500, \qquad x_{17} = 1000, \qquad x_{26} = 2500, \qquad x_{27} = 500,$$

$$x_{28} = 1000, \qquad x_{34} = 2000, \qquad x_{35} = 1500,$$

with the other shipping quantities equal to zero.

Speed

The Hungarian method revises costs repeatedly. Each revision adds at least 2 nodes to the set R of nodes that can be reached from nodes having residual supply by shipping on zero-cost chains. After finitely many iterations, the set R must include a node that has residual demand, which makes incremental shipment possible.

For an assignment problem having m source nodes and m demand nodes, the number of times that flow must be augmented cannot exceed m. Also, not more than m data revisions will be needed before each flow augmentation. Each data revision requires work proportion to m^2, at worst. Thus, a worst-case work bound for the assignment problem is m^4. That is the *square* of the number of decision variables.

7. Review

Network flow models are an enormous subject, to which this chapter is an introduction. Reading it shows you how to identify a network flow model and whether the simplex method can be guaranteed to produce an integer-valued optimal solution. In the context of a transportation problem, you have seen:

- How to use multipliers to implement the simplex method without building simplex tableaus.
- That bases correspond to spanning trees.
- That the entering variable in a simplex pivot identifies a loop whose length is negative.

With some modification, these three properties hold for all network flow problems.

If you read the starred section on the Hungarian method, you learned of an algorithm that competes with the simplex method for network flow problems and that has polynomial worst-case behavior when it is applied to the assignment problem.

8. Homework and Discussion Problems

1. Beginning with the tableau in Figure 9.12, continue pivoting until you find an optimal solution to the transportation problem whose data appear in Figure 9.6. Did you encounter any degenerate pivots?

2. From a mathematical viewpoint, the network flow model is not presented in its simplest form. It is possible to eliminate all fixed flows. How? Does eliminating them simplify the linear program? Support your answers.

3. *(faster start for the transportation problem)* The Northwest Corner rule initializes the simplex method for transportation problems, but it ignores the shipping costs. This problem illustrates one of the ways to obtain an initial spanning tree that is feasible and that accounts for the shipping costs.

   (a) Suppose 3 is subtracted from each shipping cost in the left-most column of Figure 9.6. Argue that this subtracts 6,000 from the cost of every shipping plan, hence keeps the optimal plan(s) optimal.

   (b) Subtract a number from each column in Figure 9.6. Select these numbers so that each column contains at least one 0.

   (c) Begin construction of a spanning tree by shipping the most you can through a cell whose cost equals 0. If this exhausts the amount available in the cell's row (column), ship as much as possible to the next cheapest cell in its column (row). Continue until a feasible shipping plan has been constructed.

   (d) Does the computation in part (c) provide a bound on how close to optimal it is? If so, what is it? Support your answer.

4. A swimming coach has timed her five best swimmers in each of the four strokes that are part of a relay event. No swimmer is allowed to do more than one stroke. Recorded below is the amount (possibly zero) by which each swimmer's time exceeds the minimum in each stroke.

Lap	Alice	Beth	Carla	Doris	Eva
Freestyle	0.31	0	0.09	0.41	0.25
Butterfly	0.19	0	0.03	0.16	0.18
Breaststroke	0.31	0	0.2	0.19	0.21
Backstroke	0.28	0	0.32	0.21	0.26

(a) Formulate an assignment problem that assigns swimmers to strokes in a way that minimizes the time required for the team to complete the relay.

(b) Use Solver to find an optimal solution. Interpret the values that Solver assigns to the shadow prices.

Note: The next three problems relate to the transportation problem whose data are presented in the diagram:

2	0	3	0	3	2	3	supplies
0	1	6	0	0	6	4	
0	6	0	3	4	0	2	
4	1	4	0	3	4	8	
3	3	6	2	1	2		
demands							

5. For the 4 × 6 transportation problem whose data appear above:

   (a) Use the NW corner rule to find a spanning tree with which to initialize the simplex method.

   (b) Then use tableaus like the one in Figure 9.9 of this chapter to execute two pivots of the simplex method.

6. For the 4 × 6 transportation problem whose data appear above:

   (a) Find a partial shipping plan that exhausts the top three supplies and ships only on arcs that cost 0. Which demands are fully satisfied?

   (b) Draw the analogue of Table 9.2 for this partial shipment plan.

   (c) Indicate how to alter the shipping costs so as to allow the network you drew in part (b) to be enlarged.

   (d) Repeat steps (b) and (c) until you have found an incremental network that includes a demand node that has residual demand. Alter the partial shipping plan to ship as much as possible to that node, while keeping the shipping cost equal to zero.

7. For the 4 × 6 transportation problem whose data appears above:

   (a) Formulate this problem for solution on an Excel worksheet.

   (b) Use the Options Button to help you record each basic solution that Solver encounters as it solves this linear program. Turn in a list of the basic solutions that Solver found. Were any of these solutions degenerate?

8. Perform the Hungarian method on the 5 × 5 assignment problem whose costs and partial shipment plan are given in the table below. (Shipment of one unit at zero cost occurs on each shaded arc; also, one unit of residual supply exists at node 4, and 1 unit of unsatisfied demand exists at node 9.)

		demand node					
		6	7	8	9	10	
supply node	1	1	3	0	3	4	
	2	6	5	1	7	1	
	3	9	1	8	0	10	
	4	13	11	12	14	0	1
	5	16	15	1	7	0	
					1		

Part IV–LP Theory

Part IV presents a more penetrating account of linear programming than is in earlier chapters. The theory in Part IV is important on its own, as are the algorithms to which it leads. They also prepare you for Part V (game theory) and for Part VI (nonlinear systems).

Chapter 10. Vector Spaces and Linear Programs

Chapter 10 contains the information about vector spaces that relates directly to linear programs. You may find that much of this information is familiar, and you may find that linear programming strengthens your grasp of it.

Chapter 11. Multipliers and the Simplex Method

As was noted in earlier chapters, shadow prices may not exist. The multipliers always exist. They may not be unique. Even when the multipliers are ambiguous, they are shown to account properly for the relative opportunity cost of each decision variable. The multipliers are also shown to guide the simplex method as it pivots.

Chapter 12. Duality

In this chapter, the simplex method with multipliers is used to prove the "Duality Theorem" of linear programming. This theorem shows how each linear program is paired with another. Several uses of the Duality Theorem are presented in this chapter, and other uses appear in later chapters.

Chapter 13. The Dual Simplex Pivot and Its Uses

This chapter introduces you to the "dual" simplex pivot, and it presents several algorithms that employ simplex pivots and dual simplex pivots. One of these algorithms is a one-phase "homotopy" that pivots from an arbitrary basis to an optimal basis. Another algorithm solves integer programs.

Chapter 10: Vector Spaces and Linear Programs

1. Preview

Introductory textbooks on linear algebra present a great deal of information about vector spaces. This chapter includes the information that pertains directly to linear programming. It omits the rest. You may find that what remains here is coherent and, as a consequence, particularly accessible.

It is recalled that the column space of the matrix A is the set of all linear combinations of the columns of A. Similarly, the **row space** of the matrix A is the set of all linear combinations of the rows of A. Included in this chapter are:

E. V. Denardo, *Linear Programming and Generalizations*, International Series in Operations Research & Management Science 149,
DOI 10.1007/978-1-4419-6491-5_10, © Springer Science+Business Media, LLC 2011

- A demonstration that different bases for the same vector space must contain the same number of elements.
- A matrix interpretation of pivots.
- A demonstration that the "row rank" of a matrix equals its "column rank."
- Information about "invertible" matrices.
- A "theorem of the alternative" for solutions to systems of linear equations.

Gauss-Jordan elimination is this chapter's workhorse. As is usual, examples are used to introduce and illustrate properties that hold in general.

2. Matrix Notation

Let us begin by using matrix notation to describe a linear program that has been cast in Form 1. This linear program appears below as

Program 10.1. Maximize (or minimize) z, subject to the constraints

$$\begin{aligned} cx - z &= 0, \\ Ax \quad &= b, \\ x \quad &\geq 0. \end{aligned}$$

The data in Program 10.1 are the $m \times n$ matrix A, the $m \times 1$ vector b and the $1 \times n$ vector c. Its decision variables are z and x_1 through x_n. The decision variables x_1 through x_n are required to be nonnegative, and they are arrayed into the $n \times 1$ vector x. Evidently, Program 10.1 has m equality constraints, excluding the equation that defines z as the objective value, and it has n decision variables, other than z.

Matrix notation that had been introduced in Chapter 3 is now reviewed and developed somewhat. When A is an $m \times n$ matrix, A_j denotes the jth column of A, and A^i denotes the ith row of A. When A is an $m \times n$ matrix and B is an $n \times r$ matrix, the matrix product AB can be taken, and AB is the $m \times r$ matrix whose ij^{th} element is given by

(1) $$(AB)_{ij} = A^i B_j = \sum_{k=1}^{n} A_{ik} B_{kj}.$$

Using subscripts to identify columns allows the jth column of the matrix product AB to be expressed as

(2) $$(AB)_j = AB_j = A_1B_{1j} + A_2B_{2j} + \cdots + A_nB_{nj} \qquad \text{for each j.}$$

The second equation in (2) is familiar; to see why, substitute x for the column vector B_j and observe that this equation expresses Ax as a linear combination of the columns of A.

Similarly, using superscripts to identify rows allows the ith row of the matrix product AB to be written as

(3) $$(AB)^i = A^iB = A_{1i}B^1 + A_{2i}B^2 + \cdots + A_{mi}B^m \qquad \text{for each i,}$$

If, in addition, C is a $r \times s$ matrix, the matrix products (AB)C and A(BC) are easily seen to equal each other. In other words, matrix multiplication is *associative*, so no ambiguity occurs from omitting the parentheses in ABC, as we shall do.

As before, the superscript "T" denotes transpose; A^T denotes the $n \times m$ matrix whose ji^{th} element equals A_{ij} for each i and j.

An $m \times m$ matrix I is called an **identity matrix** if $I_{ii} = 1$ for $i = 1, 2, \ldots, m$, and if $I_{ij} = 0$ for each pair (i, j) having $i \neq j$. Evidently, I is a square matrix having 1's on the diagonal from its upper-leftmost entry to its lower-rightmost entry and having 0's in all other entries. Here and throughout, the symbol I is reserved for the identity matrix.

3. The Dimension of a Vector Space

A fundamental result in linear algebra is that different bases for the same vector space must contain the same number of vectors. In earlier chapters, this result was cited without proof. A proof appears here. This proof rests squarely on Gauss-Jordan elimination. It begins with

Proposition 10.1. Each set of $r \times 1$ vectors that is linearly independent contains r or fewer vectors.

Proof. Designate as S a set of linearly independent $r \times 1$ vectors, and denote as s the number of vectors in S. To prove Proposition 10.1, we must show that $s \le r$. Assign these s vectors the labels A_1 through A_s and let A be the $r \times s$ matrix whose j^{th} column is A_j for $j = 1, 2, \ldots, s$. Apply Gauss-Jordan elimination to the equation $Ax = 0$. This equation is satisfied by setting $x = 0$, so Proposition 3.1 shows that Gauss-Jordan elimination must identify a set C of columns of A that is a basis for the column space of A. The number $|C|$ of columns in this basis equals the number of rows on which pivots have occurred, and that cannot exceed the number r of rows of A. Hence, $|C| \le r$.

Aiming for a contradiction, suppose $r < s$. Since $|C| \le r$, at least one column of A must be excluded from the basis C for the column space, and that column is a linear combination of the set C of columns, which contradicts the hypothesis that the vectors in S are linearly independent, thereby completing a proof. ■

To adapt Proposition 10.1 to a set of row vectors, apply it to their transposes. Proposition 10.1 leads directly to:

Proposition 10.2. Consider any set V of $n \times 1$ vectors that contains at least two vectors and is a vector space.

(a) The vector space V has a basis, and that basis contains not more than n vectors.

(b) Every basis for V contains the same number of vectors.

Proof. This set V must contain at least one vector v other than 0. Beginning with $S = \{v\}$, augment S repeatedly, as follows: If V contains a vector w that is not a linear combination of the vectors in S, replace S by $S \cup \{w\}$. Repeat. Proposition 10.1 guarantees that this procedure stops before S contains $n + 1$ vectors. It stops with S equal to a set of linearly independent vectors that span V, that is, with a basis for V. This proves part (a).

For part (b), consider any two bases for V. Label the vectors in one of these bases A_1 through A_r, and label the vectors in the other basis B_1 through B_s, and label these bases so that $r \le s$. Each vector in the second basis is a linear combination of the vectors in the first. Since B_j is a linear combination of the vectors A_1 through A_r, there exist scalars C_{1j} through C_{rj} such that

$$(4.j) \qquad B_j = A_1C_{1j} + A_2C_{2j} + \cdots + A_rC_{rj} \qquad \text{for } j = 1, 2, \ldots s.$$

With A as the $n \times r$ matrix whose i^{th} column equals A_i for $i=1, \ldots r$ and with C as the $r \times s$ matrix whose ij^{th} entry is C_{ij} for $i=1, \ldots, r$ and $j=1, \ldots, s$, equations (4.j) and (2) give

$$(5.j) \qquad B_j = AC_j \qquad \text{for } j = 1, 2, \ldots, s.$$

With B as the $n \times s$ matrix whose j^{th} column equals B_j for $j=1, \ldots, s$, the equations in system (5) form

$$(6) \qquad B = AC.$$

The bases have been labeled so that $r \leq s$. Aiming for a contradiction, suppose that $r < s$. In this case, the $r \times s$ matrix C has more columns than rows, so Proposition 10.1 shows that its columns are linearly dependent, hence that there exists an $s \times 1$ vector $x \neq 0$ such that $Cx=0$. Postmultiply (6) by x to obtain $Bx=ACx=A(Cx)=A0=0$, which shows that the columns of B are linearly dependent. This cannot occur because the columns of B are a basis, so it cannot be that $r<s$. Thus, $r=s$, and a proof is complete. ■

Proposition 10.2 shows that each basis for a given vector space V has the same number of vectors. The number of elements in each basis for a vector space is called the **dimension** or **rank** of that vector space. In particular, the number of vectors in each basis for the column space of a matrix A is known as the **column rank** of A. Similarly, the number of vectors in each basis for the row space of a matrix A is known as the **row rank** of A.

4. Pivot Matrices: An Example

This is the first of two sections in which matrices are used to describe pivots. In the current section, an example is used to illustrate properties that hold in general.

A familiar example

The example that was used in Chapter 3 to introduce Gauss-Jordan elimination is now revisited. In that example, a solution was sought to matrix equation $Ax=b$ in which A and b are given by

$$(7) \qquad A = \begin{bmatrix} 2 & 4 & -1 & 8 \\ 1 & 2 & 1 & 1 \\ 0 & 0 & 2 & -4 \\ -1 & 1 & -1 & 1 \end{bmatrix} \quad \text{and} \quad b = \begin{bmatrix} 4 \\ 1 \\ -4/3 \\ 0 \end{bmatrix}.$$

The initial tableau [A, b] for this example is given by

$$(8) \qquad [A, b] = \begin{bmatrix} 2 & 4 & -1 & 8 & 4 \\ 1 & 2 & 1 & 1 & 1 \\ 0 & 0 & 2 & -4 & -4/3 \\ -1 & 1 & -1 & 1 & 0 \end{bmatrix}.$$

Equation (8) omits the column headings: Reading from left to right, these column headings are x_1 through x_4 and RHS. The entries in (8) are identical to the entries in cells B2:F5 of Table 3.1.

In Chapter 3, a sequence of three pivots transformed (8) into the tableau $[\bar{A}, \bar{b}]$ given by

$$(9) \qquad [\bar{A}, \bar{b}] = \begin{bmatrix} 1 & 0 & 0 & 5/3 & 1 \\ 0 & 0 & 1 & -2 & -2/3 \\ 0 & 0 & 0 & 0 & 0 \\ 0 & 1 & 0 & 2/3 & 1/3 \end{bmatrix}.$$

This tableau in equation (9) is basic: The variables x_1, x_2 and x_3 are basic for rows 1, 4 and 2, respectively, and row 3 is trite. Columns A_1, A_2 and A_3 are a basis for the column space of A.

It will soon be seen that the tableau in (9) is obtained by premultiplying the tableau in (8) by the product P(3) P(2) P(1) of three "pivot matrices," P(j) being the pivot matrix for the jth pivot.

A pivot matrix

In Chapter 3, the first pivot made x_1 basic for the 1st row of the tableau. This pivot changed the entries in the 1st column of (8) from $[2 \ \ 1 \ \ 0 \ \ -1]^T$ to $[1 \ \ 0 \ \ 0 \ \ 0]^T$. It did so by altering the rows of the tableau in (8) in these ways:

- Row (1) was replaced by itself times 1/2.
- Row (2) was replaced by itself plus row (1) times (−1/2).

- Row (3) was replaced by itself plus row (1) times (0).
- Row (4) was replaced by itself plus row (1) times (1/2).

It will be seen that the effect of this pivot is to premultiply [A, b] by the 4×4 **pivot matrix** P(1) that is specified by

$$P(1) = \begin{bmatrix} 1/2 & 0 & 0 & 0 \\ -1/2 & 1 & 0 & 0 \\ 0 & 0 & 1 & 0 \\ 1/2 & 0 & 0 & 1 \end{bmatrix}. \tag{10}$$

To see that the matrix product P(1) A has the desired effect, note that

$$P(1)\,A_1 = \begin{bmatrix} 1/2 & 0 & 0 & 0 \\ -1/2 & 1 & 0 & 0 \\ 0 & 0 & 1 & 0 \\ 1/2 & 0 & 0 & 1 \end{bmatrix} \begin{bmatrix} 2 \\ 1 \\ 0 \\ -1 \end{bmatrix} = \begin{bmatrix} 1 \\ 0 \\ 0 \\ 0 \end{bmatrix}.$$

Evidently, premultiplying [A, b] by P(1) makes x_1 basic for row 1. The tableau that results is

$$P(1)\,[A,\ b] = \begin{bmatrix} 1 & 2 & -1/2 & 4 & 2 \\ 0 & 0 & 3/2 & -3 & -1 \\ 0 & 0 & 2 & -4 & -4/3 \\ 0 & 3 & -3/2 & 5 & 2 \end{bmatrix}. \tag{11}$$

A variable has become basic for the 1st row of the tableau, for which that reason only the 1st *column* of P(1) differs from the corresponding column of the identity matrix.

The second pivot makes x_3 basic for the row (2) of the tableau in (11). This pivot changes the entries in the 3rd column of (11) from $[-1/2\ \ 3/2\ \ 2\ \ -3/2]^T$ to $[0\ \ 1\ \ 0\ \ 0]^T$. To check that premultiplying the tableau in (11) by the matrix P(2) given in

$$P(2) = \begin{bmatrix} 1 & 1/3 & 0 & 0 \\ 0 & 2/3 & 0 & 0 \\ 0 & -4/3 & 1 & 0 \\ 0 & 1 & 0 & 1 \end{bmatrix}. \tag{12}$$

executes this pivot, we note from (11) and (12) that:

$$P(2)\ [P(1)\,A]_3 = P(2)\begin{bmatrix}-1/2\\3/2\\2\\-3/2\end{bmatrix} = \begin{bmatrix}1 & 1/3 & 0 & 0\\0 & 2/3 & 0 & 0\\0 & -4/3 & 1 & 0\\0 & 1 & 0 & 1\end{bmatrix}\begin{bmatrix}-1/2\\3/2\\2\\-3/2\end{bmatrix} = \begin{bmatrix}0\\1\\0\\0\end{bmatrix}.$$

Only the 2nd column of P(2) differs from the identity matrix. Equations (10) and (12) illustrate a property that holds in general and is highlighted below:

> The pivot matrix for a pivot on a coefficient in the kth row differs from the identity matrix only in its *k[th] column.*

Spreadsheet computation

These pivot matrices can be created on a spreadsheet, and the matrix multiplications can be done with Excel. Table 10.1 indicates how. For instance:

- The array B3:E6 is the pivot matrix P(1).
- The array G9:K12 records the matrix product P(1)[A, b].
- Function in rows 27 to 30 indicate how the pivot matrices and matrix products are computed. The product Q(3) of three pivot matrices is found by using the Excel function =**MMULT(array, array)** *recursively*. The ability to do this is a nifty feature of Excel!

Cells B21:E:24 of Table 10.1 record the entries in

$$(13) \qquad Q(3) = P(3)\,P(2)\,P(1) = \begin{bmatrix}1/3 & -1/3 & 0 & -2/3\\-1/3 & 2/3 & 0 & 0\\2/3 & -4/3 & 1 & 0\\0 & 1/3 & 0 & 1/3\end{bmatrix}.$$

This example will next be used to illustrate two properties that are shared by every sequence of pivots.

An observation

In equation (13), the 3rd column of Q(3) equals I_3, the 3rd column of the identity matrix. That is no accident. No variable been made basic for the 3rd

Table 10.1. Gauss-Jordan elimination with pivot matrices.

	A	B	C	D	E	F	G	H	I	J	K
1											
2			P(1)						[A, b]		
3		0.5	0	0	0		2	4	-1	8	4
4		-0.5	1	0	0		1	2	1	1	1
5		0	0	1	0		0	0	2	-4	-1.333
6		0.5	0	0	1		-1	1	-1	1	0
7											
8			P(2)						P(1) [A, b]		
9		1	0.333	0	0		1	2	-0.5	4	2
10		0	0.667	0	0		0	0	1.5	-3	-1
11		0	-1.33	1	0		0	0	2	-4	-1.333
12		0	1	0	1		0	3	-1.5	5	2
13											
14			P(3)						P(2) P(1) [A, b]		
15		1	0	0	-0.67		1	2	0	3	1.6667
16		0	1	0	0		0	0	1	-2	-0.667
17		0	0	1	0		0	0	0	0	0
18		0	0	0	0.333		0	3	0	2	1
19											
20			Q(3) = P(3) P(2) P(1)						P(3) P(2) P(1) [A, b]		
21		0.333	-0.33	0	-0.67		1	0	0	1.6667	1
22		-0.33	0.667	0	0		0	0	1	-2	-0.667
23		0.667	-1.33	1	0		0	0	0	0	0
24		0	0.333	0	0.333		0	1	0	0.6667	0.3333
25											
26	cell	function			cell block				array function		
27	B3	=1/G3			G9 : K12				= mmult(B3:E6 : G3:K6)		
28	B4	= - G4 / G$3			G15 : K18				= mmult(B9:E12 , G9:K12)		
29	C9	= - I9 / I$10			G21 : K24				= mmult(B15:E18 : G15:K18)		
30	C10	= 1 / I10			B21 : E24				=mmult(B15:E18,mmult(B9:E12,B3:E6))		

row of the tableau, for which reason the 3rd column of the pivot matrices P(1), P(2) and (P3) equal I_3, so repeated use of equation (2) gives

$$(14)\qquad Q(3)_3 = P(3)P(2)P(1)_3 = P(3)P(2)I_3 = P(3)P(2)_3 = P(3)I_3 = P(3)_3 = I_3.$$

This suggests – correctly – that:

> If none of the first p pivots occur on an element in row k, the kth column of Q(p) equals I_k.

A second observation

Equation (9) shows that row (3) of the tableau $[\bar{A}, \bar{b}] = Q(3)\,[A, b]$ is trite; it consists entirely of 0's. Equation (3) shows that $\bar{A}^3 = Q(3)^3 A$. Hence, from (13), we see that

(15) $$[0, 0, 0, 0] = \overline{A}^3 = Q(3)^3 A = (2/3)A^1 + (-4/3)A^2 + 1A^3.$$

Equation (15) demonstrates that the 3rd row of A is a linear combination of the 1st and 2nd rows, specifically, that

(16) $$A^3 = (-2/3)A^1 + (4/3)A^2.$$

This suggests – correctly – that:

> If a sequence of pivots causes the kth row $\overline{A}^k$ to equal 0, then no pivot has occurred any coefficient in row k, and A^k is a linear combination of the rows of A on which pivots have occurred.

5. Pivot Matrices: General Discussion

Pivot matrices are now presented in a general setting, namely, that in which a solution is sought to the equation system, Ax=b. Here, as above, A is an $m \times n$ matrix, and b is an $m \times 1$ vector. The data in this equation system array themselves into the *initial tableau* (matrix):

(17) $$[A, b] = \begin{bmatrix} A_{11} & A_{12} & \cdots & A_{1n} & b_1 \\ A_{21} & A_{22} & \cdots & A_{2n} & b_2 \\ \vdots & \vdots & \cdots & \vdots & \vdots \\ A_{m1} & A_{m2} & \cdots & A_{mn} & b_m \end{bmatrix}.$$

After any number p (including 0) of pivots have occurred, the initial tableau is transformed into the tableau,

(18) $$[\overline{A}, \bar{b}] = \begin{bmatrix} \overline{A}_{11} & \overline{A}_{12} & \cdots & \overline{A}_{1n} & \bar{b}_1 \\ \overline{A}_{21} & \overline{A}_{22} & \cdots & \overline{A}_{2n} & \bar{b}_2 \\ \vdots & \vdots & \cdots & \vdots & \vdots \\ \overline{A}_{m1} & \overline{A}_{m2} & \cdots & \overline{A}_{mn} & \bar{b}_m \end{bmatrix}.$$

A pivot matrix

Let us suppose the tableau $[\overline{A}, \bar{b}]$ in (18) has been obtained by some number p of pivots and that the $p+1^{st}$ pivot occurs on the coefficient $\overline{A}_{ij}$ in the ith row and j^{th} column of (18). This coefficient must be nonzero. It will be seen that this pivot is executed by premultiplying the array $[\overline{A}, \bar{b}]$ by the $m \times m$ matrix $P(p+1)$ that differs from the identity matrix only in its ith column and is given by

$$(19) \qquad P(p+1) = \begin{bmatrix} 1 & & -\overline{A}_{1j}/\overline{A}_{ij} & & 0 \\ \vdots & \ddots & \vdots & & \vdots \\ 0 & \cdots & 1/\overline{A}_{ij} & \cdots & 0 \\ \vdots & & \vdots & \ddots & \vdots \\ 0 & & -\overline{A}_{mj}/\overline{A}_{ij} & & 1 \end{bmatrix}.$$

If the matrix product $P(p+1)[\overline{A}, \bar{b}]$ is to make x_j basic for row i, its jth column must have a 1 in row i and have 0's elsewhere. Substitute to obtain

$$P(p+1)\,\overline{A}_j = \begin{bmatrix} 1 & & -\overline{A}_{1j}/\overline{A}_{ij} & & 0 \\ \vdots & \ddots & \vdots & & \vdots \\ 0 & \cdots & 1/\overline{A}_{ij} & \cdots & 0 \\ \vdots & & \vdots & \ddots & \vdots \\ 0 & & -\overline{A}_{mj}/\overline{A}_{ij} & & 1 \end{bmatrix} \begin{bmatrix} \overline{A}_{1j} \\ \vdots \\ \overline{A}_{ij} \\ \vdots \\ \overline{A}_{mj} \end{bmatrix} = \begin{bmatrix} 0 \\ \vdots \\ 1 \\ \vdots \\ 0 \end{bmatrix},$$

exactly as desired. Evidently, the matrix product $P(p+1)[\overline{A}, \bar{b}]$ executes the pivot.

A sequence of pivots

Let us consider the effect of beginning with the tableau [A, b] and executing *any* finite sequence of pivots. If p pivots have occurred so far, the initial tableau [A , b] has been transformed into the current tableau

$$(20) \qquad [\overline{A}, \bar{b}] = Q(p)[A, b]$$

where the $m \times m$ matrix $Q(p)$ is given by

$$(21) \qquad Q(p) = P(p)P(p-1) \cdots P(1)$$

and where $P(j)$ is the pivot matrix for the jth pivot.

The example in the prior section suggests that if none of the first p pivots occurred on a coefficient in row k, then the kth column of $Q(p)$ equals the

k^{th} column of the identity matrix. That example also suggests that if the kth row of the matrix $\overline{A}$ that results from the first p pivots consists entirely of 0's, then the kth row of A is a linear combination of the rows of A on which pivots have occurred. These suggestions are shown to be accurate by parts (a) and (b) of

Proposition 10.3. Equations (20) and (21) describe the tableau $[\overline{A}, \bar{b}]$ that results from any finite number p of pivots on an initial tableau [A, b]. Denote as R the set of rows on which these p pivots have occurred.

(a) If R excludes k, then $Q(p)_k = I_k$.

(b) If $\overline{A}^k = 0$, then R excludes k, and A^k is a linear combination of the set $\{A^i : i \in R\}$ of rows of A.

Proof. The hypothesis of part (a) is that p pivots occur on coefficients in rows other than row (k). Thus, as noted earlier, $P(j)_k = I_k$ for $j = 1, 2, \ldots, p$. Since $Q(1) = P(1)$, this guarantees $Q(1)_k = P(1)_k = I_k$. Adopt the inductive hypothesis that $Q(j-1)_k = I_k$, which has just been verified for the case $j = 2$. Since $Q(j) = P(j)Q(j-1)$, equation (2) gives $Q(j)_k = P(j)Q(j-1)_k$, and the inductive hypothesis gives $Q(j)_k = P(j)I_k = P(j)_k = I_k$. This proves part (a).

The hypothesis of part (b) is that row (k) of $\overline{A}^k$ consists entirely of zeros. Had a pivot occurred on a coefficient in row (k), row (k) would have a basic variable at each tableau thereafter, and $\overline{A}^k$ could not consist entirely of zeros. So it must be that $k \notin R$. We have $0 = \overline{A}^k = [Q(p)A]^k = Q(p)^k A$, the last from equation (3), so that

$$0 = Q(p)^k A = \sum_{i=1}^{m} Q(p)_{ik} A^i . \tag{22}$$

Since $k \notin R$, part (a) of this proposition gives $Q(p)_{kk} = 1$. Part (a) also gives $Q(p)_{ik} = 0$ for each $i \neq k$ that is not in R, Thus, from (22),

$$0 = A^k + \sum_{i \in R} Q(p)_{ik} A^i. \tag{23}$$

This proves part (b). ■

6. The Rank of a Matrix

In Chapter 3, it had been shown that application of Gauss-Jordan elimination to the equation Ax=0 constructs a basis for the column space of the matrix A. This basis consists of the columns on which pivots occur. Proposition 10.4 (below) shows the same execution of Gauss-Jordan elimination constructs a basis for the row space of A. That basis consists of the rows on which pivots occur. A bi-product of this result is that the row rank of every matrix equals its column rank.

Proposition 10.4 *(bases via Gauss-Jordan elimination).* Consider any matrix A. Apply Gauss-Jordan elimination to the equation Ax=0 and, at termination, denote as C the set of columns on which pivots have occurred, and denote as R the set of rows on which pivots have occurred. Then:

(a) The set $\{A_j : j \in C\}$ is a basis for the column space of A.

(b) The set $\{A^i : i \in R\}$ is a basis for the row space of A.

(c) The row rank of A equals the column rank of A.

Proof. The equation Ax=0 has a solution, so Gauss-Jordan elimination must terminate with a basic tableau and with $|R| = |C|$. Part (a) has been established by Proposition 3.1. Part (b) of Proposition 10.3 shows that $\{A^i : i \in R\}$ spans the row space of A. This set contains $|R|$ vectors, so Proposition 10.2 shows that the row rank of A cannot exceed $|R|$.

Since $|R| = |C|$, we have shown that *every* matrix A has row rank that does not exceed its column rank. The row rank of the transpose of A equals the column rank of A, so it must be that every matrix A has row rank equal to its column rank, which proves part (c). Moreover, since $\{A^i : i \in R\}$ spans the row space of A, it cannot be that $\{A^i : i \in R\}$ is linearly dependent, as this would imply that the row rank of A is less than $|R|$. Hence, $\{A^i : i \in R\}$ is a basis for the row space of A, which proves part (b). ■

The conclusions of Proposition 10.4 hold when the vector 0 is replaced by any $m \times 1$ vector b in the column space of A: When Gauss-Jordan elimination is implemented, the columns that become basic are a basis for the column space of A, and the rows that do not become trite are a basis for the row space of A.

Proposition 10.4 demonstrates that that the row rank of each matrix equals its column rank. This justifies a definition; the **rank** of a matrix is the number of vectors in any basis for its column space or in any basis for its row space.

7. The Full Rank Proviso

The Full Rank proviso was employed in Chapter 4. In that chapter, a linear program was said to satisfy the Full Rank proviso if *any* basic tableau for its Form 1 representation has a basic variable for each row. Program 10.1 is written in the format of Form 1.

Proposition 10.5. Program 10.1 satisfies the Full Rank proviso if and only if the rows of A are linearly independent.

Proof. The constraints of Program 10.1 are the equations $cx - z = 0$ and $Ax = b$ and the nonnegativity requirements $x \geq 0$. The equations form the linear system

$$\begin{bmatrix} c & 1 \\ A & 0 \end{bmatrix} \begin{bmatrix} -z \\ x \end{bmatrix} = \begin{bmatrix} 0 \\ b \end{bmatrix}.$$

Let us consider the $(m + 1) \times (n + 1)$ matrix F given by

$$F = \begin{bmatrix} c & 1 \\ A & 0 \end{bmatrix}.$$

No linear combination of the rows of $[A \quad 0]$ spans the row vector $[c \quad 1]$, for which reason the row rank of F exceeds the row rank of A by 1.

Suppose the Full Rank proviso is satisfied. A basic solution exists that has $m+1$ basic variables. Proposition 10.2 shows that every basic solution has $m+1$ variables, hence that the rank of A equals m.

Suppose the rank of A equals m. The row rank of F must equal $m+1$, so every basic solution must have a basic variable for each row. ■

8. Invertible Matrices

Many readers will recall that the $m \times m$ matrices B and C are each others' inverses if BC = I. Some readers will recall that the preceding statement is a theorem – an implication of a more primitive definition of the "inverse" of a matrix.

The $m \times m$ matrix B is now said to be **invertible** if there exists $m \times m$ matrices C and D such that

(24) $$CB = BD = I.$$

If B is invertible, then (24) and the fact that matrix multiplication is associative gives

(25) $$C = CI = C(BD) = (CB)D = ID = D.$$

Evidently, B is invertible if and only if there exists an $m \times m$ matrix C such that

(26) $$CB = BC = I.$$

Equation (25) also shows that at most one matrix C can satisfy (26).

If B is invertible, the unique matrix C that satisfies (26) is called the **inverse** of B. The inverse of B, if it exists, is denoted B^{-1}. Not every square matrix is invertible; if a row of B consists entirely of 0's, the matrix B cannot be invertible, for instance.

Elementary properties of inverses are recorded in

Proposition 10.6.

(a) A square matrix B can have at most one inverse.

(b) If B and D are invertible $m \times m$ matrices, their product BD is invertible, and

(27) $$(BD)^{-1} = D^{-1}B^{-1}.$$

Proof. Part (a) has been proved. For part (b) the fact that matrix multiplication is associative is used in

$$(BD)(D^{-1}B^{-1}) = B(DD^{-1})B^{-1} = BIB^{-1} = BB^{-1} = I.$$

A similar argument shows that $(D^{-1}B^{-1})(BD) = I$, which verifies (27), and completes a proof. ■

Pivot matrices

In Chapter 3, it was observed that the effects of a pivot could be undone. This suggests that pivot matrices are invertible. The $m \times m$ pivot matrix P to the right of the equal sign in equation (19) differs from the identity matrix only in its ith column. Let us consider the $m \times m$ matrix R that also differs from the identity matrix only in its ith column and is given by

$$\text{(28)} \qquad R = \begin{bmatrix} 1 & & \overline{A}_{1j} & & 0 \\ \vdots & \ddots & \vdots & & \vdots \\ 0 & \cdots & \overline{A}_{ij} & \cdots & 0 \\ \vdots & & \vdots & \ddots & \vdots \\ 0 & & \overline{A}_{mj} & & 1 \end{bmatrix}.$$

It is easily seen that $PR = RP = I$, hence that R is the inverse of P.

Permutation matrices

The $m \times m$ matrix S is called a **permutation** matrix if S contains exactly m non-zero entries, if each row of S contains exactly one 1 and if each column of S contains exactly one 1. It is easily checked that the transpose S^T of a permutation matrix S satisfies $SS^T = S^TS = I$. In other words, the transpose of a permutation matrix is its inverse.

Conditions for invertibility

The main result of this section is the characterization of invertible matrices in

Proposition 10.7. Let B be an $m \times m$ matrix. The following are equivalent:

(a) There exists an $m \times m$ matrix C that satisfies $CB = I$.

(b) The columns of B are linearly independent.

(c) The rows of B are linearly independent.

(d) There exists a product Q of pivot matrices and a permutation matrix J such that QP=J.

(e) There exists an $m \times m$ matrix C that satisfies CB=BC=I.

Proof. It will be demonstrated that $(a) \Rightarrow (b) \Rightarrow (c) \Rightarrow (d) \Rightarrow (e) \Rightarrow (a)$.

$(a) \Rightarrow (b)$: Suppose that the $m \times m$ matrix C satisfies CB=I. To show that the columns of B are linearly independent, consider any $m \times 1$ vector x for which Bx=0. Premultipy by C and then use CB=I to obtain Ix=C0, equivalently, x=0. $(b) \Rightarrow (c)$: Suppose that the columns of B are linearly independent. The column rank of B equals m. By Proposition 10.4, the row rank of B also equals m. So the rows of B are linearly independent.

$(c) \Rightarrow (d)$: Suppose the rows of B are linearly independent. Application of Gauss-Jordan elimination to the equation Bx=0 transforms the initial tableau [B, 0] into the final tableau [QB, 0] in which Q is a product of pivot matrices and in which Proposition 10.3 guarantees that the matrix J=QB is a permutation matrix.

$(d) \Rightarrow (e)$: Suppose J=QB where Q is a product of pivot matrices and J is a permutation matrix. We have seen that each pivot matrix is invertible and that each permutation matrix is invertible. Proposition 10.6 shows that the product Q of invertible matrices is invertible. Set $C=(J^{-1}Q)$. Premultiply J=QB by J^{-1} to obtain $I=J^{-1}QB=(J^{-1}Q)B=CB$. Premultiply J=QB by Q^{-1} to obtain $Q^{-1}J=B$. Postmultiply this equation by $J^{-1}Q$ to obtain $I=BJ^{-1}Q=BC$, which completes a demonstration that CB=BC=I.

$(e) \Rightarrow (a)$: Suppose CB=BC=I. Clearly, CB=I. This competes a proof. ∎

Gauss-Jordan elimination lies at the heart of the proof of Proposition 10.7.

9. A Theorem of the Alternative

A **theorem of the alternative** is a statement that exactly one of two alternatives must hold. The proposition that appears below is known as a **theorem of the alternative for linear systems.**

Proposition 10.8 *(theorem of the alternative for linear systems).* For each $m \times n$ matrix A and each $m \times 1$ vector b, exactly one of the following alternatives holds:

(a) There exists an $n \times 1$ vector x such that $Ax = b$.

(b) There exists a $1 \times m$ vector y such that $yA = 0$ and $yb \neq 0$.

Proof. The proof will show that if (a) holds, (b) cannot and that if (a) does not hold, (b) must.

(a) implies not (b). By hypothesis, there exists an $n \times 1$ vector x such that $Ax = b$. Aiming for a contradiction, suppose that (b) also holds, so there exists a $1 \times m$ vector y such that $yA = 0$ and $yb \neq 0$. Premultiply $Ax = b$ by y to obtain $yAx = yb \neq 0$. Postmultiply $yA = 0$ by x to obtain $yAx = 0\,x = 0$. This establishes the contradiction $0 = yAx \neq 0$. Thus, if (a) holds, (b) cannot.

Not (a) implies (b). By hypothesis, there exists no $n \times 1$ vector x such that $Ax = b$. Application of Gauss-Jordan elimination to the array [A, b] must result in an inconsistent row. Proposition 10.3 shows that the resulting tableau is $Q[A, b] = [QA, Qb]$ for some matrix Q. This tableau has an inconsistent row, say, the ith row. From (3), we see that $Q^iA = 0$ and that $Q^ib \neq 0$, so (b) holds with $y = Q^i$. ■

Who cares?

Of what interest is a theorem of the alternative? Suppose we wished to demonstrate that no solution can exist to the matrix equation $Ax = b$. Proposition 10.8 shows that this is equivalent to the existence of a solution to $yA = 0$ and $yb \neq 0$, which may be easier to demonstrate.

A point of logic

It's easy to stumble when trying to prove that two or more statements are equivalent. Proposition 10.8 will be used to illustrate the pitfall, along with a foolproof way to avoid it. Proposition 10.8 asserts that the two statements listed below are equivalent:

- Condition (a) holds.
- Condition (b) does not hold.

This raises a point of logic. Listed below are four implications, each of which can be part of a demonstration that the above two conditions are equivalent. Here and throughout, " $\Rightarrow$ " means "implies" and " $\Leftarrow$ " means "is implied by."

1. (a) $\Rightarrow$ not (b)
2. (b) $\Rightarrow$ not (a)
3. not (a) $\Rightarrow$ (b)
4. not (b) $\Rightarrow$ (a)

To prove an equivalence, we must demonstrate two of the four implications that are listed above, *but not any two*. Highlighted below is a rule worth memorizing.

Logic: An implication is identical to the implication that "reverses everything."

"Reversing everything" in the implication "(b) $\Rightarrow$ not (a)" produces the implication "not (b) $\Leftarrow$ (a)." Thus, implications 1 and 2 are identical to each other, and implications 3 and 4 are identical to each other. We established Proposition 10.8 by proving 1 and 3. We could have proved 2 and 3. It would not do to prove 1 and 2.

Typically, in an equivalence relation, one pair of implications is easy to establish, and the other pair is more difficult. The pitfall is to prove both members of the easy pair. Use the "reverses everything" paradigm to avoid that trap.

10. Carathéodory's Theorem

A very useful result that is due to Constantin Carathéodory (1872–1950) is presented as

Proposition 10.9 *(Carathéodory's theorem)*. Let $S \subseteq \Re^m$ be a set that contains more than $m+1$ vectors. If b is a convex combination of the vectors in S, then b is a convex combination of at most $m+1$ of these vectors.

Proof. By hypothesis, b is a convex combination of the vectors in S. Number the vectors in S that have positive weights in this convex combination v_1 through v_r. There is a set c_1 through c_r of positive numbers such that

$$b = \sum_{i=1}^{r} c_i v_i, \qquad \sum_{i=1}^{r} c_i = 1.$$

If $r \le m+1$, there is nothing to prove.

Suppose $r > m+1$. In this case, the set consisting of the vectors $(v_2 - v_1)$, $(v_3 - v_1)$, …, $(v_r - v_1)$ consists of at least $m+1$ vectors in $\Re^m$. Proposition 10.1 shows that these vectors are linearly dependent, so there exist numbers d_2 through d_r not all of which equal zero such that

$$0 = \sum_{i=2}^{r} d_i (v_i - v_1).$$

Define d_1 by $d_1 = -\sum_{i=2}^{r} d_i$, and note that

$$0 = \sum_{i=1}^{r} d_i v_i, \qquad 0 = \sum_{i=1}^{r} d_i,$$

Not all of d_1 through d_r equal zero, and they sum to 0, so at least one of them is positive. Define R by

$$R = \min \left\{ \frac{c_i}{d_i} \; : \; d_i > 0 \right\}. \tag{29}$$

Note that R must be positive. Define

$$e_i = c_i - R d_i \qquad \text{for } i = 1, \ldots, r.$$

Evidently, e_1 through e_r are nonnegative numbers that sum to 1, with $e_i = 0$ for at least one i and with $\sum_{i=1}^{r} e_i v_i = b$. Thus, b is a convex combination of fewer than r of the vectors in S. Repeating this argument reduces r to $m+1$ and completes a proof. ■

Proposition 10.9 is known as **Carathéodory's theorem**. It was proved in 1911, and it shares a feature with the simplex method; the "smallest ratio" in (29) is used to determine which vector is removed.

11. Review

Gauss-Jordan elimination is indeed the workhorse of this chapter. A single application of Gauss-Jordan elimination has been shown construct a pair of bases, one for the column space and one for the row space: Since these bases have the same number of vectors, the column rank of each matrix equals its row rank. Gauss-Jordan elimination also played a crucial role in the demonstration that square matrices B and C are each others' inverses if BC=I.

The pivot matrices that are introduced in this chapter will play a key role in Chapter 11. A generalization of the theorem of the alternative in this chapter will play a key role in Chapter 12.

12. Homework and Discussion Problems

1. This problem concerns the matrix equation Ax=b in which

$$A = \begin{bmatrix} 2 & 4 & -1 \\ 1 & 2 & 1 \\ -1 & 1 & -1 \end{bmatrix} \quad \text{and} \quad b = \begin{bmatrix} 8 \\ 1 \\ 1 \end{bmatrix}.$$

Parts (a)-(c) of this problem ask you to adapt the Excel computation in Table 10.1 to the above data.

(a) For these data, use pivot matrices to execute Gauss-Jordan elimination, pivoting at each opportunity on the left-most nonzero coefficient in the lowest-numbered row for which a basic variable has not yet been found.

(b) Compute the product Q of the pivot matrices.

(c) Is QA=J where J is a permutation matrix?

(d) Do the results of your computation resemble Table 10.1? If so, why?

2. Verify that matrix multiplication is associative by showing that (AB)C= A(BC).

3. Table 10.1 specifies pivot matrices P(1), P(2) and P(3) that transform (8) into (9). Use equation (28) to write down the inverse of P(1), of P(2) and of P(3).

4. All parts of this problem concern the 4×4 matrix B given by

$$B = \begin{bmatrix} 1 & 1 & 0 & 0 \\ 0 & 0 & 1 & 1 \\ 1 & 0 & 0 & 4 \\ 0 & 0 & 0 & 2 \end{bmatrix}.$$

(a) Show that the rows of B are linearly independent.

(b) Without doing any further calculation, determine whether or not the columns of B are linearly independent.

(c) On a spreadsheet, execute Gauss-Jordan elimination to find a product Q of pivot matrices and a permutation matrix J such that QB=J. *Remark:* One (relatively easy?) way to do this is to begin with the 4×8 tableau [BI] and pivot only on elements in the first four columns.

(d) On the same spreadsheet, compute J^T and J^TQ. *Remark:* Excel has an array function that computes the transpose of a matrix.

(e) On the same spreadsheet, verify that $J^TQ = B^{-1}$.

5. Displayed below are the 5×5 permutation matrices J and $\hat{J}$.

$$J = \begin{bmatrix} 0 & 0 & 1 & 0 & 0 \\ 0 & 0 & 0 & 0 & 1 \\ 0 & 0 & 0 & 1 & 0 \\ 1 & 0 & 0 & 0 & 0 \\ 0 & 1 & 0 & 0 & 0 \end{bmatrix} \qquad \hat{J} = \begin{bmatrix} 0 & 0 & 1 & 0 & 0 \\ 1 & 0 & 0 & 0 & 0 \\ 0 & 0 & 0 & 1 & 0 \\ 0 & 1 & 0 & 0 & 0 \\ 0 & 0 & 0 & 0 & 1 \end{bmatrix}$$

(a) Specify the inverse of J and the inverse of $\hat{J}$.

(b) Draw a directed network that includes nodes 1 through 5 and directed arc (i, j) if and only if $J_{ij} = 1$. Use this network to determine the smallest positive integer n for which $J^n = I$. (Here, $J^3 = J\,J\,J$, for instance.)

(c) Repeat part (b) for the matrix $\hat{J}$.

(d) What is the smallest positive integer n such that *every* 5×5 permutation matrix J has $J^n = I$? Why?

6. A square matrix E is called an **exchange** matrix if it can be obtained by switching exactly two rows of the identity matrix. Is the exchange matrix E invertible? If so, what is its inverse?

7. Let Q, R and S be $m \times m$ matrices. Suppose that Q and S are invertible, and that R is not invertible.

   (a) Show that the columns of QR are linearly dependent.

   (b) Show that the rows of QR are linearly dependent.

   (c) Show that QRS is not invertible.

8. Let A be an $m \times n$ matrix and let B be an $n \times m$ matrix. Suppose $m < n$. Show that BA is not invertible.

9. Let a, b, c and d be any numbers such that $(ab - cd) \neq 0$. Show that the 2×2 matrices given below are each others' inverses.

$$\begin{bmatrix} a & c \\ d & b \end{bmatrix} \qquad \frac{1}{(ab - cd)} \begin{bmatrix} b & -c \\ -d & a \end{bmatrix}.$$

10. Let A be an $m \times n$ matrix, and let b be a $m \times 1$ vector. Prove or disprove: If the equation $Ax = b$ has no solution, the rows of A are linearly dependent.

11. Is it possible to locate four ×'s and one ⋆ on a sheet of paper in such a way that the ⋆ is a convex combination of the four ×'s but is not a convex combination of fewer than four of the ×'s? If not, why not?

Chapter 11: Multipliers and the Simplex Method

1. Preview

The tableau-based simplex method in Chapter 4 masks a relationship between the data in the initial tableau and in the tableaus that result from sequences of pivots. That relationship is now brought into view. Each tableau encountered by the simplex method will be shown to have at least one set of "multipliers," one per constraint. These multipliers will be seen to:

- Serve as break-even prices, even when they are not unique.

- Guide the simplex method in its choice of pivot element.

In Chapter 12, the multipliers will emerge as the decision variables in a second linear program, which will be called the "dual."

This chapter is focused on the simplex method, as was Chapter 4. The development here is more advanced. The key ideas in this chapter are highlighted (enclosed in boxes). It might be best to focus on them first and to

E. V. Denardo, *Linear Programming and Generalizations,* International Series in Operations Research & Management Science 149,
DOI 10.1007/978-1-4419-6491-5_11,

fill in the details later. Proofs of the propositions in this chapter are starred because of their lengths.

The simplex method with multipliers was illustrated in Chapter 9. Each basis for an m×n transportation problem had multipliers u_1 through u_m and v_1 through v_n that were easy to compute because each basic variable x_{ij} has $c_{ij} = u_i + v_j$.

2. The Initial and Current Tableaus

Let us turn our attention to a general description of a linear program that has been cast in Form 1, namely:

Program 11.1. Maximize (or minimize) z, subject to

$$
\begin{array}{llcl}
(1.0) & c_1x_1 + c_2x_2 + \cdots\cdots + c_nx_n - z &=& 0 \\
(1.1) & A_{11}x_1 + A_{12}x_2 + \cdots\cdots + A_{1n}x_n &=& b_1 \\
(1.2) & A_{21}x_1 + A_{22}x_2 + \cdots\cdots + A_{2n}x_n &=& b_2 \\
 & \vdots \qquad \vdots \qquad \qquad \vdots & & \vdots \\
(1.m) & A_{m1}x_1 + A_{m2}x_2 + \cdots\cdots + A_{mn}x_n &=& b_m \\
(1.m+1) & x_1 \geq 0, \quad x_2 \geq 0, \ \ldots, \ x_n \geq 0. & &
\end{array}
$$

The format of Program 11.1 is familiar:

- The decision variable are z and x_1 through x_n.
- The integer m is the number of equations, other than the one that defines z as the objective value.
- The integer n is the number of decision variables, other than z.
- The number c_j is the coefficient of x_j in the objective.
- The number b_i is the right-hand-side (RHS) value of the ith constraint.
- The number A_{ij} is the coefficient of x_j in the ith constraint.

The initial tableau

The data for Program 11.1 array themselves into the **initial tableau** (matrix) that is depicted below.

$$
(2) \qquad \begin{matrix} \text{row } 0 \\ \text{row } 1 \\ \text{row } 2 \\ \vdots \\ \text{row } m \end{matrix} \begin{bmatrix} c_1 & c_2 & \cdots & c_n & 1 & 0 \\ A_{11} & A_{12} & \cdots & A_{1n} & 0 & b_1 \\ A_{21} & A_{22} & \cdots & A_{2n} & 0 & b_2 \\ \vdots & \vdots & & \vdots & \vdots & \vdots \\ A_{m1} & A_{m2} & \cdots & A_{mn} & 0 & b_m \end{bmatrix}
$$

As this notation suggests, the top-most row of the initial tableau is called **row 0**, and the others are called **row 1** through **row m.**

The numbers in the 1st column of the initial tableau are the coefficients of x_1 in equations (1.0) through (1.m), the numbers in the 2nd column are the coefficients of x_2 in these equations, and so forth, through the nth column. The numbers in the next-to-last column multiply $-z$, and the final column consists of the RHS values.

Admissible pivots

When applied to the initial tableau, the simplex method executes a sequence of pivots. Each of these pivots occurs on a nonzero coefficient of a decision variable in some row other than row 0. Any pivot that occurs on a non-zero coefficient of some variable in some row $i \geq 1$ is now said to be an **admissible pivot**. Each simplex pivot is admissible, but admissible pivots need not be simplex pivots.

The current tableau

A tableau that can result from *any* finite sequence of admissible pivots is now called a **current tableau.** Each current tableau is depicted as

$$
(3) \qquad \begin{matrix} \text{row } 0 \\ \text{row } 1 \\ \text{row } 2 \\ \vdots \\ \text{row } m \end{matrix} \begin{bmatrix} \bar{c}_1 & \bar{c}_2 & \cdots & \bar{c}_n & 1 & \bar{b}_0 \\ \bar{A}_{11} & \bar{A}_{12} & \cdots & \bar{A}_{1n} & 0 & \bar{b}_1 \\ \bar{A}_{21} & \bar{A}_{22} & \cdots & \bar{A}_{2n} & 0 & \bar{b}_2 \\ \vdots & \vdots & & \vdots & \vdots & \vdots \\ \bar{A}_{m1} & \bar{A}_{m2} & \cdots & \bar{A}_{mn} & 0 & \bar{b}_m \end{bmatrix} .
$$

Bars atop the entries in (3) record the fact that they can differ from the numbers in the corresponding positions of (2). The next-to-last column of tableau (3) equals that of (2) because no admissible pivot occurs on a coefficient in row 0. The entry in the upper-right-hand corner of tableau (3) is denoted $\bar{b}_0$, which need not equal 0 because each admissible pivot replaces row 0 by itself less a constant times some other row.

Matrix notation

The data in the initial and current tableaus group themselves naturally into the matrices and vectors in:

$$(4) \qquad \begin{bmatrix} c & 1 & b_0 \\ A & 0 & b \end{bmatrix} \quad \text{and} \quad \begin{bmatrix} \bar{c} & 1 & \bar{b}_0 \\ \bar{A} & 0 & \bar{b} \end{bmatrix}.$$

Here, c and $\bar{c}$ are the $1 \times n$ vectors

$$c = [c_1 \quad c_2 \quad \cdots \quad c_n], \qquad \bar{c} = [\bar{c}_1 \quad \bar{c}_2 \quad \cdots \quad \bar{c}_n],$$

and A and $\bar{A}$ are the $m \times n$ matrices

$$A = \begin{bmatrix} A_{11} & A_{12} & \cdots & A_{1n} \\ A_{21} & A_{22} & \cdots & A_{2n} \\ \vdots & \vdots & & \vdots \\ A_{m1} & A_{m2} & \cdots & A_{mn} \end{bmatrix}, \quad \bar{A} = \begin{bmatrix} \bar{A}_{11} & \bar{A}_{12} & \cdots & \bar{A}_{1n} \\ \bar{A}_{21} & \bar{A}_{22} & \cdots & \bar{A}_{2n} \\ \vdots & \vdots & & \vdots \\ \bar{A}_{m1} & \bar{A}_{m2} & \cdots & \bar{A}_{mn} \end{bmatrix},$$

and b and $\bar{b}$ are the $m \times 1$ vectors

$$b = \begin{bmatrix} b_1 \\ b_2 \\ \vdots \\ b_m \end{bmatrix}, \qquad \bar{b} = \begin{bmatrix} \bar{b}_1 \\ \bar{b}_2 \\ \vdots \\ \bar{b}_m \end{bmatrix},$$

and the RHS value b_0 of the initial tableau equals 0. Finally, the "0's" in equation (4) are $m \times 1$ vectors of 0's.

A relationship

The relationship between the initial and current tableaus is described by a vector y and a matrix Q in

Proposition 11.1. Suppose Ax = b has a solution. Consider any current tableau for Program 11.1.

(a) There exist at least one $1 \times m$ vector y and at least one $m \times m$ matrix Q such that

$$\bar{c} = c - yA, \qquad \bar{b}_0 = -yb, \tag{5}$$

$$\overline{A} = QA, \qquad \overline{b} = Qb. \tag{6}$$

(b) If the rank of A equals m, equations (5) and (6) have unique solutions y and Q. If the rank of A is less than m, equations (5) and (6) have multiple solutions.

Proof*. The proof of part (a) will be by induction on the number p of pivots that have occurred. For the case p=0, equations (5) and (6) hold with y=0 and Q=I. We adopt the inductive hypothesis that (5) and (6) hold after p pivots have occurred. These equations state, respectively, that:

- Row 0 of the current tableau equals row 0 of the initial tableau less a linear combination of rows 1 through m of the initial tableau.
- Rows 1 through m of the current tableau are linear combinations of rows 1 through m of the initial tableau.

The $(p+1)^{st}$ pivot occurs on a nonzero coefficient $\overline{A}_{ij}$ in row $i \geq 1$ of the current tableau. Row i is multiplied by the constant $1/\overline{A}_{ij}$, so it remains a linear combination of rows 1 through m of the initial tableau. Each of the other rows is replaced by itself less some constant times row i of the current tableau. Thus, each row is replaced by itself less a linear combination of rows 1 through m of the initial tableau. As a consequence, a revised matrix Q and a revised vector y satisfy the highlighted properties after p + 1 pivots have occurred. This completes an inductive proof of part (a).

For part (b), we first consider the case in which the rank of [A, b] is less than m. The rows of [A, b] are linearly dependent, so there exists a nonzero $1 \times m$ vector w such that wA=0 and wb = 0. Replacing y by (y + w) preserves a solution to (5). Similarly, with W as the $m \times m$ matrix each of whose rows equals w, replacing Q by (Q + W) preserves a solution to (6). Hence, the solutions to (5) and (6) cannot be unique.

Let us consider the case in which the rank of A equals m. The rows of A are linearly independent. Solutions y and $\tilde{y}$ to (5) satisfy $(y - \tilde{y})A = 0$, and, since the rows of A are linearly independent, this guarantees $y = \tilde{y}$. Similarly, solutions Q and $\tilde{Q}$ to (6) satisfy $(Q - \tilde{Q})A = 0$. For each i, we have $(Q - \tilde{Q})^i A = 0$. The fact that the rows of A are linearly independent (again) guarantees $Q^i = \tilde{Q}^i$ for each i, so that $Q = \tilde{Q}$. This completes a proof of part (b). ■

It was demonstrated in Proposition 10.5 that the Full Rank proviso holds if and only if the rank of A equals m. Thus, y and Q are unique if the Full Rank proviso is satisfied, and they are not unique if it is violated.

Multipliers

Each vector y that satisfies (5) is said to be a set of **multipliers** for the current tableau, and y_i is called the **multiplier** for the ith constraint. To see where the multipliers get their name, we rewrite equation (5) as

(7) $$[\bar{c} \quad 1 \quad \bar{b}_0] = [c \quad 1 \quad 0] - y[A \quad 0 \quad b].$$

Equation (7) contains the same information as does equation (5); it states that $\bar{c} = c - yA$ and that $\bar{b}_0 = -yb$. Equation (7) can be read as:

> The top row of the current tableau equals the top row of the initial tableau less the sum over each $i \geq 1$ of the constant (multiplier) y_i times the ith row of the initial tableau.

An existential result

Proposition 11.1 is not constructive. It does not tell us how to compute a vector y and a matrix Q that satisfy (5) and (6). Initially, before any pivots have occurred, equations (5) and (6) hold with y=0 and Q=I. We will soon see how to compute y and Q *recursively*, that is, by updating them in a way that accounts for each pivot.

3. Updating the Current Tableau

Presented in this section is a method for updating Q and y so as to implement a pivot. The discussion commences with an interpretation of Proposition 11.1. Part (a) asserts that there exist at least one $1 \times m$ vector y and at least one m × m matrix Q that satisfy equations (5) and (6). Equations (5) and (6) can be written succinctly as

$$\text{(8)} \qquad \begin{bmatrix} \bar{c} & 1 & \bar{b}_0 \\ \bar{A} & 0 & \bar{b} \end{bmatrix} = \begin{bmatrix} 1 & -y \\ 0 & Q \end{bmatrix} \begin{bmatrix} c & 1 & 0 \\ A & 0 & b \end{bmatrix}.$$

Please pause to check that equation (8) contains *exactly* the same information as do equations (5) and (6), for instance, that (8) gives

$$\bar{c} = 1c - yA = c - yA.$$

Equation (8) motivates the introduction of the $(m+1)\times(m+1)$ matrix $\tilde{Q}$ given by

$$\text{(9)} \qquad \tilde{Q} = \begin{bmatrix} 1 & -y \\ 0 & Q \end{bmatrix}.$$

An interpretation of equations (8) and (9) is highlighted below.

> The current tableau is obtained by premultiplying the initial tableau by the matrix $\tilde{Q}$ that is given by equation (9).

Part (b) of Proposition 11.1 shows that the $\tilde{Q}$ is unique if and only if the rank of A equals m.

Accounting for a pivot

Let us consider how to update $\tilde{Q}$ so as to account for a pivot on a nonzero coefficient $\overline{A}_{ij}$ in the current tableau. From Chapter 10, we see that this pivot premultiplies the current tableau by the $(m+1)\times(m+1)$ matrix $\tilde{P}$ that differs from the identity matrix only in its $(i+1)^{st}$ column and is given by

$$\text{(10)} \qquad \tilde{P} = \begin{bmatrix} 1 & 0 & \cdots & -\bar{c}_j/\overline{A}_{ij} & \cdots & 0 \\ 0 & 1 & \cdots & -\overline{A}_{1j}/\overline{A}_{ij} & \cdots & 0 \\ \vdots & \vdots & \ddots & \vdots & & \vdots \\ 0 & 0 & \cdots & 1/\overline{A}_{ij} & \cdots & 0 \\ \vdots & \vdots & & \vdots & \ddots & \vdots \\ 0 & 0 & \cdots & -\overline{A}_{mj}/\overline{A}_{ij} & \cdots & 1 \end{bmatrix}.$$

Note, from equations (8) and (9), that the effect of this pivot is to replace $\tilde{Q}$ by the matrix product $\tilde{P}\,\tilde{Q}$. This matrix product may look messy. It will turn out to have a simple interpretation, however. What remains of $\tilde{P}$ after removal of its top row and left-most column is the (familiar) $m\times m$ matrix P that is given by

$$\text{(11)} \qquad P = \begin{bmatrix} 1 & \cdots & -\overline{A}_{1j}/\overline{A}_{ij} & \cdots & 0 \\ \vdots & \ddots & \vdots & & \vdots \\ 0 & \cdots & 1/\overline{A}_{ij} & \cdots & 0 \\ \vdots & & \vdots & \ddots & \vdots \\ 0 & \cdots & -\overline{A}_{mj}/\overline{A}_{ij} & \cdots & 1 \end{bmatrix}.$$

With I^i as the ith row of the $m\times m$ identity matrix, the $(m+1)\times(m+1)$ matrix $\tilde{P}$ partitions itself as

$$\text{(12)} \qquad \tilde{P} = \begin{bmatrix} 1 & (-\bar{c}_j/\overline{A}_{ij})\,I^i \\ 0 & P \end{bmatrix},$$

where "0" denotes the $m \times 1$ vector each of whose entries equals 0.

With $\tilde{Q}$ given by (9) and $\tilde{P}$ given by (12), the matrix product $\tilde{P}\,\tilde{Q}$ partitions itself as

$$\text{(13)} \qquad \tilde{P}\tilde{Q} = \begin{bmatrix} 1 & (-\bar{c}_j/\overline{A}_{ij})\,I^i \\ 0 & P \end{bmatrix} \begin{bmatrix} 1 & -y \\ 0 & Q \end{bmatrix} = \begin{bmatrix} 1 & -y - \dfrac{\bar{c}_j}{\overline{A}_{ij}} Q^i \\ 0 & PQ \end{bmatrix}.$$

This discussion is summarized by

Proposition 11.2 *(updating y and Q)*. Consider a current tableau, and suppose that y and Q satisfy (5) and (6). A pivot on a nonzero coefficient $\overline{A}_{ij}$ in this tableau updates y by

$$\text{(14)} \qquad y \leftarrow y + \frac{\bar{c}_j}{\overline{A}_{ij}} Q^i,$$

and, with P given by (12), this pivot updates Q by

$$\text{(15)} \qquad Q \leftarrow PQ.$$

Proof*. As noted above, this update premultiplies $\tilde{Q}$ by $\tilde{P}$, for which reason equation (13) verifies (14) and (15). ■

The content of Proposition 11.2 is highlighted below.

> A pivot on a nonzero coefficient $\bar{A}_{ij}$ in the current tableau replaces y by itself plus the constant $(\bar{c}_j/\bar{A}_{ij})$ times the ith row of Q, and it replaces Q by PQ.

Thus, starting with Q=I and y=0 and updating Q and y with each pivot results in a matrix Q and a vector y that satisfy (5) and (6) for the current basis. This occurs even if the Full Rank proviso is violated.

4. Multipliers as Break-Even Prices

Let us recall from Chapter 5 that relative opportunity cost is *relative* to the current plan. In a linear program, each basis (set of basic variables) is a plan, and the relative opportunity cost of doing something equals the decrease in profit (increase in cost) that occurs if the resources needed to do that thing are freed up and the values of the basic variables are adjusted accordingly. Proposition 11.3 (below) shows that the multipliers can be interpreted as break-even prices even when they are not unique.

Proposition 11.3 *(multipliers as prices).* Consider any current tableau for Program 11.1 that is basic. Let y be any set of multipliers for this tableau.

(a) The basic solution for this tableau has y b as its objective value.

(b) Let d be any vector in the column space of A. Replacing b by (b+d) in the initial tableau and repeating the sequence of pivots that led to the current tableau keeps the tableau basic and changes its objective value by y d.

(c) Suppose Program 11.1 is a maximization problem. For each j, the relative opportunity cost of the resources needed to set $x_j = 1$ equals yA_j.

(d) Row i has a unique multiplier y_i if and only if the column space of A contains the $m \times 1$ vector I_i (which has 1 in its ith position and has 0's in all other positions).

Proof*. For part (a), we note that the basic solution to this tableau equates $-z$ to $\bar{b}_0$, so (5) gives $-z=-yb$ and $z=yb$.

For part (b), consider a vector y and matrix Q for which (5) and (6) hold with the RHS vector b. Let us replace b by (b + d) and then repeat the pivot sequence that led to the current tableau. This has no effect on y or Q. It has no effect on $\bar{A} = QA$ or on $\bar{c} = c - yA$. Each variable x_j that was basic remains basic. The vector $\bar{b} = Q\,b$ is replaced by $Q(b+d)$, and the number $\bar{b}_0 = -yb$ is replaced by $-y(b+d)$. By hypothesis, b and d are in the column space of A, so that any equation that was trite remains trite. The tableau remains basic, and (5)-(6) continue to hold, which proves part (b).

For part (c), note that removing the resources needed to set $x_j=1$ replaces b by $(b - A_j)$, so part (b) shows that the basic solution's objective changes by $y \times (-A_j) = -yA_j$. In a maximization problem, profit decreases by yA_j. This proves part (c).

For part (d), we first consider the case in which the column space V_c of A contains the vector I_i. The sum of two vectors in V_c is in V_c, so the vector $(b+I_i)$ is in V_c, and part (b) shows that changing the RHS vector from b to $(b+I_i)$ changes the basic solution's objective by $yI_i=y_i$. This demonstrates that y_i is unique.

Now consider the case in which the column space V_c does not contain I_i. Since b is in V_c, there does exist an $n\times 1$ vector x such that $b=Ax$. Since I_i is not in V_c, no solution exists to $Az=I_i$. Proposition 10.8 shows that there does exist a row vector v such that $vA=0$ and $vI_i \neq 0$. Premultiply $b=Ax$ by v to obtain $vb=vAx=(vA)x=0$. We have seen that $vA=0$, that $vb=0$ and that $0 \neq vI_i = v_i$. Each vector y that satisfies (5) is a set of multipliers for the current tableau. With y as such a vector, note that $(y+v)$ also satisfies (5), hence is a set of multipliers. These multipliers satisfy $(y+v)I_i = y_i + v_i \neq y_i$ because $v_i \neq 0$. Hence, the multiplier for the ith constraint cannot be unique. This completes a proof. ■

Proposition 11.3 shows that the vector y of multipliers play the role of break-even prices in these ways:

- The equation $z=yb$ shows that the multipliers are break-even prices for the *entire* bundle b of resources.

- The equation $z = y(b + d)$ shows that the multipliers are break-even prices for any vector d of perturbations of the RHS values that lies in the column space of A.

- For a maximization problem, the equation $\bar{c}_j = c_j - yA_j$ shows that yA_j is the decrease in profit that occurs if the resources needed to set $x_j = 1$ are set aside.

All of this occurs whether or not y is unique.

Shadow prices

Denote as V_c the column space of A. Let us consider two cases: If V_c contains the vector I_i, Part (d) of Proposition 11.3 shows that the multiplier y_i for the ith constraint is unique, and part (b) shows that y_i is the shadow price for the ith constraint. Alternatively, if V_c does not contain I_i, Part (d) shows that the multiplier y_i for the ith constraint is not unique. Also, since $b + I_i$ is not in V_c, the ith constraint cannot have a shadow price. As a consequence, the multiplier for a constraint is unique if and only if it is the constraint's shadow price. It is emphasized:

> Consider any basic solution to $Ax = b$. The multiplier y_i for the ith constraint is unique if and only if y_i is the shadow price for the ith constraint.

Solver and Premium Solver report shadow prices whether or not they exist. What these codes are actually reporting is a set of multipliers for the final basis. Calling these multipliers "shadow prices" follows a long-standing tradition.

Sensitivity Analysis with Premium Solver

To illustrate what occurs when the Full Rank proviso is violated, we present to Premium Solver the linear program that appears below.

Program 11.2. $z^* = \text{Maximize } \{4x_1 + 2x_2 + 4x_3\}$, subject to the constraints

row 1 $\quad 3x_1 + 2x_2 + 1x_3 = 4,$

row 2 $\quad 6x_1 + 4x_2 + 2x_3 = 8,$

row 3 $\quad 1x_1 + 2x_2 + 3x_3 = 4,$

$$x_1 \geq 0, \quad x_2 \geq 0, \quad x_3 \geq 0.$$

This linear program has three equality constraints. Its 2nd constraint is a linear multiple of its 1st constraint. Perturbing the RHS of either of these two constraints renders the linear program infeasible. Neither can have a shadow price. When Premium Solver is presented with this linear program, it reports an optimal solution that sets

$$x_1 = 1, \qquad x_2 = 0, \qquad x_3 = 1,$$

and it reports an optimal value $z^* = 8$. Its sensitivity analysis reports that reduced costs of x_1, x_2 and x_3 equal 0, –2, and 0, respectively. Its sensitivity analysis also reports the shadow prices and ranges that appear in

Table 11.1. Shadow prices and ranges for Program 11.2.

	Shadow Price	Constraint R.H. Side	Allowable Increase	Allowable Decrease
row 1	0	4	0	0
row 2	0.5	8	0	0
row 3	1	4	8	2 2/3

The "shadow prices" for rows 1 and 2 are actually multipliers. To double-check that the vector $y = [0 \;\; 0.5 \;\; 1]$ of multipliers does satisfy equation (5), we substitute and obtain

$$\bar{c} = [4 \;\; 2 \;\; 4] - (0.5)[6 \;\; 4 \;\; 2] - (1.0)[1 \;\; 2 \;\; 3] = [0 \;\; -2 \;\; 0],$$

$$z^* = (0)(4) + (0.5)(8) + (1.0)(4) = 8,$$

both of which are correct.

Table 11.1 reports that the RHS value of row 1 has 0 as its Allowable Increase and its Allowable Decrease because perturbing the RHS value of row 1 renders the linear program infeasible. The same is true for row 2. Premium Solver (correctly) reports that the basis solution remains feasible when the RHS value of row 3 is increased by as much as 8 and when it is decreased by as much as 2 2/3. Row 3 does have a shadow (break-even) price, and it does apply to changes in the RHS value of the 3rd constraint that lie between –2 2/3 and +8.

Sensitivity Analysis with Solver

As the time this book is being written, the Sensitivity Report issued by Solver differs from that in Table 11.1. Solver reports correct values of the multipliers. For this example (and others that violate the Full Rank proviso), Solver reports *incorrect ranges* of those RHS values that cannot be perturbed without rendering the linear program infeasible.

Final words

The fact that multipliers are break-even prices suggests that they ought to play a key role in the simplex method. Yet the multipliers are all but invisible in the tableau-based simplex method that was presented in Chapter 4. The next section of this chapter shows that the multipliers are crucial to a version of the simplex method that is better suited to solving large linear programs.

The tem "multiplier" abbreviates **Lagrange multiplier**. The multipliers *are* the Lagrange multipliers. And if you use an algorithm that is designed for non-linear optimization, Solver will report the values of the Lagrange multipliers.

5. The Simplex Method with Multipliers

The tableau-based simplex method is a great way to learn how the simplex method works, and it is a fine way in which to solve linear programs that have only a modest number of equations and decision variables. For really large linear programs, there is a better way.

Described in this section is the version of the simplex method that is implemented in several commercial codes. This method was originally dubbed the **revised simplex method**[1], but it has long been known as the **simplex method with multipliers**. As its name suggests, this method uses the multipliers to guide the simplex method as it pivots. The simplex method with multipliers has two main advantages:

- It is faster when the number n of decision variables is several times the number m of constraints.
- It requires careful control of round-off error on 3 m numbers, rather than on the entire tableau.

[1] Dantzig, George B. and William Orchard-Hayes, "Notes on linear programming: Part V – alternative algorithm for the revised simplex method using product form for the inverse," RM 1268, The RAND Corporation, Santa Monica, CA, November 19, 1953.

A third advantage, as is noted later in this section, is that it dovetails nicely with "column generation."

A nonterminal iteration

To describe an iteration of the simplex method with multipliers, suppose we are in Phase II and that Q and y for the current basis are known.

A simplex pivot with multipliers (i.e., with y and Q):

1. From y, compute the vector $\bar{c} = c - yA$ of reduced costs.

2. Select as the entering variable x_j as any variable whose reduced cost $\bar{c}_j$ is positive (negative) in the case of a maximization (minimization) problem. Compute $\bar{b}$ and $\bar{A}_j$ from

(16) $$\bar{b} = Qb \qquad \text{and} \qquad \overline{A}_j = QA_j.$$

3. Find a row i whose ratio $\bar{b}_i/\overline{A}_{ij}$ is closest to zero, among those rows having $\overline{A}_{ij} > 0$.

4. Replace y by itself plus the multiple $\bar{c}_j/\overline{A}_{ij}$ of the ith row Q^i of Q. Then, with the pivot matrix P given by (11), replace Q by PQ. Return to Step 1.

Proposition 11.2 shows that the update of Q and y in Step 4 executes the pivot. Phase II of keeps pivoting until it encounters an optimal solution (in Step 2) or an unbounded linear program (in Step 3). This version of the simples method requires careful control of the round-off error in the vectors y, $\bar{b}$ and $\bar{A}_j$ but not on the entire tableau.

Sparsity and the product form

When pivoting in a basis, it is a good idea to keep the product Q of the pivot matrices as sparse as possible for as long as possible. Pivot in the slack variables first. An alternative to storing the entire m×m matrix Q is to store only the column of each pivot matrix P that corresponds to the row on which the pivot occurs and to compute $\bar{b}$ and $\bar{A}_j$ and y recursively. This storage system is dubbed the **product form of the inverse.** (That usage is accurate *if* the constraint matrix A has full rank.) Finally, replacing pivots by "lower" pivots and back-substitution (as suggested in Chapter 3) retards error growth and improves sparseness.

Eventually, after enough pivots have occurred, the round-off error will have accumulated to the point at which it can no longer be dealt with. When that occurs, it becomes necessary to begin again – to "pivot in" the current basis, and then restart the simplex method. This too is easier to accomplish using the simplex method with multipliers.

Column generation

Column generation is well-suited to a linear program that has a great many columns and in which the following is true: Given any basis, a column whose reduced cost is best (e.g., largest in a maximization problem) can be found from the solution to a subordinate optimization problem whose data include the values of the multipliers for the current basis. If the reduced cost of this column is positive in a maximization problem (negative in a minimization problem), pivot and repeat. If not, stop. An example in which column generation is attractive can be found in Denardo, Feinberg and Rothblum.[2]

6. The Basis Matrix and Its Inverse

The presentation of the simplex method in this chapter is not typical. In most presentations, the relationship between the initial tableau and the current tableau is *not* established via a vector y and a matrix Q that satisfy (5) and (6). Instead, the Full Rank proviso is assumed to hold, and the "inverse of the basis matrix" is employed. In this section, this chapter's development is related to the more common one.

Commonly-used notation

This subsection deals with the case in which Program 11.1 satisfies the Full Rank proviso: Thus, the rank of A equals m, the equation $Ax=b$ is consistent, and each basic tableau has one basic variable per row.

Let us consider any basic tableau that might be encountered by the simplex method. The variable $-z$ is basic for row 0, and rows 1 through m have basic variables. These basic variables are used to identify a function β, an

[2] Denardo, E. V., E. A. Feinberg and U. G. Rothblum, "The multi-armed bandit, with constraints," submitted for publication.

m×m matrix B and a 1×m vector c_B by the following procedure: For i=1, 2, …, m.

- The decision variable $x_{\beta(i)}$ is basic for row i of this tableau.
- The 1×m vector c_B has $c_{\beta(i)}$ as its ith entry.
- The m×m matrix B has $A_{\beta(i)}$ as its ith column.

In the literature, the matrix B that is prescribed by these rules is called a **basis matrix.** The ith column of B is the column $A_{\beta(i)}$ of coefficients of the variable $x_{\beta(i)}$ that is basic for row i. The matrix B is square (because the Full Rank proviso is satisfied), and B is invertible (because its columns are linearly independent).

An example

To illustrate this notation, we reconsider the linear program that was used in Chapter 4 to introduce the simplex method. Table 11.2 reproduces its initial and final tableaus. Its decision variables (previously x, y and s_1 through s_4) are now labeled x_1 through x_6, however. This example satisfies the Full Rank proviso because the variables x_3 through x_6 are basic for rows 1 through 4 of the initial tableau.

Table 11.2. Initial and final tableaus for a maximization problem.

	A	B	C	D	E	F	G	H	I
1									
2		x1	x2	x3	x4	x5	x6	- z	RHS
3	row 0	2	3					1	0
4	row 1	1		1					6
5	row 2	1	1		1				7
6	row 3	0	2			1			9
7	row 4	-1	3				1		9
8									
9		x1	x2	x3	x4	x5	x6	- z	RHS
10	row 0				-2 1/4		- 1/4	1	-18
11	row 1			1	- 3/4		1/4		3
12	row 2	1			3/4		- 1/4		3
13	row 3				- 1/2	1	- 1/2		1
14	row 4		1		1/4		1/4		4

The tableau in rows 10-14 of Table 11.2 is basic. The variables that are basic for rows 1 through 4 of this tableau are x_3, x_1, x_5 and x_2, respectively. For each i, $\beta(i)$ identifies the variable that is basic for row i of this tableau, and

$$\beta(1) = 3, \quad \beta(2) = 1, \quad \beta(3) = 5, \quad \beta(4) = 2,$$

$$c_B = \begin{bmatrix} c_3 & c_1 & c_5 & c_2 \end{bmatrix} = \begin{bmatrix} 0 & 2 & 0 & 3 \end{bmatrix},$$

(17) $$B = \begin{bmatrix} A_3 & A_1 & A_5 & A_2 \end{bmatrix} = \begin{bmatrix} 1 & 1 & 0 & 0 \\ 0 & 1 & 0 & 1 \\ 0 & 0 & 1 & 2 \\ 0 & -1 & 0 & 3 \end{bmatrix}.$$

Each column of B is a column of A, but these columns do not appear in their natural order. For instance, column A_1 appears as the 2nd column of B (not the 1st) because x_1 is basic for row 2. The columns of B are linearly independent, so B is invertible. Its inverse is recorded as:

(18) $$B^{-1} = \begin{bmatrix} 1 & -3/4 & 0 & 1/4 \\ 0 & 3/4 & 0 & -1/4 \\ 0 & -1/2 & 1 & -1/2 \\ 0 & 1/4 & 0 & 1/4 \end{bmatrix},$$

This matrix B^{-1} appears in Table 11.2. To see why, recall that x_3 through x_6 are the slack variables for the original tableau.

When the Full Rank proviso is satisfied

If the Full Rank proviso is satisfied, equations (5) and (6) are satisfied by a unique vector y and a unique matrix Q. Their relationship to each basic tableau's basis matrix B and to its vector c_B are the subject of

Proposition 11.4. Suppose that Program 11.1 satisfies the Full Rank proviso, and consider any current tableau that is basic. The matrix Q and vector y that satisfy (5) and (6) are

(19) $$y = c_B B^{-1} \quad \text{and} \quad Q = B^{-1}.$$

Furthermore, this current tableau relates to the initial tableau through

(20) $$\overline{A} = B^{-1}A, \qquad \bar{b} = B^{-1}b, \qquad \bar{c} = c - c_B B^{-1}A,$$

and the current tableau's basic solution has objective value z that is given by

(21) $$z = c_B B^{-1} b.$$

Proof*. Preparing to verify the left-hand equation in (19), we denote as I_i the ith column of the $m \times m$ identity matrix. In the current tableau, the variable $x_{\beta(i)}$ is basic for row i, so $\overline{A}_{\beta(i)} = I_i$, and (6) gives $\overline{A}_{\beta(i)} = QA_{\beta(i)}$. The matrix B has $B_i = A_{\beta(i)}$. We have shown that

$$I_i = \overline{A}_{\beta(i)} = QA_{\beta(i)} = QB_i.$$

This equation holds for each i, so $I = QB$. Hence, $Q = B^{-1}$.

The variable $x_{\beta(i)}$ is basic for row i, so the reduced cost $\bar{c}_{\beta(i)}$ of this variable equals 0. Thus, (5) gives $c_{\beta(i)} = yA_{\beta(i)}$. Since $A_{\beta(i)} = B_i$, we have demonstrated that $c_{\beta(i)} = yB_i$. This equation holds for each i, so $c_B = yB$. Postmultiply this equation by B^{-1} to obtain $c_B B^{-1} = y$. This verifies (19). Equations (20) and (21) are immediate from (6), (5) and the fact that the basic solution equates $-z$ to the RHS value $\bar{b}_0$ of row 0. This completes a proof. ■

The gist of Proposition 11.4 is highlighted below:

> When the Full Rank proviso is satisfied, the matrix Q and the vector y satisfy $Q = B^{-1}$ and $y = c_B B^{-1}$.

When the Full Rank proviso is violated

Suppose, however, that the Full Rank proviso is *not* satisfied. The rank of A is less than m, so each basis for the column space of A consists of fewer than m columns. The "basis matrix" has fewer columns than rows. It is not square, and it cannot have an inverse. Results that are stated in terms of B^{-1} cannot be valid. These results become correct, however, when B^{-1} is replaced by any matrix Q that satisfies (6) and when $c_B B^{-1}$ is replaced by any vector y that satisfies (5). It is highlighted:

> When the Full Rank proviso is violated, results that are stated in terms of B^{-1} and $c_B B^{-1}$ become correct when Q replaces B^{-1} and y replaces $c_B B^{-1}$.

In brief, the more standard development coalesces with ours when B^{-1} is replaced by the product Q of the pivot matrices that led to the current tableau and when $c_B B^{-1}$ is replaced by the vector y of multipliers.

7. Review

Proposition 11.1 relates the current tableau to the initial tableau via a row vector y and a square matrix Q that satisfy equations (5) and (6). This vector y and matrix Q are unique if and only if the Full Rank proviso is satisfied. Proposition 11.2 shows how to compute solutions y to (5) and Q to (6) recursively, by accounting for each pivot. Proposition 11.3 shows how the vector y of *multipliers* plays the role of break-even prices, even when y is not unique. Proposition 11.4 relates the development in this chapter to the more typical one, in which the Full Rank proviso is satisfied. In addition, the multipliers have been shown to be key to the "revised" simplex method, which is better-suited to solving large linear programs.

In concert, the results in this chapter show that the multipliers play a crucial role in linear programming. In Chapter 12, it will be seen that the multipliers play yet another role – they are the decision variables in a second linear program, which is known as the "dual."

8. Homework and Discussion Problems

1. Suppose that the equation Ax=b is consistent and that its ith row includes a slack variable (that converted an inequality into an equation.)

   (a) Show that the system Ax=b remains consistent when the RHS value of the ith constraint is perturbed.

   (b) Does the ith constraint have a shadow price?

   (c) In every basic tableau, is y_i unique?

   (d) In every basic tableau, is y_i a shadow price?

2. (*nonnegative column*) For a maximization problem in Form 1, the following tableau has been encountered. In it, * stands for an unspecified data element. Prove there exist no values of the unspecified data for which it is optimal to set A > 0. *Hint:* If it did, could you reduce A and increase ________?

A	B	C	D	E	F	-z	RHS
-3		*		*		1	*
4	1	*		*			*
2		*	1	*			*
0		*		*	1		*

3. Table 11.1 reports the optimal solution and sensitivity report to Program 11.2 that was obtained with Premium Solver. Obtain comparable reports from Solver. Do you see any differences?

4. Suppose that a constraint in a linear program lacks a shadow price. Premium Solver reports one any how. What is it reporting? Does Premium Solver provide a clue that no shadow price exists?

5. Consider a linear program that is written in Form 1 and for which the sum of the rows of A equals the $1 \times n$ vector $0 = (0 \;\; 0 \;\; \cdots \;\; 0)$. Does this linear program have shadow prices? Support your answer.

6. Cells D11:G14 of Table 11.2 contain the inverse of the matrix B that is given by (17). This is not an accident. Why? *Hint:* cells D4:G7 of the same table give the entries in the matrix $I = BB^{-1}$ and the product Q of the pivot matrices that produce the tableau in Table 11.2 equals B^{-1}.

7. Suppose that a linear program's constraints are $Ax \le b$ and $x \ge 0$, so that its Form 1 representation includes a slack variable for each "≤" constraint. Support your answers to parts (a)-(c).

   (a) Does this linear program satisfy the Full Rank proviso?

   (b) In each basic tableau, where can the shadow prices be found?

   (c) In each basic tableau, where can the inverse of the basis matrix be found?

8. This problem concerns the linear program that appears below. Which of its constraints have shadow prices and which do not? Support your answer.

$$\text{Maximize } \{2x_1 + 1x_2 + 2x_3\}, \text{ subject to}$$

$$3x_1 + 2x_2 + 1x_3 = 4,$$

$$1x_1 + 2x_2 + 3x_3 = 6,$$

$$9x_1 + 6x_2 + 3x_3 = 12,$$

$$x_1 \geq 0, \quad x_2 \geq 0, \quad x_3 \geq 0.$$

9. Consider a basic tableau for a linear program that is written in the format: Minimize c x, subject to the constraints $Ax \geq b$ and $x \geq 0$. Where can the shadow prices for this tableau be found? Support your answer.

Chapter 12: Duality

1. Preview

In Chapter 11, each current tableau was seen to have at least one vector y of *multipliers* that determine its vector $\bar{c}$ of reduced costs and its objective value z via $\bar{c} = c - yA$ and $z = yb$. It was also noted that these multipliers, if unique, are the shadow prices. A method was presented for computing a vector y of multipliers, whether or not they are unique.

In the current chapter, these multipliers emerge as the decision variables in a second linear program, which is known as the "dual." It will be demonstrated that a linear program and its dual have these properties:

E. V. Denardo, *Linear Programming and Generalizations*, International Series in Operations Research & Management Science 149,
DOI 10.1007/978-1-4419-6491-5_12, © Springer Science+Business Media, LLC 2011

- If a linear program is unbounded, its dual cannot be feasible.
- A linear program cannot be feasible if its dual is unbounded.
- If a linear program and its dual are feasible, then
  - These two linear programs have the same optimal value.
  - Application of the revised simplex method to either linear program terminates with an optimal solution to it and with a vector y of multipliers that is optimal for the dual.

Duality is a potent tool. In this chapter, duality is used to:

- Prove a classic and important result of Farkas (1896).
- Analyze a model that compares the efficiency of different units of an organization.
- Characterize the "no-arbitrage tenet" of financial economics.
- Establish a result that is known as "strong complementary slackness."

In later chapters, duality will be used to:

- Construct a general equilibrium in a simplified model of an economy (see Chapter 14).
- Construct an equilibrium in a competitive game (see Chapter 14).
- Characterize optimal solutions of nonlinear programs (see Chapter 20).

The role played by duality in constrained optimization is comparable to the role played by Gaussian elimination in linear algebra. Both are fundamental, and equally so.

2. Dual Linear Programs

Duality will be introduced in the context of

Program 12.1. $z^* = \text{Max } c\, x$, subject to

$$A x = b,$$

$$x \geq 0.$$

Program 12.1 particularizes Form 1 by requiring that the objective function be maximized, not minimized. Let us recall that Form 1 is a canonical form. Program 12.1 is a also canonical form because a minimization problem can be converted into an equivalent maximization problem by multiplying each coefficient in its objective function by –1.

The dual of Program 12.1

Program 12.1^D (below) has the same data as does Program 12.1. Its decision variables form the $1 \times m$ (row) vector y.

Program 12.1^D. $z_* =$ Min y b, subject to

$$yA \geq c,$$

$$y \text{ is free.}$$

Program 12.1^D is called the **dual** of Program 12.1. Since Program 12.1 is a canonical form, this defines the dual of every linear program.

The "D" in Program 12.1^D is intended to connote "dual." The superscripted optimal value z^* of Program 12.1 reminds us that something is being maximized (made high), and the subscripted optimal value z_* in Program 12.1^D reminds us that something is being minimized (made low). The word "dual" suggests – correctly, as we shall see – that taking the dual of the dual brings us back to the linear program with which we started.

An unwieldy definition?

This definition of the dual linear program is unambiguous, but it can be unwieldy. For instance, to take the dual of Program 12.1^D, we would need to cast it in the format of Program 12.1. That is unnecessary. Later in this chapter, a recipe will be provided for taking the dual of any linear program, *without* first writing it in the format of Program 12.1.

3. Weak Duality

This chapter's analysis of Program 12.1 and its dual begins with an easy-to-prove result that is known as "weak duality."

Proposition 12.1 *(weak duality).* Suppose that x is feasible for Program 12.1 and that y is feasible for Program 12.1^D.

(a) Then

(1) $$yb \geq z_* \geq z^* \geq cx.$$

(b) Also, each inequality in (1) holds as an equation if and only if x and y satisfy

(2) $$\left(\sum_{i=1}^{m} y_i A_{ij} - c_j\right)(x_j) = 0 \quad \text{for } j = 1, 2, \ldots, n.$$

Proof. Feasibility of x and y gives

$$Ax = b, \quad x \geq 0, \quad \text{and} \quad yA \geq c.$$

Premultiply $Ax = b$ by y to obtain $yAx = yb$. Write $yA \geq c$ as $yA = c + t$ with $t \geq 0$ and then postmultiply this equation by x to obtain $yAx = cx + tx$. Equate the two expressions for yAx and then substitute $yA - c$ for t in

(3) $$\begin{aligned} yb &= cx + tx \\ &= cx + (yA - c)x \\ &= cx + \sum_{j=1}^{n} (yA_j - c_j)x_j. \end{aligned}$$

Feasibility of x and y guarantees $x_j \geq 0$ and $(yA_j - c_j) \geq 0$ for each j. Thus, (3) shows that $yb \geq cx$. This inequality holds for each feasible solution x, so it holds when c x is maximized; $yb \geq z^* \geq cx$. This inequality holds for every feasible solution y, so it holds when y b is minimized, $yb \geq z_* \geq z^* \geq cx$. This proves part (a).

For part (b), we first suppose that feasible solutions x and y satisfy (2). In this case, (3) gives $yb = cx$, so every inequality in (1) must hold as an equation.

Finally, suppose, that feasible solutions x and y satisfy $yb = cx$. In this case, (3) gives $0 = \sum_{j=1}^{n} (yA_j - c_j)x_j$. Feasibility of x and y assure us that each term on the RHS of this equation is nonnegative, and the fact that the sum of nonnegative terms equals 0 guarantees (2). ■

Weak duality?

The name, **weak duality**, stems from the fact that the optimal value z^* of a maximization problem cannot exceed the optimal value z_* of the minimization problem that is its dual.

It will soon be demonstrated that if Program 12.1 is feasible and bounded, its optimal value z^* and the optimal value z_* of its dual equal each other. In certain pairs of *nonlinear* programs, the optimal value z^* of the maximization problem can lie below the optimal value z_* of the minimization problem, in which case the difference $(z^* - z_*)$ is known as the "duality gap."

Unbounded linear programs

A linear program is said to be **unbounded** if it is feasible and if the objective value of its feasible solutions can be improved without limit. If Program 12.1 is unbounded, it has $z^* = +\infty$, and Proposition 12.1 guarantees that Program 12.1^D can have no feasible solution. Similarly, if Program 12.1^D is unbounded, it has $z_* = -\infty$, and Proposition 12.1 guarantees that Program 12.1 cannot be feasible. It is emphasized:

If a linear program is unbounded, its dual must be infeasible.

If a linear program is unfeasible, must its dual be unbounded? No. Examples exist in of a linear program and its dual in which *both* are infeasible. Problem 4 hints at how to construct such an example.

4. Strong Duality

Proposition 12.2 (below) shows that if a linear program is feasible and bounded, so is its dual, and the two have the same optimal value. This proposition also shows that the simplex method constructs optimal solutions to both linear programs.

Proposition 12.2 *(strong duality)***.** The following are equivalent:

(a) The optimal value of Program 12.1 equals the optimal value of Program 12.1^D, and both are finite.

(b) Programs 12.1 and 12.1^D are feasible.

(c) Program 12.1 is feasible and bounded.

(d) Application of the simplex method with multipliers and with an anti-cycling rule to Program 12.1 terminates with a basis whose basic solution x is optimal for Program 12.1 and with a vector y of multipliers that is optimal for Program 12.1^D.

Proof. *(a) ⇒(b):* Linear programs whose optimal values are finite must be feasible.

(b) ⇒ (c): Immediate from Proposition 12.1.

(c)⇒(d): Suppose Program 12.1 is feasible and bounded. Application to Program 12.1 of the simplex method with multipliers and with an anti-cycling rule terminates finitely. By hypothesis, it terminates with a basis whose basic solution x is feasible, has $cx = z^*$, and has vector $\bar{c}$ of reduced costs that satisfies $\bar{c} \leq 0$. Also, from Propositions 11.2 and 11.3, the simplex method with multipliers constructs a vector y of multipliers such that

$$\bar{c} = c - yA, \qquad z^* = yb.$$

From $\bar{c} = c - yA$ and $\bar{c} \leq 0$, we see that $yA \geq c$, hence that y is feasible for Program 12.1^D. Weak Duality shows that $yb \geq z_* \geq z^*$, which couples with the equation $z^* = yb$ that is displayed above to show that $yb = z_*$, so y is optimal for Program 12.1^D.

(d)⇒(a): By hypothesis, x is feasible for Program 12.1, and y is feasible for Program 12.1^D. Also, since x is optimal for Program 12.1, Part (a) of Proposition 11.3 shows that $cx = yb$. That $z^* = z_*$ is immediate from (1). This completes a proof. ■

The term, **strong duality**, describes conditions under which the inequality $z_* \geq z^*$ holds as an equation. The portion of Proposition 12.2 that does not mention the simplex method is called the "Duality Theorem."

> The **Duality Theorem** states that if a linear program is feasible and bounded, so is its dual, and these two linear programs have the same optimal value.

The proof provided here of Proposition 12.2 rests on the simplex method. Once you know the role played by multipliers, the method of proof is straightforward – examine the conditions that cause the simplex method to terminate.

5. A Recipe for Taking the Dual

It is not necessary to cast a linear program in the format of Program 12.1 before taking its dual. There's a recipe for taking the dual. A glimpse of this recipe appears below:

- The dual of a maximization problem is a minimization problem, and conversely.
- The RHS values of a linear program become the objective coefficients of its dual.
- The objective coefficients of a linear program become the RHS values of its dual.
- The column of coefficients of each variable in a linear program become the data in a constraint of its dual.

To illustrate the last of these points, we observe that the coefficients of the variable x_j in Program 12.1 are c_j and column vector A_j; these coefficients are the data in the constraint $yA_j \geq c_j$ in Program 12.1^D.

Complementary variables and constraints

The first step in a recipe for taking the dual of a linear program is to assign to each non-sign constraint in that linear program a "complementary" decision variable in its dual. The senses of these complementary variables and constraints are determined by Table 12.1, below.

Table 12.1. Senses of complementary variables and constraints.

Row	Maximization	Minimization
1	non-sign constraint ≤ RHS	variable ≥ 0
2	non-sign constraint = RHS	variable is free
3	non-sign constraint ≥ RHS	variable ≤ 0
4	variable ≥ 0	non-sign constraint ≥ RHS
5	variable is free	non-sign constraint = RHS
6	variable ≤ 0	non-sign constraint ≤ RHS

When taking the dual of a maximization problem, Table 12.1 is read from left to right. When taking the dual of a minimization problem, Table 12.1 is read from right to left. That's why Table 12.1 is dubbed the **cross-over table**.

To illustrate Table 12.1, consider Program 12.1 and its dual. Program 12.1 has equality constraints; from row 2 of the cross-over table, we see that its dual has variables that are free (unconstrained as to sign). Program 12.1 has nonnegative decision variables, and row 4 of the cross-over table shows that its dual has constraints that are "≥" inequalities.

A memory aid

Table 12.1 is easy to remember if you interpret the complementary variables as shadow prices. For instance:

- Row 1, when read from left to right, states that the complementary variable (shadow price) for a "≤" constraint in a maximization problem is nonnegative. That must be so because increasing the constraint's RHS value can increase the optimal value, but cannot decrease it.
- Row 4, when read from right to left, states that the complementary variable (shadow price) for a "≥" constraint in a minimization problem is nonnegative. That must be so for the same reason.
- Rows 2 and 5 state that the complementary variable (shadow price) for an "=" constraint can have any sign. That is so because increasing the RHS value of an equality constraint can cause the optimal value to increase or decrease.

The recipe

The recipe that appears below constructs the dual of every linear program. This recipe is wordy, but an example will make everything clear.

Recipe for taking the dual of any linear program:

1. The dual of a maximization problem is a minimization problem, and conversely.
2. To each non-sign constraint in a linear program is assigned a **complementary** variable in the dual, and the sense of that variable is determined from Table 12.1.

3. The objective of the dual is found by summing the product of each constraint's RHS value and that constraint's complementary variable.

4. To each variable in a linear program is assigned a **complementary** constraint in the dual, and this constraint is formed as follows:

   - This constraint's sense is determined from Table 12.1.
   - This constraint's RHS value equals the coefficient of this variable in the objective.
   - This constraint's LHS value is found by summing the product of this variable's coefficient in each constraint and the complementary variable for that constraint.

An example

This recipe will be illustrated by using it to take the dual of:

Program 12.2. Minimize $1a - 2b + 3c$, subject to the constraints

$$\begin{aligned} x:&\quad -4a + 5b - 6c \geq 7, \\ y:&\quad 8a + 9b + 10c = -11, \\ z:&\quad 12a + 13b - 14c \leq 15, \\ &\quad a \geq 0, \quad b \text{ is free}, \quad c \leq 0. \end{aligned}$$

Step 1 of the recipe states that the dual of Program 12.2 is a maximization problem. The non-sign constraints in Program 12.2 have been assigned the complementary variables x, y and z (any labels other than a, b and c would do). Step 2 determines the senses of x, y and z from rows 4, 5 and 6 of the cross-over table. Evidently,

$$x \geq 0, \qquad y \text{ is free}, \qquad z \leq 0.$$

Step 3 states that the objective function of the dual is $7x - 11y + 15z$.

For each decision variable in Program 12.2, Step 4 creates a complementary constraint. We will verify that the constraint that is complementary to the variable a is

$$a: \quad -4x + 8y + 12z \leq 1.$$

Since a is nonnegative, row 1 shows that the above constraint is a "≤" inequality. The RHS value of this constraint equals the coefficient of a in the objective, namely, 1. The LHS value of this constraint equals the sum of the coefficient of a in each constraint times that constraint's complementary variable, as above. Similarly, the constraints that are complementary to b and c are

$$\text{b:} \qquad 5x + 9y + 13z = -2,$$

$$\text{c:} \qquad -6x + 10y - 14z \geq 3.$$

In brief, the complete dual to Program 12.2 is:

Program 12.2^D. Maximize $7x - 11y + 15z$, subject to

$$\text{a:} \qquad -4x + 8y + 12z \leq 1,$$

$$\text{b:} \qquad 5x + 9y + 13z = -2,$$

$$\text{c:} \qquad -6x + 10y - 14z \geq 3,$$

$$x \geq 0, \qquad y \text{ is free}, \qquad z \leq 0.$$

Note that the RHS values of Program 12.2 become the objective coefficients of its dual, and the objective coefficients of Program 12.2 become the RHS values of its dual. Note also that the column of coefficients of each variable in Program 12.2 become the row of coefficients of its complementary constraint.

You are urged to take the dual of Program 12.2^D and see that it is Program 12.2.

Two types of constraints?

The recipe for taking the dual treats the constraints on the signs of the decision variables differently from the so-called "non-sign" constraints. This seems arbitrary. What happens if we don't?

To find out, let's reconsider Program 12.2. This time, we interpret a as a free variable and $a \geq 0$ as a non-sign constraint. Being a non-constraint, $a \geq 0$ is assigned a complementary (dual) variable by Step 2. Let's label that variable s_1. So Program 12.2 now appears as

Program 12.2. Minimize $1a - 2b + 3c$, subject to the constraints

x:	$-4a + 5b - 6c \geq 7,$
y:	$8a + 9b + 10c = -11,$
z:	$12a + 13b - 14c \geq 15,$
s_1:	$1a \geq 0,$

$$a \text{ is free}, \quad b \text{ is free}, \quad c \leq 0.$$

The dual's objective now includes the addend $0\,s_1$, which equals zero. Because s_1 is complementary to a "≥" constraint, row 4 of the cross-over table shows that $s_1 \geq 0$. Because the variable a is now free, row 2 shows that its complementary constraint is

a: $$-4x + 8y + 12z + 1s_1 = 1.$$

Evidently, treating $a \geq 0$ as a non-sign constraint inserts a slack variable in the constraint that is complementary to a. This has no material effect on Program 12.2^D.

No proof?

No proof has been provided that the recipe works. To supply a proof, we would need to show that using the recipe has the same effect as forcing a linear program into the format of Program 12.1 and then taking its dual. Such a proof would be cumbersome, it would provide no insight, and it is omitted.

Weak and strong duality

Weak and strong duality had been established in the context of Program 12.1 and its dual. These results apply to any pair of linear programs, however. That is so because:

- Casting a maximization problem in the format of Program 12.1 has no effect on its feasibility or on its optimal value.
- Casting a minimization problem in the format of Program 12.1^D has no effect on its feasibility or on its optimal value.

For instance, if a maximization problem is unbounded, its dual must be infeasible. Also, if any linear program is feasible and bounded, then so is its dual, and they have the same optimal value.

6. Complementary Slackness

A set of values of the decision variables for a linear program and for its dual is said to satisfy **complementary slackness** if the following conditions are satisfied:

- If a variable for either of these linear programs is not zero, its complementary constraint holds as an equation.
- If a constraint for either of these linear programs holds as a strict inequality, the complementary variable equals zero.

This definition violates the principle of parsimony; either of the above conditions implies the other (see Problem 5).

Let us consider the implications of complementary slackness for Program 12.1 and its dual. The decision variable x_j in Program 12.1 and the constraint

$$\sum_{i=1}^{m} y_i A_{ij} \geq c_j$$

in Program 12.1^D are complementary to each other. Complementary slackness holds if the values taken by the decision variables x and y have this property: if $x_j \neq 0$ then $\sum_{i=1}^{m} y_i A_{ij} = c_j$. Put another way, complementary slackness is the requirement that

$$(x_j)\left[\sum_{i=1}^{m} y_i A_{ij} - c_j\right] = 0 \qquad \text{for each } j. \tag{4}$$

The next three subsections describe facets of complementary slackness. Each of these facets is familiar.

Complementary slackness and basic tableaus

Complementary slackness is inherent in basic tableaus. To see how, consider any basic tableau for Program 12.1. This tableau has a basic solution x, and it has at least one vector y such that $\bar{c} = c - yA$. If x_j is not zero, then x_j must be basic, and its reduced cost $\bar{c}_j$ must equal 0, which guarantees $c_j = yA_j$. In brief:

Consider any basic tableau for Program 12.1. Its basic solution x and each vector y of its multipliers satisfy complementary slackness.

Complementary slackness and weak duality

Complementary slackness is also familiar from weak duality. Equations (2) and (4) are identical. Thus, Proposition 12.1 (weak duality) states that:

> Feasible solutions to Program 12.1 and its dual are optimal if and only if they satisfy complementary slackness.

Complementary slackness and shadow prices

Complementary slackness is familiar in a third way. If a linear program has an inequality constraint, that constraint has a shadow price. Moreover, if a basic solution causes that constraint to be slack (to hold as a strict inequality), its break-even (shadow) price must equal zero, *exactly* as is required by complementary slackness.

Pivot strategies

The simplex method pivots from basic tableau to basic tableau. Each basic tableau has a basic solution x and at least one vector y of multipliers. Listed below are conditions that are necessary and sufficient for x and y to be optimal for a linear program and its dual:

(i) x is feasible for the linear program.

(ii) y is feasible for the dual linear program.

(iii) x and y satisfy complementary slackness.

Each basic tableau has a basic solution and multipliers that satisfy (iii). The simplex method pivots to preserve (i) and (iii). It aims to improve the basic solution's objective value with each pivot. It stops as soon as it encounters a tableau whose multipliers satisfy (ii).

Is there a variant of the simplex method that pivots as to preserve conditions (ii) and (iii) and stops when it satisfies condition (i)? Yes, there is. It is called the "dual simplex method," and it is discussed in Chapter 13.

Is there a variant of the simplex method that pivots to preserve condition (iii) and stops when it attains conditions (i) and (ii)? Yes, there is. It is called the "parametric self dual method," and it too is discussed in Chapter 13.

7. A Theorem of the Alternative

Chapter 10 includes a theorem of the alternative for the matrix equation $Ax = b$. Presented below is a theorem of the alternative for *nonnegative* solutions to this equation.

Proposition 12.3 *(Farkas).* Consider any $m \times n$ matrix A and any $m \times 1$ vector b. Exactly one of the following alternatives occurs:

(a) There exists an $n \times 1$ vector x such that

$$Ax = b, \quad x \geq 0.$$

(b) There exists a $1 \times m$ vector y such that

$$yA \leq 0, \quad yb > 0.$$

Proof. *(a) implies not (b):* Suppose that (a) holds, so that a solution x exists to $Ax = b$ and $x \geq 0$. Aiming for a contradiction, suppose that (b) also holds, i.e., there exists a solution y to $yA \leq 0$ and $yb > 0$. Premultiply $Ax = b$ by y and obtain $yAx = yb > 0$. Postmultiply $yA \leq 0$ by the nonnegative vector x and obtain $yAx \leq 0$. The contradiction $0 < yAx \leq 0$ shows that (b) cannot hold if (a) does.

Not (a) implies (b): Suppose that (a) does not hold. Let us consider the linear program and its dual that are specified by:

LP: min $\{0x\}$, subject to	Dual: max $\{yb\}$, subject to
$Ax = b, \quad x \geq 0.$	$yA \leq 0.$

Since (a) does not hold, LP is infeasible. The $1 \times m$ vector $y = 0$ is feasible for Dual. If its optimal value equaled 0, the Duality Theorem would imply that the optimal value of LP equals zero. This cannot occur, by hypothesis, so a solution must exist to (b). This completes a proof. ■

Proposition 12.3 is a **theorem of the alternative** because it demonstrates that exactly one of two alternatives holds. Proposition 12.3 is known as **Farkas's lemma** in honor of the Hungarian mathematician, Gyula Farkas, who published it in 1896.

A recipe for theorems of the alternative

Proposition 12.3 illustrates a handy way to construct and prove theorems of the alternative. Suppose you wish to construct a theorem of the alternative for a particular set of linear constraints. Proceed as follows:

- Set up a linear program that maximizes (minimizes) 0, subject to this set of constraints.
- Use the cross-over table to construct the dual linear program.
- Observe that the dual has a feasible solution whose objective value is negative (positive) if and only if the constraint system has no solution.

An illustration

This recipe is useful enough to illustrate. Should you wish to determine a theorem of the alternative for the inequality system,

$$Ax \leq b, \qquad x \geq 0. \tag{5}$$

The handiest linear program is to maximize {0x}, subject to (5). The dual linear program minimizes {yb} subject to the constraints

$$yA \geq 0, \qquad y \geq 0. \tag{6}$$

The dual is feasible because setting $y = 0$ satisfies (6). The Duality Theorem guarantees that no solution exists to (5) if and only if a solution exists to (6) that has $yb < 0$. This proves

Proposition 12.4 *(Farkas).* Consider any $m \times n$ matrix A and any $m \times 1$ vector b. Exactly one of the following alternatives occurs:

(c) There exists an $n \times 1$ vector x that satisfies (5).

(d) There exists a $1 \times m$ vector y that satisfies (6) and $yb < 0$.

Any result that can be proved this way is dubbed a "Farkas."

Farkas's lemma?

Perhaps Farkas's lemma once was a lemma, that is, a step toward an important result. This "lemma" is now recognized as one of the most fundamental theorems in constrained optimization. Needless to say, perhaps, Farkas did not prove it via the Duality Theorem, which came six decades later. In Chapter 17, a generalization of Farkas's lemma will be presented, and the Bolzano-Weierstrass theorem will be used to prove it.

8. Data Envelopment*

This is the first of three self-contained sections. Each of these three sections uses a linear program *and* its dual to analyze an issue. These sections are starred because they can be read independently of each other and because the information in them is not used in later chapters.

The subject of the current section is the efficiency of different units of an organization. These units might be branches of a bank, offices of a group medical practice, hospitals in a region, or academic departments in a university. The characteristic feature of the model that is under development is that the inputs and outputs of each unit are easy to measure but are difficult to place values upon.

A theorem of the alternative

Let's focus on a particular unit, say, unit B. It will be demonstrated that exactly one of these two alternatives holds:

(a) There exist values of the outputs and costs of the inputs such that unit B has a benefit-to-cost ratio that is at least as large as any of the other units.

(b) There exists a nonnegative linear combination of the other units that produces more of each output and consumes less of each input than does unit B.

If condition (a) holds, unit B is said to be "potentially efficient." If condition (b) holds, the data of unit B are said to be "enveloped." For the latter reason, the situation we are probing is called **data envelopment.** The pioneering work on this model was a done in 1978 by Charnes, Cooper and Rhodes[1].

---

[1] Charnes, A., W. Cooper, and E. Rhodes, "Measuring the efficiency of decision-making units," *European Journal of Operational Research*, V. 2, pp. 429-444, 1978.

Describing the units

An example will be used to introduce data envelopment. In this example, there are three units, and they are labeled A, B and C. There are three outputs, which are labeled 1, 2 and 3. There are two inputs, which are labeled 1 and 2. Table 12.2 specifies the amount of each input that is required by each unit, as well as the amount of each output that is produced by each unit. This table shows that unit A consumes 15 units of input 2 and produces 3.5 units of output 2, for instance.

Table 12.2. Inputs and outputs of units A, B and C.

	input 1	input 2	output 1	output 2	output 3
unit A	10	15	20	3.5	10
unit B	24	30	25	7	20
unit C	21	24	20	6	25

A schedule of prices

The decision variables in this model are the prices (values) that are placed on the inputs and on the outputs. Let us designate

p_i = the price per unit of output i for i = 1, 2, 3,

q_j = the price per unit of input j for j = 1, 2.

These prices are required to be nonnegative. Each schedule of prices assigns to each unit a *benefit* of its outputs and a *cost* of its inputs. The cost of the inputs to unit B equals $24q_1 + 30q_2$, for instance.

Benefit-to-cost ratios

Given a schedule of prices, the benefit-to-cost ratio ρ_B of unit B is determined by the data in its row of Table 12.2 and is given by

$$\rho_B = \frac{25p_1 + 7p_2 + 20p_3}{24q_1 + 30q_2}.$$

Units A and C have similar benefit-to-cost ratios, ρ_A and ρ_C. Implicit in this definition is the requirement that at least one of the inputs has a price that is positive. (Otherwise, the denominator would equal 0.)

A potentially-efficient unit

A unit is potentially-efficient if there is a schedule of prices such that its benefit-to-cost ratio is at least as large as the others. In particular, unit B is potentially-efficient if there exist prices for which

$$(7) \qquad \rho_B \geq \rho_A \qquad \text{and} \qquad \rho_B \geq \rho_C.$$

The inequalities in (7) entail the comparison of ratios. This seems to present a difficulty; the requirement $\rho_B \geq \rho_A$ cannot be represented as a linear inequality. Note, however, that multiplying the price p_i of each output by the same constant θ multiplies each ratio by θ and preserves (7). Thus, if a solution exists to (7), then a solution exists in which ρ_B is at least 1 and in which ρ_A and ρ_C are at most 1. In other words, a schedule of prices satisfies (7) if and only if a (possibly different) schedule of prices satisfies

$$(8) \qquad \rho_B \geq 1, \qquad \rho_A \leq 1, \qquad \rho_C \leq 1.$$

Clearing the denominators in (8) produces linear inequalities.

A linear program

Multiplying the ratios in (8) by their respective denominators produces the first three inequalities in:

Program 12.3. Maximize $\{p_1 + p_2 + p_3 + q_1 + q_2\}$ subject to

$$y_A: \qquad 20p_1 + 3.5p_2 + 10p_3 \leq 10q_1 + 15q_2,$$

$$y_B: \qquad 24q_1 + 30q_2 \leq 25p_1 + 7p_2 + 20p_3,$$

$$y_C: \qquad 20p_1 + 6p_2 + 25p_3 \leq 21q_1 + 24q_2,$$

$$v: \qquad p_1 + p_2 + p_3 + q_1 + q_2 \leq 10$$

$$p_i \geq 0, \qquad q_j \geq 0, \qquad \text{each i and j}.$$

Let us interpret Program 12.3:

- Its 1st constraint keeps the benefit of the outputs of unit A from exceeding the cost of its inputs, thereby enforcing $\rho_A \leq 1$.
- Its 2nd constraint keeps the cost of the inputs to unit B from exceeding the benefit of its outputs, thereby enforcing $\rho_B \geq 1$.

- Its 3rd constraint keeps the benefit of the outputs of unit C from exceeding the cost of its inputs, thereby enforcing $\rho_C \leq 1$.
- Its objective seeks a schedule of prices for which unit B is potentially efficient. Its 4th constraint imposes upper bounds on these prices. (With that constraint omitted, Program 12.3 would either have 0 or + ∞ as its optimal value.)

Program 12.3 is feasible – setting each variable equal to zero satisfies its constraints. If the optimal value of Program 12.3 is positive, it reports a schedule of prices for which unit B has the highest benefit-to-cost ratio. If the optimal value of Program 12.3 is zero, no such prices exist.

Duality will be used in the analysis of Program 12.3. For that reason, each of its non-sign constraints has been assigned a complementary dual variable; y_A is complementary to the 1st constraint, for instance.

Solver says

Solver reports that Program 12.3 has 0 as its optimal value. No schedule of prices exists for which unit B has the highest benefit-to-cost ratio. Solver reports an optimal solution having $p_i = 0$ and $q_j = 0$ for each i and j, and it reports shadow prices y_A, y_B and y_C and v that are given below.

(9) $$y_A = 2.875, \quad y_B = 3.9375, \quad y_C = 3.0833, \quad v = 0.$$

Proposition 12.2 shows that these shadow prices (multipliers) are an optimal solution to the dual of Program 12.3. From the dual, we will learn that unit B is enveloped by the combination of units A and B whose weights w_A and w_B are given by

(10) $$w_A = \frac{y_A}{y_B} = \frac{2.875}{3.9375} = 0.730,$$

(11) $$w_C = \frac{y_C}{y_B} = \frac{3.0833}{3.9375} = 0.783.$$

The dual of Program 12.3

The dual of Program 12.3 appears below, where it is labeled

Program 12.3 ^{D}. Minimize {10 v}, subject to the constraints

p_1: $20y_A + 20y_C + v \geq 1 + 25y_B,$

p_2: $3.5y_A + 6y_C + v \geq 1 + 7y_B,$

p_3: $10y_A + 25y_C + v \geq 1 + 20y_B,$

q_1: $24y_B + v \geq 1 + 10y_A + 21y_C,$

q_2: $30y_B + v \geq 1 + 15y_A + 24y_C,$

$$y_A \geq 0, \quad y_B \geq 0, \quad y_C \geq 0, \quad v \geq 0.$$

Program 12.3 is feasible and bounded (its optimal value equals 0). Proposition 12.2 shows that Program 12.3^D is also feasible and bounded, moreover, that the values of the multipliers given by (9) are an optimal solution to Program 12.3^D. This optimal solution has $v=0$ and it has $y_B > 0$.

Let us interpret the optimal solution to Program 12.3^D. Dividing its second constraint by y_B and noting that $v=0$ shows that

$$3.5\frac{y_A}{y_B} + 6\frac{y_C}{y_B} > 7.$$

The LHS of this inequality is a weighted combination of the typ-2 outputs of units A and C. This weighted combination exceeds the type-2 output of unit B. Similarly, dividing the 5th constraint by y_B shows that

$$30 > 15\frac{y_A}{y_B} + 24\frac{y_C}{y_B}$$

This inequality shows that unit B consumes more of input 2 than does the same weighted combination of units A and C. The pattern is evident; unit B consumes more of each input and produces less of each output than does the weighted combination of units A and C with weights w_A and w_C given by (10) and (11). Unit B is enveloped.

The general result

The preceding line of analysis holds in general. If a general analysis were presented as a proposition, its conclusion would be that the following are equivalent:

- There do not exist prices on the inputs and outputs for which unit k has a benefit-to-cost ratio that is at least as large as that of any other unit.
- The analogue of Program 12.3 has 0 as its optimal value, and its optimal solution has a shadow price y_i for each unit i that is nonnegative, with $y_k > 0$.
- The dual of that linear program is feasible and bounded, and it has 0 as its optimal value.
- With $w_i = y_i/y_k$, the dual shows that unit k consumes more of each input and produces less of each output than does the nonnegative linear combination of the other units in which, for each i, in the inputs and outputs of unit i are multiplied by w_i.

It suffices for this result that the each of the inputs be positive and that each of the outputs be nonnegative. (If an input equaled 0, a ratio could have 0 as its denominator.)

9. The No Arbitrage Tenet of Financial Economics*

In financial economics, "arbitrage" describes a situation in which the possibility exists of earning a profit with no possibility of incurring a loss. A tenet of financial economics is an arbitrage opportunities are fleeting; if one emerges, investors flock to it, and by so doing alter the prices so as to eliminate it.

A link exists between arbitrage and duality. This link holds in general, but it is established here in the context of a family of *one-period* investment opportunities. Dealing with a one-period model lets us focus on the investor's asset position at the end of the period, and this simplifies the discussion.

A risk-free asset

It is assumed that the opportunity exists to invest and to borrow at a known **risk-free rate** of r per period. Think of a bank. Each dollar deposited in the bank at the start of the period returns $(1 + r)$ dollars at the period's end. Similarly, for each dollar that one borrows from the from the bank at the start of the period, one must repay $(1 + r)$ dollars at the period's end.

Risky assets

Each risky asset has a fixed market price at the start of the period. Each risky asset's price at the end of the period depends on the "state" that occurs then. These states are mutually exclusive and collectively exhaustive – one of them will occur.

An example

Table 12.3 describes a model having one risk-free asset, three risky assets (that are labeled 1, 2 and 3), and four states (that are labeled a, b, c and d).

Table 12.3. Net return for three risky assets.

	A	B	C	D	E	F	G
1							
2	risk-free rate r =		0.03				
3							
4			price at end of period				
5	asset	a	b	c	d		cost
6	1	107	109	95	107		100
7	2	215	200	212	197		200
8	3	305	321	305	293		300
9							
10			net return				
11	asset	a	b	c	d		
12	1	4	6	-8	4		
13	2	9	-6	6	-9		
14	3	-4	12	-4	-16		
15							
16			= E8 - $G8 * (1 + C2)				

Think of assets 1, 2 and 3 as shares of common stock in different companies. From row 6 of Table 12.3 we see that the start-of-period price of stock #1 is $100 per share and that the end-of-period price depends on the state that occurs then. If state b occurs, the end-of-period price is $109, for instance.

The *net return* from investing in one unit of an asset is the amount of money (possibly negative) that remains from borrowing the price of that asset at the start of the period, purchasing that asset at that time, selling it at the end of the period, and repaying the loan and the accrued interest. For instance, if state b occurs, the net return for investing in one share of stock #1 equals $\$109 - \$100 \times (1 + 0.03) = \$6$. The formula in cell E14 shows how to compute the net return for investing in one unit of each asset.

A portfolio

For this example, a portfolio is a 3-tuple $x = [x_1 \ \ x_2 \ \ x_3]$ in which x_1, x_2 and x_3 are the number of units of asset 1, 2 and 3 in which one invests. A positive value of x_j is called a "long." and a negative value of x_j is called a "short." To *short* one unit of asset 2 is to borrow one unit of that asset at the start of the period, sell it at that time, invest the money obtained in the bank for the period, withdraw the money from the bank at the end of the period, buy the asset at that time, and repay the loan. Shorting one share of stock #1 has a net return of –\$6 if state b occurs because $-6 = 100 \times (1 + 0.03) - 109$.

An arbitrage opportunity

An **arbitrage opportunity** exists if there exists a portfolio that has a nonnegative net return under every state and has a positive net return under at least one state.

Rows 12-14 of Table 12.3 show that no single asset presents an arbitrage opportunity. Asset 1 does not because row 12 contains a negative entry (going long could lose) and a positive entry (going short could lose). The same is true of rows 13 and 14. At issue is whether there exists any portfolio $x = [x_1 \ \ x_2 \ \ x_3]$ that creates an arbitrage opportunity.

A linear program

In Program 12.4 (below), the decision variables x_1, x_2 and x_3 specify the number of units of asset 1, 2 and 3 in the portfolio, and the numbers w_a through w_d specify the net return at the end of the period if states a through d occur. The objective seeks a portfolio that achieves positive net return under at least one state with nonnegative net return under all states. If the inequality constraint

$$w_a + w_b + w_c + w_d \leq 1$$

were omitted from this linear program, it would have 0 or $+\infty$ as its optimal value. With that constraint included, the optimal solution to the linear program exhibits a portfolio that creates an arbitrage opportunity, if one exists.

Program 12.4. Maximize $\{w_a + w_b + w_c + w_d\}$, subject to the constraints

q_a: $$w_a = 4x_1 + 9x_2 - 4x_3,$$

q_b: $$w_b = 6x_1 - 6x_2 + 12x_3,$$

q_c: $$w_c = -8x_1 + 6x_2 - 4x_3,$$

q_d: $$w_d = 4x_1 - 9x_2 - 16x_3,$$

θ: $$w_a + w_b + w_c + w_d \le 1,$$

$$w_a \ge 0, \quad w_b \ge 0, \quad w_c \ge 0, \quad w_d \ge 0.$$

No arbitrage

Solver reports that Program 12.4 has 0 as its optimal value, so this set of investment opportunities presents no arbitrage opportunity. Solver reports these values of the shadow prices:

$$q_a = 1, \quad q_b = 28/9, \quad q_c = 31/9, \quad q_d = 11/9, \quad \theta = 0.$$

A probability distribution over the states

To construct a probability distribution from q_a through q_d, we equate K to the sum of q_a through q_d (this gives K = 79/9) and define p_a through p_d as q_a/K through q_d/K, respectively, getting

$$p_a = 9/79, \quad p_b = 28/79, \quad p_c = 31/79, \quad p_d = 11/79.$$

This probability distribution has a property that you might surprise you.

Expected net return

Given this probability distribution over the states, let us compute the expectation of the net return for each unit invested in asset 1. Row 12 of Table 12.3 contains the net return of asset 1 for each state. From that row and the probability distribution given above, we learn that expected net return for asset 1 is given by

$$\frac{1}{79} \times (9 \times 4 + 28 \times 6 - 31 \times 8 + 11 \times 4) = 0.$$

Repeating this computation with the data in row 13 of Table 12.3 verifies that the expected net return of asset 2 equals 0, as does the expected net return of asset 3.

A theorem of the alternative

It is no accident that each risky asset has 0 as its expected net return. A theorem of the alternative is at work. This theorem demonstrates that exactly one of the following alternatives holds:

- A risk-free asset and a set of risky assets offer an arbitrage opportunity.
- There is a probability distribution over the states such that the probability of each state is positive and such that the expected net return of each risky asset equals 0.

This result will be seen to follow directly from duality. It will be established in a general setting, rather than for the data in Table 12.3.

A general model

Let there be m assets, and let them be numbered 1 through m. Let there be n states of nature, and let them be numbered 1 through n. Exactly one of these states will occur at the end of the period. For each i such that $1 \le i \le m$ and for each j such that $1 \le j \le n$, the number A_{ij} has this interpretation:

A_{ij} equals the net return at the end of the period per unit invested in asset i at the start of the period if state j occurs at the end of the period.

The A_{ij}'s are known to the investor. For the data in Table 12.3, the A_{ij}'s form the 3×4 array in cells B12:E14.

A pair of linear programs

Program 12.5 (below) seeks to determine whether or not an arbitrage opportunity exists. Interpret x_i as the number (possibly negative) of units of asset i that are purchased at the beginning of the period and sold at the end of the period. Interpret w_j as the net profit at the end of the period if state j occurs at that time. The constraints require the net profit w_j to be nonnegative for each state j, and the objective seeks a portfolio that has at least one w_j positive. With the constraint on the sum of the w_j's deleted, the optimal value of Program 12.5 would be 0 or $+\infty$. Including that constraint causes it to exhibit a portfolio that have an arbitrage opportunity, if one exists.

Program 12.5. Maximize $\sum_{j=1}^{n} w_j$, subject to the constraints

$$q_j: \quad w_j = \sum_{i=1}^{m} x_i A_{ij} \quad \text{for } j = 1, \ldots, n,$$

$$\theta: \quad \sum_{j=1}^{n} w_j \le 1,$$

$$w_j \ge 0 \quad \text{for } j = 1, \ldots, n.$$

Program 12.5 is feasible because setting the x_i's and the w_j's equal to zero satisfies its constraints. If no arbitrage opportunity exists, the optimal value of Program 12.5 equals zero. To see what this implies, we investigate the dual of Program 12.5, which appears below as:

Program 12.5^D. Minimize θ, subject to

$$x_i: \quad (-1)\left[\sum_{j=1}^{n} A_{ij} q_j\right] = 0 \quad \text{for } i = 1, \ldots, m,$$

$$w_j: \quad q_j + \theta \ge 1 \quad \text{for } j = 1, \ldots, n.$$

If the optimal value of Program 12.5 is zero, Strong Duality shows that the Program 12.5^D is feasible and that its optimal value equals 0, hence that there exists a set of q_j's that satisfy $q_j \ge 1$ for each j and that satisfy

$$\sum_{j=1}^{n} A_{ij} q_j = 0 \quad \text{for } i = 1, \ldots, m. \tag{12}$$

With $K = \sum_{j=1}^{n} q_j$, set

$$p_j = q_j / K \quad \text{for } j = 1, \ldots, n. \tag{13}$$

The p_j's satisfy

$$p_j > 0 \quad \text{for } j = 1, \ldots, n, \tag{14}$$

$$\sum_{j=1}^{n} p_j = 1. \tag{15}$$

Divide equation (12) by $-K$ to verify that the p_j's also satisfy

(16) $$\sum_{i=1}^{m} A_{ij} p_j = 0 \qquad \text{for } i = 1, \ldots, m.$$

A risk-neutral probability distribution

A probability distribution over the states is said to be **risk neutral** if the probability of each state is positive and if the expected net return on each asset equals zero. The p_j's that are constructed from (13) are a risk-neutral probability distribution. These probabilities are positive, and they assign each asset an expected net profit of 0.

The general result

The line of development that is underway leads to

Proposition 12.5. For each i such that $1 \le i \le m$ and for each j such that $1 \le j \le n$, let A_{ij} equal the net return that at that the end of the period if state j is observed then and if one unit is invested in asset i at the start of the period. The following are equivalent.

(a) No arbitrage opportunity exists.

(b) Program 12.5 has 0 as its optimal value.

(c) Program 12.5^D has 0 as its optimal value.

(d) There exists a risk neutral probability distribution.

Proof. That (a) $\Rightarrow$ (b) $\Rightarrow$ (c) $\Rightarrow$ (d) has been established. That (b) $\Rightarrow$ (a) is immediate. Showing that (d) $\Rightarrow$ (b) will complete a proof.

(d) $\Rightarrow$ (b): Suppose that there exists numbers p_1 through p_n that satisfy (14), (15) and (16). Write (16) in matrix notation as $0 = Ap$. Program 12.5 is feasible because equating all of its decision variable so zero satisfies its constraints. Consider any feasible solution to Program 12.5. Write its equality constraints in matrix form as $w = xA$. Postmultiply this equation by p to obtain $wp = xAp = x0 = 0$. Since w is feasible, $w_i \ge 0$ for each i. Since $p_i > 0$ for each i, the only way to obtain $\sum_{i=1}^{m} w_i p_i = 0$ is to have $w_i = 0$ for each i, so the optimal value of Program 12.5 equals 0. This completes a proof. ■

The proof technique in this section is similar to that in the prior section. It is to set up a linear program whose objective value can be made positive if and only if a condition holds and examine its dual. This technique will be used again in the next section.

10. Strong Complementary Slackness*

Let us consider a linear program that is feasible and bounded. We have seen that its dual is feasible and bounded (Proposition 12.2), that they have the same optimal value (also Proposition 12.2), and that feasible solutions to the linear program and its dual are optimal if and only if they satisfy complementary slackness (Proposition 12.1).

Every optimal solution to a linear program and every optimal solution to its dual must satisfy complementary slackness. In this section, it is shown that there exist optimal solutions to a linear program and its dual that satisfy a somewhat stronger condition. This result is established for Program 12.1 and its dual with the latter rewritten as:

Program 12.1^D. Min y b, subject to the constraints

x: $$yA - t = c,$$
$$t \geq 0.$$

The surplus variables that convert the inequalities in $yA \geq c$ to equations are shown explicitly in this representation of Program 12.1^D.

Feasible solutions to Program 12.1 and Program 12.1^D are optimal if and only if they satisfy the complementary slackness conditions,

(17) $$x_j t_j = 0 \qquad \text{for } j = 1, 2, \ldots, n.$$

Complementary slackness requires at least one of each pair $\{x_j, t_j\}$ to equal zero. Complementary slackness allows *both* members of the pair $\{x_j, t_j\}$ to equal zero. By contrast, **strong complementary slackness** requires that only one member of each pair $\{x_j, t_j\}$ equals zero. Thus, feasible solutions to Program 12.1 and Program 12.1^D satisfy strong complementary slackness if (17) holds and, in addition, if

(18) $$x_j + t_j > 0 \qquad \text{for } j = 1, 2, \ldots, n.$$

The goal of this section is to show if Program 12.1 is feasible and bounded, there exist optimal solutions to it and its dual that satisfy (18). The first – and main – step in the proof is to show that there exist optimal solutions that satisfy $x_j + t_j > 0$ for a particular j.

Proposition 12.6. Suppose Program 12.1 is feasible and bounded. Consider any single integer j between 1 and n inclusive. Program 12.1 and Program 12.1^D have optimal solutions for which $(x_j + t_j) > 0$.

Proof. The Duality Theorem guarantees Programs 12.1 and 12.1^D have the same optimal value and that feasible solutions x and y to Program 12.1 and Program 12.1^D, respectively, are optimal if $cx \geq yb$.

The feasible solutions to Program 12.6 (below) are the optimal solutions to Programs 12.1 and 12.1^D. Program 12.6 seeks optimal solutions to Programs 12.1 and 12.1^D that have $x_j + t_j > 0$.

Program 12.6. Max θ, subject to the constraints

$$\begin{array}{llcl}
\alpha: & \theta \quad - x_j - t_j & \leq & 0, \\
v: & Ax & = & b, \\
w: & -yA \quad + t & = & -c, \\
\lambda: & yb - cx & \leq & 0, \\
 & x \geq 0, \qquad t \geq 0. & &
\end{array}$$

To the left of the constraints of Program 12.6 are the names that have been assigned to its complementary variables. Note that v is a $1 \times m$ vector, that w is an $n \times 1$ vector and that α and λ are scalars.

Aiming for a contradiction, we assume that Proposition 12.6 is false, hence that 0 is the optimal value of Program 12.6. The Duality Theorem shows that 0 is also the optimal value of its dual. The objective of the dual is to minimize (vb – cw). An optimal solution to the dual equates this objective to 0, so it satisfies the constraints.

$$\begin{array}{lll}
 & vb - cw & = 0, \\
\theta: & \alpha & = 1, \\
x: & vA \quad - \alpha I^j - \lambda c & \geq 0, \\
y: & -Aw \quad + \lambda b & = 0, \\
t: & w - \alpha I_j & \geq 0, \\
 & \alpha \geq 0, \quad \lambda \geq 0. &
\end{array}$$

In the above, I^j and I_j denote the row and column vectors having 1 in their j^{th} entry and 0's elsewhere.

The hypothesis (which will be refuted) is that values of v, w, α and λ exist that satisfy the above constraints. Evidently, $\alpha = 1$. There are two cases to consider.

Case #1: In this case, $\lambda = 0$, The above constraints include $vA \geq I^j$ and $Aw = 0$ and $w \geq I_j$. Postmultiply $vA \geq I^j$ by w to obtain $vAw \geq w_j \geq 1$. But $Aw = 0$, so $vAw = 0$, and we have obtained the contradiction $0 = vAw \geq 1$. Case #1 cannot occur.

Case #2: In this case, $\lambda > 0$. Set $y = v(1/\lambda)$ and set $x = w(1/\lambda)$. From the constraints that are displayed above,

$$\begin{aligned} yb &= cx, \\ yA - I^j/\lambda &\geq c, \\ Ax &= b, \\ x &\geq I_j/\lambda. \end{aligned}$$

Of these four constraints, the last requires $x \geq 0$ with $x_j > 0$ and the second requires $yA - t = c$ with $t \geq 0$ and $t_j > 0$. Since $yb = cx$ and since $Ax = b$, optimal solutions to Program 12.6 and its dual have been constructed. But these solutions have $x_j > 0$ and $t_j > 0$, so they violate complementary slackness, hence cannot be optimal. Case #2 cannot occur either. The proof is complete. ■

Proposition 12.7 (strong complementary slackness). Suppose Program 12.1 is feasible and bounded. Then there exist optimal solutions to Program 12.1 and its dual that have

$$x_j + t_j > 0 \qquad \text{for } j = 1, 2, \ldots, n\,.$$

Proof. Proposition 12.6 holds for each particular value of j. The average of n optimal solutions to a linear program is optimal. Denote as $\hat{x}$ and as $(\hat{y}, \hat{t})$ the average of the optimal solutions found from Proposition 12.6 for the values $j = 1, 2, \ldots, n$. Note that $\hat{x}$ is optimal for Program 12.1, that $(\hat{y}, \hat{t})$ is optimal for Program 12.1^D and that $\hat{x}_j + \hat{t}_j > 0$ for each j. This proves the theorem. ■

11. Review

It is now clear that the simplex method attacks a linear program and its dual. A variety of issues can be addressed by studying a linear program and its dual. Three of them are studied in the starred sections of this chapter. Others appear in later chapters. In Chapter 14, for instance, duality will be used to develop a (simplified) model of an economy in general equilibrium.

Prominent in this chapter is a "lemma" that Farkas published in 1896. Very few results in linear algebra predate linear programming and deal with inequalities. That has changed. Farkas's lemma is now understood to be central to constrained optimization. Proposition 12.3 obtains Farkas's lemma as an immediate consequence of the Duality Theorem. The converse is also true; see Problem 10.

12. Homework and Discussion Problems

1. *(taking the dual)* Use the recipe to take the dual of Program 12.2^D. Indicate where and how you used the cross-over table.

2. *(the dual of Program 12.1^D)*

   (a) Recast Program 12.1^D in the format of Program 12.1.Specifically, rewrite Program 12.1^D as a maximization problem with nonnegative decision variables and equality constraints. Did z_* get replaced by $-z_*$?

   (b) Take the dual of the program you constructed in part (a).

   (c) Recast the dual as a maximization problem.

   (d) True or false: You have demonstrated Program 12.1 is the dual of Program 12.1^D.

3. *(diet)* An individual is willing to consume any of n foods in quantities that are sufficient to meet or exceed his minimum daily requirement of each of m nutrients.This person takes no pleasure in eating; he wishes to minimize the cost of feeding himself. The foods are numbered 1 through n, and the nutrients are numbered 1 through m. For $j = 1, \ldots, n$, each unit of food j costs c_j and, for $i = 1, \ldots, m$, provides the amount A_{ij} of nutrient i.

For $i = 1, \ldots, m$, his minimum daily requirement of nutrient i is for b_i units. Food quantities are divisible (he can purchase fractional units).

(a) Formulate his diet problem as a linear program.

(b) A food chemist has found a way to create each of the nutrients directly, rather than from foods. She wishes maximize the revenue she receives from selling nutrients to him. Formulate her problem as a linear program.

(c) Is there a relationship between the linear programs you have created and, if so, what is it?

(d) When eating food, might it be optimal for him to consume more than b_i units of a particular nutrient? If so, what price must she set on that nutrient? Why?

4. *(infeasibility)* This problem concerns a linear program that is written in the format: Max {cx}, subject to $Ax \le b$ and $x \ge 0$. For the 2×2 matrix A given below, find vectors b and c such that neither the linear program nor its dual is feasible.

$$A = \begin{bmatrix} 1 & -1 \\ -1 & 1 \end{bmatrix}$$

5. Two *complementary slackness* conditions are presented in the beginning of Section 6, along with the assertion that they are equivalent. Is this assertion true? Support your answer.

6. *(a self-dual linear program)* This problem concerns a linear program that is written as: Maximize cx, subject to the constraints $Ax \le b$ and $x \ge 0$. With y as a $1 \times m$ vector, consider the "related" linear program: Maximize $(cx - yb)$, subject to the constraints

$$\begin{bmatrix} A & 0 \\ 0 & -A^T \end{bmatrix} \begin{bmatrix} x \\ y^T \end{bmatrix} \le \begin{bmatrix} b \\ -c^T \end{bmatrix}, \qquad \begin{bmatrix} x \\ y^T \end{bmatrix} \ge \begin{bmatrix} 0 \\ 0 \end{bmatrix}.$$

(a) Assume the original linear program is feasible and bounded. Is the "related" linear program feasible and bounded? What is its optimal value?

(b) Take the dual of the related linear program.

(c) A linear program is said to be **self-dual** if it and its dual are identical.

(d) True or false: Each linear program that is feasible and bounded can be written as a linear program whose optimal value equals zero.

7. *(bounded variables)* Consider the linear program: Maximize {cx} subject to the constraint $Ax \leq b$ and $0 \leq x \leq u$. The data in this linear program are the matrix A, the column vector $b \geq 0$, the row vector c, and the column vector $u \geq 0$.

   (a) Is this linear program feasible? Is it bounded?

   (b) What is the dual of this linear program?

   (c) Is the dual feasible and bounded?

8. *(LP without optimization)* Suppose Program 12.1 is feasible and bounded.

   (a) Use only the data of Program 12.1 (specifically, A, b and c) to write a system of linear constraints whose solution includes an optimal solution to Program 12.1.

   (b) What can you say about the constraints you formed in part (a) if Program 12.1 is not feasible and bounded?

9. This problem concerns Program A: Max cx, subject to $Ax \leq b$ and $x \geq 0$. Suppose that Program A is feasible.

   (a) Suppose a solution $\hat{x}$ exists to $A\hat{x} \leq 0$, $\hat{x} \geq 0$ and $c\hat{x} > 0$. Show that Program A is unbounded.

   (b) Suppose that Program A is unbounded. Can there exist a row vector y that satisfies $yA \geq c$ and $y \geq 0$? If there cannot, must there exist a column vector $\hat{x}$ that satisfies the constraints in part (a)? Support your answers.

   (c) Complete the following sentence: If Program A is feasible, it is unbounded if and only if there is a column vector that satisfies the constraints ____________. Support your answer.

10. *(duality from Farkas)* Suppose that a solution x exists to $Ax \leq b$ and $x \geq 0$ and that a solution y exists to $yA \geq c$ and $y \geq 0$. *Remark:* The hypothesis of this problem is that a linear program and its dual are feasible.

(a) Is it true that the vectors x and y in that satisfy the constraints in part (a) have $cx \le yb$? Support your answer.

(b) Suppose (this will be refuted) that no solution exists to

$$\begin{aligned} \hat{y}: &\quad Ax \le b, \\ \hat{x}: &\quad -yA \le -c, \\ \theta: &\quad yb \le cx, \\ &\quad x \ge 0, \quad y \ge 0. \end{aligned}$$

Use Farkas to show that a solution does exist to

$$(*) \quad \hat{y}A \ge \theta c, \quad A\hat{x} \le b\theta, \quad \hat{y}b < c\hat{x}, \quad \hat{y} \ge 0, \quad \hat{x} \ge 0, \quad \theta \ge 0.$$

(c) Can a solution to (*) exist that has $\theta = 0$? Support your answer. (*Hint:* Refer to part (a) of the preceding problem.)

(d) Can a solution to (*) exist that has $\theta = 1$? What about $\theta > 0$? Support your answers.

(e) Have you obtained the Duality Theorem as a consequence of Farkas? Support your answer.

11. *(bounded feasible regions)* This problem concerns Program A: Max {cx}, subject to the constraints $Ax \le b$ and $x \ge 0$.

(a) Suppose that Program A is feasible and that its feasible region is bounded. Show that a row vector $\hat{y}$ satisfies $\hat{y} \ge 0$ and $\hat{y}A \ge e$, where e is the $1 \times n$ vector each of whose entries equals 1. Show that the dual of Program A has an unbounded feasible region.

(b) Suppose that the dual of Program A is feasible and that its feasible region is bounded. Show that a column vector $\hat{x}$ satisfies $\hat{x} \ge 0$ and $A\hat{x} \le -e$ where e is now the $m \times 1$ vector of 1's. Conclude that the feasible region of Program A is unbounded.

12. *(unbounded feasible regions)* This problem concerns Program A: Max {cx}, subject to the constraints $Ax \le b$ and $x \ge 0$. Can it occur that this linear program *and* its dual have unbounded feasible regions. *Hint*: With

$$A = \begin{bmatrix} 1 & -1 \\ -1 & 1 \end{bmatrix},$$

see whether you can find vectors b and c such that both feasible regions are unbounded.

13. *(data envelopment)* This problem concerns the data envelopment model whose data are in Table 12.2.

    (a) Formulate a linear program that determines whether or not unit A is potentially efficient. Solver it. Find a set of prices for which unit A is potentially efficient or a nonnegative linear combination of the other units that envelops unit A.

    (b) Redo part (a) for unit C.

14. *(data envelopment)* Consider a data envelopment model that compares three academic departments each of which has 2 inputs and 3 outputs. The positive number C_{ij} denotes the type j output of department i, and the positive number D_{ik} denotes the type k input of department i. (For the data in Table 12.2, the matrix D consists of the two columns to the left, and the matrix C consists of the three columns to the right.) Suppose that department 2 is enveloped, hence, that there exist nonnegative numbers α and γ such that

$$C^2 < \alpha C^1 + \gamma C^3 \quad \text{and} \quad D^2 > \alpha D^1 + \gamma D^3,$$

and suppose that department 1 is also enveloped.

    (a) Does department 3 envelope each of the others?

    (b) For the data in Table 12.2, unit B is enveloped. Can you determine whether or not unit A is enveloped without solving a linear program? If so, how?

15. *(the no-arbitrage tenet)* This problem concerns a variant of the no-arbitrage model. Let us assume that investors cannot go "short;" equivalently, that each portfolio x must have $x_j \geq 0$ for $j = 1, 2, \ldots, n$. The no-arbitrage tenet remains unchanged, but the definition of a portfolio is more restrictive.

    (a) What changes occur to Program 12.5?

    (b) What changes occur to Program 12.5^D?

    (c) What changes occur to the statement of Proposition 12.5?

16. *(strong complementary slackness)* State and prove the variant of Proposition 12.7 for the variant of Program 12.1 in which $Ax = b$ is replaced by $Ax \le b$. *Hint:* Might it suffice to apply Proposition 12.6 and Proposition 12.7 to the linear program in which A is replaced by [A, I]?

17. *(a matrix game)* Suppose that you and I know the entries in the $m \times n$ matrix A. You pick a row. Simultaneously, I pick a column. If you pick row i and I pick column j, I pay you the amount A_{ij}. Suppose you choose a randomized strategy p (a probability distribution over the rows) that maximizes your smallest expected payoff over all columns I might choose.

    (a) Interpret the constraints and the objective of the linear program:

$$
\begin{array}{lll}
 & \text{Max } v, \text{ subject to} & \\
q_j: & v \le \sum_{i=1}^{m} p_i A_{ij} & \text{for } j = 1, 2, \ldots, n, \\
w: & \sum_{i=1}^{m} p_i = 1, & \\
 & p_i \ge 0 & \text{for } i = 1, 2, \ldots, m.
\end{array}
$$

    (b) Is this linear program feasible and bounded?

    (c) Write down and interpret the dual of this linear program. Is it feasible and bounded?

    (d) What does complementary slackness say about the optimal solutions of the two linear programs?

Chapter 13: The Dual Simplex Pivot and Its Uses

1. Preview

The simplex method pivots to preserve feasibility and seek optimality (nonpositive reduced costs in a maximization problem). By contrast, the "dual" simplex method pivots to preserve optimality and seek feasibility.

The dual simplex method is presented in this chapter, as are three of its uses. To introduce the first of these uses, we recall that the simplex method consists of two phases. The "parametric self-dual method" is a pivot scheme that has only one phase. It uses simplex pivots and dual simplex pivots to move from a basic tableau to an optimal tableau. How it works is discussed here.

The dual simplex method is well-suited to re-optimizing a linear program after a constraint has been added to it. For that reason, the dual simplex method plays a key role in two different schemes for solving integer programs. Both schemes are discussed in this chapter. Each of them solves a sequence of linear programming "relaxations;" and each relaxation after the first adds a constraint to a linear program that has been solved previously.

E.V. Denardo, *Linear Programming and Generalizations*, International Series in Operations Research & Management Science 149,
DOI 10.1007/978-1-4419-6491-5_13, © Springer Science+Business Media, LLC 2011

2. The Dual Simplex Method

The "dual simplex method" will be introduced in the context of:

Problem 13.A. Minimize $\{6a + 7b + 9c + 9d\}$, subject to the constraints

$$
\begin{array}{llr}
x: & 1a + 1b \qquad - 1d \geq 2, \\
y: & 1b + 2c + 3d \geq 3, \\
 & a \geq 0, \quad b \geq 0, \quad c \geq 0, \quad d \geq 0.
\end{array}
$$

Does Problem 13.A look familiar? Its dual is the linear program that was used in Chapter 4 to introduce the simplex method. The initial steps of the dual simplex method are also familiar. They are to:

- Use a slack or surplus variable to convert each inequality constraint to an equation.
- Introduce an equation that defines z as the objective value.
- Pivot to create a basis.

Executing the first two of these steps casts Problem 13.A in the format of

Program 13.1. Minimize $\{z\}$, subject to the constraints

$$
\begin{array}{ll}
(1.0) & 6a + 7b + 9c + 9d \qquad - z = 0, \\
(1.1) & 1a + 1b \qquad - 1d - t_1 \qquad = 2, \\
(1.2) & 1b + 2c + 3d \qquad - t_2 \qquad = 3, \\
 & a \geq 0, \quad b \geq 0, \quad c \geq 0, \quad d \geq 0, \quad t_1 \geq 0, \quad t_2 \geq 0.
\end{array}
$$

In Program 13.1, the variables t_1 and t_2 are not basic because their coefficients in equations (1.1) and (1.2) equal −1. Multiplying these equations by −1 produces a basic tableau and places Program 13.1 in the equivalent form,

Program 13.1. Minimize $\{z\}$, subject to the constraints

$$
\begin{array}{ll}
(2.0) & 6a + 7b + 9c + 9d \qquad - z = \ \ 0, \\
(2.1) & -1a - 1b \qquad + 1d + t_1 \qquad = -2, \\
(2.2) & -1b - 2c - 3d \qquad + t_2 \qquad = -3, \\
 & a \geq 0, \quad b \geq 0, \quad c \geq 0, \quad d \geq 0, \quad t_1 \geq 0, \quad t_2 \geq 0.
\end{array}
$$

The variables t_1, t_2 and $-z$ are a basis for system (2). This basis satisfies the optimality conditions for a minimization problem because the reduced costs of the nonbasic variables are nonnegative. The basic solution is not feasible because it sets

$$t_1 = -2, \qquad t_2 = -3, \qquad z = 0,$$

which violates the nonnegativity constraints on t_1 and t_2.

Phase II

Program 13.1 is actually being used to introduce **Phase II** of the dual simplex method. Phase II is initialized with a basic tableau that satisfies the optimality condition but whose basic solution is not feasible.

The simplex method in reverse

The simplex method pivots so as to preserve feasibility and improve the objective value. It terminates when the optimality condition is satisfied. This and practically everything else is *reversed* in Phase II of the **dual simplex method,** for which:

- Each pivot preserves the optimality condition and aims to worsen the basic solution's objective value.
- Pivoting terminates when the basic solution becomes feasible.
- In each pivot, the departing variable (pivot row) is chosen *first*, before the entering variable (pivot column) is selected.
- The row in which the pivot element occurs has a RHS value that is *negative.*
- Ratios determine the pivot *column.*
- A ratio is computed for each column whose entry in the pivot row is *negative*, and this ratio equals the column's reduced cost (top-row coefficient) divided by its coefficient in the pivot row.
- The column in which the pivot element occurs has a ratio that is closest to zero.

The dual simplex method will be executed twice, first by hand and then on a spreadsheet.

A dual simplex pivot

When the dual simplex method is applied to system (2), the first pivot could occur a coefficient in equation (2.1) or (2.2) because both of their RHS values are negative. We mimic Rule A of Chapter 4 and select the equation whose RHS value is most negative, equation (2.2) in this case. Ratios are computed for the variables b, c and d because their coefficients in equation (2.2) are negative. These ratios equal 7/(−1), 9/(−2) and 9/(−3), respectively. The first **dual simplex pivot** occurs on the coefficient of d in row (2.2) because its ratio is closest to zero. Arrows record the selection of the pivot row and column.

$$
\begin{array}{llr}
(2.0) & 6a + 7b + 9c + 9d \qquad\qquad - z = & 0, \\
(2.1) & -1a - 1b \qquad + 1d + t_1 \qquad\qquad = & -2, \\
(2.2) & \quad - 1b - 2c - 3d \qquad + t_2 \qquad = & -3, \leftarrow \\
\text{ratios:} & \quad -7 \quad -4.5 \quad -3 & \\
 & \qquad\qquad\qquad \uparrow &
\end{array}
$$

Executing this pivot recasts Program 13.1 as:

Program 13.1. Minimize {z}, subject to the constraints

$$
\begin{array}{llr}
(3.0) & 6a + \quad 4b + \quad 3c \qquad\qquad\qquad + 3t_2 - z = & -9, \\
(3.1) & -1a - (4/3)b - (2/3)c \qquad + t_1 \qquad\qquad = & -3, \\
(3.2) & \qquad (1/3)b + (2/3)c + d \qquad - (1/3)t_2 \quad = & +1,
\end{array}
$$

$$a \geq 0, \quad b \geq 0, \quad c \geq 0, \quad d \geq 0, \quad t_1 \geq 0, \quad t_2 \geq 0.$$

This pivot has preserved the optimality condition for a minimization problem, i.e., nonnegative reduced costs. The optimality condition would not have been preserved if the pivot had occurred on the coefficient of any other variable in equation (2.2). The basic solution to system (3) equates the basic variables to the values

$$z = 9, \qquad t_1 = -3, \qquad d = 1.$$

The first pivot has increased the basic solution's objective value from 0 to 9; this worsened the objective value because Program 13.1 is a minimization problem.

A second dual simplex pivot

The next dual simplex pivot occurs on a coefficient in equation (3.1) because only its RHS value is negative. The variables a, b and c have ratios of 6/(−1), 4/(−4/3) and 3/(−2/3), respectively. The variable b has ratio of −3, which is closest to zero, so the second dual simplex pivot occurs on the coefficient of b in equation (3.1). This pivot casts Program 13.1 in the equivalent form:

Program 13.1. Minimize {z}, subject to the constraints

$$\begin{array}{llr}
(4.0) & 3a + 3c + 3t_1 + 4t_2 - z = & -18, \\
(4.1) & (3/4)a + b + (1/2)c - (3/4)t_1 - (1/4)t_2 = & 9/4, \\
(4.2) & -(1/4)a + (1/2)c + d + (1/4)t_1 - (1/4)t_2 = & 1/4, \\
 & a \ge 0, \quad b \ge 0, \quad c \ge 0, \quad d \ge 0, \quad t_1 \ge 0, \quad t_2 \ge 0. &
\end{array}$$

The basic solution to system (4) is an optimal solution to Program 13.1. That is so because it is feasible and because the reduced costs of the nonbasic variables satisfy the optimality condition for a minimization problem. This optimal solution equates the basic variables to the values

$$z = 18, \qquad b = 9/4, \qquad d = 1/4.$$

Pivoting on a spreadsheet

The dual simplex method is easy to execute on a spreadsheet. Table 13.1 shows how. Its format is familiar, except that a row (rather than a column) is set aside for each tableau's ratios.

Table 13.1. Dual simplex pivots.

	A	B	C	D	E	F	G	H	I
1									
2		a	b	c	d	t1	t2	-z	RHS
3	(1.0)	6	7	9	9			1	0
4	(1.1)	1	1		-1	-1			2
5	(1.2)		1	2	3		-1		3
6									
7		a	b	c	d	t1	t2	-z	RHS
8	(2.0)	6	7	9	9	0	0	1	0
9	(2.1)	-1	-1	0	1	1	0	0	-2
10	(2.2)	0	-1	-2	-3	0	1	0	-3
11	ratio	---	-7	-4.5	-3	---	---		
12							= IF(E10 < 0, E8/E10 , "---")		
13									
14		a	b	c	d	t1	t2	-z	RHS
15	(3.0)	6	4	3	0	0	3	1	-9
16	(3.1)	-1	-1.333	-0.667	0	1	0.3333	0	-3
17	(2.2)	-0	0.3333	0.6667	1	-0	-0.333	-0	1
18	ratio	-6	-3	-4.5	---	---	---		
19							= IF(E16 < 0, E15/E16 , "---")		
20									
21		a	b	c	d	t1	t2	-z	RHS
22	(4.0)	3	0	1	0	3	4	1	-18
23	(4.1)	0.75	1	0.5	0	-0.75	-0.25	0	2.25
24	(4.2)	-0.25	0	0.5	1	0.25	-0.25	-0	0.25

In Table 13.1, rows 8-10 reproduce the information in system (2). Cell I10 is shaded because its RHS value is the most negative. The functions in row 11 compute the ratios. Cell E11 is shaded because its ratio is closest to zero. Evidently, the coefficient in cell E10 is the pivot element. Table 13.1 omits the array function =**pivot(E10, B8:I10)** that executes the pivot and creates the array in the block B15 : I17 of cells. This block corresponds to system (3). The next pivot occurs on the coefficient in cell C16, and it creates the tableau in cells B22 : I24. That tableau is optimal.

A coincidence?

This application of the dual simplex method required two pivots, and it encountered tableaus whose basic solutions have objective values of 0, 9 and 18. When the simplex method was introduced Chapter 4, two pivots were needed, and the same sequence of objective values was encountered. There are other similarities (for instance, in the ratios), and they are not a coincidence.

Problem 13.A is the dual of the linear program that was used in Chapter 4 to introduce the simplex method, and it's a fact that:

> Application of the dual simplex method to a linear program is equivalent – pivot for pivot – to the application of the simplex method to the dual linear program, provided that comparable rules are used to resolve the ambiguity in the pivot element.

The dual simplex method is aptly named. It amounts to applying the simplex method to the dual of the linear program.

A disappointment?

This equivalence suggests that the dual simplex method is nothing new, that it is not useful. That is incorrect! Three uses of the dual simplex method are presented in this chapter, and each of these uses is important.

A bit of the history

Carlton E. Lemke (1920-2004) made several major contributions to the development of operations research. Two of these are the dual simplex method[1] (1954) and the complementary pivot method (1964). The former has just been introduced. The latter appears in Chapters 15 and 16, where it plays a vital role in the development of techniques for computing economic equilibria.

Like many in his generation, Carl Lemke interrupted college to serve in the military. He joined the 82nd Airborne Division as a paratrooper in 1940 and remained until 1946. In addition to his other duties, Lemke was a standout on the division's boxing team. His keen interest in mathematics, his athleticism, his droll humor, and his passion for the outdoors endured for a lifetime.

3. The Parametric Self-Dual Method

The simplex method has two phases, as does the dual simplex method. By contrast, the "parametric self-dual" method has one phase. It uses simplex

[1] C. Lemke, "The dual method of solving linear programming problems," *Naval Research Logistics Quarterly*," V. 1, pp. 36-47.

pivots and dual simplex pivots to move from a basic tableau to an optimal tableau. It will be introduced in the context of

Problem 13.B. Maximize $\{4p + 1q + 2r\}$, subject to the constraints

$$-1p + 1q + 2r \geq 6, \tag{5.1}$$

$$1p - 3.5q - 3r = -10, \tag{5.2}$$

$$-2p - 3q \leq 0, \tag{5.3}$$

$$p \geq 0, \quad q \geq 0, \quad r \geq 0. \tag{5.4}$$

Getting started

The first steps of the parametric self-dual method are familiar. They are to:

- Convert each inequality constraint to an equation by inserting a slack or surplus variable.
- Introduce an equation that defines z as the objective value.
- Pivot so as to create a basic tableau.

In Program 13.2 (below), the surplus variable s_1 and the slack variable s_3 have been used to convert constraints (5.1) and (5.3) to equations, and z has been defined as the objective value in the usual way.

Program 13.2. Maximize z, subject to the constraints

$$4p + 1q + 2r - z = 0, \tag{6.0}$$

$$-1p + 1q + 2r - 1s_1 = 6, \tag{6.1}$$

$$1p - 3.5q - 3r = -10, \tag{6.2}$$

$$-2p - 3q + 1s_3 = 0, \tag{6.3}$$

$$p \geq 0, \quad q \geq 0, \quad r \geq 0, \quad s_1 \geq 0, \quad s_3 \geq 0. \tag{6.4}$$

It remains to pivot so as to create basic variables for the equations that lack them. Let's pivot on the coefficient of s_1 in equation (6.1) and then on the coefficient of p in equation (6.2). This places Program 13.2 in the equivalent format:

Program 13.2. Maximize z, subject to the constraints

$$
\begin{array}{llr}
(7.0) & 15q + 14r \qquad\qquad - z = & 40, \\
(7.1) & 2.5q + 1r + 1s_1 \qquad\qquad = & 6, \\
(7.2) & 1p - 3.5q - 3r \qquad\qquad = & -10, \\
(7.3) & -10q - 6r \qquad + 1s_3 \qquad = & -20. \\
(7.4) & p \geq 0, \quad q \geq 0, \quad r \geq 0, \quad s_1 \geq 0, \quad s_3 \geq 0. &
\end{array}
$$

The basic solution to system (7) is not feasible because the RHS values of equations (7.2) and (7.3) are negative. This basic solution also violates the optimality condition for a maximization problem because q and r have positive the reduced costs.

System (7) should look familiar. It is. Problem 13.B is identical to the linear program that was used in Chapter 6 to introduce Phase I of the simplex method, and (7) is the basic tableau with which Phase I was initiated.

A homotopy

The fancy-sounding word "homotopy" describes a parametric scheme for solving a relatively difficult problem. A **homotopy** introduces a parameter α and creates a family of related problems, one for each value of α, with these properties:

- Setting α large enough produces a problem that is easy to solve.
- Setting α equal to zero produces the problem whose solution is sought.
- If the problem has been solved for a particular value of α, it is relatively easy to find the solution for a somewhat lower value of α.

Homotopies have a great many uses in optimization. The version of Phase I in Chapter 6 employed a homotopy! The decision variable α was incorporated into system (7), and simplex pivots were used to keep the basic solution feasible as α was reduced from 20 to 0.

Correcting non-optimality

Our current agenda is to introduce a parameter α and use simplex pivots *and* dual simplex pivots to obtain optimal solutions as α is reduced from a large value to 0. To get started:

- Subtract α from each reduced cost that is positive.
- Add α to each RHS value that is negative, but not from the equation for which −z is basic.

Program 13.2 becomes:

Program 13.2α. Maximize z, subject to the constraints

(8.0)		$(15-\alpha)q$	$+(14-\alpha)r$		$-z=$	$40,$
(8.1)		$2.5q$	$+1r+s_1$		$=$	$6,$
(8.2)	p	$-3.5q$	$-3r$		$=$	$-10+\alpha,$
(8.3)		$-10q$	$-6r$	$+s_3$	$=$	$-20+\alpha.$
(8.4)	$p\geq 0,$	$q\geq 0,$	$r\geq 0,$	$s_1\geq 0,$	$s_3\geq 0.$	

The basic solution to system (8) is optimal for all values of α that satisfy $\alpha\geq 20$. When α is slightly below 20, the RHS value of equation (8.3) becomes negative, and a dual simplex pivot is needed to restore optimality. This pivot occurs on a negative coefficient in equation (8.3), and that coefficient is determined by the ratios in:

Table 13.2. Ratios for system (8) with α = 20.

decision variable	ratio
q	$\frac{15-\alpha}{-10}=\frac{15-20}{-10}=0.5$
r	$\frac{14-\alpha}{-6}=\frac{14-20}{-6}=1$

The first pivot

In Table 13.2, the variable q has the ratio that is closer to zero, so the 1st pivot occurs on the coefficient of q in equation (8.3). Executing this pivot casts Program 13.2α in the equivalent form,

Program 13.2α. Maximize z, subject to the constraints

$$\begin{array}{llllll}
(9.0) & & & (5-0.4\alpha)r & +(1.5-0.1\alpha)s_3 - z & = 10 + 3.5\alpha + 0.1\alpha^2, \\
(9.1) & & & -0.5r + s_1 & +0.25s_3 & = -1 + 0.25\alpha, \\
(9.2) & p & & -0.9r & -0.35s_3 & = -3 + 0.65\alpha, \\
(9.3) & & q & +0.6r & -0.10s_3 & = 2 - 0.10\alpha, \\
(9.4) & \multicolumn{5}{l}{p \geq 0, \quad q \geq 0, \quad r \geq 0, \quad s_1 \geq 0, \quad s_3 \geq 0.}
\end{array}$$

We seek the range on α for which the basic solution to system (9) is optimal, namely, the range on α for which the reduced costs of r and of s_3 satisfy

$$5 - 0.4\alpha \leq 0, \qquad 1.5 - 0.1\alpha \leq 0$$

and for which the RHS values of constraints (9.1)-(9.3) satisfy

$$-1 + 0.25\alpha \geq 0, \quad -3 + 0.65\alpha \geq 0, \quad 2 - 0.10\alpha \geq 0.$$

These five inequalities hold when α lies in the interval $15 \leq \alpha \leq 20$.

As α decreases to 15, the reduced cost of s_3 increases to 0. As α decreases below 15, the reduced cost of s_3 becomes positive, and the basic solution to system (9) is no longer optimal. A simplex pivot is called for, with s_3 as the entering variable. This pivot occurs on the coefficient of s_3 in equation (9.1) because it is the only positive coefficient of s_3.

Dependence on α

System (9) indicates which coefficients can depend on α, and how they depend on α. After any number of pivots, only the reduced costs and the RHS values can depend on α, and the dependence is linear, except for the RHS value of the equation for which −z is basic, for which the dependence can be quadratic.

A spreadsheet

Executing the parametric self-dual method by hand is tedious and error-prone. Table 13.3 uses a spreadsheet to record the information in system (8) and determine the first pivot element.

Table 13.3. The optimal tableau for $\alpha \geq 20$ and the 1st pivot element.

	A	B	C	D	E	F	G	H	I	J	K	L
1												
17												
18				α =	20			= H23 + E$18 * I23				
19												
20		p	q	r	s1	s3	-z	RHS	α			
21	(8.α)		-1	-1								
22	(8.0)	0	15	14	0	0	1	40			net	
23	(8.1)	0	2.5	1	1	0	0	4			4	
24	(8.2)	1	-3.5	-3	-0	-0	-0	-10	1		10	
25	(8.3)	0	-10	-6	-0	1	-0	-20	1		0	
26												
27	net	0	-5	-6	0	0		= F22 + F21 * $E18				
28	ratio	---	0.5	1	---	---		= IF(F25 < 0, F27/F25, "---")				

To interpret Table 13.3:

- Compare rows 21 and 22 with the reduced costs in equation (8.0). Note that row 21 contains the coefficients of α and that row 22 contains the coefficients that are independent of α
- Compare columns H and I with the RHS values in system (8). Note that column I contains the coefficients of α and that column H contains the coefficients that are independent of α.
- Cell E18 contains a value of α.
- The functions in cells B27:F27 compute the *net* reduced costs for this value of α.
- The functions in cell K23:K25 compute the *net* RHS values of equations (8.1)-(8.3) for the same value of α.
- Evidently, decreasing α to 20 reduces the net RHS value of row 25 to zero. Cell K25 is shaded because a dual simplex pivot will occur on a coefficient in row 25.
- The ratios are computed in row 28. Column C has the ratio that is closest to zero, and cell C28 is shaded to record this fact.
- The 1st pivot will occur on the coefficient in cell C25, which lies at the intersection of the pivot row and the pivot column. This coefficient is shaded to record its selection.

This pivot updates the entries in the entire tableau, *including* the row and column that depend on α. The array function =**pivot(C25, B21:I25)** executes the first pivot and produces the array in the block B32 : I36 of cells in Table 13.4, below. That table contains the same information as does system (9). In particular, rows 32 and 33 describe the equation

$$(5 - 0.4\alpha)r + (1.5 - 0.1\alpha)s_3 - z = 10 + (2 + 1.5)\alpha - 0.1\alpha^2,$$

which is equation (9.0). The format of Table 13.4 is similar to that of Table 13.3. To reduce clutter, the functions that compute the net reduced costs and net RHS values have not been recorded.

Table 13.4. The optimal tableau for $15 \leq \alpha \leq 20$ and the 2nd pivot element.

	A	B	C	D	E	F	G	H	I	J	K	L
1												
29				α =	15							
30												
31	α	p	q	r	s1	s3	-z	RHS	α			
32	(9.α)	-0	0	-0.4	0	-0.1	0	2	-0.1			
33	(9.0)	0	0	5	0	1.5	1	10	1.5		net	ratio
34	(9.1)	0	0	-0.5	1	0.25	0	-1	0.25		2.75	11
35	(9.2)	1	0	-0.9	0	-0.35	0	-3	0.65		6.75	---
36	(9.3)	-0	1	0.6	0	-0.1	0	2	-0.1		0.5	---
37												
38	net	0	0	-1	0	0						

The 2nd pivot

Table 13.4 indicates that as α decreases to 15, the reduced cost of the nonbasic variable s_3 increases to zero. The 2nd pivot will be a simplex pivot with s_3 as the entering variable. The coefficient on which this pivot occurs is in cell F34. The array function =**pivot(F34, B32:I36)** executes the 2nd pivot.

The 3rd tableau and its pivot element

Table 13.5 contains the tableau that results from the 2nd pivot. Its format is identical to that of Table 13.4. Evidently, as α decreases to 13 1/3, the reduced cost of r increases to 0. The 3rd pivot will be a simplex pivot, with r as the entering variable. This pivot will occur on the coefficient of r in cell D47 because only it is positive.

Table 13.5. The optimal tableau for 13 1/3 ≤ α ≤ 15 and the 3rd pivot element.

	A	B	C	D	E	F	G	H	I	J	K	L
1												
40				α =	13.3							
41												
42	α	p	q	r	s1	s3	-z	RHS	α			
43		0	0	-0.6	0.4	0	0	1.6	0			
44		0	0	8	-6	0	1	16	0		net	ratio
45		0	0	-2	4	1	0	-4	1		9.33	---
46		1	0	-1.6	1.4	0	0	-4.4	1		8.93	---
47		0	1	0.4	0.4	0	0	1.6	0		1.6	4
48												
49	net	0	0	0	-0.67	0						

The final tableau

Executing the 3rd pivot produces the tableau in Table 13.6.

Table 13.6. The optimal tableau for 0 ≤ α ≤ 13 1/3.

	A	B	C	D	E	F	G	H	I	J	K
1											
51				α =	0						
52											
53	α	p	q	r	s1	s3	-z	RHS	α		
54		0	1.5	0	1	0	0	4	0		
55		0	-20	0	-14	0	1	-16	0		net
56		0	5	0	6	1	0	4	1		4
57		1	4	0	3	0	0	2	1		2
58		0	2.5	1	1	0	0	4	0		4
59											
60	net	0	-20	0	-14	0					

The basic solution to rows 54-58 of Table 13.5 is optimal when $\alpha = 0$. This basic solution equates the nonbasic variables q and s_1 to 0, and it equates the basic variables to the values

$$z = 16, \quad p = 2, \quad r = 4, \quad s_3 = 4.$$

This basic solution is an optimal solution to Program 13.2.

Recap

The parametric self-dual method did indeed execute a homotopy. It was initialized with a basic tableau that is optimal for values of α that satisfy

$\alpha \geq 20$. The 1st pivot produced a basis that is optimal for all α between 15 and 20. The 2nd pivot produced a basis that is optimal for all α between 13 1/3 and 15. The final pivot produced a basis that is optimal for all α between 0 and 13 1/3.

Speed

It seems reasonable that solving a linear program by a one-parameter homotopy would run quickly. As had been noted in Chapter 6, Robert J. Vanderbei provided empirical evidence that this method requires roughly $(m+n)/2$ pivots to solve a linear program that has m equations and n nonnegative decision variables.

4. Branch-and-Bound

Let us recall from Chapter 1 that an **integer program** differs from a linear program in that one or more of the decision variables must be integer-valued. An example of an integer program appears below as

Problem 13.C. Maximize $\{3a + 2b + 1.5c\}$, subject to the constraints

$$\begin{aligned} 1a + 1b \quad\quad &\leq 5, \\ 1a - 1b \quad\quad &\leq 0, \\ 2a \quad\quad + 5c &\leq 7, \\ a \geq 0, \quad b \geq 0, \quad c &\geq 0, \end{aligned}$$

a, b and c are integer-valued.

Problem 13.C is easy to solve by trial and error. Two candidate solutions are

$$a = 2, \quad b = 3, \quad c = 0, \quad \text{objective} = 12,$$
$$a = 1 \quad b = 4, \quad c = 1, \quad \text{objective} = 12.5,$$

and the latter is optimal.

This example will be used to introduce two different methods for solving integer programs. Both of these methods solve a sequence of linear programs. Each linear program after the first differs from a linear program that had been solved previously by the inclusion of one extra inequality constraint. For that reason, each linear program other than the first is well-suited to solution by the dual simplex method.

The method that's known as branch-and-bound is introduced in this section. The method of cutting planes is introduced in the next section.

The LP relaxation

A **relaxation** of an optimization problem is what one obtains by weakening or removing one or more of its constraints. Nearly every method for finding a solution to an integer program begins by solving its **LP relaxation,** namely, the relaxation in which the requirement that the decision variables be integer-valued is removed. The LP relaxation of Problem 13.C appears below as Program 13.3, where it has been placed in Form 1.

Program 13.3. Maximize {z}, subject to the constraints

$$\begin{array}{llr}
(10.0) & 3a + 2b + 1.5c \qquad\qquad - z = 0 & \\
(10.1) & 1a + 1b \qquad + s_1 \qquad\qquad = 5, & \\
(10.2) & 1a - 1b \qquad\qquad + s_2 \qquad = 0, & \\
(10.3) & 2a \qquad + 5c \qquad\qquad + s_3 \quad = 7, & \\
 & a \geq 0, \quad b \geq 0, \quad c \geq 0, \quad s_i \geq 0 \quad \text{for} \quad i = 1, 2, 3. &
\end{array}$$

Three simplex pivots solve this relaxation and produce the tableau in Table 13.7.

Table 13.7. Optimal tableau for the LP relaxation of Program 13.3.

	A	B	C	D	E	F	G	H	I	J
1										
18			a	b	c	s1	s2	s3	-z	RHS
19			0	0	0	-2.2	-0.2	-0.3	1	-13.1
20			0	1	0	0.5	-0.5	0	0	2.5
21			1	0	0	0.5	0.5	0	0	2.5
22			0	0	1	-0.2	-0.2	0.2	0	0.4

The basic solution to the tableau in Table 13.7 sets

$$a = 2.5, \qquad b = 2.5, \qquad c = 0.4, \qquad z = 13.1.$$

If the solution to a relaxation happens to satisfy the constraints that had been relaxed or omitted, it solves the original problem. That did not occur in this case because the solution to the relaxation violates the integrality constraints on a, b and c.

Divide and conquer

The branch-and-bound method is a "divide-and-conquer" scheme that consists of three constituents, which are dubbed "branching," "bounding" and "pruning." Each constituent will be described and illustrated using Program 13.3.

Branching

To **branch** is to pick a decision variable whose value in the optimal solution to a linear program violates an integrality constraint and:

- Replace that linear program by two others, each with one added constraint. One of these new linear programs requires the decision variable to be not greater than the next lower integer. The other requires the decision variable to be not smaller than the next larger integer.
- Solve these two linear programs.

The optimal solution to the LP relaxation of Program 13.3 violates the integrality constraints on a, on b and on c. We could branch on any one of these decision variables. Let's branch on a. This optimal solution sets $a=2.5$. To branch on a is to replace the LP relaxation by two others, one with the added constraint $a \le 2$, the other with the added constraint $a \ge 3$, and to solve these two linear programs.

Bounding

The **bound** is the best (largest in the case of maximization) of the objective values of the feasible solutions to the integer program that have been found so far. The initial bound is $-\infty$ in the case of a maximization problem and is $+\infty$ in the case of a minimization problem. If the bound is finite, the **incumbent** is a feasible solution to the integer program whose objective value equals the bound.

Pruning

A linear program is **pruned** (eliminated) if either of these conditions obtains:

- The linear program has no feasible solution.
- The linear program is feasible, but its optimal value fails to improve on the prior bound.

No linear program that is pruned can have a feasible solution that satisfies the integrality constraints and has an objective value that improves the incumbent's.

A branch-and-bound tree

Branching, bounding and pruning creates a series of relaxations of the integer program. These relaxations organize themselves into a **branch-and-bound tree.** Figure 13.1 exhibits a branch-and-bound tree for Problem 13.C. The linear program at Node 1 of this tree has already been solved, and the others will be.

Figure 13.1. A branch-and-bound tree for Problem 13.C.

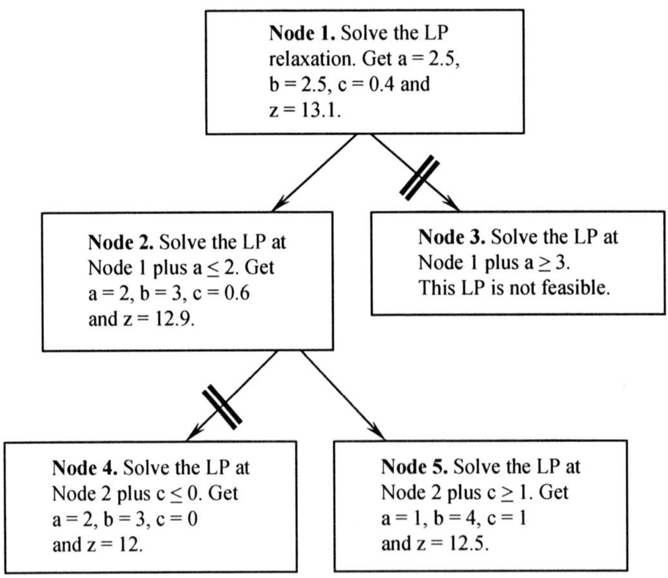

Each node (square box) in Figure 13.1 describes a linear program. Node 1 describes the LP relaxation, whose optimal solution has been found and is presented in Table 13.7. This optimal solution violates the integrality constraints on the decision variables a, b and c.

Nodes (linear programs) 2 and 3 are obtained from Node 1 by branching on the decision variable a. The linear program at Node 3 is infeasible, so Node 3 can be pruned.

The optimal solution to the linear program at Node 2 sets $c = 0.6$, and Nodes 4 and 5 have the added constraint $c \leq 0$ and $c \geq 1$, respectively.

Let's suppose that the linear program at Node 4 is solved next. Its optimal solution is integer-valued and its optimal value equals 12. This optimal solution becomes the incumbent, and 12 becomes the bound. Any other node whose optimal value is 12 or less can be pruned because adding constraints to its linear program cannot improve (increase) its optimal value.

The linear program at Node 5 must be solved. Its optimal solution is also integer-valued, and its optimal value equals 12.5. This improves on the current bound. So the optimal solution to the linear program at Node 5 becomes the new incumbent, and 12.5 becomes the new bound. Each node whose optimal value does not exceed 12.5 is pruned. In this instance, Node 4 is pruned. Only Node 5 remains, so its optimal solution solves the integer program.

Executing branch-and-bound

One could solve the linear program at each node of the branch-and-bound tree from scratch. An attractive alternative is start with the optimal solution to the node that is one level up and use the dual simplex method to account for the new (inequality) constraint. How that is accomplished is discussed next.

Solving the LP at Node 2

Table 13.8 shows how to solve the linear program at Node 2.

Table 13.8. Node 2: the LP plus a ≤ 2.

	A	B	C	D	E	F	G	H	I	J
25										
26		s4	a	b	c	s1	s2	s3	-z	RHS
27			0	0	0	-2.2	-0.2	-0.3	1	-13.1
28			0	1	0	0.5	-0.5	0	0	2.5
29			1	0	0	0.5	0.5	0	0	2.5
30			0	0	1	-0.2	-0.2	0.2	0	0.4
31		1	1							2
32										
33		s4	a	b	c	s1	s2	s3	-z	RHS
34		-0	0	0	0	-2.2	-0.2	-0.3	1	-13.1
35		-0	0	1	0	0.5	-0.5	0	0	2.5
36		0	1	0	0	0.5	0.5	0	0	2.5
37		-0	0	0	1	-0.2	-0.2	0.2	0	0.4
38		1	0	-0	-0	-0.5	-0.5	-0	-0	-0.5
39										
40		s4	a	b	c	s1	s2	s3	-z	RHS
41		-0.4	0	0	0	-2	0	-0.3	1	-12.9
42		-1	0	1	0	1	0	0	0	3
43		1	1	0	0	0	0	0	0	2
44		-0.4	0	0	1	0	0	0.2	0	0.6
45		-2	-0	0	0	1	1	0	0	1

Rows 27-30 of Table 13.8 contain the optimal tableau for the LP relaxation. Row 31 models the new constraint, a ≤ 2. It does so by introducing a slack variable s_4 that converts this constraint to the equation $a + s_4 = 2$. The slack variable s_4 is basic for row 31, but a is no longer basic for row 29. Pivoting on the coefficient of a in row 29 restores the basis, preserves the optimality condition, and produces the tableau in rows 34-38. This tableau's basic solution sets $s_4 = -0.5$. A dual simplex pivot is called for. It occurs on the coefficient of s_2 in row 38. This pivot produces an optimal tableau. (In general, more than one dual simplex pivot may be needed to produce an optimal tableau). Evidently, the optimal solution to the LP at Node 2 sets

$$a = 2, \qquad b = 3, \qquad c = 0.6, \qquad z = 12.9.$$

This optimal solution is not integer-valued. It may be necessary to branch on c, depending on what happens when the linear program at node 3 is solved.

Solving the LP at node 3

To solve the linear program at Node 3, we append to the optimal tableau for the LP relaxation the constraint $a \geq 3$, restore the basis, and execute the dual simplex method. Table 13.9 shows what occurs. The constraint $a - s_5 = 3$ has been multiplied by -1, thereby making s_5 basic for row 54. A pivot on the coefficient of a in row 52 restores the basis and makes the RHS value of row 61 negative. A dual simplex pivot is called for.

Table 13.9. Node 3: the LP plus $a \geq 3$.

	A	B	C	D	E	F	G	H	I	J
25										
49		s5	a	b	c	s1	s2	s3	-z	RHS
50			0	0	0	-2.2	-0.2	-0.3	1	-13.1
51			0	1	0	0.5	-0.5	0	0	2.5
52			1	0	0	0.5	0.5	0	0	2.5
53			0	0	1	-0.2	-0.2	0.2	0	0.4
54		1	-1							-3
55										
56		s5	a	b	c	s1	s2	s3	-z	RHS
57		-0	0	0	0	-2.2	-0.2	-0.3	1	-13.1
58		-0	0	1	0	0.5	-0.5	0	0	2.5
59		0	1	0	0	0.5	0.5	0	0	2.5
60		-0	0	0	1	-0.2	-0.2	0.2	0	0.4
61		1	0	0	0	0.5	0.5	0	0	-0.5

But no coefficient in row 61 is negative, so there is nothing to pivot upon. Row 61 represents the equation

$$s_5 + 0.5s_1 + 0.5s_2 = -0.5.$$

The decision variables s_5, s_1 and s_2 are required to be nonnegative, so the left-hand side of the above constraint can not be negative. The RHS value of this constraint is negative, so it can have no solution. This demonstrates that the LP at Node 3 is infeasible. That node can be pruned.

Node 4

Node 4 depicts the LP at Node 2 with the added constraint $c \leq 0$. Mimicking the procedure used at node 2 solves this LP. (Its solution can be found in the spreadsheet that accompanies this chapter). Its optimal solution is

$$a = 2, \quad b = 3, \quad c = 0, \quad z = 12.$$

Because this optimal solution is integer-valued, it is the initial incumbent, and 12 becomes the current bound. Any other node whose optimal value is 12 or less (there are no such nodes at the moment) can be pruned.

Node 5

Node 5 depicts the LP at Node 2 with the added constraint $c \geq 1$. Mimicing the procedure used at node 3 solves this linear program and produces the optimal solution

$$a = 1, \quad b = 4, \quad c = 1, \quad z = 12.5.$$

which is integer-valued. This solution improves on the incumbent because its objective value exceeds the current bound. It's optimal solution becomes the incumbent, and the bound increases to 12.5. Node 4 is pruned because its optimal value is below 12.5.

The general pattern

The branch-and-bound tree is a family of linear programs, each of which is a relaxation of the linear program that one wishes to solve. Each linear program in this tree differs from the LP "above" it by an inequality constraint on one decision variable. Each linear program is easily solved by starting with the optimal tableau for the linear program above it and:

- Making the slack or surplus variable for the new constraint basic for that constraint.
- Making the decision variable that is being bounded basic for the equation for which it had been basic. (This does not affect the reduced costs, so the optimality condition is preserved.)
- Solving the linear program via dual simplex pivots.

Typically, only a few dual simplex pivots are needed to solve a particular linear program in the tree. But the number of nodes in the branch-and-bound tree could be enormous. More about this later.

A bit of the history

The branch-and-bound method is remarkable for its simplicity, and it is equally remarkable for its usefulness. In 1960, Land and Doig[2] published their classic paper on branch-and-bound.

The simplex method vs. interior-point methods

As a subroutine in a scheme for solving integer programs, the simplex method has an advantage over interior-point methods. This is so because the simplex method gives an extreme point as its solution. If there are multiple optimal solutions (and there often are), interior-point methods will not report an extreme point as best. They will give fractional answers, rather than 0's and 1's.

Pure and mixed integer programs

A **pure** integer program differs from a linear program in that *every* decision variable is required to be integer-valued. A **mixed** integer program differs from a linear program in that at one or more – but not all – of its decision variables are required to be integer-valued.

Branch-and-bound works for pure integer programs, and it works for mixed integer programs provided, of course, that one branches on the decision variables that are required to be integer-valued.

Program 13.3 seems to be a mixed integer program because its slack variables are not required to be integer-valued. But any solution to equations (10.1)-(10.3) that sets a, b and c to integer values also sets s_1, s_2 and s_3 to integer values. So nothing is lost by regarding Program 13.3 as a pure integer program.

5. The Cutting Plane Method

Program 13.3 will now be solved by the "cutting plane" method. This method can be made to work for the case of a mixed integer program. To simplify the discussion, it is presented for the case of a pure integer program, for

---

[2] A. H. Land and A. G. Doig, "An automatic method for describing discrete programming problems," *Econometrica*, V. 28, pp. 497-520, 1960.

which all decision variables are required to be integer-valued. As just noted, Program 13.3 can be regarded as a pure integer program.

A sequence of linear programs

The cutting-plane method solves a *sequence* of linear programs, rather than a tree of linear programs. The optimal solution to each linear program in this sequence is used to introduce an inequality constraint in the next. The first linear program in the sequence is (you guessed it) the LP relaxation.

Table 13.7 gives the optimal tableau for the LP relaxation of Program 13.3. This tableau is reproduced in dictionary format as.

(11.0) $z = 13.1 - 2.2s_1 - 0.2s_2 - 0.3s_3$

(11.1) $b = 2.5 - 0.5s_1 + 0.5s_2$

(11.2) $a = 2.5 - 0.5s_1 - 0.5s_2$

(11.3) $c = 0.4 + 0.2s_1 + 0.2s_2 - 0.2s_3$

The basic solution to system (11) is not feasible for Program 13.3 because it fails to equate a, b and c to integer values.

A cutting plane

Each iteration of the cutting plane method selects any variable whose value in the optimal solution violates an integrality constraint and uses the equation for which it is basic to create one new constraint. Let us select the variable a, as was done in the previous section. The basic solution to system (11) sets $a=2.5$, which violates the integrality constraint on a. We write 2.5 as $2+0.5$ and write equation (11.2) as

(11.2) $a = 2 + [0.5 - 0.5s_1 - 0.5s_2].$

The term "[…]" in (11.2) cannot exceed 0.5, so it must be that $a \le 2$. This leads us to the linear program having the added constraint

(11.4) $s_4 + a = 2,$

where s_4 is a slack variable.

A second cutting plane

The second linear program mazimizes z, subject to equations (11.0) through (11.4) and to the nonnegativity constraints on all decision variables other than z. This linear program has been solved. Rows 41-45 of Table 13.8 report its optimal tableau, which appears below as system (12).

(12.0) $z = 12.9 - 2.0s_1 - 0.3s_3 - 0.4s_4$

(12.1) $b = 3.0 - 1.0s_1 + 1.0s_4$

(12.2) $a = 2.0 - 1.0s_4$

(12.3) $c = 0.6 - 0.2s_3 + 0.4s_4$

(12.4) $s_2 = 1.0 - 1.0s_1 + 2.0s_4$

The basic solution to system (12) equates the variables a, b and s_2 to integer values. Only c is equated to a fraction, so equation (12.3) must be used to produce a cutting plane. The addend 0.4 s_4 on the right-hand-side of (12.3) seems to present a slight difficulty, but substituting $0.4 = 1.0 - 0.6$ lets (12.3) be written as

$$c = 0 + [0.6 - 0.2s_3 - 0.6s_4] + 1s_4,$$

The term "[...]" in the above cannot be larger than 0.6, so each integer-valued solution to (123) satisfies $c \le s_4$. This generates the cutting plane

(12.5) $$s_6 + c - s_4 = 0,$$

where s_6 is a new slack variable.

A spreadsheet

Thus, the third linear program maximizes z, subject to equations (12.0) through (12.5) along with nonnegativity constraints on all variables other than z. The constraints of this linear program appear in rows 3-8 of Table 13.10. Pivoting on the coefficient of c in row 6 restores the basis, preserves the optimality condition, and produces the tableau in rows 11-16. The RHS value of row 16 is negative. A dual simplex pivot is called for. Ratios are computed for columns C and I, and this pivot occurs on the coefficient of s_4 in row 16.

Table 13.10. The 2nd cutting plane.

	A	B	C	D	E	F	G	H	I	J	K
1											
2		s6	s4	a	b	c	s1	s2	s3	-z	RHS
3			-0.4	0	0	0	-2	0	-0.3	1	-12.9
4			-1	0	1	0	1	0	0	0	3
5			1	1	0	0	0	0	0	0	2
6			-0.4	0	0	1	0	0	0.2	0	0.6
7			-2	-0	0	0	1	1	0	0	1
8		1	-1			1					0
9											
10		s6	s4	a	b	c	s1	s2	s3	-z	RHS
11		-0	-0.4	0	0	0	-2	0	-0.3	1	-12.9
12		-0	-1	0	1	0	1	0	0	0	3
13		-0	1	1	0	0	0	0	0	0	2
14		0	-0.4	0	0	1	0	0	0.2	0	0.6
15		-0	-2	-0	0	0	1	1	0	0	1
16		1	-0.6	-0	-0	0	-0	-0	-0.2	-0	-0.6
17	ratio		0.667						1.5		
18											
19		s6	s4	a	b	c	s1	s2	s3	-z	RHS
20		-0.67	0	0	0	0	-2	0	-0.17	1	-12.5
21		-1.67	0	0	1	0	1	0	0.333	0	4
22		1.667	0	1	0	0	0	0	-0.33	0	1
23		-0.67	0	0	0	1	0	0	0.333	0	1
24		-3.33	0	0	0	0	1	1	0.667	0	3
25		-1.67	1	0	0	-0	0	0	0.333	0	1

Rows 20-25 of Table 13.10 describe a basic tableau that satisfies the optimality condition for a maximization problem (the reduced costs are nonpositive) and that equates the basic variables to nonnegative integer values. An optimal solution to Problem 13.C has been found, and it is

$$b = 4, \qquad a = 1, \qquad c = 1, \qquad \text{objective equals } 12.5.$$

Strong cuts

The first cut used equation (11.4) to impose the constraint $a \leq 2$ and the second cut used equation (12.5) to impose the constraint $c \leq s_4$. These were (as it will turn out) the "strong" cuts of their type. Precise mathematical definition of a **strong cut** requires notation that is slightly involved, but the example

$$a = 3.8 + 1.8b - 3.4c \tag{13}$$

will make everything clear. In this example, a is a basic variable, and b and c are nonbasic. Substituting $3.8 = 3 + 0.8$ and $1.8 = 2 - 0.2$ and $-3.4 = -3 - 0.4$ into (13) rewrites it as

$$a = 3 + [0.8 - 0.2b - 0.4c] + 2b - 3c. \tag{14}$$

The expression "[...]" cannot exceed 0.8, so the integer-valued solutions to (14) satisfy

$$a \leq 3 + 2b - 3c. \tag{15}$$

Note that (15) was obtained by rounding down or up as needed to guarantee that the term in braces is below 1. (Doing this precisely would require notation to describe the "floor" and "ceiling" of a number x).

A bit of the history

In 1958, Ralph E. Gomory introduced the cutting plane method and showed how to solve pure integer programs[3] and mixed integer programs[4] with finitely many cuts. For this and other path-breaking work, he was made an IBM Fellow in 1964. Gomory proved equally adept as an administrator; he served as IBM's Director of Research from 1980 to 1986, as IBM's Senior Vice President for Science and Technology from 1986 to 1989, and then as President of the Alfred P. Sloan foundation, where he continued his mathematical research.

What's best?

The cutting plane method is surely elegant, but it tends to run slowly on practical problems. The branch-and-bound method is supremely inelegant, but it runs surprisingly quickly on practical problems. Really big integer programs – such as routing fleets of aircraft – are being solved. They are solved by

---

[3] Gomory, R. E., "Outline of an algorithm for integer solutions to linear programs," *Bull. Amer. Math. Soc.*, V. 64, pp. 275-278, 1958.

[4] Gomory, R. E., "An algorithm for integer solutions to linear programs," *Princeton-IBM Research Technical Report Number 1*, Nov. 17, 1958. Reprinted as pp. 269-302 of *Recent Advances in Mathematical Programming*, R. L. Graves and P. Wolfe (eds.), McGraw-Hill, NY, 1963.

artfully designed hybrid algorithms that begin with cutting planes and switch to branch-and-bound.

6. Review

This chapter completes our account of the simplex method. It and its twin, the dual simplex method, are so fast that they can be used as subroutines in algorithms that solve optimization problems that are not linear. Their speed and the fact that they report extreme points as optimal solutions account for the fact that rather large integer programs can be solved – and are solved – fairly quickly.

7. Homework and Discussion Problems

1. Execute the dual simplex method on Program 13.1, but arrange for the first pivot to occur on a coefficient in equation (2.1), rather than in equation (2.2).

2. Write the dual of Problem 13.A, labeling the variables that are complementary to its constraints as x and y and its slack variables as s_1 through s_4.

    (a) Is the linear program you just constructed identical to a linear program in Chapter 4?

    (b) What pairs of variables (one from Problem 13.A and the other from the dual that you have just constructed) are complementary?

    (c) Fill in the blanks: At each iteration of the application of the dual simplex method to Problem 13.A, the reduced cost of each decision variable equals the value of the ___ in the comparable iteration of __________.

3. Is it possible to initiate Phase II of the dual simplex method in the case of a linear program that is unbounded? If not, why not?

4. *(cycling and Bland's rule)* Suppose that Phase II of the dual simplex method is being used to solve a linear program whose objective is maximization

and that a basic tableau has been found that has nonpositive reduced costs. Fill in the blanks:

(a) The analogue of Rule A for the simplex method picks the pivot row as follows: The pivot row has the most ____ RHS value; ties, if any, are broken by picking the row whose basic variable is listed farthest to the ____.

(b) The analogue of Rule A for the simplex method picks the pivot column as one whose ratio is ____ to zero; ties, if any, are broken by picking the column that is _____.

(c) The rule chosen in parts (a) and (b) will cycle when it is used to solve the dual of the linear program that appears in this book as _____.

(d) Cycling in the dual simplex method is precluded by using the analogue of Bland's rule, which resolves the ambiguity in the pivot element as follows _____________.

5. In branch-and-bound, can it occur that all nodes get pruned? If so, under what circumstances does it occur?

6. In equation (13), we could write $3.8 = 3 + 0.8$ and $1.8 = 3 - 1.2$ and $-3.4 = -3 - 0.4$ obtain $a = 3 + [0.8 - 1.2b - 0.4c] + 3b - 3c$. The term in braces cannot exceed 0.8, so the integer-valued solutions to this equation must satisfy $a \leq 3 + 3b - 3c$. Do you prefer (15)? If so, why?

7. Solve Problem 13.C (on page 427) by branch-and-bound, but branch first on the variable c, rather than a. Construct the analogue of Figure 13.1. (An optimal tableau for the LP relaxation appears as Table 13.7 on the spreadsheet for this chapter.)

8. Solve Problem 13.C (on page 427) by the cutting plane method, but begin with a cutting plane for equation (11.3) rather than (11.2). (An optimal tableau for the LP relaxation appears as Table 13.7 on the spreadsheet for this chapter.)

Part V–Game Theory

Linear programs deal with a single decision maker who acts in his or her best interest. Game theory deals with multiple decision makers, each of whom acts in his or her own best interest. At first glance, these subjects seem to have nothing in common. But there are two strong connections – one through the Duality Theorem, the other through the simplex method.

Chapter 14. Introduction to Game Theory

Game theory has a wide variety of solution concepts and applications. Three different solution concepts are described and illustrated in this chapter. Several famous games are discussed. The Duality Theorem is used to construct optimal strategies for von Neumann's matrix game and to construct a general equilibrium for a stylized model of an economy.

Chapter 15. The Bi-Matrix Game

The bi-matrix game is not a zero-sum game. The Duality Theorem provides no insight into it. But the simplex method does. Feasible pivots are used to construct an equilibrium.

Chapter 16. Fixed Points and Equilibria

An economic equilibrium has long been understood to be a "Brouwer fixed point," namely, a vector x = f(x) where f is a continuous map of a closed bounded convex set C into itself. A deft adaptation of the pivot scheme in Chapter 15 constructs an equilibrium for an n-player competitive game. The same method provides a constructive proof of Brouwer's fixed-point theorem.

Chapter 14: Introduction to Game Theory

1. Preview

Prior chapters of this book have been focused on the search by a single decision maker (individual or firm) for a strategy whose net benefit is largest, equivalently, whose net cost is smallest.

Game theory is the study of models in which two or more decision makers must select strategies and in which each decision maker's well being can be affected by his or her strategy *and* by the strategies of the other participants. Being mindful of the interests of the other participants lies at the heart of game theory.

It is emphasized:

> To decide what you should do in a game-theoretic situation, examine it from the viewpoint of each participant and, possibly, each coalition.

E. V. Denardo, *Linear Programming and Generalizations*, International Series in Operations Research & Management Science 149,
DOI 10.1007/978-1-4419-6491-5_14,

Game theory is an enormous subject. It is an important part of several academic disciplines, which include political science, economics, and operations research. It provides insight into business, politics, military affairs, and public life. Terminology from game theory has entered every-day usage. A **win-win** situation is one in which each participant benefits (no one loses). A **zero-sum** situation is one in which the sum of the net benefits of the participants equals zero.

Game theory encompasses many different models, and it includes several different concepts of effective behavior. Any introduction to game theory must be very selective. The models in this chapter illustrate three different solution concepts and emphasize the connections between game theory and linear programming.

2. Three Solution Concepts

In game theory, each of several players selects his or her strategy. As mentioned above, the benefit each player receives can depend on that player's strategy and on the strategies of the other players. Listed below are three different performance criteria:

- An individual player's strategy is said to be **dominant** for that player if this strategy performs best for that player, independent of the strategies of the other players.
- A set of strategies, one per player, is said to be **stable**[1] if no group of players can all benefit by changing their strategies, given that the players who are not in the group do not change their strategies.
- A set of strategies, one per player, is said to be an **equilibrium**[2] if no single player can benefit by changing his or her strategy, given that the other players do not change their strategies.

---

[1] This usage of "stable" is not uniformly agreed upon. Some writers have used "strong equilibrium" instead, but that usage never caught on. Other writers describe a set of *outcomes* (rather than strategies) as stable if no group of participants can all get outcomes they prefer by changing their strategies simultaneously.

[2] In the literature on economics, what we are calling an equilibrium is sometimes referred to as a *Nash equilibrium*; this distinguishes it from a *general equilibrium*, this being a Nash equilibrium in which the "market clears."

A dominant strategy is a *best response* to every set of strategies of the other players. Only a few games have dominant strategies. If a game presents you with a dominant strategy, it is in your self-interest to employ it. The benefit that you garner can depend on the actions of the other players, but – given any set of strategies of the other players – you have nothing to gain by deviating from your dominant strategy.

Dominance is a property of an individual player's strategy. The other two solution concepts (stable and equilibrium) are properties of the strategies of all of the players. In particular, in an equilibrium, each player's strategy is a best response to the strategies that all of the other players are *currently* using. Equilibrium is a central theme in economic reasoning. But be warned: In general, there can be several equilibria, and there need not be an equilibrium that is best for all of the participants.

This chapter presents a model for which each of these solution concepts is germane. The "sealed-bid auction" in Section 3 has a dominant strategy. The "marriage game" in Section 4 has stable strategies. The matrix game in Section 5 has an equilibrium, as does the simplified model of an economy in Section 6.

3. A Sealed-Bid Auction

Auctions are used to sell commodities as diverse as art, government bonds, oil rights, telecommunication bandwidth, and used automobiles. At the major auction houses, art is sold by an *ascending bid* or *English* auction in which:

- The auctioneer announces a starting price. If someone bids that price, the auctioneer announces a somewhat higher price and asks whether a different person will bid that price.
- If so, the auctioneer increases the price and asks again.
- This process is repeated until no bidder is willing to improve upon the price offered most recently.
- The artwork is sold to the most recent (and highest) bidder at the price that he or she had bid.

Several other auction mechanisms are used. At the Amsterdam flower market, a *descending price* or *Dutch* auction is employed. Each bidder has a button. An opening price for a lot of flowers is displayed visually. This price begins to decrease at a constant rate that is visible to all bidders. The bidder who first pushes his or her button buys the lot at the currently-displayed price.

A Vickery auction

In 1961, a paper by William Vickery[3] created a sensation with an analysis of an auction of an indivisible item that proceeds according to these rules:

- Each bidder submits a sealed bid for the item.
- The bidders are precluded from sharing any information about their bids.
- The bids are opened, and the item is purchased by the person whose bid is highest. The price that person pays is the *second* highest bid.

Such an auction has long been known as a **Vickery auction.** For it and related work, Vickery (also spelled Vickrey) was awarded a share of the 1996 Nobel Prize in Economics. Like many good ideas, this one has deep roots. Stamps had been auctioned in this way since the 1890's. To illustrate this type of auction, we consider:

Problem 14.A. A Vickery auction will be used to sell a home. You are willing to pay as much as $432,000 for this home. Others will bid on it. You have no idea what they will bid. What do you bid?

A dominant strategy

Should you bid less than $432 thousand? Suppose you do. For purposes of discussion, suppose you bid $420 thousand. If you win the auction, you pay exactly the same price that you would have paid if you had bid $432 thousand. But suppose the high bid was $425 thousand. By bidding low, you lost. Had you bid $432 thousand, you would have gotten the home for $7 thousand less than you would have been willing to pay for it. Evidently, you should not bid less than $432 thousand.

---

[3] Vickery, William, "Counterspeculation, auctions and competitive sealed tenders," *Journal of Finance*, V. 16, pp 8-37, 1961.

Should you bid more than \$432 thousand? Suppose you do. If you win the auction, you pay the second highest price. If that price is below \$432 thousand, you pay the same amount you would if you had bid \$432 thousand. If that price is above \$432 thousand, you pay more than the value of you placed on the house.

Evidently, you should bid exactly \$432 thousand. This strategy is dominant because it is best for you, *independent* of what the other players do. In a Vickery auction, it is optimal for each bidder to bid the value that he or she places on the item. The winning bidder earns a profit equal to the difference between that person's bid and the second highest bid.

4. Matching in a Two-Sided Market

The next illustration of game theory is to a market in that has two classes of participant, say, class A and class B. Each member of class A seeks an association with a member of class B, and conversely. It is assumed that:

- Each member of class A has a strict ranking over some or all of the members of class B.
- Each member of class B has a strict ranking over some or all of the members of class A.

If member α of either class omits member β of the other class from his/her ranking, it means that member α prefers being unassociated to being paired with member β. To illustrate this type of two-sided market, consider:

> *Graduates and firms:* Each college graduate has a strict ranking over the various positions that he or she might fill, and each firm with an open position has a strict ranking over the college graduates who might fill it.
>
> *Medical school graduates and hospitals:* Each medical school graduate has a strict preference over the internships that she or he might wish to fill. Each hospital with one or more open internships has a strict preference over the graduates who might fill each position.

Note that if a firm or a hospital has more than one open position, it is assumed to rank candidates for its open positions independently.

A matching

In this context, a **matching** is a set of pairs, each pair consisting of one member of class A and one member of class B with the property that no member of either class is paired twice. In the first example, graduates are paired with positions. In the second example, MDs are paired with internships.

Unstable matchings

A matching of medical school graduates to internships can be unstable in these ways:

- It can match an internship to a graduate that the hospital did not rank.
- It can match a graduate to an internship that the graduate did not rank.
- It can fail to assign graduate α to internship β in a case for which graduate α would prefer internship β to his/her assigned status and in which the hospital would prefer to fill the internship with graduate α to this internship's assigned status.

In the first example, the hospital prefers to leave the internship vacant. In the second, the graduate prefers no internship. In the third, the graduate and hospital prefer to abandon their current assignments and pair up with each other.

A matching of graduates to residencies is *stable*, as defined earlier in this chapter, if none of these instabilities occur. These questions present themselves:

- Does there exist a stable matching?
- If a stable matching exists, how can it be found?
- Can there be more than one stable matching? If so, how do they compare?

In 1962, David Gale and Lloyd Shapley coauthored a lovely paper[4] that posed this matching problem and answered the questions that are listed above.

[4] D. Gale and L. Shapley, "College admissions and the stability of marriage," *American Mathematical Monthly*, V. 69, pp 9-15, 1962.

The dance competition

To illustrate this model, we turn our attention to:

Problem 14.B *(the dance competition).* For Saturday night's ballroom dance competition, each of four women can be paired any of five men. The four women are labeled A through D, and the five men are labeled v through z. Each woman has a strict preference over the men, and each man has a strict preference over the women. The preferences are listed in Table 14.1. This table indicates that woman A's first choice is man w, her second choice is man x, and so forth. This table also indicates that man z's 1st choice is woman A, his 2nd choice is woman B, his 3rd choice is woman D, and he prefers to stay home than be partnered with woman C.

Table 14.1. Preferences of each woman and of each man.

woman	preference	Man	preference
A	w x v y z	v	A B C D
B	x v w y z	w	D B C A
C	z y v w x	x	D C A B
D	v y z w x	y	A D C B
		z	A B D

Let us consider whether a matching that includes the pairs (A, v) and (B, x) can be stable. Woman B and man v are ranked with their 1st choice partners. They are can do no better. What about woman A and man x? Woman A prefers man x to her assigned partner. Man x prefers woman A to his assigned partner. Given the option, they will break their dates and go to the dance competition with each other. This matching is not stable.

DAP/M

The procedure that is described below is has the acronym **DAP/M**, which is short for "deferred acceptance procedure with men proposing." The bidding process is as follows:

1. In the first round, each man proposes to the woman he ranks as best.

2. Each woman who has multiple offers rejects all but the one she ranks best. (No woman has yet accepted any offer.)

3. In the next round, each man who was rejected in Step 2 and has not exhausted his preference list proposes to the woman he ranks just below the one who just rejected him. Return to Step 2.

The bidding process terminates when no woman rejects an offer or when each man who is rejected has proposed to every woman that he ranked.

Table 14.2 shows what happens when DAP/M is applied to the data in Table 14.1. In the first round, woman A receives three offers and woman D receives two. Woman A rejects offers from men y and z because she prefers man v. Woman D rejects man x because she prefers man w. In the second round, men x, y and z propose to the women they rank as second. Proposals continue for four rounds. In Round 4, only man z proposes, and at the end of this round he has been rejected by each woman he wishes to dance with. At that point, DAP/M establishes the matching:

(A, v) (B, w) (C, x) (D, y),

with man z unmatched.

Table 14.2. Four rounds of DAP/M.

man	Round 1	Round 2	Round 3	Round 4
v	A	A	A	A
w	D	D no	B	B
x	D no	C	C	C
y	A no	D	D	D
z	A no	B	B no	D no

Proof of stability

To demonstrate that DAP/M creates a stable matching, we consider a woman α and a man ω who are not matched to each other by DAP/M. Might they prefer each other to their current assignments? Well:

- If woman α received an offer from man ω, she rejected him in some round in which she had an offer she preferred. She still has that offer or something she prefers even more. So she prefers her current assignment to man ω.

- If woman α received no offer from man ω, he prefers his current assignment to her.

Evidently, woman α and man ω do not prefer each other to their current assignments. The matching created by DAP/M is stable.

DAP/W

DAP/W describes the same deferred acceptance procedure, but with the women doing the proposing. For the data in Table 14.1, the women rank different men first, but woman C ranks man z highest, and he would rather stay home, so she proposes to man y in the 2nd round. At this point, no man has more than one offer, so the bidding process terminates. It produces the matching

(A, w) (B, x) (C, y) (D, v),

with man z unmatched. This too is a stable matching, and for the same reason.

Different stable matchings

The stable matchings produced by DAP/M and DAP/W are compared in Table 14.3. This table shows that woman A gets the partner she ranks 3rd when the men are proposing and that she gets the man she ranks 1st when the women are proposing. For this example, every woman is better off when the women are proposing. Similarly, with one exception, every man is better off when the men are proposing. The exception is man z, who stays home no matter who is proposing. Table 14.3 suggests (correctly, as we shall see) that it is better to be part of the group that is doing the proposing.

Table 14.3. The rank that each player places on her/his partner under DAP/M and DAP/W.

process	A	B	C	D	v	w	x	y	z
DAP/M	3	3	5	2	1	2	2	2	4
DAP/W	1	1	2	1	4	4	3	4	4

The marriage problem

The general matching problem that we've illustrated is known as the **marriage problem** and is as follows:

- Sets M and W have finitely many members and are disjoint.
- Each member of set M has a strict preference ranking over some or all of the members of W.
- Each member of W has a strict preference ranking over some or all of the members of M.

A **matching** is a set of pairs, each consisting of one member of M and one member of W, with no member of M or of W being paired more than once. A matching is **stable** if no member of M or W is assigned a partner that the individual has not ranked and if there do not exist a member m of M and a member w of W who prefer each other to their current assignments.

DAP/M describes the matching procedure that is illustrated above with the members of M doing the bidding. DAP/W is the same procedure with the members of W doing the proposing. A stable matching is said to be **best** for a participant if no other stable matching produces an outcome that is preferred by this participant.

Proposition 14.1. For the marriage problem, DAP/M and DAP/W produce stable matchings. Indeed:

(a) The stable matching produced by DAP/M is best for each member of M and is worst for each member of W.

(b) The stable matching produced by DAP/W is best for each member of W and is worst for each member of M.

Proof. The proof that DAP/M (and hence DAP/W) produce stable matchings is identical to the proof given for Problem 14.B. An inductive proof of parts (a) and (b) can be found on page 32 of a book by Roth and Sotomayor.[5] ■

A bit of the history

Pages 2-8 of the aforementioned book by Roth and Sotomayor describe the market for medical interns in the era between 1944 and 1951. It was chaotic, and increasingly so. The chaos ended with the introduction in 1951 of a process that is virtually identical to the deferred acceptance procedure devised a decade later by Gale and Shapley, with the hospitals doing the proposing!

---

[5] Roth, Alvin E. and Oliveira Sotomayor, *Two-sided matching: a study in game-theoretic modeling and analysis*, Cambridge University Press, Cambridge, England, 1990.

5. A Zero-Sum Two-Person Matrix Game

Presented in this section is a game that had analyzed decades before the advent of linear programming by the great 20th century mathematician, John von Neumann. To introduce this game, we consider:

Problem 14.C. You and I know the data in the matrix A whose entries are given in equation (1). You choose a row of this matrix. Simultaneously, I choose a column. I pay you the amount at the intersection. Each of us prefers more money to less.

$$(1) \qquad A = \begin{bmatrix} 3 & 5 & -2 \\ 6 & 7 & 4 \end{bmatrix}.$$

Problem 14.C is a *zero-sum* game because you win what I lose. For the payoff matrix A that is given by (1), it is easy to see how to play this game. You prefer row 2 to row 1 because each entry in row 2 is larger than the corresponding entry in row 1. Playing row 2 is dominant for you; it is better for you than row 1 no matter what I do.

Similarly, I prefer column 3 to the others because each entry in column 3 is lower than the corresponding entries in columns 1 and 2. Playing column 3 is dominant for me. In brief, the dominant strategies are:

- You play row 2.
- I play column 3.

This pair of strategies is also an equilibrium: If I play column 3, you have no motive to deviate from row 2. Similarly, if you play row 2, I have no motive to deviate from column 3. The amount that I must pay you if we both choose equilibrium strategies is called the **value** of this game. With the payoff matrix A in equation (1), the value of this game equals 4.

A minor complication

Let's reconsider the same game with a different payoff matrix A, namely, with

$$(2) \qquad A = \begin{bmatrix} 9 & 1 & 3 \\ 6 & 5 & 4 \\ 0 & 8 & 2 \end{bmatrix}.$$

With payoff matrix (2), you no longer have a dominant row, and I no longer have a dominant column.

To see what you should do, think about the *least* you can win if you pick each row. If you pick row 1, the least you can win is 1. If you pick row 2, the least you can win is 4. If you pick row 3, the least you can win is 0. Evidently, picking row 2 maximizes the least you can win.

Similarly, think about the most I could lose. If I choose column 1, 2 or 3, the most I can lose is 9, 8 and 4, respectively. The payoff matrix A that's given by equation (2) is a case in which the largest of the "row mins" *equals* the smallest of the "column maxes." The equilibrium strategy remains:

- You play row 2 (which has the largest minimum).
- I play column 3 (which has the smallest maximum).

If you pick row 2, I have no motive to deviate from column 3. Similarly, if I pick column 3, you have no motive to deviate from row 2. The value of the game still equals 4.

A less minor complication

The payoff matrices given by equations (1) and (2) did not make this game famous. Let us now turn our attention to the 3×4 payoff matrix A given as

$$(3) \qquad A = \begin{bmatrix} 5 & 2 & 6 & 4 \\ 2 & 3 & 1 & 2 \\ 1 & 4 & 7 & 6 \end{bmatrix}.$$

The "row mins" equal 2, 1 and 1. The "column maxes" equal 5, 4, 7 and 6. The largest of the row mins (namely 2) is *less* than the smallest of the column maxes (namely 4). This is enough to guarantee that there can be no equilibrium in which you play a particular row and I play a particular column.

Randomized strategies

To reestablish the notion of equilibrium, we need to relax the performance criteria and enrich the set of strategies. Your goal is now to maximize the *expectation* of the amount you will earn, and my goal is to minimize the *expectation* of the amount I will lose. We continue to implement our strategies simultaneously.

Aiming to maximize your expected payoff, you consider the possibility of playing a **randomized strategy** that picks rows 1, 2 and 3 with probabilities p_1, p_2 and p_3, respectively. Being probabilities, these numbers are nonnegative, and they sum to 1.

Similarly, I aim to minimize my expected payout, and I play a randomized strategy that picks columns 1 through 4 with probabilities q_1 through q_4, these being nonnegative numbers whose sum equals 1.

Matrix notation

It proves handy to represent your strategy as a 1×3 (row) vector p and mine as a 4×1 (column) vector q, so that

$$p = [p_1 \quad p_2 \quad p_3] \quad \text{and} \quad q^T = [q_1 \quad q_2 \quad q_3 \quad q_4].$$

Let us recall that the jth column of the matrix A is denoted A_j and that its ith row is denoted A^i. For the 3×4 payoff matrix A given by expression (3), the matrix products pA_j and A^iq have interesting interpretations:

pA_j equals your expected payoff if you choose strategy p and I play column j,

A^iq equals my expected payout if I choose strategy q and you play row i.

In particular, if you choose strategy p and I play column 2, your expected payoff (and my expected payout) equals

$$pA_2 = 2p_1 + 3p_2 + 4p_3.$$

Similarly, if I choose strategy q and you play row 3, your expected payoff (and my expected payout) equals

$$A^3q = 1q_1 + 4q_2 + 7q_3 + 6q_4.$$

Particular strategies

Without (yet) revealing how these strategies were selected, we consider a particular randomized strategy p^* for you and a particular randomized strategy q^* for me, namely,

$$p^* = [1/2 \quad 0 \quad 1/2],$$

$$(q^*)^T = [1/3 \quad 2/3 \quad 0 \quad 0].$$

With strategy p^*, you pick row 1 with probability of 1/2, you pick row 3 with probability of 1/2, and you avoid row 2. Similarly, with strategy q^*, I pick column 1 with probability 1/3, I pick column 2 with probability 2/3, and I avoid the other two columns.

The entries in the matrix product $p^* A$ equals my expected payout if you choose strategy p^* and I choose the corresponding column of A.

$$\text{(4)} \qquad p^* A = [1/2 \quad 0 \quad 1/2] \begin{bmatrix} 5 & 2 & 6 & 4 \\ 2 & 3 & 1 & 2 \\ 1 & 4 & 7 & 6 \end{bmatrix} = \begin{bmatrix} 3 & 3 & 6.5 & 5 \end{bmatrix}$$

Evidently, if you play strategy p^*, the least I can expect to lose is 3. Moreover, my expected payout equals 3 if I randomize in any way over columns 1 and 2 and play columns 3 and 4 with probability 0. Strategy q^* randomizes in this way, in which sense it is a *best response* to strategy p^*.

Similarly, the entries in the matrix product Aq^* equal your expected payoff if I play strategy q^* and you choose the corresponding row of A.

$$\text{(5)} \qquad Aq^* = \begin{bmatrix} 5 & 2 & 6 & 4 \\ 2 & 3 & 1 & 2 \\ 1 & 4 & 7 & 6 \end{bmatrix} \begin{bmatrix} 1/3 \\ 2/3 \\ 0 \\ 0 \end{bmatrix} = \begin{bmatrix} 3 \\ 2.67 \\ 3 \end{bmatrix}$$

Evidently, if I play strategy q^*, the most you can expect to win is 3, and you win 3 if you randomize in any way over rows 1 and 3 and avoid row 2. Strategy p^* randomizes in this way, so it is a *best response* to strategy q^*.

An equilibrium

We have seen that if you choose strategy p^*, my expected payout equals 3 if I choose strategy q^*, and this is the least I can expect to lose, so

$$\text{(6)} \qquad 3 = p^* A q^* \le p^* A q \qquad \text{for each } q.$$

We have also seen that if I choose strategy q^*, your expected payoff equals 3 if you choose strategy p^*, and this is the most you can expect to win, so

(7) $$3 = p^* A q^* \geq p A q^* \qquad \text{for each p.}$$

Expressions (6) and (7) show that the pair p^* and q^* of strategies is an equilibrium; each of these strategies is a best response to the other. These expressions also show that 3 is the value of the game.

A "maximin" problem

This (rhetorical) question remains: How are p^* and q^* determined? This question will be answered for the general case, that is, for an $m \times n$ payoff matrix A whose entries are known to both players.

If the row player chooses strategy p and the column player chooses column j, the row player's expected payoff equals pA_j, and it seems natural from expression (4) that the row player aims to maximize the smallest such payoff. In other words, the row player seeks a randomized strategy p^* that solves:

Program 14.1. Maximize $_p \{\min_j(pA_j)\}$, subject to

$$p_1 + p_2 + \ldots + p_m = 1,$$

$$p_i \geq 0 \qquad \text{for } i = 1, \ldots, m.$$

As written, this "maximin" problem is not a linear program. But consider

Program 14.2. Max {v}, subject to

(8) $$v \leq pA_j \qquad \text{for } j = 1, \ldots, n,$$

$$p_1 + p_2 + \ldots + p_m = 1,$$

$$p_i \geq 0 \qquad \text{for } i = 1, \ldots, m.$$

As was noted in Chapter 1, maximizing the smallest of the quantities pA_1 through pA_n is equivalent to maximizing v subject to (8) and the other constraints that p needs to satisfy.

Program 14.2 is easily seen to be feasible and bounded. To construct a feasible solution, take $p_i = 1/m$ for each i, and equate v to the smallest entry in A. To see that Program 14.2 is bounded, note that each feasible solution equates v to a value that does not exceed the largest entry in A.

A minimax problem

Similarly, if the column player chooses randomized strategy q and the row player picks row i, the column player's expected payout equals $A^i q$, and expression (5) suggests that the column player should aim to minimize the largest such payout. In other words, the column player should seek a solution to:

Program 14.3. Minimize $_q$ $\{\max_i(A^i q)\}$, subject to

$$q_1 + q_2 + \ldots + q_n = 1,$$

$$q_j \geq 0 \qquad \text{for } j = 1, \ldots, n.$$

Program 14.3 converts itself into the linear program,

Program 14.4. Min $\{w\}$, subject to

$$\text{(9)} \qquad w \geq A^i q \qquad \text{for } i = 1, \ldots, m,$$

$$q_1 + q_2 + \ldots + q_n = 1,$$

$$q_j \geq 0 \qquad \text{for } j = 1, \ldots, n.$$

Program 14.4 is easily seen to be feasible and bounded.

Dual linear programs

We have seen that Program 14.2 and Program 14.4 are feasible and bounded. Hence, both linear programs have optimal solutions.

- Each optimal solution to Program 14.2 prescribes the value v^* of its objective and a vector p^* of probabilities.
- Each optimal solution to Program 14.4 prescribes the value w^* of its objective and a vector q^* of probabilities.

You may have guessed that duality will be used to prove

Proposition 14.2. Optimal solutions to Programs 14.2 and 14.4 exist and have these properties: $v^* = w^*$ and the pair p^* and q^* form an equilibrium.

Remark: The proof of Proposition 14.2 is starred because it rests on the Duality Theorem of linear programming, which appears in Chapter 12.

Proof*. It is left to you (see Problem 11) to verify that Program 14.2 and Program 14.4 are each others' duals. These linear programs are feasible and bounded; the Duality Theorem (Proposition 12.2) demonstrates that they have the same optimal value. So $v^* = w^*$. From constraint (8) and (9), we see that the optimal solutions to Programs 14.2 and 14.4 satisfy

(10) $$v^* \le p^* A_j \qquad \text{for } j = 1, \ldots n,$$

(11) $$v^* \ge A^i q^* \qquad \text{for } i = 1, \ldots m.$$

Multiply inequality (10) by the nonnegative number q_j^* and then sum over j. Since the q_j^*'s sum to 1, this gives $v^* \le p^* A q^*$. Similarly, multiply (11) by p_i^* and then sum over i to obtain $v^* \ge p^* A q^*$. Thus,

(12) $$v^* = p^* A q^*.$$

Next, consider any strategy q for the column player. Multiply (10) by q_j and then sum over j to obtain

(13) $$v^* \le p^* A q.$$

Finally, consider any strategy p for the row player. Multiply (11) by p_i and then sum over i to obtain

(14) $$v^* \ge p A q^*.$$

Expressions (12)-(14) show that the pair p^* and q^* fare an equilibrium and that v^* equals the value of the game. ■

It is easy to see that Programs 14.2 and 14.4 satisfy the Full Rank proviso, hence that each basic solution to either assigns a shadow price to each constraint. The shadow prices for the optimal basis are an optimal solution to the dual (Proposition 12.2), so it is only necessary to solve one of these linear programs. The shadow prices that Solver reports for either linear program are an optimal solution to the other linear program.

The minimax theorem

With a $m \times n$ matrix A, the analogue of Program 14.2 and its dual find a unique number v* and strategies p* and q* that attain the optima in

$$v^* = \max_p\{\min_j(pA_j)\},$$

$$v^* = \min_q\{\max_i(A^iq)\}.$$

It's easy to see that

$$\min_j(pA_j) \le pAq \qquad \text{for every strategy } q,$$

$$\max_i(A^iq) \ge pAq \qquad \text{for every strategy } p.$$

Thus, the analysis of Program 14.2 and its dual have proved:

Proposition 14.3 *(the minimax theorem).* For every payoff matrix A, there exists a unique number v* such that

(15) $$v^* = \max_p\{\min_q(pAq)\} = \min_q\{\max_p(pAq)\}.$$

Proposition 14.3 is the celebrated **minimax theorem** of John von Neumann. His proof of this theorem employed Brouwer's fixed-point theorem and was existential; it did not show how to find the best strategies.

An historic conversation

In a reminiscence[6], Dantzig described his visit with John von Neumann in the latter's office at the Institute for Advanced Study on October 3, 1947. This visit occurred a few months after Dantzig developed the simplex method. Shortly into their conversation, two startling insights occurred:

- Dantzig was surprised to learn that the simplex method solves two linear programs, the one under attack and its dual.
- von Neumann was surprised to learn that the simplex method is the natural weapon with which to prove the minimax theorem and to compute solutions to matrix games.

Their conversation initiated a decades-long process of replacing existential arguments in game theory with constructive arguments that are based on the simplex method and its generalizations.

---

[6] George B. Dantzig, "Linear Programming," *Operations Research*, V. 50, pp. 42-47, 2002.

6. An Economy in General Equilibrium

In this section, the concept of an economy in general equilibrium is discussed, a stylized (simplified) model of an economy is described, and a general equilibrium is constructed from the optimal solution to a linear program and its dual.

Aggregation

Models of an economy are highly aggregated. A large number of commodities are grouped into relatively few types of **good**, which might include capital, labor, land, steel, energy, foodstuff, and so forth. The many production processes are grouped into relatively few **technologies**, which might include steel production, agriculture, automobile manufacture, steel capacity expansion, and so forth.

Agents

An **economy** consists of two types of agents, which are known as "consumers" and "producers." These agents interact with each other through a "market" at which goods can change hands. An economy differs from a typical linear program in two important ways:

- There are multiple agents, each acting in his or her self-interest.
- The prices of the goods are endogeneous, which is to say that they are set within the model.

By contrast, in a linear program, there is only one decision-maker, and that person has no influence on the prices of the goods that he or she buys or sells.

The **consumers** in an economy have these characteristics:

- The consumers own all of the goods. Each consumer begins with an endowment of each good.
- At the market, each consumer sells the goods that he or she owns but does not wish to consume and buys the goods that he or she wishes to consume, but does not own.

- Each consumer faces a budget constraint, namely, that the market value of the goods that the consumer buys cannot exceed the market value of the goods that the consumer sells.
- Each consumer trades at the market in order to maximize the value (to that consumer) of the bundle of goods that he or she consumes.

The producers own nothing. All they do is to operate "technologies." The **technologies** have these properties:

- Each technology transforms one bundle of goods into another.
- If a technology is operated, the goods it consumes must be purchased at the market, and the goods it produces must be sold at the market.
- Producers who operate technologies aim to maximize the profit they earn by so doing.

The market for a good is said to **clear** if the quantity of the good that is offered for sale is at least as large as the quantity that is demanded. (The model that is under development allows for free disposal of unwanted goods.) Whether or not the market clears can depend on the price; if the price of a good is too low, its demand may exceed its supply. These prices are *endogenous*, which is to say that they are determined within the model.

A general equilibrium

An economy is said to be in **general equilibrium** if the following conditions are satisfied:

- Each consumer maximizes his or her welfare, given the actions of the other participants.
- Each producer maximizes profit, given the actions of the other participants.
- Each good is traded at the market and at a price that is set within the model.
- The market for each good clears.

A great bulk of theoretical and applied economics rests on the assumption that an economy has a general equilibrium. In this section, duality will be used to construct a general equilibrium.

A simplification

The model of an economy that is being developed is simplified in the ways that are listed below.

- Only a single-period is studied.
- The technologies are assumed to have constant returns to scale.
- The economy has only *one* consumer.
- This consumer has a linear utility function on quantities consumed.

The approach that is under development accommodates multiple periods, decreasing returns to scale on production, and decreasing marginal utility on consumption. This approach does *not* generalize to multiple consumers who have different utility functions. To obtain a general equilibrium for the case of multiple consumers, we would need to switch tools, from linear and nonlinear programming to fixed-point methods.

The data

The data that describe this one-period model are listed below:

- There are m goods, and these goods are numbered 1 through m.
- There are n technologies, and these technologies are numbered 1 through n.
- For $g = 1, \ldots, m$, the consumer possesses the amount e_g of good g at the start of the period.
- For $g = 1, \ldots, m$, the consumer obtains the benefit u_g from each unit of good g that the consumer consumes during the period.
- For each good g and each technology t, the quantity A_{gt} equals the *net* output of good g per unit level of technology t.

Mnemonics are in use: The letter "g" identifies a good, the letter "t" identifies a technology, the number e_g is called the consumer's **endowment** of good g, and the number u_g is called the consumer's **utility** of each unit of good g. Goods exist in nonnegative quantities, so these endowments (the e_g's) are nonnegative numbers. The consumer owns all of the assets. If, for instance, good 7 is steel, then e_7 equals the number of units of steel that the consumer possesses at the start of the period. In this (linear) model, the per-unit utility can vary with the good, but not with the quantity that is consumed.

"Net output" can have any sign. If A_{gt} is positive, good g is an output of technology t. If A_{gt} is negative, good g is an input to technology t. The $m \times n$ array A of net outputs describes a linear **activity analysis** model of the sort that had been discussed in Chapter 7.

The central issue

The motivating questions are posed as:

Problem 14.D *(general equilibrium).* For this model, is there a general equilibrium? If so, how can it be found?

For the model that is under development, a linear program and its dual will be used to demonstrate that a general equilibrium exists and to construct one.

The decision variables

The decision variables in these linear programs are of three types – the level at which each technology is operated, the market price for each good, and the amount of each good that the consumer consumes. For $t = 1, \ldots, n$ and for $g = 1, \ldots, m$, the model includes the decision variables:

x_t = the level at which the producers operate technology t during the period,

p_g = the market price of good g,

z_g = the amount of good g that the consumer consumes during the period.

The production levels and the consumption levels must be nonnegative, so the levels of the decision variables must satisfy

$$x_t \geq 0, \qquad t = 1, \ldots, n,$$

$$z_g \geq 0, \qquad g = 1 \ldots m.$$

A general equilibrium will appear as constraints on these decision variables.

Net production

The *net* production of good g during the period is given by the quantity

$$\sum_{t=1}^{n} A_{gt} x_t$$

because operating technology t at the level x_t produces net output $A_{gt}\, x_t$ of good g. Net production of a good can be negative, and net production is negative in the case of goods (such as capital or labor) that are inputs to every technology.

Net profit

A producer who operates technology t must buy its inputs at the market and sell its outputs at the market. Thus

(16) $\sum_{g=1}^{m} p_g A_{gt}$ = the net profit per unit level of technology t.

This sum equals the revenue received from the outputs of the technology less the price paid for its inputs. Capital is an input, so this sum is positive if the producer earns an *excess* profit, that is, a profit that is above the market rate of return on capital.

A producers' equilibrium

In this model of an economy, the producers own no assets. All they can do is to operate the technologies. A **producers' equilibrium** is a set $x = (x_1, \ldots, x_n)$ of levels of the technologies and a set $p = (p_1, \ldots, p_m)$ of prices that satisfy the constraints:

$$\begin{aligned} x_t &\geq 0 && \text{for } t = 1, \ldots, n, \\ \sum_{g=1}^{m} p_g A_{gt} &\leq 0 && \text{for } t = 1, \ldots, n, \\ x_t \left(\sum_{g=1}^{m} p_g A_{gt} \right) &= 0 && \text{for } t = 1, \ldots, n. \end{aligned}$$

A producers' equilibrium requires the that each production level be non-negative, that no technology operates at a profit, where "profit" means profit in excess of the rate of return on capital, and that no technology is operated if it would incur a loss. Today, these conditions seem natural and obvious.

But they were unnoticed for decades after Walras's seminal work (1884) on general equilibrium.

Tjalling G. Koopmans published these conditions in 1951, early in the history of linear programming. Koopmans' conditions demonstrate that the existence of each technology imposes a constraint on the market prices. To illustrate, suppose the economy is in equilibrium and then a new technology emerges, say, technology 9. If the prices that existed before this technology emerged violate the inequality $\sum_{g=1}^{m} p_g A_{g9} \leq 0$, the prices will have to shift in order for a producers' equilibrium to be restored.

Market clearing

The **market clearing** constraint for good g states that the amount z_g of good g that is consumed by the consumer cannot exceed the sum of the consumer's endowment e_g and the net production of good g. Expression (16) specifies the net production of good g, so market clearing requires

$$z_g \leq e_g + \sum_{t=1}^{n} A_{gt} x_t, \qquad \text{for } g = 1, 2, \ldots, m.$$

Free disposal accounts for the fact that these constraints are inequalities, not equations. If a good g were noxious (think of slag), its market clearing constraint would be an equation, rather than an inequality.

The prices do not appear in the market clearing constraints. Whether or not the market clears will depend on the prices, however. If the price of a good is too low, the demand for that good will exceed the supply, and the market for that good will not clear.

A consumer's equilibrium

The consumer faces a **budget constraint,** namely, that the market value of the bundle of goods that the consumer consumes cannot exceed the market value of the consumer's endowment. At a given set of prices, a **consumer's equilibrium** is any trading and consumption plan that maximizes the utility of the bundle of goods that the consumer consumes, subject to the consumer's budget constraint. Our model has only one consumer, and the satisfaction that the consumer receives from each good is linear in the amount of that good that is consumed. For our model, a consumer's equilibrium is an optimal solution to the linear program,

Consumer's LP. Maximize$_z$ $\sum_{g=1}^{m} u_g z_g$, subject to the constraints

$$\sum_{g=1}^{m} p_q z_g \leq \sum_{g=1}^{m} p_g e_g,$$
$$z_g \geq 0, \qquad \text{for } g = 1, \ldots, m.$$

In this LP, the subscript "z" to the right of "Maximize" is a signal that the z_g's are its decision variables. The prices (the p_g's) are fixed because the consumer has no direct effect on the prices. The objective of this linear program measures the consumer's level of satisfaction (utility) with the bundle of goods that he or she consumes. Its constraint keeps the market value of the bundle of goods that the consumer consumes from exceeding the market value of the consumer's endowment.

A linear program

We are now poised to answer the questions posed in Problem 14.D. For the case of a single (canonical) consumer, a general equilibrium will be constructed from the optimal solutions to Program 14.5 (below) and its dual.

Program 14.5. $u^* =$ Maximize$_{z,x}$ $\sum_{g=1}^{m} u_g z_g$, subject to the constraints

$$p_g: \qquad z_g - \sum_{t=1}^{n} A_{gt} x_t \leq e_g, \qquad \text{for } g = 1, \ldots, m,$$
$$x_t \geq 0, \qquad \text{for } t = 1, \ldots, n,$$
$$z_g \geq 0, \qquad \text{for } g = 1, \ldots, m.$$

The subscripts z and x on "Maximize" indicate that the decision variables in Program 14.5 are the consumption quantities (the z_g's) and the levels at which the technologies are operated (the x_t's). The objective of Program 14.5 measures the utility to the consumer of the bundle of goods that he or she consumes. Its constraints keep the consumption of each good g from exceeding its net supply.

A curious feature of Program 14.5 is that the producers are altruistic; they set their production levels so as to maximize the consumer's level of satisfaction, with no regard for their own welfare.

Is Program 14.5 feasible? Yes. The endowments (the e_g's) are nonnegative numbers, so it is feasible to equate each decision variable to zero. Program 14.5 enforces the market clearing constraints. It omits these facets of a general equilibrium:

- The consumer's budget constraint.
- The market prices.
- The requirement that the producers maximize their profits.

The notation hints that the optimal values of the dual variables will serve as market prices, that a general equilibrium will be constructed from optimal solutions to Program 14.5 and its dual.

The dual linear program

In Program 14.5, the market clearing constraint on good g has been assigned the complementary dual variable, p_g. Each decision variable in Program 14.5 gives rise to a complementary constraint in its dual. This dual appears below as

Program 14.6. $u^* = \text{Minimize} \sum_{g=1}^{m} e_g p_g$, subject to the constraints

$$
\begin{array}{llr}
x_t: & \sum_{g=1}^{m} p_g(-A_{gt}) \geq 0, & \text{for } t = 1, \ldots, n, \\
z_g: & p_g \geq u_g, & \text{for } g = 1, \ldots, m, \\
 & p_g \geq 0, & \text{for } g = 1, \ldots, m.
\end{array}
$$

Each non-sign constraint in either linear program is labeled with the variable to which it is complementary.

Constructing a general equilibrium

In order for an equilibrium to exist, it must be assumed that the consumer cannot be made infinitely well off. That is part of the hypothesis of

Proposition 14.4 *(general equilibrium).* Assume that Program 14.5 is bounded. Then:

(a) Program 14.5 and Program 14.6 have optimal solutions and have the same optimal value.

(b) Each optimal solution $x = (x_1, \ldots, x_n)$ and $z = (z_1, \ldots, z_m)$ to Program 14.5 and each optimal solution $p = (p_1, \ldots, p_m)$ to Program 14.6 form a general equilibrium. Moreover, if z_g is positive, then $p_g = u_g$.

Remark: The proof of Proposition 14.4 rests on the Duality Theorem for linear programming and is starred for that reason.

Proof*. To see that Program 14.5 is feasible, note that endowment e_g of each good is nonnegative, so equating each decision variable to 0 satisfies its constraints. By hypothesis, Program 14.5 is bounded. Application of the simplex method to Program 14.5 constructs an optimal solution $x = (x_1, \ldots, x_n)$ and $z = (z_1, \ldots, z_m)$. The Duality Theorem (Proposition 12.2) guarantees that Program 14.6 is feasible and bounded, that it has an optimal solution $p = (p_1, \ldots, p_m)$, that these optimal solutions have the same objective value, u^* and that they satisfy the complementary slackness conditions:

$$(17) \qquad x_t\left(\sum_{g=1}^{m} p_g A_{gt}\right) = 0, \qquad \text{for } t = 1, \ldots, n,$$

$$(18) \qquad z_g(p_g - u_g) = 0, \qquad \text{for } g = 1, \ldots, m.$$

It will be demonstrated that these values of the decision variables form a general equilibrium, with p_1 through p_m as market prices.

To see that this is a producers' equilibrium, note that the constraints of Program 14.5 keep the production quantities nonnegative, that the constraints of Program 14.6 guarantee that no technology operates at a profit, and that (17) guarantees that no technology is used if it operates at a loss.

System (18) states that, if z_g is positive, then $p_g = u_g$, exactly as is asserted in the theorem. Let us re-write system (18) as

$$(19) \qquad u_g z_g = p_g z_g, \qquad \text{for } g = 1, \ldots, m.$$

It remains to show that this is a consumer's equilibrium, equivalently, that the consumer's budget constraint is satisfied. The fact that Programs 14.6 and 14.5 have the same optimal value u^* couples with system (19) to give

$$u^* = \sum_{g=1}^{m} p_g e_g = \sum_{g=1}^{m} u_g z_g = \sum_{g=1}^{m} p_g z_g.$$

This equation demonstrates that the market value $\sum_{g=1}^{m} p_g e_g$ of the consumer's endowment equals the market value $\sum_{g=1}^{m} p_g z_g$ of the bundle of goods that the consumer consumes. In other words, this is a consumer's equilibrium, and the proof is complete. ■

Recap

This section shows how LP duality provides insight into a fundamental concept in economics. The decision variables in one linear program are the consumption quantities and the production quantities. The decision variables in the dual linear program are the market prices. The Duality Theorem shows that these linear programs construct a general equilibrium.

The key to this analysis has been the assumption of a single (canonical) consumer. With one consumer, the content of this theorem remains valid for the case of decreasing marginal returns on production and consumption. For this more general case, the Lagrange multipliers that Solver reports *are* the market prices.

Program 14.5 and its dual can have alternative optimal, but they have a unique optimal value, u^*. Thus, the consumers' optimal consumption bundle need not be unique, but the benefit obtained by the consumer is unique.

When the model includes several consumers with different preferences, there can be more than one equilibrium, and different equilibria can have differing benefits to the consumers.

7. A Bi-Matrix Game

A **bi-matrix game** is a game with two players, one of whom chooses a row, and the other chooses a column. If the row player chooses row i and the column player chooses column j, the row player loses the (possibly negative) amount A_{ij} in the $m \times n$ matrix A and the column player loses the (possibly negative) amount B_{ij} in the $m \times n$ matrix B. The bi-matrix game simplifies to the matrix game of von Neumann if $A + B = 0$, that is, if one player wins what the other loses. In Chapter 15, we will see that every bi-matrix game has at least one equilibrium in randomized strategies *and* that a clever "tweak" of the simplex method will find an equilibrium. A famous bi-matrix game appears here as

Problem 14.E *(the prisoner's dilemma).* You and I have been arrested and have been placed in separate cells. The district attorney calls us into her office and tells us that she is confident that we committed a major crime, but she only has enough evidence to get us convicted of a minor crime. If neither of us squeals on the other, each of us will do 1 year in jail on the minor crime. If only one of us squeals, the squealer will not go to jail and the other will serve 7 years for the major crime. If both of us squeal, each will go to jail for 5 years for the major crime. She tells us that we must make our decisions independently, and then sends us back to our respective cells. She visits each of our cells and asks each of us to squeal. Each of us prefers less time in the slammer to more. How shall we respond?

The players in this game are being treated symmetrically. If both players clam, both serve 1 year. If one clams and the other squeals, the squealer serves no time and the person who was squealed on serves 7 years. If both squeal, both serve 5 years. Table 14.4 displays the cost to each player under each pair of strategies.

Table 14.4. The cost of each strategy pair, my cost at the left, yours at the right

	you clam	you squeal
I clam	(1, 1)	(7, 0)
I squeal	(0, 7)	(5, 5)

If you clam, it is better for me to squeal (I serve 0 years instead of 1). If you squeal, it is better for me to squeal (I serve 5 years instead of 7). Squealing is dominant for me. Squealing is also dominant for you, and for the same reason. Squealing is dominant for both of us, and it causes each of us to spend 5 years in jail. But squealing is it is not stable. If both of us clammed, each of us would serve 1 year rather than 5!

This example questions the solution concepts: It is dominant for each of us to squeal. It is an equilibrium for each of us to squeal. But it is better for each of us to clam.

8. Review

This introduction to game theory is brief and selective. It includes five of the many models of game theory and three of the many solution concepts. The most cogent of the ideas in this chapter may be that of the best response.

In an equilibrium, each player's strategy is a best response to the current strategies of the other players. A single player's strategy is dominant if it is the best response to any strategies that the other players might choose.

This chapter has illustrated a connection between linear programming and game theory. The Duality Theorem has provided insight into von Neumann's zero-sum matrix game and into a stylized model of an economy in general equilibrium.

In the next two chapters, a different connection between linear programming and game theory will be brought into view. In Chapter 15, a variant of the simplex method that is known as "complementary pivoting" will be used to compute an equilibrium in a bimatrix game. In Chapter 16, an algorithm that is strikingly similar to complementary pivoting will be used to approximate fixed points, including those of an economic (Nash) equilibrium.

9. Homework and Discussion Problems

1. *(a Vickery auction)* In Problem 14.A, suppose you (the seller) are allowed to place a sealed bid on the property that is for sale. Let V denote the value you place on the property. What do you bid? Do you have a dominant strategy? Under what circumstance will you earn a profit, and how large will it be?

2. *(the marriage game)* Suppose all rankings are as in Table 14.1, except that man z's ranking is A B D C. What matching would DAP/M produce? What ranking would DAP/W produce? Would the same man stay home in both of these rankings?

3. *(the marriage game)* How could you determine whether or not a particular instance of the marriage game has a unique stable matching?

4. In Problem 14.B (*the marriage game*), the men doing the proposing and the true preferences are as given in Table 14.1. Can the women to misrepresent their preferences in such a way that DAP/M yields the stable matching that they would attain under DAP?W? If so, how?

5. *(lunch)* Each of six students brings a sandwich to school. These six students meet for lunch in the school cafeteria. Each student has a strict preference

over the sandwiches. The students are labeled A through F. The sandwiches they bring are labeled a through f, respectively. Their preferences are as indicated below. For instance, student A's 1st choice is sandwich c (which was brought by student C), her 2nd choice is sandwich e, her 3rd choice is sandwich f, her 4th choice is the sandwich she brought, and so forth. Each of these students will eat a sandwich, at lunch but not necessarily the one that he or she brought.

student			preference			
A	c	e	f	a	b	d
B	b	a	c	e	f	d
C	e	f	c	a	d	b
D	c	a	b	e	d	f
E	d	c	b	f	e	a
F	b	d	e	f	a	c

(a) Design a procedure that creates a stable matching of students to sandwiches in which no student eats a sandwich the she likes worse than the one she brought. *Hint:* Begin by drawing a network having 6 nodes and 6 directed arcs; each node represents a sandwich, and arc (x, y) is included if student X has sandwich y as her 1st choice.

(b) Who eats which sandwich?

6. *(lunch, continued).* For the preferences in the preceding problem, find an allocation of lunches to students in which each student gets the lunch that she ranks 1st or 2nd. *Hint:* This can be set up as an assignment problem. Is this allocation stable? If so, why?

7. Some matrix games can be solved by "eyeball." The matrix game whose payoff matrix A is given by (3) is one of them. Let's see how:

   (a) I (the column player) look at A and observe that playing rows 1 and 3 with probability of 0.5 is at least as good for you as is playing row 2. What does this tell me about your best strategy?

   (b) If you choose the randomization in part (a), my expected payout for my four pure strategies form the vector [3 3 6.5 5]. What does this tell me about the columns I should avoid?

(c) Has this game been boiled down to the 2×2 payoff matrix

$$\begin{bmatrix} 5 & 2 \\ 1 & 4 \end{bmatrix}?$$

(d) If so, what strategy for me causes your payoff to be independent of what you do? And what strategy for you causes my payout to be independent of what I do?

(e) Have you constructed an equilibrium? If so, why?

8. True or false: In a zero-sum matrix game, each player has a dominant strategy. Support your answer.

9. Consider a two-person zero-sum matrix game whose $m \times n$ payoff matrix A has the property that

$$\max_i(\min_j A_{ij}) = \min_j(\max_i A_{ij}).$$

Show that this game has an equilibrium in pure strategies.

10. Use a linear program to find an equilibrium for the zero-sum matrix game with payoff matrix A that is given by

$$A = \begin{bmatrix} 1 & 2 & 3 & 4 \\ 6 & 3 & 0 & -3 \\ 2 & 8 & -5 & -1 \end{bmatrix}.$$

11. Using the cross-over table, take the dual of Program 14.2. Interpret the linear program that you obtained. Are both linear programs feasible and bounded? What does complementary slackness say about their optimal solutions?

12. **(general equilibrium with production capacities)** Suppose that each technology t has a finite production capacity, C_t, so that the levels at which the technologies are operated must satisfy the constraints $x_t \le C_t$ for $t = 1, \ldots, n$. The consumer owns all of the assets in the economy, including the production capacities. What changes, if any, occur in Programs 14.5 and 14.6? What changes, if any, occur in statement and proof of Proposition 14.4? Why?

13. Find an equilibrium for the bimatrix game in which your cost matrix A and my cost matrix B are given below. *Hints:* Can I pick a strategy such that what you lose is independent of the row you choose? Can you do something similar?

$$A = \begin{bmatrix} 1 & 5 \\ 3 & 0 \end{bmatrix}, \quad B = \begin{bmatrix} 4 & 1 \\ 3 & 6 \end{bmatrix}$$

14. Dick and Harry are the world's best sprinters. They will place 1st and 2nd in each race that they both run, whether or not they take a performance-enhancing drug. Dick and Harry are equally likely to win a race in which neither of them take this drug or if both take it. Either of them is certain to win if only he takes it. They are in it for the money. Each race pays K dollars to the winner, nothing to anyone else.

There is a test for this drug. The test is perfect (no false positives or false negatives), but it is expensive to administer. If an athlete is tested at the conclusion of the race and is found to have taken the drug, he is disqualified from that race and is fined 12 K dollars.

(a) Without drug testing, are there dominant strategies? If so, what are they?

(b) Now, suppose that with probability p Harry and David are both tested for this drug at the conclusion of the race. Are there values of p that are large enough to that neither cheats? If so, how large must p be?

(c) Redo part (b) for the case in which Harry and David have product endorsement contracts that are worth 10 K and that each of their contracts has a clause stating that payment will cease if he tests positive for performance-enhancing drugs.

15. You are a contestant in the TV quiz show, *Final Jeopardy*. Its final round is about to commence. Each of three contestants (yourself included) has accumulated a certain amount of "funny money" by answering questions correctly in prior rounds. The rules for the final round are these:

- The program's host announces the category of a question that he will pose.
- Knowing the category but not the question, each contestant wagers part or all of his/her funny money by writing that amount on a slate that is not visible to anyone else.

- Next, the question is posed, and each contestant writes his/her answer on the same slate.
- Then, the slates are made visible. Each contestant who had the correct (incorrect) answer has his/her funny money increased (decreased) by the amount of that person's wager.
- The contestant whose final amount of funny money is largest wins that amount in dollars. The others win nothing.

Having heard the final category, you are confident that you will be able to answer the question correctly with probability q that exceeds 0.5 . Your goal is to maximize the expectation of the amount that you will win.

(a) Denote as y your wealth position in funny money at the *end* of the final round, and denote as f(y) the probability that you will win given y. Can f(y) decrease as y increases?

(b) Denote as x your wealth position in funny money as the *start* of the final round and, given a wager of w (necessarily, $w \le x$), denote as e(x, w) the expectation of your winnings. Argue that

$$\begin{aligned} e(x, w) &= f(x + w)q(x + w) + f(x - w)(1 - q)(x - w) \\ &\le f(2x)\{qx + (1 - q)x + w[(2q - 1]\} \\ &\le f(2x)[x + x(2q - 1)] = e(x, x). \end{aligned}$$

(c) For the final round, do you have a dominant strategy? If so, what is it, and why is it dominant?

16. On the web, look up the definition of a "cooperative game" and of its "core." What are they?

Chapter 15: A Bi-Matrix Game

1. **Preview**

The zero-sum matrix game of von Neumann was studied in Chapter 14. The current chapter is focused on a non-zero sum generalization. This generalization, which is known as a **bi-matrix game**, is described below:

- You and I know the entries in the $m \times n$ matrices A and B.
- You pick a row. Simultaneously, I pick a column.
- If you choose row i and I choose column j, you lose A_{ij} and I lose B_{ij}.
- You wish to minimize the expectation of your loss, and I wish to minimize he expectation of my loss.

The bi-matrix game reduces to the zero-sum matrix game if $A + B = 0$.

E.V. Denardo, *Linear Programming and Generalizations*, International Series in Operations Research & Management Science 149,
DOI 10.1007/978-1-4419-6491-5_15, © Springer Science+Business Media, LLC 2011

Pivot strategies

In Chapter 14, we saw that the zero-sum matrix game has an equilibrium in randomized strategies, moreover, that this equilibrium can be found by using the simplex method to solve a linear program. Here, we will see that the bi-matrix game also has an equilibrium in randomized strategies. This equilibrium will not be constructed by the simplex method, however. In its place, an equilibrium will be found by application of a related procedure that is known as the "complementary pivot method."

Payoff matrices

The bi-matrix game has been introduced in the context of a pair A and B of cost matrices. The entire discussion is adapted in Section 6 to the case in which A and B are payoff matrices, rather than cost matrices.

Cooperation and side payments

Section 7 of this chapter touches briefly on the topic of cooperation in a bi-matrix game. It probes a pair of questions: How shall two players act so as to obtain the largest possible total reward? How shall they divvy it up?

Significance

The bi-matrix game is important in its own right, and the complementary pivot method has several other uses, which include the solution of convex quadratic programs. The most amazing feature of the complementary pivot method may be that it leads directly to a method for approximating a "Brouwer fixed point," as will be seen in Chapter 16.

2. Illustrations

In this section, a pair of examples is used to probe the bi-matrix game and to suggest a pattern of analysis that will be developed in subsequent sections.

An equilibrium in pure strategies

Let us begin with the particularly simple instance of the bi-matrix game whose cost matrices are

$$A = \begin{bmatrix} 1 & 5 \\ 3 & 0 \end{bmatrix} \qquad B = \begin{bmatrix} 4 & 1 \\ 3 & 2 \end{bmatrix}.$$

For these matrices, you do not have a dominant row, but I have a dominant column. Each entry in the 1st column of B is larger than the corresponding entry in the 2nd column of B. I will pick column 2 because it costs me less than column 1, independent of the row that you choose. Knowing that I will pick column 2, you choose row 2 because $A_{22} = 0 < A_{21} = 5$. For these matrices, the bi–matrix game has an equilibrium in nonrandomized strategies, namely:

- You choose row 2.
- I choose column 2.

Each of these strategies is a best response to the other: If I choose column 2, you have no economic motive to deviate from row 2. And if you choose row 2, I have no economic motive to deviate from column 2. As was mentioned above, column 2 is a dominant strategy for me, but row 2 is not dominant for you.

An equilibrium in randomized strategies

A somewhat more representative example has payoff matrices A and B that are given by

(1) $$A = \begin{bmatrix} 2 & 5 & 7 \\ 5 & 7 & 3 \end{bmatrix}, \qquad B = \begin{bmatrix} 7 & 5 & 2 \\ 2 & 1 & 6 \end{bmatrix}.$$

To establish an equilibrium for these matrices, you and I will need to employ randomized strategies. As was the case in Chapter 14, it proves convenient to represent a randomized strategy for you as a row vector p and a to represent a randomized strategy for me as a column vector q.

$$p = \begin{bmatrix} p_1 & p_2 \end{bmatrix} \qquad q^T = \begin{bmatrix} q_1 & q_2 & q_3 \end{bmatrix}$$

Here, p_i is the probability that you play row i, so p_1 and p_2 are nonnegative numbers whose sum equals 1. Similarly, q_j is the probability that I play column j, so q_1, q_2 and q_3 are nonnegative numbers whose sum equals 1.

Solution by eye

For the cost matrices in (1), it is fairly easy to construct an equilibrium. You observe that I will avoid column 1 because column 2 costs me less than column 1 independent of the row you choose. Being rational, I will set $q_1 = 0$.

Let us suppose that you randomize over the rows so that I am indifferent between columns 2 and 3. In other words, you choose p_1 and $p_2 = (1 - p_1)$ so that

$$5p_1 + 1(1 - p_1) = 2p_1 + 6\,(1 - p_1).$$

This gives $p_1 = 5/8$, which determines the randomized strategy,

$$\text{(2)} \qquad p = \begin{bmatrix} 5/8 & 3/8 \end{bmatrix},$$

for you, the row player. Each entry in the matrix product p B equals my expected loss if I play the corresponding column. For the randomized strategy p given above,

$$\text{(3)} \qquad p\,B = \begin{bmatrix} 41/8 & 28/8 & 28/8 \end{bmatrix}.$$

Evidently, for this strategy p, my expected loss equals 7/2 if I randomize in any way over columns 2 and 3, and my expected loss exceeds 7/2 if I choose column 1 with positive probability.

Now let us suppose that I randomize in a way that that avoids column 1 and makes you indifferent between rows 1 and 2. In other words, I pick q_2 and $q_3 = (1 - q_2)$ so that

$$5q_2 + 7(1 - q_2) = 7q_2 + 3(1 - q_2).$$

This equation is satisfied by $q_2 = 2/3$, so $q_3 = (1 - q_2) = 1/3$. My randomized strategy is

$$\text{(4)} \qquad q^T = \begin{bmatrix} 0 & 2/3 & 1/3 \end{bmatrix}.$$

For this randomized strategy q, each entry in the matrix product A q equals your expected loss if you choose the corresponding row, and

$$\text{(5)} \qquad A\,q = \begin{bmatrix} 17/3 & 17/3 \end{bmatrix}^T.$$

Hence, if I play use the randomized strategy q given by (4), your expected loss equals 17 / 3, independent of which row you choose.

Equation (3) and the fact that $q_1 = 0$ shows that q is a best response to the strategy p given by (2). Equation (5) shows that p is a best response to q. Evidently, an equilibrium in randomized strategies has been constructed.

Empathy

The prior analysis of the cost matrices in (1) illustrate a principle that can help in the construction of an equilibrium.

- You (the row player) figure out which columns I should avoid and select your randomized strategy p so that I am indifferent between those columns that I should not avoid.
- I (the column player) ascertain which rows you should avoid and select my randomized strategy q so that you are indifferent between the rows you should not avoid.

With larger and more complicated cost matrices, it can be difficult for the players to "eyeball" strategies p and q that have these properties.

3. An Equilibrium

In this section, our attention turns from the payoff matrices in equation (1) to the general situation. Presented in this section is a set of equations that prescribe an equilibrium in randomized strategies.

The general situation

To develop this equation system, we turn our attention to a bi-matrix game whose cost matrices A and B have m rows and n columns. A randomized strategy for you (the row player) is represented as a $1 \times m$ vector p whose i^{th} entry p_i equals the probability that you play row i. A randomized strategy for me (the column player) is represented as an $n \times 1$ vector q whose j^{th} entry q_j is the probability that I play column j. It is noted that:

- The probability that you choose row i *and* I choose column j equals the product $p_i q_j$ because our choices are made independently.

- If you choose row i and I choose column j, then you lose A_{ij} and I lose B_{ij}.
- If you choose randomized strategy p and I choose randomized strategy q, then your expected loss equals p A q and my expected loss equals p B q because

$$\text{p A q} = \sum_{i=1}^{m} \sum_{j=1}^{n} p_i A_{ij} q_j,$$
$$\text{p B q} = \sum_{i=1}^{m} \sum_{j=1}^{n} p_i B_{ij} q_j.$$

This notation facilitates a succinct description of an equilibrium. Strategy p is a best response to q if

(6) $$\text{p A q} \le \hat{\text{p}}\text{ A q} \qquad \text{for every strategy } \hat{\text{p}}.$$

Expression (6) states that if I choose strategy q, you cannot reduce your expected loss below p A q, no matter what strategy $\hat{p}$ you choose. Similarly, strategy q is a best response to p if

(7) $$\text{p B q} \le \text{p B } \hat{\text{q}} \qquad \text{for every strategy } \hat{\text{q}}.$$

Conditions (6) and (7) describe an equilibrium when it is understood that p and $\hat{p}$ are $1 \times m$ vectors of probabilities and that q and $\hat{q}$ are $n \times 1$ vectors of probabilities.

The requirements that p, $\hat{p}$, q and $\hat{q}$ be probability distributions are easily expressed in terms of linear equations and nonnegativity requirements. But (6) and (7) contain nonlinear addends (such as $p_i A_{ij} q_j$ and $p_i B_{ij} q_j$), which makes them less easy to deal with. The nonlinearities in equations (6) and (7) will soon be replaced by complementarity conditions.

A convenient simplification

For the 2×3 example whose cost matrices are specified by (1), let us ask ourselves what happens if the number 10 is subtracted from every entry in B. Numerically, this replaces the data in (1) with

(8) $$A = \begin{bmatrix} 2 & 5 & 7 \\ 5 & 7 & 3 \end{bmatrix}, \qquad B = \begin{bmatrix} -3 & -5 & -8 \\ -8 & -9 & -4 \end{bmatrix}$$

This subtracts 10 from p B q for every pair p and q of randomizes strategies. It has no effect on the relative desirability of strategy q over strategy $\hat{q}$, so it preserves the set of equilibria. Similarly, adding 10 to each entry in A has no effect on the set of equilibria.

In brief, the set of equilibria is preserved if a constant is subtracted from each entry in B, and if another constant is added to each entry in A. Doing so allows us to construct equilibria for cost matrices A and B that satisfy

$$(9) \qquad A_{ij} > 0 \quad \text{and} \quad B_{ij} < 0 \qquad \text{for each i and j.}$$

Thus, no loss of generality occurs if we work with matrices A and B such that you (the row player) incur a positive cost and I (the column player) incur a negative cost, no matter what strategies we choose.

Linear constraints

Imposing (9) entails no loss of generality, and it lets us deduce the requirement that probabilities sum to 1. That requirement is absent from the system of equations and nonnegativity conditions in

$$(10) \qquad 1 + s_i = \textstyle\sum_{j=1}^{n} A_{ij}\, y_j \qquad \text{for } i = 1, \ldots, m,$$

$$(11) \qquad -1 + t_j = \textstyle\sum_{i=1}^{m} x_i\, B_{ij} \qquad \text{for } j = 1, \ldots, n,$$

$$(12) \qquad x_i \geq 0, \quad s_i \geq 0 \qquad \text{for } i = 1, \ldots, m,$$

$$(13) \qquad y_j \geq 0, \quad t_j \geq 0 \qquad \text{for } j = 1, \ldots, n.$$

Each solution to (10)-(13) is shown to lead to a pair of randomized strategies in

Proposition 15.1. Suppose that each entry in the cost matrix A is positive and that each entry in the cost matrix B is negative. Consider any solution to (10)-(13). Then the numbers $\rho = \sum_{i=1}^{m} x_i$ and $\sigma = \sum_{j=1}^{n} y_j$ are positive, and the pair p and q of randomized strategies given by

$$(14) \qquad p_i = x_i/\rho \qquad \text{for } i = 1, \ldots, m,$$

$$(15) \qquad q_j = y_j/\sigma \qquad \text{for } j = 1, \ldots, n,$$

Satisfy

$$(16) \qquad (1 + s_i)/\sigma = A^i q \qquad \text{for } i = 1, \ldots, m,$$

$$(17) \qquad (-1 + t_j)/\rho = p\, B_j \qquad \text{for } j = 1, \ldots, n.$$

Proof. Consider any solution to (10)-(13). Since each entry in A is positive, (10) cannot hold if $y = 0$. Similarly, since each entry in B is negative, (11) cannot hold if $x = 0$. Thus, $\rho = \sum_{i=1}^{m} x_i$ and $\sigma = \sum_{j=1}^{n} y_j$ are positive, and equations (14) and (15) define, respectively, a randomized strategy p for the row player and a randomized strategy q for the column player. Dividing (10) by σ produces (16), and dividing (11) by ρ produces (17). This completes a proof. ■

Proposition 15.1 constructs a pair p and q of strategies from (10)-(13). This pair of strategies need not form an equilibrium. It will form an equilibrium *if* a set of "complementarity" conditions is also satisfied.

Complementary variables

In system (10)-(11), the variables x_i and s_i are now said to be **complementary** to each other, and the variables y_j and t_j are said to be **complementary** to each other.

A solution to (10)-(11) is now said to be **complementary** if this solution satisfies the nonnegativity conditions in (12)-(13) and if it also satisfies

$$(18) \qquad x_i\, s_i = 0 \qquad \text{for } i = 1, \ldots, m,$$

$$(19) \qquad y_j\, t_j = 0 \qquad \text{for } j = 1, \ldots, n.$$

Conditions (18) and (19) require one member of each complementary pair to equal zero. These conditions are reminiscent of the complementary slackness conditions of linear programming. Their role is apparent in

Proposition 15.2. Suppose that each entry in the cost matrix A is positive and that each entry in the cost matrix B is negative. Consider any complementary solution to (10)-(11). The pair p and q of randomized strategies specified by Proposition 15.1 form an equilibrium.

Proof. Consider any complementary solution to (10)-(11). Proposition 15.1 shows that the randomized strategies p and q satisfy (16) and (17). Multiply (16) by p_i and then sum over i. The p_i's sum to 1 and (18) gives $p_i\, s_i = 0$ for each i, so

$$1/\sigma = p\,A\,q. \tag{20}$$

Similarly, multiply (17) by q_j and then sum over j. The q_j's sum to 1, and (19) gives $t_j\, q_j = 0$ for each j, so

$$-1/\rho = p\,B\,q. \tag{21}$$

To see that p is a best response to q, consider any randomized strategy $\hat{p}$ for the row player. Multiply (16) by $\hat{p}_i$, note that $\hat{p}_i\, s_i \geq 0$, and then sum over i to obtain $1/\sigma \leq \hat{p}\,A\,q$, which combines with (20) to give $p\,A\,q = 1/\sigma \leq \hat{p}\,A\,q$. Similarly, consider any randomized strategy $\hat{q}$ for the column player. Multiply (17) by $\hat{q}_j$, note that $t_j\, \hat{q}_j \geq 0$, and then sum over j to obtain $-1/\rho \leq p\,B\,\hat{q}$, which combines with (21) to give $p\,B\,q = -1/\rho \leq p\,B\,\hat{q}$, for every randomized strategy $\hat{q}$ for the column player. Thus, the pair p and q form an equilibrium, which completes the proof. ■

Propositions 15.1 and 15.2 show how to construct an equilibrium from a solution to the linear equations in (10) and (11) that is complementary.

4. Complementary Pivots

A sequence pivots will be used to construct a complementary solution to (10)-(11). The first of these pivots will be reminiscent of Phase I of the simplex method. Subsequent pivots will use ratios to preserve feasibility, but – unlike simplex pivots – they will not strive to improve an objective.

Complementary bases

The hypothesis of Proposition 15.2 includes the $m+n$ linear equations in (10) and (11). Each of these linear equations includes a slack variable, so the rank of system (10)-(11) equals $m+n$. Each basis for (10)-(11) consists of exactly $m+n$ variables. A basis for (10)-(11) is now said to be **complementary** if:

- For each i, the basis includes exactly one member of the pair $\{x_i, s_i\}$.
- For each j, the basis includes exactly one member of the pair $\{y_j, t_j\}$.
- Its basic solution equates each decision variable to a value that is nonnegative.

Proposition 15.2 shows how to construct an equilibrium from a complementary basis. Exhibited here is a method for constructing a complementary basis.

One artificial variable

To prepare for pivoting, each equation in system (10)-(11) is rewritten with the constant on the right and the decision variables on the left:

$$s_i - \sum_{j=1}^{n} A_{ij}\, y_j = -1 \qquad \text{for } i = 1, \ldots, m, \tag{22}$$

$$t_j - \sum_{i=1}^{m} x_i\, B_{ij} = 1 \qquad \text{for } j = 1, \ldots, n. \tag{23}$$

The set $\{s_1, \ldots, s_m, t_1, \ldots, t_n\}$ of decision variables is a basis for (22)-(23), and this set does include exactly one member of each complementary pair, but its basic solution is not feasible because it sets $s_i = -1$ for each i.

The infeasibility of s_1 through s_m, can be corrected by the insertion of an **artificial variable** α on the LHS of each equation in (22). This variable α must have a negative coefficient on the LHS of each equation in (22). It would do to assign α the coefficient of –1 in each of these equations, but an upcoming argument simplifies if the coefficients of α are different from each other. Let us replace (22) by

$$s_i - \sum_{j=1}^{n} A_{ij}\, y_j - i\alpha = -1 \qquad \text{for } i = 1, \ldots, m. \tag{24}$$

Now, when the nonbasic variable α is set equal to 1, the basic solution has $s_1 = 0$, and it equates s_2 through s_m to positive values.

The initial pivot

A sequence of pivots will occur on the system consisting of (23) and (24). The initial pivot will occur on the coefficient of α in the equation for which s_1 is basic. The variable α enters the basis, and s_1 departs. The resulting basis consists of the set $\{\alpha, s_2, \ldots, s_m, t_1, \ldots, t_n\}$ of decision variables. Its basic solution sets

$$\alpha = 1,$$

$$s_i = i - 1 \qquad \text{for } i = 2, 3, \ldots, m,$$

$$t_j = 1 \qquad \text{for } j = 1, 2, \ldots, n,$$

This basic solution satisfies the complementarity conditions in (18)-(19), and it satisfies the nonnegativity conditions in (12)-(13), but it equates the artificial variable α to a positive value.

An almost complementary basis

A basis for system (23) and (24) is said to be **almost complementary** if this basis:

- Includes the artificial variable α;
- Includes exactly one member of every complementary pair but one;
- Has a basic solution that equates all decision variable (including α) to nonnegative values.

The initial pivot has produced an almost complementary basis.

A pivot sequence

The **complementary pivot method** begins with the system consisting of equations (23) and (24), and it executes the following sequence of pivots:

- The initial pivot occurs on the coefficient of α in the equation for which s_1 is basic.
- The entering variable in each pivot after the first is the complement of the variable that departed on the preceding pivot, and the row on which that pivot occurs is determined by the usual ratios, which keep the basic solution nonnegative.
- Pivoting stops when α is removed from the basis.

Since s_1 leaves on the 1st pivot, its complement, x_1, will enter on the 2nd pivot. The usual ratios determine the coefficient of x_1 to pivot upon. The variable that departs in the 2nd pivot is the basic variable for the row on which that pivot occurs. Its complement enters on the 3rd pivot. And so forth – until a pivot occurs on a coefficient in the row for which α is basic.

An Illustration

To illustrate the complementary pivot method, we return to the matrices A and B that are specified by equation (8). The tableau in Table 15.1 describes the first two pivots. It is noted that:

- Rows 3-4 of this tableau mirror equation (24). Rows 5-7 mirror equation (23).
- In each pivot, the entering variable's label is lightly shaded, the pivot element is shaded more darkly, and the departing variable's label is surrounded by dashed lines.
- The 1st pivot occurs on the coefficient of α in the row for which s_1 is basic.
- Since s_1 departs on the 1st pivot, its complement x_1 enters on the 2nd pivot.
- Column N displays the usual ratios. The 2nd pivot occurs on the coefficient of x_1 row 14 because its ratio is smallest. The variable t_3 is basic for the equation modeled by that row. Hence, the complement y_3 of t_3 will enter on the 3rd pivot.

Table 15.1. The first two pivots for the matrices A and B in (8).

	A	B	C	D	E	F	G	H	I	J	K	L	M	N
1														
2		s1	s2	t1	t2	t3	x1	x2	y1	y2	y3	α	RHS	
3		1							-2	-5	-7	-1	-1	
4			1						-5	-7	-3	-2	-1	
5				1			3	8					1	
6					1		5	9					1	
7						1	8	4					1	
8														
9		s1	s2	t1	t2	t3	x1	x2	y1	y2	y3	α	RHS	ratios
10		-1							2	5	7	1	1	
11		-2	1						-1	3	11		1	
12				1			3	8					1	0.333
13					1		5	9					1	0.2
14						1	8	4					1	0.125

The complementary pivot method is executed on the spreadsheet that accompanies this chapter. The pivot sequence is as follows:

- On the 1st pivot, α enters and s_1 departs.
- On the 2nd pivot, x_1 enters and t_3 departs.
- On the 3rd pivot, y_3 enters and s_2 departs.
- On the 4th pivot x_2 enters and t_2 departs.
- On the 5th pivot, y_2 enters and α departs.

The 5th pivot produces a complementary basis. From the spreadsheet that accompanies this chapter, we see that its basic solution sets

$$x_1 = 5/52, \quad x_2 = 3/52, \quad y_1 = 0, \quad y_2 = 2/17, \quad y_3 = 1/17,$$

which is converted by (14) and (15) into the randomized strategies

$$p = \begin{bmatrix} 5/8 & 3/8 \end{bmatrix} \qquad q = \begin{bmatrix} 0 & 2/3 & 1/3 \end{bmatrix}.$$

Proposition 15.2 demonstrates that these strategies form an equilibrium. This is the same equilibrium that was constructed by "eyeball" in the prior section.

Failure?

What can go wrong? Like any pivot scheme, the complementary pivot method could cycle (revisit a basis). And it could call for the introduction of a variable that can be made arbitrarily large without reducing the values of any of the basic variables. In the next section, a routine perturbation argument will preclude cycling, and each entering variable will be shown to have a coefficient on which to pivot.

5. The Guarantee

In this section, it is demonstrated that the complementary pivot method must end with a pivot to a complementary basis. This will show – constructively – that every bi-matrix game has an equilibrium in mixed strategies.

The structure of the argument used here is strikingly similar to the argument that will be used in Chapter 16 to construct a Brouwer fixed-point. In both chapters, the pivot scheme can be described as taking a "walk" through the "rooms" of a mansion.

Rooms

Let us call each almost complementary basis a **green room**, and let us call each complementary basis a **blue room**. The **mansion** is the set of all green and blue rooms. Thus, each room in the mansion identifies a complementary basis or an almost complementary basis. Doors between certain rooms will soon be created.

Degeneracy

When the complementary pivot procedure is executed, there could exist a tie for the smallest ratio. Should two rows tie for the smallest ratio, we have a choice as to the pivot element and, hence, as to the departing variable. That possibility is eliminated by the:

> **Nondegeneracy Hypothesis.** In every basic feasible tableau for system (23) and (24), no entering variable has identical ratios for two or more rows.

This hypothesis guarantees a unique pivot sequence.

Perturbation

The Nondegeneracy Hypothesis may seem awkward, but it can be guaranteed by perturbing the RHS value of each equation in (23) and (24). With ε as a small positive number:

- For i = 1, …, m, subtract ε^i from the RHS value of the ith equation in (24).
- For j = 1, …, m, add ε^{m+j} to the RHS value of the jth equation in (23).

A standard argument, which is due to A. Charnes, shows that no two ratios can be tied for all sufficiently small positive values of ε, hence that one has no choice as to the pivot row. And a standard technique (also due to Charnes) breaks ties when ε = 0 so as pivot as though ε were miniscule, but positive.

Reversibility of pivots

We shall need to use a property of feasible pivots that is nearly self-evident but which has not been mentioned previously. This property is not tied to the bi-matrix game. To describe it in general terms, we consider the equation system,

$$C\,w = b,$$

whose data are the entries in the $r \times s$ matrix C and in the $r \times 1$ vector b and whose decision variables form the $s \times 1$ vector w.

Let us suppose we have at hand a basic solution to C w = b, and suppose that we elect to pivot on the coefficient of the nonbasic variable w_j in the row for which w_k is basic. This pivot produces a new basis and a new basic solution. To *reverse* it, pivot on the coefficient of w_k in the row for which w_j has become basic; this restores the original basis and the original basic solution. If the first pivot kept the basic solution nonnegative, so must the second.

Doors

Each green room (almost complementary basis) has two **doors**, each door corresponding to selecting as the entering variable one member of the pair that is not represented in this basis. Each blue room (complementary basis) has a single **door**, which corresponds to selecting α as the entering variable.

Each door is **labeled** with the entering variable to which it corresponds. If you are standing in a blue room, you see one door, and its label is α. If you are standing in a green room, you see two doors, and their labels are complementary – if one door's label is s_k, the other's label is t_k, for instance.

A door is said to lead **outside** the mansion if there is nothing to pivot upon, equivalently, if none of the basic variables decrease as the entering variable becomes positive. Alternatively, if one or more of the basic variables decreases as the entering variable becomes positive, the door leads to the room that results from pivoting on the row with the smallest ratio. With perturbed data, that row is unique.

Each label identifies the entering variable for the pivot that corresponds to walking through that door. Each door between two rooms is labeled on *both* sides. The label on the other side of the door you are looking at is that of the variable that will depart as you walk through the door. Because pivots are reversible, after walking through a door into a new room, you could turn around and walk through the same door back to the room at which you started.

At least one door to the outside

Let us observe that the mansion *does* have a door to the outside. To exhibit such a door, we consider the green room (almost complementary basis) that results from the initial pivot. That basis is the set $\{\alpha, s_2, \ldots, s_m, t_1, \ldots, t_n\}$ of decision variables. This pivot removed s_1 from the basis. Let us take s_1 as the entering variable. Setting s_1 positive alters the values of the basic variables like so:

$$\alpha = 1 + s_1,$$

$$s_2 = 1 + 2s_1, \qquad s_3 = 2 + 3s_1, \qquad \ldots, s_m = (m-1) + ms_1,$$

$$t_1 = 1, \qquad t_2 = 1, \qquad \ldots, t_n = 1.$$

As s_1 becomes positive, none of the basic variables decrease in value. The solution remains nonnegative no matter how large s_1 becomes. This basis and entering variable correspond to a door to the outside.

We have seen that the mansion has a door between a green room (almost complementary basis) and the outside. Suppose – and this will be demonstrated – that:

Only one door leads to the outside of the mansion.

It will soon be clear that if there is only one door to the outside, the complementary pivot method must terminate with a pivot to a blue room (complementary basis).

Complementary paths

The complementary pivot method is initialized with a particular pivot. But suppose we allow ourselves to start in any room or outside the mansion and walk through a sequence of doors subject to the requirement that we leave each green room via the door other than the one through which we just entered it. Any such walk is said to follow a **complementary path**. The complementary pivot method follows one such path.

Types of complementary path

Figure 15.1 displays the types of paths we can encounter if there is only one door to the outside. The mansion (set of rooms) is enclosed in dashed lines, and the door to the outside is represented by a line segment between a green room and the outside.

Figure 15.1. Types of complementary path.

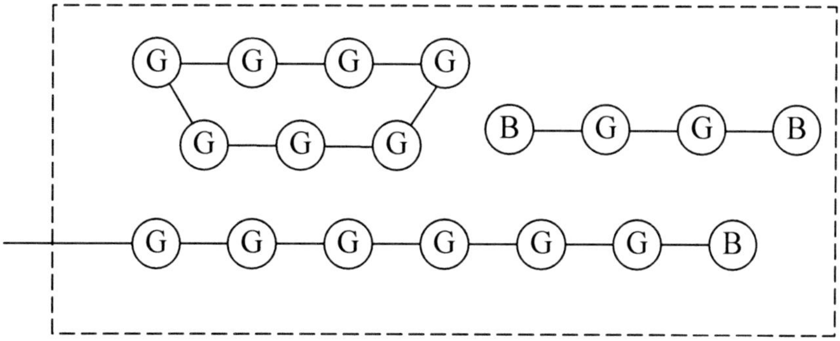

Figure 15.1 illustrates the three types of complementary path that can occur if there is only one door to the outside:

- *Type 1:* If we start in a green room (almost complementary basis), we could return to that room, thereby encountering a cycle.
- *Type 2:* Suppose we enter the mansion and then follow the complementary path. We cannot revisit a room – the first room we revisit would need to have at least three doors, and none do. We cannot leave the mansion – there is only one door to the outside, and we would need to revisit the room we entered before we left. We *must* end at a blue room (complementary basis).
- *Type 3:* Suppose we begin at a blue room other than the one at the end of the Type-2 path. We cannot join the Type-2 path because it has no unused doors. We cannot leave the mansion because there is only one door to the outside. We must end at another blue room.

The complementary pivot method follows the Type-2 path. This path must lead to a complementary basis – hence to an equilibrium in randomized strategies – if we can show that there is only one door to the outside.

Something extra

The Nondegeneracy Hypothesis gives us something extra: The number of complementary bases is odd. It equals 1 or 3 or 5, and so forth.

Are cycles a good thing?

If the complementary pivot method is initialized at an arbitrarily selected green room, it can cycle. Is the possibility of cycling good news or bad news? Well, suppose the number of almost complementary bases is large, e.g., 10,000. We would like the path from the outside the mansion to the blue room (complementary basis) to have only a few edges because we will need to execute one pivot for each green room in that path. It will be short if the preponderance of the green rooms are parts of cycles. The hope is that the vast majority of green rooms are parts of cycles.

No pivot row

Two propositions will be used to show that the bi-matrix game, as specified by equations (23) and (24), has only one door to the outside. The first of these propositions is not particular to the bi-matrix game – it relates an entering variable that can be made arbitrarily large to a "homogeneous" equation system. Because it holds in general, the first proposition is presented in the context of the system,

$$C\,w = b \qquad \text{and} \qquad w \geq 0,$$

whose data are the entries in the $r \times s$ matrix C and in the $r \times 1$ vector b and whose decision variables form the $s \times 1$ vector w. As is usual, C_j denotes the jth column of the matrix C. This proposition studies the case in which an entering variable x_k can be made arbitrarily large without causing any of the basic variables to become negative.

Proposition 15.3. Let $\{C_j : j \in \beta\}$ be a basis for the column space of C. Suppose that for some $k \notin \beta$, the equation system,

$$\sum_{j\in\beta} C_j w(\theta)_j + \theta C_k = b, \tag{25}$$

has a solution $w(\theta) \geq 0$ for all $\theta \geq 0$. Then there exists a non-zero nonnegative solution to

$$\sum_{j\in\beta} C_j v_j + C_k v_k = 0. \tag{26}$$

Proof. Subtracting (25) with the value $\theta = 0$ from (25) with a positive value θ produces

$$\sum_{j\in\beta} C_j[w(\theta)_j - w(0)_j] + \theta C_k = 0. \tag{27}$$

Since θ is positive, a nonnegative non-zero solution to (26) will be exhibited by (27) if we can show that $w(\theta)_j - w(0)_j \geq 0$ for each $j \in \beta$.

Since $\{C_j : j \in \beta\}$ is a basis for the column space of C and since C_k is in that space, there exist a set $\{z_j : j \in \beta\}$ of numbers such that

$$C_k = \sum_{j\in\beta} C_j(-z_j).$$

Substituting the above expression for C_k into (27) produces

$$\sum_{j\in\beta} C_j[w(\theta)_j - w(0)_j - \theta\, z_j] = 0$$

Being a basis, the set $\{C_j : j \in \beta\}$ of columns is linearly independent, so each coefficient in the expression that is displayed above must equal zero:

$$(28) \qquad w(\theta)_j - w(0)_j = \theta z_j \qquad \text{for each } j \in \beta.$$

By hypothesis, $w(\theta)_j \geq 0$ for all positive values of θ, so (28) gives $z_j \geq 0$ for each $j \in \beta$. Thus, $w(\theta)_j - w(0)_j \geq 0$ for each $j \in \beta$. This completes a proof. ■

The homogeneous equation

Proposition 15.3 concerns the feasible region of any linear program that has been cast in Form 1. Its feasible region consists of the vectors w that satisfy $C\,w = b$ and $w \geq 0$.

The matrix equation $C\,w = 0$ is said to be **homogeneous** because its RHS values equal zero. Proposition 15.3 states that if an entering variable can be made arbitrarily large without making any basic variables negative, then there exists a vector $w > 0$ such that $C\,w = 0$ and such that the value of the entering variable is positive and the values of the other nonbasic variables equal zero.

Exactly one door to the outside

The crux of the argument that the complementary pivot method cannot cycle appears below as

Proposition 15.4 *(one door).* Suppose that each entry in the cost matrix A is positive and that each entry in the cost matrix B is negative. Consider an almost complementary basis for system (23) and (24) and suppose that perturbing its basic solution by equating one of the missing pair of variables to a positive value causes no basic variable to decrease in value. Then the basis is $\{\alpha, s_2, \ldots, s_m, t_1, \ldots, t_n\}$ and the entering variable is s_1.

Proof*. The proof of this proposition earns a star because it is a bit intricate.

By hypothesis, there exists an almost complementary solution to

$$(29) \qquad s_i - \sum_{j=1}^{n} A_{ij}\, y_j - i\alpha = -1 \qquad \text{for } i = 1, \ldots, m,$$

$$(30) \qquad t_j - \sum_{i=1}^{m} x_i\, B_{ij} = 1 \qquad \text{for } j = 1, \ldots, n.$$

Since the entering variable can be made arbitrarily large, Proposition 15.3 show that there exists a solution to the homogeneous version of these equations, namely, a nonnegative solution to

$$\hat{s}_i - \sum_{j=1}^{n} A_{ij}\,\hat{y}_j - i\hat{\alpha} = 0 \qquad \text{for } i = 1, \ldots, m, \tag{31}$$

$$\hat{t}_j - \sum_{i=1}^{m} \hat{x}_i\,B_{ij} = 0 \qquad \text{for } j = 1, \ldots, n. \tag{32}$$

Further, in this homogeneous solution only the entering variable and the basic variables can be positive. Since the basis is almost complementary and since the complement of the entering variable is not basic, these solutions satisfy

$$(s_i + \hat{s}_i) \cdot (x_i + \hat{x}_i) = 0 \qquad \text{for } i = 1, \ldots, m, \tag{33}$$

$$(t_t + \hat{t}_j) \cdot (y_j + \hat{y}_j) = 0 \qquad \text{for } j = 1, \ldots, n. \tag{34}$$

The total number of decision variables in equations (29)-(34) equals $2(m+n)+1$. All of these decision variables are nonnegative. It will be seen that many of them must equal 0. Since each entry in B is negative, (32) guarantees

$$\hat{t}_j = 0 \qquad \text{for each } j,$$

$$\hat{x}_i = 0 \qquad \text{for each } i.$$

All of the variables in (32) equal zero, not all of the variables in (31) can equal zero. Each coefficient A_{ij} is positive, so (31) guarantees

$$\hat{s}_i \geq 0 \qquad \text{for each } i.$$

Thus, $\hat{s}_i$ is either basic or is the entering variable. In either case, (33) guarantees

$$x_i = 0 \qquad \text{for each } i.$$

Thus, $x = 0$. Hence, (30) gives

$$t_j = 1 \qquad \text{for each j.}$$

Hence, t_j is basic for each j. So (34) guarantees

$$y_j = 0 \qquad \text{for each j.}$$

Every basis for (23) and (24) contains $(m + n)$ variables. By hypothesis, the current basis is almost complementary, so it includes α. It includes t_1 through t_n, and it includes $(m - 1)$ of the variables s_1 through s_m. It must exclude exactly one of the variables s_1 through s_m.

Aiming for a contradiction, we suppose this basis excludes s_k with $k > 1$. The basic solution has $y = 0$, so the kth constraint in (29) gives $\alpha = 1/k$, and the 1st constraint in (29) shows that the basic variable $s_1 = \alpha - 1 = 1/k - 1 < 0$. This cannot occur because the basic solution is feasible. Hence, s_1 is the entering variable, and the basis consists of the set $\{\alpha, s_2, \ldots, s_m, t_1, \ldots, t_n\}$ of variables. This completes a proof. ■

The clincher

Complementary pivoting cannot leave the mansion by the door through which it entered. And Proposition 15.4 demonstrates that the only door to the outside is the one by which it entered. It must terminate, and it must terminate at a blue room, that is, at an equilibrium.

A bit of the history

In 1964, Lemke and Howson published a stunning paper[1] that introduced the complementary pivot method and demonstrated that it computes an equilibrium in mixed strategies for a bi-matrix game. Prior to their work, the bi-matrix game was known to have an equilibrium in mixed strategies, but the proof rested on Brouwer's fixed-point theorem, and no method was known for computing the equilibrium.

---

[1] Carleton E. Lemke and J. T. Howson, "Equilibrium Points of Bi-Matrix Games," SIAM Journal, V. 12, pp. 413-423, 1964.

6. Payoff Matrices

The entire development in this chapter has concerned a bi-matrix game with a pair A and B of $m \times n$ *cost* matrices. The variant with **payoff matrices** is as follows:

- You and I know the entries in the $m \times n$ matrices A and B.
- You pick a row. Simultaneously, I pick a column.
- If you choose row i and I choose column j, you earn A_{ij} and I earn B_{ij}.
- You wish to maximize the expectation of your earnings, and I wish to maximize the expectation of my earnings.

Expected net cost is the negative of expected net profit. A perfectly satisfactory way to treat the problem with payoff matrices is to is to replace A by $-A$ and B by $-B$ and solve the resulting bi-matrix game problem with cost matrices.

This conversion is *exactly* equivalent to replacing A by $-A$ in (10) and B by $-B$ in (11) and executing the prior analysis under the hypothesis that each entry in A is negative and each entry in B is positive.

7. Cooperation and Side Payments

Until this point, our discussion of bi-matrix games has been focused on competitive behavior. This section concerns a model that includes cooperation. Let us consider this situation:

- You and I know the entries in the $m \times n$ matrices A and B.
- You pick a row. Simultaneously, I pick a column.
- If you choose row i and I choose column j, you receive A_{ij} dollars and I receive B_{ij} dollars.
- You and I can engage in a contract that governs how the total $A_{ij} + B_{ij}$ of the amounts we receive is to be allocated between us.

To have a concrete example to work with, let's suppose that

(35) $$A = \begin{bmatrix} 4 & 1 \\ -3 & 0 \end{bmatrix}, \qquad B = \begin{bmatrix} 1 & 2 \\ 3 & 2 \end{bmatrix}.$$

The matrix A + B measure the total of the rewards we can attain, and the matrix A – B represents the difference For the data in (35),

(36) $$A + B = \begin{bmatrix} 5 & 3 \\ 0 & 2 \end{bmatrix}, \qquad A - B = \begin{bmatrix} 3 & -1 \\ -6 & -2 \end{bmatrix}.$$

The competitive solution

To assess the potential benefit of cooperation, let's see what happens if we do not. I observe that you will pick row 1 because it pays you more than does row 2, independent of what I do. So I play column 2. You receive \$1. I receive \$2. That is well short of the \$5 payoff we could attain by working together.

A threat

To get me to play column 1, you will need to compensate me. I can threaten to play column 2. In this case – and in general – a reasonable measure of the threat is the value of the game A – B. For the data in (36), that value equals –1. (It has an equilibrium in pure strategies; you play row 1, and I play column 2).

Dividing the spoils

In general, the numbers α and β are defined by

$$\alpha = \max_{i,j}\{A_{ij} + B_{ij}\}, \qquad \beta = \text{val}\{A - B\},$$

where "val(A – B)" is short for the value (to the row player) of the game whose payoff matrix is A – B. A reasonable division of the spoils is that the row player receives $(\alpha+\beta)/2$ and the column player receives $(\alpha-\beta)/2$.

For the data in (35), we have

$$\alpha = \$5.00, \quad \beta = -\$1.00.$$

The total of \$5.00 is divided like so; you receive \$1.50 and I receives \$3.50. Interestingly, although the cooperative solution awards you (the row player)

$4.00 and me (the column player) $1.00, I have enough bargaining power to come out ahead, by this account.

8. Review

Except for a brief foray into cooperative behavior, this chapter has focused on competition. The complementary pivot method has been used to find an equilibrium, that is, a pair of strategies in which each player responds optimally to the actions of the other.

In two respects, the complementary pivot method resembles the simplex method. The initial pivot is akin to the version of Phase I in Chapter 6. Subsequent pivots use ratios to keep the basic solution feasible.

In two respects, it does not. The complementary pivot method strives for a solution to the complementarity conditions. (Virtually without exception, the algorithms that preceded it pivoted had striven to improve an objective.) Second, the complementary pivot method arranges for a unique pivot sequence that can only end in one way – with a complementary basis and, hence, with an equilibrium in mixed strategies. These innovations will be used in Chapter 16 to approximate a solution to a fixed-point equation.

9. Homework and Discussion Problems

1. On a spreadsheet, execute the complementary pivot method for the data in Table 15.1 (on page 491), but with the numbers in cells L3 and L4 exchanged.

2. By eye, find all equilibria for the bi-matrix game whose cost matrices A and B are given below. *Remark:* Barring "degeneracy," the number of equilibria is odd.

$$A = B = \begin{bmatrix} 0 & 6 \\ 6 & 0 \end{bmatrix}$$

3. By eye, find all equilibria for the bi-matrix game whose cost matrices A and B are given below. *Remark:* Barring "degeneracy," the number of equilibria is odd.

$$A = B = \begin{bmatrix} 0 & 6 & 0 \\ 0 & 0 & 6 \\ 6 & 0 & 0 \end{bmatrix}$$

4. Construct 2×2 cost matrices A and B for which the bi-matrix game has three equilibria with these properties: amongst these equilibria, one has the best (unique smallest) expected cost for the row player and the worst (unique largest) expected cost for the column player,. and another has the worst (unique largest) expected cost for the row player and the best (unique smallest) expected cost for the column player.

5. Find all equilibria for the bi-matrix game whose cost matrices A and B are given below.

$$A = \begin{bmatrix} 1 & 3 \\ 4 & 0 \end{bmatrix}, \qquad B = \begin{bmatrix} 0 & 0 \\ 0 & 0 \end{bmatrix}$$

6. Without using the complementary pivot method, find an equilibrium in non-randomized strategies for the bi-matrix game whose payoff matrices A and B are given below. *Hint:* Will the column player use a column if another column has a larger entry in some row and as large an entry in each row? With these "dominated" columns deleted, what will the row player avoid?

$$A = \begin{bmatrix} 2 & 7 & 5 & 6 \\ 7 & 7 & 2 & 6 \\ 7 & 0 & 4 & 4 \\ 1 & 7 & 5 & 2 \end{bmatrix}, \qquad B = \begin{bmatrix} 9 & 4 & 5 & 6 \\ 5 & 3 & 11 & 3 \\ 8 & 4 & 4 & 8 \\ 2 & 0 & 7 & 6 \end{bmatrix}$$

7. When comparing vectors x and y that have the same size, the inequality $x < y$ means $x \leq y$ and $x \neq y$. (Thus, $x < y$ if $x \leq y$ and if at least one of these inequalities is strict.) In a bi-matrix game, column k of the cost matrix B is said to be *weakly dominated* if there exists an $n \times 1$ vector q whose entries are nonnegative, whose entries sum to 1, and that satisfy $Bq < B_k$ and $q_k = 0$.

(a) Suppose column k is dominated. Argue that at least one pair p and q of strategies that is an equilibrium has $q_k = 0$.

(b) Devise a linear program that determines whether or not column k of B is weakly dominated.

8. A bi-matrix game is said to be a **constant-sum** matrix game if there exists a number c such that its cost matrices A and B have $A_{ij} + B_{ij} = c$ for each i and j.

(a) An equilibrium for a constant-sum matrix game can be found by solving a linear program. Exhibit a linear program for which this the preceding statement is true and demonstrate that it works.

(b) A constant-sum matrix game can have multiple equilibria, but they all have the same expected ____________________. Complete the sentence and indicate why it is true.

9. This problem concerns the bi-matrix game whose payoff matrices A and B are given by

$$A = \begin{bmatrix} 1 & 3 \\ 5 & 2 \end{bmatrix}, \qquad B = \begin{bmatrix} -2 & -1 \\ 4 & 0 \end{bmatrix}.$$

(a) Find the equilibrium and its expected payoff to each player. *Hint:* Each player can arrange for the other's expected payoff to be independent of his strategy.

(b) What is the largest total amount α that the two players could receive from this game?

(c) Find the value β of the matrix game whose payoff matrix is A – B. Can the row player guarantee a minimum payoff that exceeds that of the column player by β?

(d) Describe a procedure that provides the row player with a payoff of $(\alpha + \beta)/2$ and the column player with a payoff of $(\alpha - \beta)/2$.

Chapter 16: Fixed Points and Equilibria

1. Preview

In 1909, L. E. J. Brouwer proved a *fixed point* theorem that is illustrated by this scenario: At dawn, the surface of an oval swimming pool is perfectly still. Then a breeze begins to blow. The wind is strong enough to create waves, but not breakers. At dusk, the wind dies down, and the surface becomes still again. Each point on the surface of the pool may have shifted continuously during the day, but each point that began on the surface remains there throughout the day. Brouwer's theorem guarantees that at least one point on the surface ends up where it began.

For decades, all proofs of Brouwer's fixed-point theorem were existential; they demonstrated that a fixed point exists, but offered no clue as to how to locate it. That has changed. An analogue of the complementary pivot scheme

E. V. Denardo, *Linear Programming and Generalizations*, International Series in Operations Research & Management Science 149,
DOI 10.1007/978-1-4419-6491-5_16, © Springer Science+Business Media, LLC 2011

in the prior chapter approximates a fixed point of a continuous map of a simplex into itself. This development has had a profound impact on several fields. It has become possible to compute economic equilibria, for instance.

This chapter introduces fixed-point computation and its use in the calculation of economic equilibria. The development begins with a statement of Brouwer's fixed- point theorem. Next, the problem of finding an equilibrium is formulated for solution as a Brouwer fixed-point problem. The complementary pivot method is then adapted to approximate Brouwer fixed-points, first on the "unit simplex," then on any simplex, and finally on any closed convex subset of $\Re^n$. The chapter is closed with a discussion of the related literature. This chapter is meaty, but geometric diagrams will help us to grapple with its key ideas.

2. Statement of Brouwer's Theorem

Brouwer's theorem concerns a continuous function f that maps a subset C of $\Re^n$ into itself. A vector x in C is said to be a **fixed point** of f if f(x) = x, that is, if the function f maps the n-vector x into itself. Brouwer provided conditions under which a continuous function f has at least one fixed point in its domain.

A few examples

A continuous function that maps a subset C of $\Re^n$ into itself need not have a fixed point within that set. The following three examples suggest what can keep this from happening.

Example 16.1. Let $C = \{x \in \Re : 0 < x \leq 1\}$, and let f(x) = x/2. This function f is continuous, and it maps C into itself, but no number x in C has f(x) = x.

Example 16.2. Let $C = \{x \in \Re : 1 \leq x < \infty\}$, and let f(x) = 1 + x. The function f is a continuous map of C into itself, but no number x in C has f(x) = x.

Example 16.3. Let $C = \{x \in \Re^2 : 1 \leq {x_1}^2 + {x_2}^2 \leq 4\}$, and let $f(x_1, x_2) = (-x_1, -x_2)$. This function f is continuous on C, but no vector x in C has f(x) = x.

In Example 16.1, the set C is not closed. In Example 16.2, the set C is closed but not bounded. In Example 16.3, the set C is closed and bounded, but it has a "hole."

A sufficient condition

The difficulties illustrated by these examples are circumvented by assuming that the set C is closed, bounded, and convex.

Proposition 16.1. *(Brouwer's fixed-point theorem).* Let the subset C of $\Re^n$ be closed, bounded and convex, and let f be a continuous function that maps C into itself. Then there exists at least one n-vector x in C such that f(x) = x.

Proposition 16.1 is known as **Brouwer's fixed-point theorem**. A proof of this proposition appears in later in this chapter. The main tool in this proof is a result that is of great interest in itself.

3. Equilibria as Fixed Points

Before Brouwer's theorem is proved, an important use of it is described. In Chapter 14, the problem of finding an equilibrium in a zero-sum matrix game was formulated for solution by the simplex method. In Chapter 15, the problem of finding an equilibrium in a bi-matrix game was posed for solution by complementary pivoting. In this section, the problem of finding an equilibrium in an n-person competitive game will be formulated as a Brouwer fixed-point problem. The line of analysis that is exhibited here works for any number n of players. The analysis requires less notation when there are only two players, so we focus on that case.

Two players

As in Chapter 15, the data for the game we shall play are the entries in the m×n matrices A and B. You (the row player) select a row. Simultaneously, I (the column player) select a column. If you pick row i and I pick column j, you receive the amount A_{ij}, and I receive the amount B_{ij}. You and I know the entries in the matrices A and B. You prefer a larger expected payoff to a smaller one, and so do I.

A randomized strategy for you (the row player) is represented as a 1×m vector p whose i^{th} entry p_i equals the probability that you play row i. A randomized strategy for me (the column player) is represented as an $n \times 1$ vector q whose j^{th} entry q_j is the probability that I play column j. The set C of our randomized strategies consists of every $1 \times m$ vector p and every $n \times 1$ vector q that satisfy

$$p \geq 0, \quad \sum_{i=1}^{m} p_i = 1, \quad q \geq 0, \quad \sum_{j=1}^{n} q_j = 1.$$

This set C is closed, bounded and convex. It is, in fact, a polyhedron.

An equilibrium

If you and I choose randomized strategies p and q, respectively, the expectation of your gain equals p A q, and the expectation of my gain equals p B q. Let us recall that the pair (p, q) of strategies is an **equilibrium** if every randomized strategy $\bar{p}$ for the row player and every randomized strategy $\bar{q}$ for the column player have

$$\bar{p}\, A\, q \leq p\, A\, q \qquad \text{and} \qquad p\, B\, \bar{q} \leq p\, A\, q.$$

Evidently, the strategies that form an equilibrium are *best responses* to each other.

Your potential for improvement

It will soon be evident that the equilibria are fixed-points of a continuous map of C into itself. Let us consider any pair p and q of strategies, and let us suppose that the inequality

(1) $$A^i\, q > p\, A\, q.$$

holds when i=3. This inequality states that if I play strategy q, you can do better by playing row 3 than you can by playing strategy p. This connotes that you ought to *increase* the probability p_3 that you play row 3. More generally, you benefit by increasing p_i for each row i for which (1) holds. To do so, you will need to reduce the probability that you play rows for which (1) does not hold.

An adjustment mechanism that accomplishes this is easily designed. For every pair of strategies p and q and fo each row i, let us define the quantity

(2) $$\alpha_i(p,\ q) = \max\,\{0,\ A^i\, q - p\, A\, q\}.$$

Note that $\alpha_i(p, q)$ is positive if and only if (1) holds. Consider the numbers $\hat{p}_1$ through $\hat{p}_m$ that are defined by

$$(3) \qquad \hat{p}_i = \frac{p_i + \alpha_i(p, q)}{1 + \sum_{k=1}^{m} \alpha_k(p, q)}, \qquad \text{for } i = 1, 2, \cdots, m.$$

Being a probability distribution, p_1 through p_m are nonnegative numbers that sum to 1, and (3) is easily seen to guarantee that $\hat{p}_1$ through $\hat{p}_m$ are nonnegative numbers that sum to 1.

My potential for improvement

The same argument works for me, the column player. Let us suppose that the inequality

$$(4) \qquad p\, B_j > p\, B\, q.$$

holds for some j. This says that if you play strategy p, I can do better by playing column j than by playing strategy q. This suggests that I should increase the probability q_j that I play column j. This observation leads to the functions,

$$(5) \qquad \beta_j(p, q) = \max\,\{0,\; p\, B_j - p\, B\, q\} \qquad \text{for } j = 1, 2, \ldots, n,$$

and to the adjustment mechanism

$$(6) \qquad \hat{q}_j = \frac{q_j + \beta_j(p, q)}{1 + \sum_{k=1}^{n} \beta_k(p, q)} \qquad \text{for } j = 1, 2, \ldots, n.$$

A fixed point

Equations (3) and (6) adjust the strategies of the row and column players through the function f that is given by

$$(7) \qquad f(p, q) = (\hat{p},\; \hat{q}),$$

where $\hat{p}_i$ and $\hat{q}_j$ are specified by the right-hand sides of (3) and (6), respectively. A pair (p, q) of probability distributions is said to be a **stable distribution** if

$$(p, q) = f(p, q).$$

Proposition 16.2, below, shows that at least one stable distribution exists, moreover, that each stable distribution is an equilibrium.

Proposition 16.2. *(A Nash equilibrium).* Consider any bi-matrix game.

(a) There exists at least one pair (p, q) of stable probability distributions.

(b) Each pair (p, q) of stable distributions is an equilibrium.

Proof. The set C of pairs (p, q) of probability distributions over the rows and columns is closed, bounded and convex. The function f that maps the pair (p, q) into $(\hat{p}, \hat{q})$ via (3) and (6) is a continuous map of C into itself. Brouwer's theorem (Proposition 16.1) shows that this function has at least one fixed point, hence that there exists at least one pair (p, q) of probability distributions for which (3) and (6) hold with $\hat{p} = p$ and $\hat{q} = q$. This proves part (a).

For part (b), let us first consider *any* pair (p, q) of strategies. Set v=pAq. It will be argued by contradiction that there exists at least one row i for which $p_i > 0$ and $v \geq A^i q$. Suppose not. Multiply $v < A^i q$ by p_i and sum over i to obtain the contradiction, $v < pAq = v$.

Now, let the pair (p, q) of strategies be stable. As noted above, there exists an i for which $p_i > 0$ and $A^i q \leq v = pAq$. Since $\alpha_i(p, q) = \max\{0, A^i q - pAq\} = 0$, this pair of stable strategies has

$$p_i = \frac{p_i + 0}{1 + \sum_{k=1}^{m} \alpha_k(p, q)}.$$

Since p_i is positive, clearing the denominator in the above equation shows that $\alpha_k(p, q) = 0$ for each k. This demonstrates that

$$A^k\, q \leq p\, A\, q \qquad \text{for } k = 1, 2, \cdots, m.$$

Let $\hat{p}$ be any randomized strategy for the row player. Premultipy the inequality that is displayed above by $\hat{p}_k$ and sum over k to obtain $\hat{p}\, A\, q \leq p\, A\, q$, which shows that p is a best response to q. A similar argument demonstrates that $\beta_j(p, q) = 0$ for each j and, consequently, that q is a best response to p. This completes a proof. ■

A prize-winning result

In 1950, John Nash[1] published the content of Proposition 16.1, with the same proof, in a brief, elegant and famous paper. In the economics literature, an equilibrium for an n-player game is often referred to as a **Nash equilibrium**. This distinguishes it from a **general equilibrium**, in which the market clearing conditions are also satisfied. John Nash shared in the 1994 Nobel Prize in Economics, which was awarded for "pioneering analysis of equilibria in the theory of noncooperative games."

Nash's paper was exceedingly influential, but a case can be made that the pioneering work in this area had been done by John von Neumann and Oskar Morgenstern. In 1928, von Neumann[2] had introduced the matrix game and had made the same use of Brouwer's fixed point theorem in his demonstration that a pair of randomized strategies has the "minimax" property that is described in Chapter 14. Von Neumann and Morgenstern had shown in their 1944 book[3] that a zero-sum game has an equilibrium in randomized strategies.

4. Affine Spaces

Brouwer's theorem will be proved in a way that facilitates the computation of fixed points and, consequently, of Nash equilibria. Our first step toward a proof is to generalize, somewhat, the notion of a vector space.

A vector space

A classic definition of a vector space was presented in Chapter 10. A subset V of $\Re^n$ is called a vector space if:

- V is not empty, and
- V contains the vector $(u + \alpha v)$ for every pair u and v of elements of V and for every real number α.

---

[1] Nash, John F (1950) "Equilibrium points in n-person games" *Proceedings of the National Academy of Sciences* v. 36, pp. 48-49.

[2] von Neumann, John (1928), "Zur Theorie der Gesellschaftsspiele." *Math. Annalen.* v. 100, pp. 295-320.

[3] von Neumann, John and Oskar Morgenstern (1944, reprinted in 2007), *Theory of Games and Economic Behavior*, Princeton University press, Princeton, NJ.

It's easy to see that each vector space V must contain the n-vector 0 (In the above, take u=v and $\alpha=-1$). An equivalent way in which to describe a vector space is presented in

Proposition 16.3. A subset W of $\Re^n$ is a vector space if and only if:

- W contains the vector 0, and
- W contains the vector $[(1-\alpha)u+\alpha v]$ for every pair u and v of vectors in W and for every real number α.

Proof of Proposition 16.3 is left to you, the reader. Evidently, if a vector space W contains the vectors u and $v \neq u$, then W contains the *line* through u and v.

An affine space

Proposition 16.3 motivates a modest generalization. A subset X of $\Re^n$ is now called an **affine space** if:

- X is not empty, and
- X contains the vector $[(1-\alpha)u+\alpha v]$ for every pair u and v of elements of X and for every real number α.

Evidently, if an affine space contains distinct vectors u and v, it contains the line through u and v. An affine space need not contain the origin. If an affine space does contain the origin, it is a vector space.

Affine spaces appear elsewhere in this book, though they are not labeled as such. The hyperplanes in Chapter 17 are affine spaces. The relative neighborhoods in Chapter 19 are defined in the context of the an affine space.

Example 16.4. Consider the subset X of $\Re^3$ that consists of all vectors $x=(x_1, x_2, x_3)$ that have $x_1+x_2+x_3=1$. This set X is easily seen to be an affine space.

Affine combinations

The sum of several vectors in a vector space is a vector that lies in that space. More generally, a vector space is closed under linear combinations. An affine space may not be closed under linear combinations. In Example 16.4, the sum of the vectors (1, 0, 0) and (0, 0, 1) is not in X, for instance.

Affine spaces are closed under a related operation. Let the subset X of $\Re^n$ be an affine space that contains the set $\{v^1, v^2, \ldots, v^k\}$ of vectors. For each set $\{\alpha_1, \alpha_2, \ldots, \alpha_k\}$ of numbers such that

$$\alpha_1 + \alpha_2 + \cdots + \alpha_k = 1,$$

the vector,

$$\alpha_1 v^1 + \alpha_2 v^2 + \cdots + \alpha_k v^k,$$

is called an **affine combination** of the vectors v^1 through v^k. The coefficients α_1 through α_k in an affine combination must sum to 1, but these coefficients need not be nonnegative. Thus, every convex combination of a set of vector is an affine combination of the same vectors, but not conversely.

Proposition 16.4. Let the subset X of $\Re^n$ be an affine space. The set X contains every affine combination of every set $\{v^1, v^2, \ldots, v^k\}$ of vectors in X.

Proposition 16.4 states that affine spaces are closed under affine combinations. This proposition can be proved by induction on k. The details are left to you, the reader.

Linear independence

A nonempty set of vectors in $\Re^n$ is linearly independent if the only way to obtain the vector 0 as a linear combination of these vectors is to multiply each of them by the scalar 0 and take the sum. A characterization of linear independent sets that contain at least two vectors is presented in

Proposition 16.5. A set of two or more vectors in $\Re^n$ is linearly independent if and only if none of these vectors is a linear combination of the others.

Proof of Proposition 16.5 is also left to you, the reader. Let us recall (from Chapter 10) that every set $\{v^1, v^2, \ldots, v^k\}$ of vectors in $\Re^n$ that is linearly independent must have $k \le n$.

Affine independence

A nonempty set of vectors in $\Re^n$ is now said to be **affinely independent** if none of these vectors is an affine combination of the others.

Example 16.5. The vectors (1, 0) and (0, 1) and (1, 1) in $\Re^2$ are affinely independent because the line that includes any two of these vectors excludes the third.

Consider a set $\{v^1, v^2, \ldots, v^k\}$ of affinely independent vectors in $\Re^n$. Example 16.5 suggests (correctly) that k can be as large as $n+1$. It is not difficult to show (see Problem 4) that k cannot exceed $n+1$.

5. A Simplex

A key step in the proof of Brouwer's theorem that is under development is a continuous map of a *simplex* into itself. The term "simplex" appears many times in this book, but has not yet been defined. Dantzig coined the phrase "simplex method" to describe the family of algorithms he devised to solve linear programs. His usage has been adopted by nearly every writer on linear programming. Simplexes turn out to play a peripheral role in the analysis of linear programs, however.

Simplexes do play a key role in fixed-point computation. Consider any nonempty set $\{v^1, v^2, \ldots, v^k\}$ of affinely independent vectors in $\Re^n$; the set of all *convex* combinations of these vectors is called a **simplex**. Thus, every simplex S can be written as

(8) $$S = \{ v = \textstyle\sum_{j=1}^k \alpha_j v^j \text{ such that } \alpha \geq 0 \text{ and } \sum_{j=1}^k \alpha_j = 1\}$$

where it is understood that the vectors v^1 through v^k are affinely independent. In brief:

A subset S of $\Re^n$ is a simplex if S equals the set of all convex combinations of a set of affinely independent vectors.

Extreme points

Equation (8) defines a simplex S as a subset of $\Re^n$ that consists of all convex combinations of a set $\{v^1, v^2, \ldots, v^k\}$ of affinely independent vectors. Because the set $\{v^1, v^2, \ldots, v^k\}$ of vectors is affinely independent, each vector in the simplex S can be written in *exactly one way* as a convex combination of v^1 through v^k. In particular, no vector in $\{v^1, v^2, \ldots, v^k\}$ is a convex combination of the others. In other words, each vector in $\{v^1, v^2, \ldots, v^k\}$ is an *extreme point* of S.

Vertexes, facets and faces

With the simplex S defined by (8), the vector v^j is called the **jth vertex** of S, and the subset of S that has $\alpha_j = 0$ is called the **jth facet** of S. The jth vertex of S is sometimes called **vertex j**, and the jth facet of S is sometimes called **facet j.**

Again, consider a simplex S that is the set of all convex combinations of a set $\{v^1, v^2, \ldots, v^k\}$ of affinely independent vectors. The set of all convex combinations of any nonempty subset T of $\{v^1, v^2, \ldots, v^k\}$ is called a **face** of S. If a simplex has four vertices, the line segment connecting any two of its vertices is a face, but it is neither a facet nor a vertex.

Simplexes in 3-space

Example 16.6. Each subset S of $\Re^3$ that is a simplex takes one of these forms:

- A point.
- A line segment.
- A triangle.
- A tetrahedron.

Each simplex in 3-space is a convex set that has not more than 4 extreme points. (Its extreme points must be affinely independent, and no set of 5 or more vectors in $\Re^3$ can be affinely independent.) In particular, a pyramid is *not* a simplex because it has 5 extreme points, which cannot be affinely independent.

A tetrahedron has seven faces that are neither facets nor vertices. Can you identify them?

Simplexes and linear programs

The simplex method pivots from extreme point to extreme point. Must the portion of the feasible region that lies close to an extreme point resemble a simplex? No. Consider a feasible region C that has the shape of a large pyramid. The portion of C whose altitude is not more than one millimeter below the apex of the pyramid has 5 extreme points, and they cannot be affinely independent.

The apex of a pyramid is the intersection of four planes, rather than three, for which reason its basic solution is degenerate. The pyramid connotes (correctly) that the neighborhood of a *degenerate* extreme point need not resemble a simplex.

6. A Simplicial Subdivision

A simplex will be expressed as the union of smaller simplexes. This will be accomplished in two different ways. Each method has its advantages. The "simplicial subdivision" method in this section is easy to visualize for a simplex S that has not more than 3 vertices, but its generalization to higher dimensions would entail a foray into algebraic topology. The "primitive set" method in the next section is slightly harder to illustrate on the plane, but it generalizes with ease to higher dimensions.

Let us consider a simplex S in $\Re^n$ whose extreme points are the vectors v^1 through v^k. A collection Ω of finitely many subsets of S is called a **simplicial subdivision** of S if Ω satisfies the three conditions that are listed below:

- S equals the union of the sets in Ω.
- Each member of Ω is a simplex that has k distinct vertices.
- If two members of Ω have a nonempty intersection, that intersection is a face of each.

This definition is a bit technical, but an example will clear the air.

Subdividing a triangle

Let us consider a simplex U^3 that is easily drawn on the plane. This simplex consists of each 3-vector whose entries are nonnegative numbers that sum to 1. Specifically,

$$U^3 = \{x \in \Re^3 \quad \text{such that} \quad x \geq 0 \text{ and } x_1 + x_2 + x_3 = 1\}.$$

This simplex has three extreme points, which are (1, 0, 0) and (0, 1, 0) and (0, 0, 1). Figure 16.1 presents two different subdivisions of U^3.

Figure 16.1. Two simplicial subdivisions of U^3.

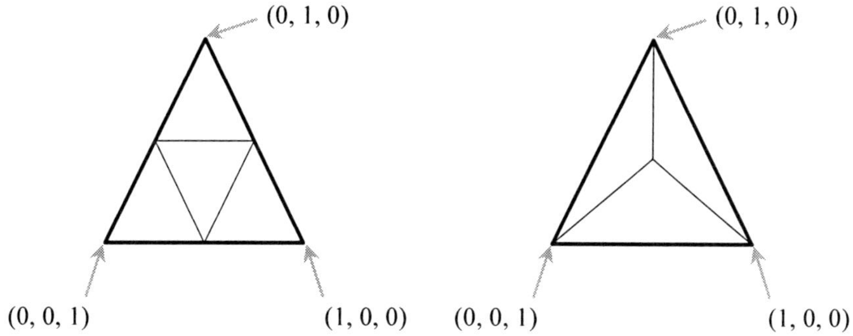

The subdivision on the left is found by placing a "dot" at the center of each edge of U^3 and then "connecting the dots." This results in four smaller simplexes, each of which has the same shape as does U^3, but not all of which have the same orientation. Please check that the intersection of any pair of these smaller simplexes is a face of both.

The subdivision on the right is found by placing a "dot" at the center (average of the vertices) of the larger simplex and then "connecting the dots." Again, the intersection of each pair of smaller simplexes is a face of both.

Repeated subdivision

Each type of subdivision can be iterated. Repeating the pattern on the left produces a sequence of smaller simplexes whose areas approach zero and whose perimeters also approach zero. Repeating the pattern on the right produces a sequence of simplexes whose areas approach zero but whose perimeters do not all approach zero. The pattern on the left is preferable.

The unit simplex in n-space

Let us turn, briefly, from 3-space to n-space. For $j = 1, 2, \ldots, n$, the symbol e^j denotes the n-vector that has 1 in its jth position and has 0's in all other positions. The **unit simplex** U^n in $\Re^n$ is the set of all convex combinations of the n-vectors e^1 through e^n, and it can be described succinctly as

(9) $$U^n = \{x \in \Re^n \quad \text{such that} \quad x \geq 0 \text{ and } x_1 + x_2 + \cdots + x_n = 1\}.$$

As a memory aide, the symbol U^n is henceforth reserved for the unit simplex in $\Re^n$. The jth vertex of U^n is the vector e^j, and the jth facet of U^n is the set of vectors in U^n that have $x_j = 0$.

The case n = 3

Let us return to the case n=3, in which case the unit simplex is the set of all convex combinations of the vectors $e^1 = (1, 0, 0)$, $e^2 = (0, 1, 0)$ and $e^3 = (0, 0, 1)$. Figure 16.2 displays the vertexes and facets of U^3, along with information that will help us to subdivide it.

Figure 16.2. The unit simplex U^3.

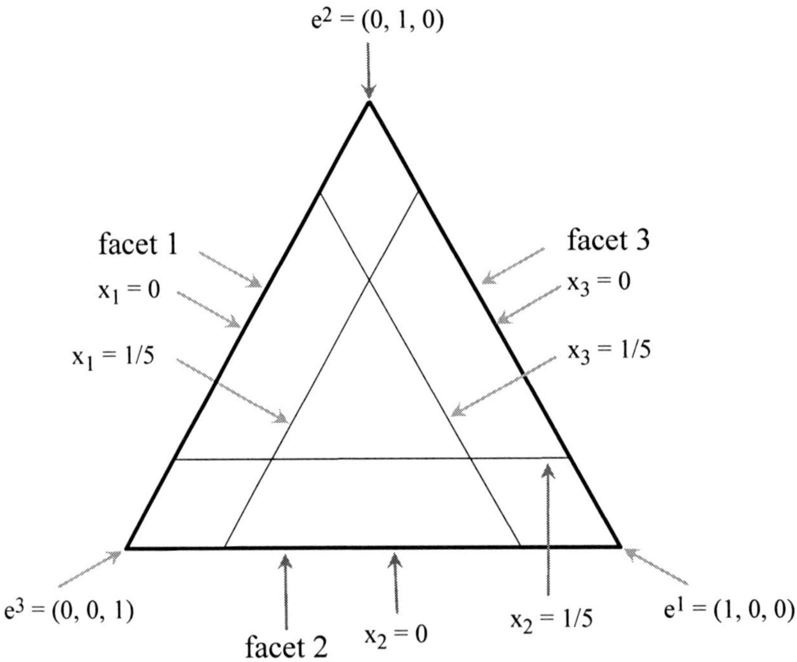

Figure 16.2 depicts U^3 as the triangle whose boundary is outlined thickly. The vertex $e^1 = (1, 0, 0)$ is located in the lower right-hand corner of U^3, across from facet 1, which is the line segment on which $x_1 = 0$. The vertexes $e^2 = (0, 1, 0)$ and $e^3 = (0, 0, 1)$ appear at the other two corners of this triangle, across from the facets on which $x_2 = 0$ and on which $x_3 = 0$. A line segment in Figure 16.2 identifies the points in U^3 for which $x_1 = 1/5$. Other

line segments in Figure 16.2 identify the points in U^3 that have $x_2 = 1/5$ and $x_3 = 1/5$.

A subdivision

Figure 16.3 uses a *grid* to subdivide this unit simplex by expressing it as the union of 25 smaller simplexes. This grid is formed by the line segments in U^3 in which each variable equals the values 1/5, 2/5, 3/5 and 4/5.

Figure 16.3. Subdividing the unit simplex U^3.

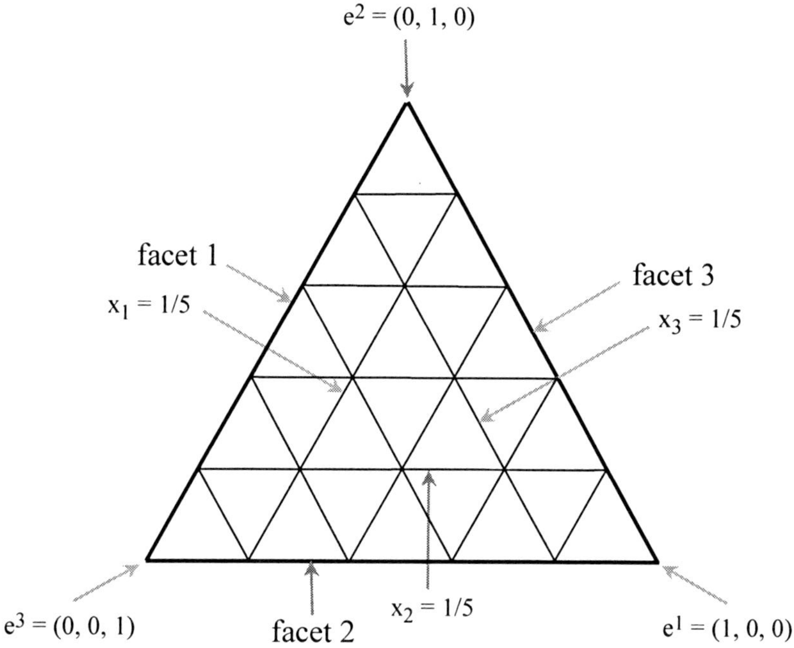

One of the 25 smaller simplexes in Figure 16.3 is the subset of U^3 in which $x_1 \geq 1/5$ and $x_2 \geq 1/5$ and $x_3 \geq 2/5$. Can you see which of the smaller simplexes it is?

Consider any two of these 25 smaller simplexes. Their intersection may be empty. If their intersection is not empty, it is a face of both.

Each of the 25 small simplexes in Figure 16.4 has three vertexes, but many of these vertexes are shared. The 25 small simplexes have a total of 21 vertexes because $21 = 1 + 2 + \cdots + 6$.

Each of the 25 small simplexes in Figure 16.4 has three facets, but many of these facets are shared. These small simplexes have a total of 45 facets. Exactly 15 of these facets lie on the boundary of U^3, and each of the other facets is shared by *exactly two* of the small simplexes.

Labeling vertexes

Let us return to the general definition of a simplex. The simplex S that is given by (8) is the set of all convex combinations of a collection $\{v^1, v^2, \ldots, v^k\}$ of affinely independent vectors in $\Re^n$. This set S is convex. Its extreme points are the vectors v^1 through v^k. *Facet j* of this simplex consists of each point x in S for which (8) holds with $x_j = 0$.

Denote as Ω a simplicial subdivision of this simplex, S. A **labeling** of Ω is an assignment of an integer between 1 and k, inclusive, to each vertex of each simplex in Ω. This labeling is arbitrary, except that it is required to satisfy the

> **Border Condition** for a simplicial subdivision: If a vertex of a simplex in Ω lies on the boundary of S, the label of that vertex equals the label of a facet of S that contains it.

Figure16.4 exhibits a labeling that satisfies this Border Condition. Each of its 25 vertexes is labeled with the integer 1, 2 or 3. The vertex at the lower right is in facets 2 and 3 of the unit simplex (the line segments $x_2 = 0$ and $x_3 = 0$), so its label could be 2 or 3, but not 1. Similarly, the vertex at the top could be labeled 1 or 3, but not 2. In this case and in general, the vertices that are not in facets of the unit simplex can have any labels.

Paths

A simplex in Ω is said to be **completely labeled** if its vertexes have different labels. An argument that is familiar from Chapter 15 will be used to show that at least one of the small simplexes in Figure 16.4 must be completely labeled.

The small simplexes in the corners of Figure 16.4 are missing one label apiece. We could focus on any one of the corners. Let's choose the small simplex in the lower right-hand corner. Its vertices omit the label 1. To identify a path to a completely-labeled simplex, in Ω, we:

Figure 16.4. A labeling that satisfies the Border Condition.

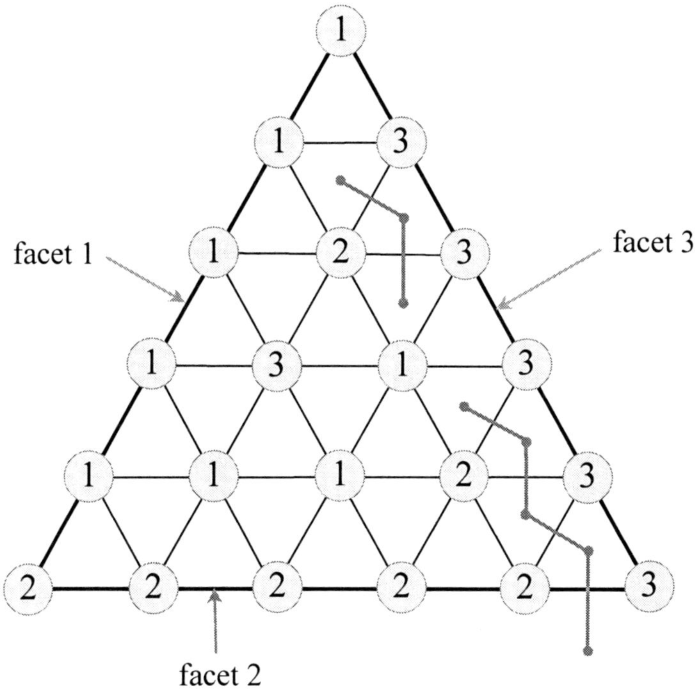

- Call each of the 25 small simplexes a *room*. The *mansion* is the set of all rooms.
- Color a room *blue* if its vertexes have the labels 1, 2 and 3.
- Color a room *green* if its vertexes have labels 2 and 3, but not 1.
- If a facet of a room contains the labels 2 and 3, create a *door* in that facet.
- Note that there is only one door to the outside of the mansion.
- Note that each green room has 2 doors.
- Note that each blue room has 1 door.
- Begin outside the mansion, enter it through its only door to the outside and leave each green room by the door through which you did not enter. This must lead you to a blue room.

- If there is a second blue room, leave it by its only door and then leave each green room by the door through which you did not enter. This must lead to a third blue room. And so forth.

The preceding argument demonstrates that Ω contains an odd number of completely-labeled simplexes. In particular, at least one simplex in Ω is completely labeled. This argument is *constructive*; it shows how to find a completely labeled simplex. Displayed in Figure 16.4 is the path from the green room in the lower right-hand corner to a completely-labeled subsimplex.

The same argument works if we start in any corner. The room at the top of Figure 16.4 has vertexes that are labeled 1 and 3, but not 2. To start there, paint each room green if its vertices have the labels 1 and 3, but not 2. Create a door in each facet of each room that bears the labels 1 and 3. Then enter the mansion as before and leave each green room by the door through which you did not enter. This must lead to a blue room. For the labeling in Figure 16.4, this walk leads to a different blue room (completely labeled simplex).

This argument is startlingly familiar. It was used by Lemke and Howson in their analysis of a bi-matrix game. This argument leads directly to algorithms that approximate Brouwer fixed points. There is a complication, however.

A complication

This general situation is more subtle than the example in Figure 16.4 suggests. A grid has been used to subdivide the unit simplex in $\Re^3$ into smaller simplexes. What about the unit simplex in $\Re^4$? A grid can be used to partition the unit simplex in $\Re^4$ into smaller simplexes, but the partitioning is not unique. The difficulties are compounded in higher dimensions. To partition the simplex in $\Re^n$, we would need to delve into algebraic topology. That can be avoided, and we shall.

The unit simplex in 4-space*

For the reader who is curious about subdivisions in higher dimensions, this starred subsection is offered. The unit simplex in $\Re^3$ has three extreme points. It is a triangle, and it can be represented on the plane.

The unit simplex in $\Re^4$ has four extreme points. It is a tetrahedron. It can be represented in 3-space. Figure 16.5 depicts the unit simplex in $\Re^4$. Its vertices are numbered 1 through 4, and each of its six edges is represented as a thick line segment. Figure 16.5 also sets the stage for subdividing this tetrahedron by bisection. The mid-points of its edges are labeled a through f.

Figure 16.5. Subdividing a tetrahedron, step 1.

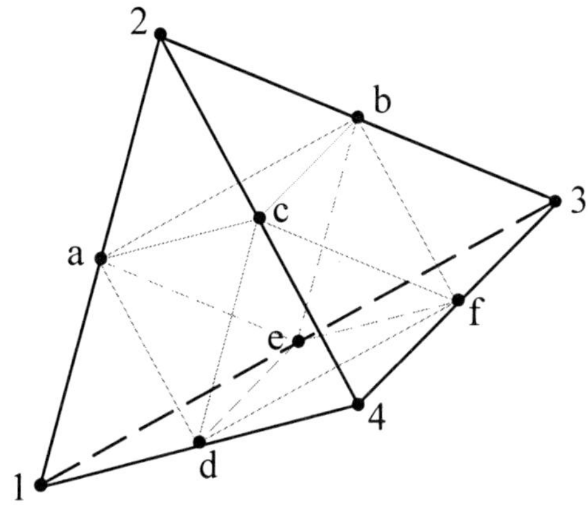

Our goal is to subdivide the tetrahedron in Figure 16.5 into smaller tetrahedrons. Figure 16.5 identifies four of these smaller tetrahedrons, one in each "corner." One of these small tetrahedrons has the vertex set {2, a, b, c}. Another has the vertex set {1, a, d, e}. A third has vertex set {3, b, e, f}. There is one other. The thin grey lanes in Figure 16.5 indicate what remains after the removal of these four small tetrahedrons.

Figure 16.6 reproduces what remains after removal of the four "corner" tetrahedrons. The convex body in Figure 16.6 has eight facets, rather than four. It is an octahedron, not a tetrahedron.

Figure 16.6. Subdividing a tetrahedron; Step 2.

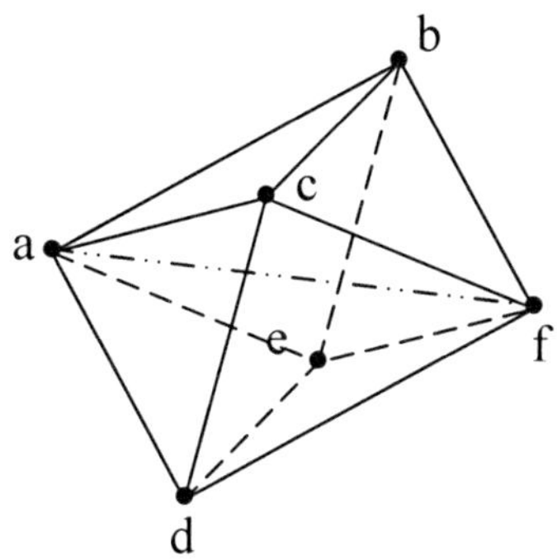

To partition the octahedron in Figure 16.6 into tetrahedrons, we could "connect" any pair of vertices that do not have an edge in common. There are three such pairs. In Figure 16.6, vertices a and f have been connected by a dashed line segment. This effects a partition of the octahedron into *four* tetrahedrons; each of which has a and f among its vertices: The vertices of one of these tetrahedrons are the set {a, f, b, c}. Another tetrahedron has the set {a, f, b, e} of vertices. Can you identify the other two?

Bisection has expressed the unit simplex in $\Re^4$ as the union of eight smaller simplexes. The set Ω that consists of these eight small simplexes is a subdivision because the intersection of any two of the small simplexes either is empty or is a face of both. Rather than subdividing simplexes in n-space, we will pursue a different approach.

7. Subdivision via Primitive Sets

A different subdivision expresses a simplex as the union of finitely many "primitive sets." These primitive sets *overlap*, rather than share boundaries. A label is assigned to each *facet* of each primitive set, rather than to each vertex. A familiar algorithm will be used to find a primitive set whose facets have different labels.

Primitive sets will be introduced in the context of the unit simplex, U^n. This simplex consists of each n-vector x whose entries are nonnegative numbers that sum to 1. This simplex has n vertexes and n facets. As in the prior section, *facet j* is the subset of U^n on which $x_j = 0$.

A set of distinguished points

Let us begin by placing a (potentially-large) number J of **distinguished points** in this unit simplex. These distinguished points are numbered w^1 through w^J. Each of them is (and must be) an n-vector whose entries are nonnegative numbers whose sum equals 1. The locations of these distinguished points are arbitrary, except that they are required to satisfy the

> **Nondegeneracy Hypothesis:** All coordinates of the points w^1 through w^J and positive, and no two points have any coordinate in common.

The Nondegeneracy Hypothesis guarantees that no distinguished point lies in any facet of the unit simplex, U^n, and that *at most one* of them lies in the line on which $x_j = c$ for any integer j between 1 and n and for any constant c between 0 and 1. The Nondegeneracy Hypothesis should be regarded as an expository device. In this setting, as in so many others, degeneracy can be worked around.

Primitive sets

The set $\{w^1, w^2, \ldots, w^J\}$ of distinguished points will be used to express the unit simplex as the union of smaller simplexes. A subset T of the unit simplex U^n is now called a **primitive set** if the following three properties are satisfied:

- With $a = (a_1, a_2, \ldots, a_n)$ as an n-vector whose entries a_1 through a_n are nonnegative numbers whose sum is less than 1, the set T is given by

$$T = \{x \in U^n : x_i \geq a_i \text{ for } i = 1, \ldots, n\}. \tag{10}$$

- No distinguished point lies in the interior of T.

- If a_k is positive, exactly one distinguished point lies in the kth facet of T, and it lies in the interior of this facet.

An illustration

The definition of a primitive set is wordy, but a picture will make everything clear. Figure 16.7 represents the unit simplex U^3 as a triangle. In this figure, six distinguished points are represented as black dots. There may be other distinguished points, but the others appear in the unshaded parts of U^3, and their dots are omitted from Figure 16.7. Each of the shaded triangles in Figure 16.7 has these properties:

- No distinguished point lies inside any of these triangles.

- Exactly one distinguished point lies inside each facet of each shaded triangle that does not lie in a facet of the unit simplex.

As a consequence, each of the shaded triangles is a primitive set. Note that each primitive set has the same shape and orientation as does the unit simplex.

Figure 16.7. Three primitive sets in the unit simplex, U^3.

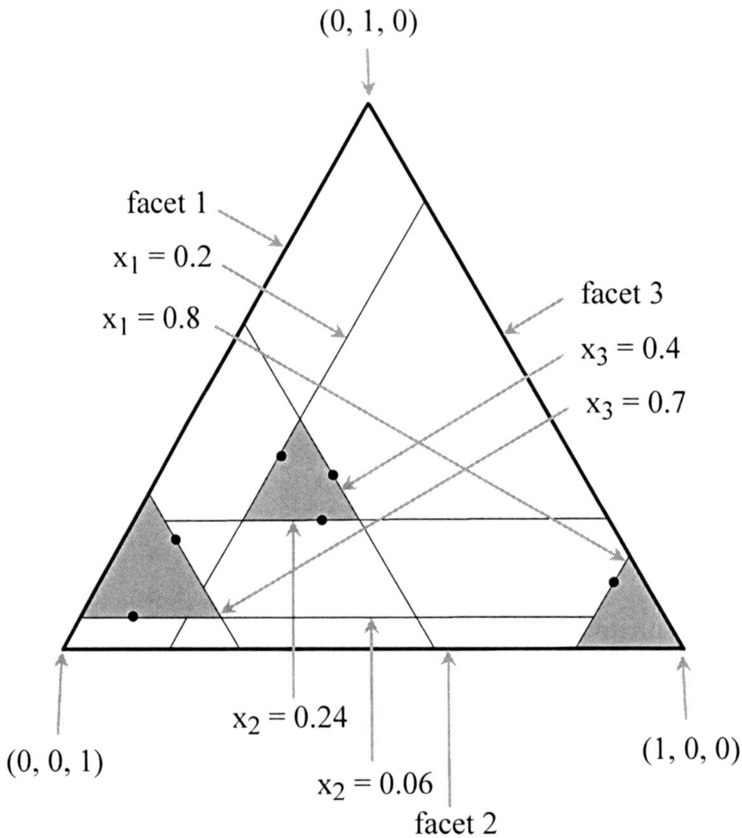

A proper labeling

A **label** is now assigned to each of the J distinguished points; each of these labels is an integer that lies between 1 and n, and each of these labels is arbitrary. The labels of the distinguished points are used to assign a label to each facet of each primitive set by these rules:

- If a facet of a primitive set lies within the kth facet of the unit simplex U^n, that facet receives the label k.
- If a facet of a primitive set includes a distinguished point, that facet receives the label of that distinguished point.

A labeling that conforms to the above rules is said to be **proper**. A proper labeling is required to satisfy the

Border Condition for primitive sets: A facet of a primitive set that is contained in the kth facet of the unit simplex U^n receives the label k.

A pivot scheme

A primitive set is said to be **completely labeled** if its facets bear the labels 1 through n, equivalently, if no two of its facets have the same label. A familiar argument will demonstrate that each proper labeling has a completely labeled primitive set. This argument will identify a path from a corner of U^n to a completely labeled primitive set. Each pivot will shift from one primitive set to another. The facets of each primitive set that is encountered prior to termination will bear all but one of the labels 1 through n. The *same* label will be missing from each primitive set that is encountered prior to termination. The pivot scheme will terminate when it encounters a completely-labeled primitive set, i.e., a primitive set whose facets bear all n labels.

Initialization

This pivot scheme is now illustrated for the case in which n=3. Each primitive set T that it encounters is specified by (10) with specific values of a_1, a_2, and a_3. The pivot scheme will be initialized at the shaded triangle in the lower right-hand corner of Figure 16.7. To accomplish this:

- Begin with $a_2 = 0$ and $a_3 = 0$.
- Find the distinguished point x having the *largest* value of x_1, and equate a_1 to its value of x_1.
- With these values of a_1, a_2 and a_3, the set T defined by (10) is a primitive set. The facet of T on which $x_1 = a_1$ is called the **entering** facet.

For the data in Figure 16.7, this initialization step sets $a_1 = 0.8$, and it produces the shaded primitive set T in the lower right-hand corner of the figure. If the entering facet had 1 as its label, we would have encountered a primitive set that has all three labels, and the algorithm would terminate. To see how the pivot scheme proceeds, we suppose that the entering facet does not have 1 as its label. For specificity, we suppose its label equals 2, rather than 3.

A pivot

In general, prior to termination, the entering facet creates a duplication: The facet whose label it duplicates will depart on the next pivot. Its departure will prepare for another facet to enter, and another pivot to occur. How the first pivot occurs will be illustrated in the context of Figure 16.8.

Figure 16.8. The first pivot.

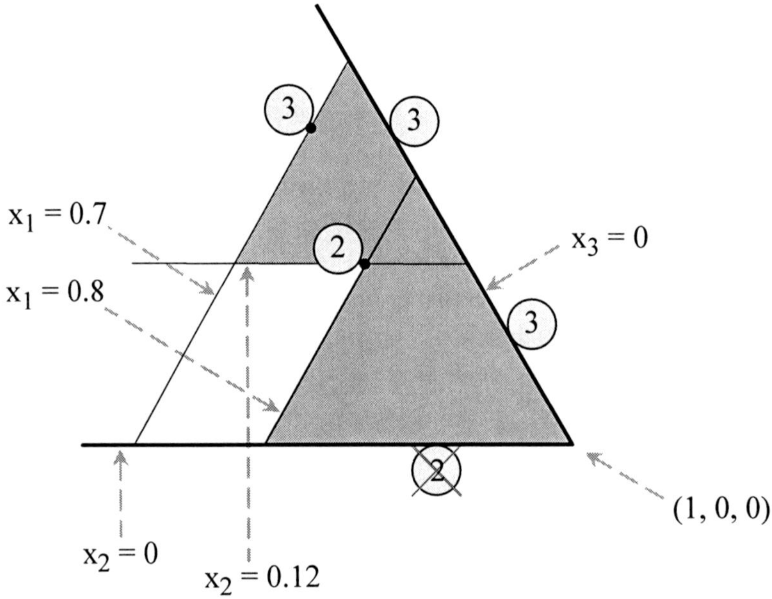

The shaded portion of Figure 16.8 consists of two overlapping triangles. This pivot scheme has been initialized at the shaded triangle T at the lower right portion of Figure 16.8. The first pivot will occur to the shaded triangle at the upper left of this figure.

The leaving facet

The initial primitive set T is given by expression (10) with $a_2 = 0, a_3 = 0$ and $a_1 = 0.8$. The entering facet lies in the interval $x_1 = 0.8$, and it bears the label 2. The other facet that bears the label 2 is $x_2 = a_2$ (currently, $a_2 = 0$), and this facet will **leave.** (In Figure 16.8, the leaving facet has an $\times$ on its label.) To cause this facet to leave:

- *Increase* x_2 from a_2 to the smallest number c that exceeds a_2 and for which the line segment $x_2 = c$ includes a point in a facet of T.
- Equate a_2 to this value of c. (For the data in Figure 16.8, the value of a_2 increases from 0 to 0.12.)

The orientation of the facet containing the point $x=(0.8, 0.12, 0.08)$ has just *shifted* from the facet having $x_1=0.8$ to the facet having $x_2=0.12$.

The entering facet

Two facets of the primitive set that results from the first pivot have now been identified. These facets lie in the intervals $x_2 = a_2$ and $x_3 = a_3$ with $a_2 = 0.12$ and $a_3 = 0$. A facet on which x_1 equals some constant has yet to be specified. This is accomplished in a familiar way:

- Among those points x having $x_2 > a_2$ and $x_3 > a_3$, find the distinguished point having the *largest* value of x_1, equate a_1 to its value of x_1.
- Denote as T the resulting primitive set, and label the facet of T on which facet $x_1=a_1$ as the new **entering** facet.

This pivot results in the shaded triangle in the upper left portion of Figure 16.8. The entering facet bears the label 3. The other facet that bears the label 3 has $x_3=0$, and it will leave on the 2nd pivot.

A later pivot

Figure 16.9 illustrates a subsequent pivot. The primitive sets encountered before and after this pivot overlap and are shaded. The primitive set that was encountered before this pivot appears toward the bottom. Just prior to this pivot, the triangle T is given by (10) with $a_1=0.2$, with $a_2=0.24$ and with $a_3=0.4$, and facet that entered had $x_1=0.2$. This facet's label is 3, and the other facet having the label 3 is crossed out, to record the fact that it will leave. This pivot occurs just like the first one:

- The *leaving facet* is in the interval on which $x_2 = 0.24$, and its departure causes a_2 to increase from 0.24 to the smallest value c for which the interval $x_2 = c$ includes a distinguished point in a facet of T. For the data in Figure 16.9, a_2 increases to 0.36.
- The facet that includes this distinguished point has shifted its orientation from the interval in which $x_3 = 0.4$ to the interval in which $x_2 = 0.36$.
- Since the facet on which $x_3 = 0.4$ has departed, we search among the distinguished points x having $x_1 > a_1$ (currently, $a_1 = 0.2$) and $x_2 > a_2$

(currently, $a_2 = 0.36$) for the one having the largest value of x_3. For the data in Figure 16.9, this distinguished point has $x_3 = 0.25$. The *entering facet* lies in the interval on which $x_3 = 0.25$. This facet bears the label 3, so the other facet having the label 3 will leave on the next pivot.

Figure 16.9. A subsequent pivot.

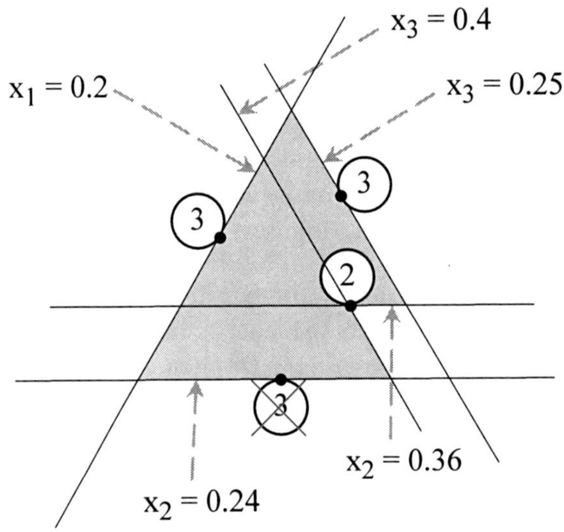

General discussion

Our attention turns to the unit simplex U^n in $\Re^n$. Pick any set of distinguished points in U^n that satisfies the Nondegeneracy Assumption. Label each distinguished point with an integer between 1 and n, inclusive, and consider the proper labeling of the facets of the primitive sets that is determined by these labels.

We could begin in any "corner" of U^n. For specificity, let us begin with the primitive set whose entering facet contains the point having the largest value of x_1 and whose other facets are contained in facets 2 through n of the unit simplex. To assure that at least one pivot occurs, we assume that the label of the facet that contains the largest value of x_1 does not equal 1. This primitive set has no facet whose label equals 1. It is the *only* primitive set that intersects the boundary of U^n and does not have a facet whose label equals 1.

The label of the facet that includes x_1 duplicates the label of some other facet in the initial primitive set. That facet will leave on the first pivot. The facet that will enter on each pivot is found by the rule that has just been illustrated. If the label of the entering facet equals 1, pivoting stops. If not, this primitive set has no facet whose label equals 1, the label of the entering facet duplicates one other label, and that label departs on the next pivot.

Termination

Let us ask ourselves the rhetorical question, "What can happen when this pivot scheme is executed?"

Proposition 16.6. Given the unit simplex U^n in $\Re^n$, consider the family of primitive sets that is determined by any set $\{w^1, w^2, \ldots, w^J\}$ of vectors in U^n that satisfies the Nondegeneracy Hypothesis. For any proper labeling of the facets of these primitive sets, the pivot scheme that is described in the prior subsections terminates after finitely many iterations, and it terminates by finding a primitive set that is completely labeled.

Proof. Let us call each primitive set a room. Let us color a room (primitive set) blue if its facets contain all n labels. Let us color a room green if its facets contain the labels 2 through n but not 1.

Begin with the room whose facets are contained in facets 2 through n of U^n. If that room is blue, there is nothing to prove. Suppose it is green. It is the only green room that intersects the boundary of U^n. Call two colored rooms *adjacent* if it is possible to shift from one to the other with a single pivot. The green room at which pivoting begins is the only green room that is adjacent to one other room. Every green room other than it is adjacent to exactly two other rooms. Pivoting cannot revisit a room because the first room revisited would need to be adjacent to three others, and none are. There are finitely many rooms, so pivoting must stop. It must stop by encountering a blue room, namely, a room whose facets bear the labels 1 through n. That room (primitive set) is completely labeled. ■

The proof of Proposition 16.6 is identical in structure to the proof in Chapter 15 that the complementary pivot scheme works. Here, as there, it is vital that pivoting begin at a particular spot. If pivoting were initiated at a room that is not in a corner, a cycle could result.

8. Fixed-Point Theorems

Proposition 16.6 will soon be used to demonstrate that a continuous map f of a closed bounded convex subset C of $\Re^n$ into itself must have a fixed point. This will be done in three stages – first for the case in which C is a unit simplex, then for the case in which C is any simplex, and then for the general case.

The unit simplex

Let f be a continuous map of the unit simplex U^n into itself. Each vector x in U^n has $x \geq 0$ and has $x_1 + x_2 + \cdots + x_n = 1$. The fact that f(x) is in U^n guarantees $f(x) \geq 0$ and $f(x)_1 + f(x)_2 + \cdots + f(x)_n = 1$. Of necessity, the inequality

$$f(x)_j \geq x_j$$

is satisfied by at least one j.

Labeling distinguished points

Let us consider any set $\{w^1, w^2, \ldots, w^J\}$ of distinguished points in U^n. Since f maps U^n into itself, each distinguished point w^k has at least one coordinate j for which the inequality

$$f(w^k)_j \geq (w^k)_j \tag{11}$$

is satisfied. Let us assign to each distinguished point w^k the label $L(w^k)$ whose value is an integer j for which (11) holds.

Labeling primitive sets

Each facet of each primitive set is now assigned a **monotone label** by this rule:

- If a facet of a primitive set is contained in facet j of U^n, it receives the label j.
- Alternatively, if this facet contains distinguished point w^k, it receives the label $L(w^k)$.

These labels are "monotone" because each facet of each primitive set is labeled with an integer j such that the inequality $f(x)_j \geq (x)_j$ is satisfied by at least one vector x in that facet. These labels are proper because they satisfy the Border Condition.

Consider what is accomplished when Proposition 16.6 is applied to a labeling that is monotone and proper. Its algorithm ends with a primitive set T that is completely labeled. For each j, at least one vector x on the boundary of T satisfies the inequality $f(x)_j \geq (x)_j$. This set T *approximates* a fixed point of f.

Proposition 16.7. Let f be a continuous function that maps the unit simplex U^n in $\Re^n$ into itself. Then there exists at least one vector x in U^n such that $f(x) = x$.

Remark: The proof of this proposition uses material from real analysis and is starred.

Proof*. Let us consider any dense sequence $\{w^1, w^2, \ldots, w^k, \ldots\}$ of vectors in U^n such that all coordinates of each vector w^k are positive and such that no two vectors have any coordinate in common. Label each vector w^k with an integer $L(w^k)$ such that (11) holds with $j = L(w^k)$.

For each positive integer J, set $W^J = \{w^1, \cdots, w^J\}$ satisfies the Nondegeneracy Condition. The distinguished points in W^J induce a set Ω^J of primitive sets on U^n. The proper labeling that satisfies (11) is monotone. Proposition 16.6 demonstrates that there exists a primitive set T^J in Ω^J whose facets bear the labels 1 through n. Thus, for each j, there is a vector x in a facet of T^J that has $f(x)_j \geq x_j$. Denote as $\bar{x}^J$ the center (average of the vertexes) of T^J. Evidently, $\bar{x}^J$ is in U^n.

Since U^n is a closed and bounded set, the Bolzano-Weierstrass property (Proposition 17.2) shows that there exists an increasing sequence $\{m_1, m_2, \ldots\}$ of positive integers and a vector y in U^n such that the sequence $\{\bar{x}^{m_1}, \bar{x}^{m_2}, \ldots\}$ converges to y. The fact that $\{w^1, w^2, \ldots\}$ is dense guarantees that the sequence $\{T^{m_1}, T^{m_2}, \ldots\}$ of primitive sets also converges to y. For each J, the facets of T^J are completely labeled. A routine continuity argument shows that $f(y)_j \geq y_j$ for each j. The entries in y sum to 1, as do the entries in f(y), which guarantees $f(y) = y$. ■

At the heart of this proof of Proposition 16.7 lies the scheme in the prior section for finding a completely-labeled primitive set. This method offers the promise of finding an approximate fixed point quickly, by examining a small fraction of the primitive sets.

An issue

The proof of Proposition 16.7 does expose a computational issue, however. To improve upon an approximate fixed point $\bar{x}^J$, one increases J and re-applies the pivot scheme. This is accomplished by *setting aside* the current approximation, $\bar{x}^J$, imposing a finer partition, going "back to the corner," and repeating the algorithm. The information obtained with one partition is ignored when a finer partition is imposed. Remedies for this defect have been devised, and two of them are mentioned in the next section of this chapter.

A simplex

Proposition 16.7 is stated in terms of a function that maps the unit simplex U^n into itself. What about a function that maps some other simplex into itself?

Proposition 16.8. Let f be a continuous function that maps a simplex S in $\Re^k$ into itself. Then there exists at least one vector x in S such that $f(x) = x$.

Proof. This simplex S has some number n of vertexes. Label these vertexes v^1 through v^n. Because S is a simplex, each vector in S can be written in a unique way as a convex combination of the vectors v^1 through v^n. Write each $x \in S$ as the convex combination

$$x = x_1 v^1 + x_2 v^2 + \cdots + x_n v^n.$$

With S expresses in this way, the prior discussion of primitive sets applies, as does the proof of Proposition 16.7. ■

A closed bounded convex set

Proposition 16.8 applies to simplexes, rather than to closed bounded convex sets, but it lies at the heart of the proof of Brouwer's fixed point theorem that appears below.

Proof of Proposition 16.1*. As stated, Brouwer's theorem concerns a continuous function f that maps a closed, bounded convex subset C of $\Re^k$ into itself. Since C is bounded, there exists a simplex S in $\Re^k$ such that C is contained in S.

Aiming to use Proposition 16.8, we will extend the domain of f to from C to S in a way that preserves its continuity and guarantees that its range is in C. For each point $x \in S \backslash C$ and each point $y \in C$, define the function g(y, x) by

$$g(y, x) = \sum_{i=1}^{k} (y_i - x_i)^2.$$

This function g is continuous. Since C is closed and bounded, the Extreme Value theorem (Proposition 17.3) shows that C contains a point $\theta(x)$ such that $g[\theta(x), x] \le g[y, x]$ for all y in C. Since C is convex, the point $\theta(x)$ is unique, and the function $\theta(x)$ is continuous in x. Evidently, if x is in C, then $\theta(x) = x$. Finally, extend f from C to S by setting $f(x) = f[\theta(x)]$ for each x in S. This function f is continuous on S.

Proposition 16.8 guarantees the existence of a vector x in S such that f(x) = x. And, since f(x) is in C, the theorem is proved. ■

From a computational viewpoint, one part of the above proof is troublesome. That part is the extension of f from C to S. Fortunately, several important applications of Brouwer's theorem are to situations in which the function f is *defined* on a simplex, and not on a closed bounded convex set that can be embedded in a simplex.

Proposition 16.8 with simplicial subdivisions

Proposition 16.8 was based on primitive sets. An analogue can be made to work with simplicial subdivisions. When dealing with subsimplexes, "monotone" labels are assigned to the vertices, rather than to the faces. One needs to subdivide in such a way that the sequence of subsimplexes converges to a point.

9. A Bit of the History

In 1910, the Dutch mathematician and philosopher L. E. J. Brouwer (1881-1960) published the fixed-point theorem that bears his name. Later in his career, Brouwer lamented the fact his proof provided no way to compute or approximate the fixed point.

Sperner's lemma

In 1928, Emanuel Sperner (1905-1980) established a result that has long been known as **Sperner's lemma.** This result is that a simplicial subdivision that satisfies a border condition much like the one in this chapter must have a completely-labeled subsimplex. Sperner's proof of this result was existential. It offered no way to find a completely labeled subsimplex, short of enumeration. It did not respond to Brouwer's self-criticism.

Primitive sets

In 1965, Herbert Scarf introduced primitive sets and indicated how one could start in a corner and follow a path to a completely-labeled primitive set. In his 1973 monograph, written in collaboration with Terje Hansen, Scarf[4] acknowledged his debt to Lemke and Howson. It seems amazing, even now, that a method devised to find a complementary solution to a *linear* system would adapt so naturally to the distinctly nonlinear a problem of approximating a Brouwer fixed point.

Impact

Scarf's work was seminal. It opened several new avenues of exploration, three of which are briefly mentioned here. First, Curtis Eaves[5], O. H. Merrrill[6] and others provided methods for improving an approximation to the fixed point without starting over in the corner.

Second, Harold Kuhn indicated how a simplicial subdivision method he had devised in 1960 could substitute for primitive sets. It was later discerned that Kuhn's method was algorithmically identical to the method devised by Terje Hansen to circumvent the Nondegeneracy Hypothesis in Section 7 of this chapter.

---

[4] Scarf, H. E., with T. Hansen, *The Computation of Economic Equilibria*, Yale University Press, New Haven, CT (1973).

[5] Eaves, B. C., "Homotopies for computation of fixed points," *Mathematical Programming*, V. 3, pp 1-22 (1972).

[6] Merrill O. H., *Applications and extensions of an algorithm that computes fixed points of certain upper semi-continuous point to set mappings*, Ph.D. Thesis, University of Michigan, Ann Arbor, MI. (1972).

The third avenue of research responds to a shortcoming of the methodology that has been presented in this chapter. In Proposition 16.2, a Nash equilibrium was described as a fixed point of a map f of a closed bounded convex set C into itself; this set C is the *direct product* of n simplexes, each of which consists of the set of randomized strategies of a particular player. Embedding C in a simplex S and extending the map f to S is awkward. In 1982, Gerard van der Laan and Dolf Talman[7] showed how to adapt Scarf's methods to the direct product of simplexes, without embedding. By doing so, they provided an efficient way to approximate a Nash equilibrium.

10. Review

In brief, prior to the work of Lemke and Scarf, the connection between economic equilibria and fixed points had been theoretical. Equilibria could be shown to exist, but no method for computing them existed. Economists needed to rely on arguments of the sort that Brouwer had spurned. That is no longer the case. Equilibria can now be computed and studied.

11. Homework and Discussion Problems

1. Prove Proposition 16.3.

2. Prove Proposition 16.4. *Hint*: try an induction on k.

3. Prove Proposition 16.5.

4. Consider a set $\{v^1, v^2, \ldots, v^k\}$ of affinely independent vectors in $\Re^n$.

   (a) Show that the vectors $\{v^2 - v^1), \{v^3 - v^1), \{v^k - v^1)$ are linearly independent.

   (b) Conclude that $k \le n + 1$.

---

[7] van der Laan, G and A. J. J. Talman [1982], "On the computation of fixed points in the product space of unit simplices and an application to noncooperative N person games," *Mathematics of Operations Research*, V. 7, pp. 1-13.

5. Let the subset C of $\Re^3$ be a pyramid. How many vertices does it have? Describe its vertices. Which of these vertices is a linear combination of the others? Which of its vertices is not a convex combination of the others?

6. Let the subset S of $\Re^2$ be a triangle. Draw a picture that expresses S as the union of three smaller triangles that have these two properties: First, no two of the smaller triangles have a shared interior. Second, the three smaller triangles are *not* a simplicial subdivision of S.

7. Consider a tetrahedron whose vertices are labeled a, b, c and d. This tetrahedron, like any other, has 15 faces. Identify them. Which faces are vertices? Which faces are facets? Which faces are neither?

8. The octahedron in Figure 16.6 can be partitioned into 4 non-overlapping tetrahedra. each of which has b and d among its vertices. What are they?

9. Alter the labels in Figure 16.4 so that the Border Condition is satisfied, but a path leads back to the room at which it began.

10. Which of the following diagrams depict a subdivision of a simplex?

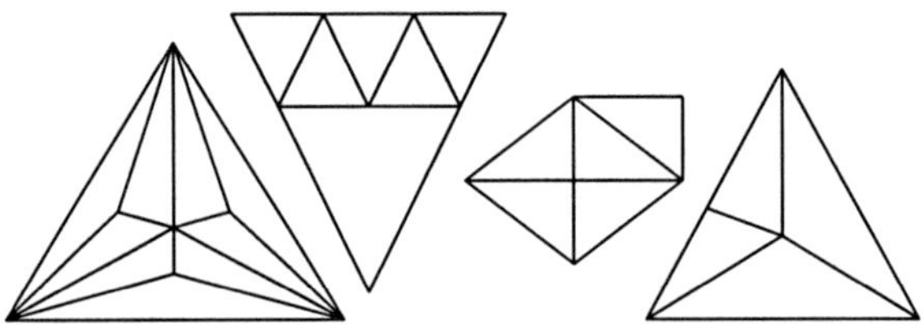

11. In Figure 16.4, create different system of doors – one in each facet of each small simplex that omits only the label 2. Does the path-from-the-outside argument continue to work? Does it lead to a different blue room?

12. In the context of Figure 16.4, suppose you start outside the mansion and follow a path to a blue room. Devise a scheme that might lead from that blue room to a different blue room.

13. *(partitioning a tetrahedron)* A tetrahedron is a three-dimensional polygon having four vertices, with an edge connecting each pair of vertices.

    (a) Draw a tetrahedron, using dashed lines to display its edges.

    (b) Bisect each edge. Identify any one of the vertices of the original tetrahedron. Use solid lines to connect the points that bisect each edge that touches this vertex. Repeat for the other three vertices.

(c) Describe the object you constructed in solid lines. How many vertices does it have? How many edges?

(d) Pick a pair of its vertices that are *not* connected by an edge. Connect them. Did you just execute a subdivision of a tetrahedron? If so, how many smaller subdivisions did you obtain?

(e) Was the partition you achieved in part (d) unique?

14. In the context of Figure 16.9, describe the next pivot. (You may wish to postulate the location of one or more points.)

15. *(A 3-player analogue of the bi-matrix game)* Suppose player 1 has m options, that player 2 has n options, and that player 3 has p options. Suppose that if players 1, 2 and 3 choose options i, j and k, they lose A_{ijk}, B_{ijk} and C_{ijk}, respectively. Each player knows the data, each player selects a randomized strategy, and each player aims to minimize the expectation of his loss.

    (a) Describe an equilibrium.

    (b) Describe an analogue of the improvement mechanism given by equations (1)-(6).

    (c) Does the proof of Proposition 16.2 adapt to this game? If so, how?

Part VI–Nonlinear Optimization

This section introduces you to the analysis of optimization problems whose objectives and constraints can be nonlinear. Even an introductory account of nonlinear programs draws upon material from multi-variable calculus, linear algebra, real analysis, and convex analysis. A coherent account of this background material appears in Chapters 17-19. Nonlinear programming is one use of this material. There are others.

Chapter 17. Convex Sets

This chapter begins with concepts that are fundamental to analysis and to constrained optimization – the dot product of two vectors, the norm of a vector, the angle between two vectors, neighborhoods, open sets, closed sets, convex sets, and continuous functions. Two of the key results in this chapter are the "extreme value theorem" and the "supporting hyperplane theorem."

Chapter 18. Differentiation

This chapter is focused on the derivative of a function of two or more variables. A differentiable function is shown to be "well-approximated" by a plane or, in higher dimensions, by a hyperplane. The gradient of a differentiable function is introduced and is shown to point in the "uphill" direction, if it is not zero.

Chapter 19. Convex Functions

In this chapter, convex functions are defined, and ways in which to recognize a convex function are described. A key result in this chapter is that a convex function has a supporting hyperplane at each point on the interior of its domain.

Chapter 20. Nonlinear Programs

A set of optimality conditions for a linear program are re-interpreted as the "Karush/Kuhn/Tucker" conditions (or KKT conditions) for a nonlinear program. For a nonlinear program satisfies a particular "constraint qualification," a feasible solution is shown to be a global optimum if and only if it satisfies the KKT conditions. Weaker constraint qualifications are shown to lead to weaker results. This chapter includes a sketch of the Generalized Reduced Gradient (or GRG) method, which is used by Solver and by Premium Solver to find solutions to nonlinear programs.

Chapter 17: Convex Sets

1. Preview

This chapter is focused on the properties of convex sets that are particularly relevant to nonlinear programming. Presented here are:

- Basic information about the dot product, the norm of a vector, the angle between two vectors, neighborhoods, open and closed sets, and limit points.
- The "extreme value theorem," which demonstrates that a continuous function on a closed and bounded set attains its maximum and its minimum.
- A theorem of the alternative for "closed convex cones."
- The "separating hyperplane" theorem and the "supporting hyperplane" theorems.

Throughout, geometric reasoning is used to motivate the analysis.

E. V. Denardo, *Linear Programming and Generalizations*, International Series in Operations Research & Management Science 149,
DOI 10.1007/978-1-4419-6491-5_17, © Springer Science+Business Media, LLC 2011

2. Preliminaries

The chapter begins with topics that may be familiar to you. These include the dot product of two vectors, the norm of a vector, the angle between two vectors, open and closed sets, neighborhoods, and continuous functions.

The dot product

Each vector in $\Re^n$ is dubbed an **n-vector**, and the n-vector x is denoted $x = (x_1, x_2, \ldots, x_n)$, so that x_i is the value taken by the ith entry in x. For each pair x and y of n-vectors, the **dot product** of x and y is denoted $x \cdot y$ and is defined by

(1) $$x \cdot y = x_1y_1 + x_2y_2 + \cdots + x_ny_n.$$

There is nothing new about the dot product: When A is an m×n matrix and x is an n×1 vector, the ith element in the matrix product A x equals the dot product of the ith row of A and x.

The norm

For each vector x in $\Re^n$, the **norm** of x is denoted $||x||$ and is defined by

(2) $$||x|| = \sqrt{x \cdot x} = \sqrt{(x_1)^2 + (x_2)^2 + \cdots + (x_n)^2}.$$

The norm of x can be interpreted as the **length** of the line segment between the vectors 0 and x. This definition harks back to the time of Euclid.

The angle between two n-vectors

When we speak of the **angle** between the n-vectors x and y, what is meant is the angle θ between their respective line segments, as is illustrated in Figure 17.1.

Figure 17.1. The angle θ between the n-vectors x and y.

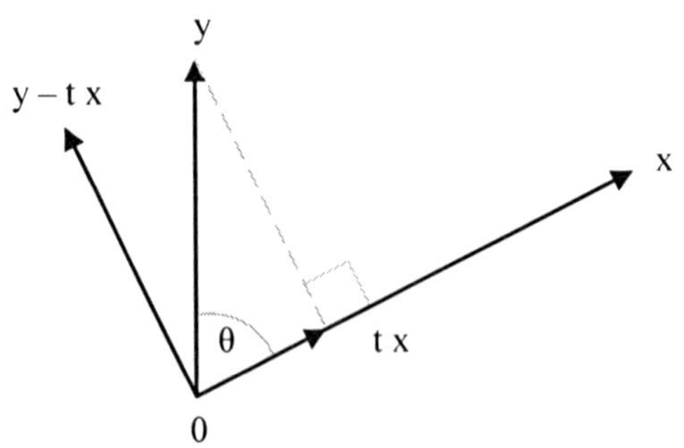

Equation (4), below, shows that cos θ is determined by the dot product of x and y and their norms. As Figure 17.1 suggests, this result is established by selecting the value of t for which the vectors tx and y – tx are perpendicular.

Proposition 17.1. Let x and y be non-zero n-vectors. For the scalar t given by

$$t = \frac{x \cdot y}{x. \cdot x}, \tag{3}$$

the vectors tx and y – tx are the sides of a right triangle, and the angle θ between the vectors x and y has

$$\cos\theta = \frac{x \cdot y}{||x.||\ ||y||}. \tag{4}$$

Proof. The number t specified by equation (3) satisfies

$$x \cdot y = tx \cdot x. \tag{5}$$

The identity y = tx + (y – tx) makes it clear that the three vectors tx and (y – tx) and y are the sides of a triangle. The Pythagorean theorem will verify that this is a right triangle whose hypotenuse is y. To employ it, we take the sum of the squares of the lengths of the vectors tx and (y – tx) and use (5) repeatedly in

$$\begin{aligned}(t\,x) \cdot (t\,x) + (y - t\,x) \cdot (y - t\,x) &= t^2(x \cdot x) + (y - t\,x) \cdot (y),\\ &= t\,x \cdot y + y \cdot y - t\,x \cdot y = y \cdot y,\end{aligned}$$

which demonstrates that y is the hypotenuse.

It remains to verify (4). To do so, we first consider the case in which t ≥ 0. In this case, cos θ equals the length of the vector tx divided by the length of y, and (3) is used in

$$\cos\theta = \frac{||t\,x||}{||y||} = \frac{t\,||x||}{||y||} = \frac{(x \cdot y)\,||x||}{(x \cdot x)||y||} = \frac{(x \cdot y)}{||x||\ ||y||}.$$

If t is negative, the above argument applies when "–1" is inserted to on the RHS of the left-most equation. ■

A key implication of Proposition 17.1 is displayed below as:

> The angle between the non-zero n-vectors x and y is:
> - Acute if $x \cdot y$ is positive.
> - Obtuse if $x \cdot y$ is negative.
> - Ninety degrees if $x \cdot y$ equals 0.

The sign of the dot product $x \cdot y$ motivates much of the algebra in this chapter.

Neighborhoods

When x is an n-vector and ε is a positive number, the symbol $B_\varepsilon(x)$ denotes the set of all n-vectors y such that the norm of y−x is less than ε. In brief,

(6) $$B_\varepsilon(x) = \{y \in \Re^n : ||y - x|| < \varepsilon\}.$$

For each positive number ε, the set $B_\varepsilon(x)$ is called a **neighborhood** of x. When n equals 3, the set $B_\varepsilon(x)$ is a "ball" that consists of those vectors y whose distance from x is below ε.

Open sets

A subset S of $\Re^n$ is said to be **open** if S contains a neighborhood of each member of S. The subset S of $\Re$ given by $S = \{x : 0 < x < 1\}$ is open. The subset T of $\Re$ given by $T = \{x : 0 < x \leq 1\}$ is not open because T contains 1 but does not contain any neighborhood of 1. The set $\Re^n$ is open, and the empty subset of $\Re^n$ is open.

Limit points and convergent sequences

A sequence $\{v^1, v^2, \ldots, v^m, \ldots\}$ of n-vectors is said to **converge** to the n−vector v if $||v^m - v|| \to 0$ as $m \to \infty$. Not every sequence of n-vectors converges, of course.

If the sequence $\{v^1, v^2, \ldots, v^m, \ldots\}$ of n-vectors converges to v, then v is called a **limit point** of this sequence. Similarly, if a sequence $\{v^1, v^2, \ldots, v^m, \ldots\}$ of n−vectors has a limit point, this sequence is said to be **convergent**. It is easy to see that a convergent sequence of vectors can have only one limit point.

Closed sets

A set S of n-vectors is said to be **closed** if every convergent sequence of elements of S converges to a vector v that is in S. In the vernacular:

A closed set contains its limit points.

The subset S of $\Re$ given by $S = \{x : 0 \leq x \leq 1\}$ is closed. The subset T of $\Re$ given by $T = \{x : 0 < x \leq 1\}$ is not closed because the sequence $\{v^1, v^2, \ldots\}$ with $v^m = 1/m$ converges to 0, which is not in T. The set $\Re^n$ is closed, and the empty subset of $\Re^n$ is closed.

A closed set is the natural environment for optimization. In Chapter 1, we optimized over closed sets. Minimizing f(x) subject to $x > 0$ is not a well-defined problem when $f(x) = 2x$, for instance.

Continuous functions

Our attention now turns to functions of n variables. Let us consider a function f that assigns to each vector v in a subset S of $\Re^n$ a number f(v). This function f is said to be **continuous** on S if $f(v) = \lim_{m \to \infty} f(v^m)$ for every sequence $\{v^1, v^2, \ldots\}$ of elements of S that converges to any vector v in S.

The definition of continuity is a bit tricky in that it is *relative* to the set S. For an example with $n = 1$, take $S = \{x : 0 < x \leq 1\}$ and let $f(x) = 1$ if x is in S and $f(x) = 0$ otherwise. This function is continuous on S, even though it is not continuous on $\Re$.

Bounded sets

A set S of n-vectors is said to be **bounded** if a number K exists such that each member v of S has $||v|| \leq K$.

Notation for n-vectors

An n-vector is described as a member of $\Re^n$ rather than as a row vector or as a column vector. The convention as concerns subscripts and superscripts of n-vectors is as follows:

- Subscripts identify the entries in an n-vector; so x_i denotes the ith entry in the n-vector x.
- Superscripts identify different n-vectors, so v^j denotes the jth n-vector, and v_i^j denotes the ith entry in the n-vector v^j.

The symbol e^j is reserved for the n-vector whose j^{th} entry equals 1 and whose other entries equal 0, so

$$e_k^j = \begin{cases} 1 & \text{if } k = j \\ 0 & \text{if } k \neq j \end{cases}.$$

Throughout this chapter, the symbol n is reserved for the number of entries in each n–vector.

3. The Extreme Value Theorem

The results in this section make use of the material that has just been discussed. The first of these results is

Proposition 17.2 *(Bolzano-Weierstrass).* Let S be a bounded subset of $\Re^n$. Every sequence $(v^1, v^2, \ldots,)$ of n-vectors in S has a subsequence that converges to some n–vector v.

Remark: This result has several proofs, none of which is truly brief. The theme of the proof offered here is to construct a nested sequence $(T_1, T_2, \ldots)$ of subsets of $\Re^n$, each of which is a closed "cube," each of which contains infinitely many members of the sequence $(v^1, v^2, \ldots,)$ and each of which has half the "width" of its predecessor.

Proof. With w as a fixed positive number and with u as a fixed n-vector, the subset T of $\Re^n$ that is defined by

$$T = \{x \in \Re^n : \max_i |x_i - u_i| \leq w/2\}$$

is called a *cube* whose *width* is w and whose *center* is u. Being bounded, S is contained in some cube T_1. Express T_1 as the union of 2^n sub-cubes each having half the width of T_1. Because there are finitely many these sub-cubes, at least one of them contains infinitely many members of $(v^1, v^2, \ldots,)$; label that sub-cube T_2. Express T_2 as the union of 2^n cubes each having half of its width, note that one of them must contain infinitely many members of $(v^1, v^2, \ldots,)$, label that cube T_3, and repeat.

Each sub-cube is closed, and the intersection of any number of closed sets is closed. Being nested, the intersection of the closed sets T_1, T_2, …, is non-empty. Because the width of T_i approaches 0, there exists exactly one n-vector v such that $\{v\} = \bigcap_{i=1}^{\infty} T_i$.

A subsequence $(v^{n(1)}, v^{n(2)}, \ldots,)$ of $(v^1, v^2, \ldots,)$ is constructed like so: Take n(1)=1. Recursively, for i=2, 3, …, pick n(i) as any element of $(v^1, v^2, \ldots,)$ that is in T_i and has $n(i) > n(i-1)$. This is possible because T_i contains infinitely many of the members of $(v^1, v^2, \ldots,)$. Note that $(v^{n(1)}, v^{n(2)}, \ldots,)$ converges to v, which completes a proof. ■

Proposition 17.2 is known as the **Bolzano-Weierstrass theorem**. It was proved by Bernard Bolzano (1781-1848) and independently (but much later) by Karl Weierstrass (1815-1897). It has many uses, which include

Proposition 17.3 *(the Extreme Value theorem).* Let S be a closed and bounded subset of $\Re^n$, and let the function f be continuous on S. Then S contains a vector v such that

$$f(v) = \min\{f(x) : x \in S\}. \tag{7}$$

Before proving the theorem, we pause to indicate what can go wrong when its hypothesis is not satisfied.

Example 17.1 *(why S must be closed).* The function f(x)=x on the open set $S = \{x : 0 < x < 1\}$ attains neither its minimum nor its maximum.

Example 17.2 *(why S must be bounded).* The function g(x)=1/x on the closed set $T = \{x : x \geq 1\}$ does not attain its minimum.

Proof. Proposition 17.2 and the continuity of f guarantee that the quantity $z^* = \inf\{f(x) : x \in S\}$ is finite. Proposition 17.2 also guarantees that there exists a convergent sequence $\{v^1, v^2, \ldots\}$ of n-vectors in S for which $f(v^m) \to z^*$. Since S is closed, there exists an element v of S such that $\{v^1, v^2, \ldots\}$ converges to v. Because f is continuous, $f(v) = z^*$. ■

Applying Proposition 17.3 to the function –f shows that a continuous function on a closed and bounded set contains its maximum as well as its minimum. In brief, the extreme value theorem demonstrates that:

A function that is continuous on a closed and bounded set S attains its largest and smallest values.

4. Convex Cones and Polar Cones

The extreme value theorem will soon be used to generalize the theorem of the alternative that was presented in Chapter 12. This generalization concerns convex cones and their "duals."

Convex cones

A subset C of $\Re^n$ is called a **convex cone** if C is nonempty and if

(8) $$(\alpha u + \beta v) \in C \qquad \text{for all } u \in C, v \in C, \alpha \geq 0, \beta \geq 0.$$

Condition (8) guarantees that this set C:

- Is convex. (Take α between 0 and 1 and $\beta = 1 - \alpha$.)
- Contains the vector 0. (By definition, C contains at least one vector u, so take $v = u$ and $\alpha = \beta = 0$.)
- Contains all nonnegative multiples of each vector in C. (Take $\beta = 0$ and α as any nonnegative number.)

This definition is widely used, but it is not quite standard. An authoritative text by Rockafellar[1] uses a slightly different definition that allows a convex cone to exclude the origin.

Examples

A convex cone need not be closed. The subset C of $\Re^2$ consisting of the origin and each vector $v = (v_1, v_2)$ having $v_1 > 0$ and $v_2 \geq 0$ is a convex cone but is not closed.

---

[1] R. Tyrrell Rockafellar, *Convex Analysis*, Princeton University Press, Princeton, NJ, 1970.

The subsets of $\Re^2$ that are closed convex cones take one of six forms, five of which are illustrated in Figure 17.2. These five are: (1) the origin, (2) a half-line through the origin, (3) a wedge-shaped region that includes the origin, (4) a line through the origin, and (5) a half-space that includes the origin. Not represented in Figure 17.2 is $\Re^2$ itself, which is a closed convex cone.

Figure 17.2. Five subsets of the plane that are closed convex cones.

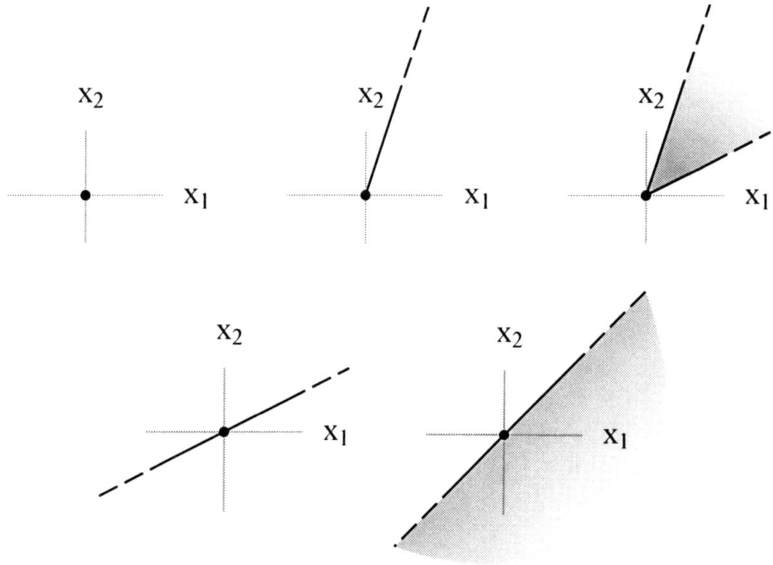

Polyhedral cones

Let A be an m×n matrix, and consider the set C of m×1 vectors given by

$$\text{(9)} \qquad C = \{Ax : x \in \Re^{n\times 1} \text{ and } x \geq 0\}.$$

This set C consists of all nonnegative linear combinations of the columns of A, and C is easily seen to be a closed convex cone. For any matrix A, the set C defined by (9) is called a **polyhedral** cone.

Polyhedral cones have a role to play in linear programming. Note that the m×1 vector b is in C if and only if there exists an n×1 vector x such that $Ax=b$ and $x\geq 0$.

Non-polyhedral cones

Figure 17.2 suggests – correctly – that each closed convex cone C in $\Re^2$ is a polyhedral cone, moreover, that it is the set of all nonnegative linear combinations of at most 3 vectors. Not every closed convex cone is polyhedral, however. Consider:

Example 17.3 *(an ice-cream cone).* The set C given by

$$C = \{x \in \Re^3 : x_3 \geq 0, (x_1)^2 + (x_2)^2 \leq (x_3)^2\}$$

is a closed convex cone, but C is not the set of nonnegative linear combinations of finitely many vectors. (The subset of C in which $x_3 \leq 6$ has the shape of an ice cream cone.)

The polar cone

Let the subset C of $\Re^n$ be a convex cone: its **polar cone** C^* is defined by

(10) $$C^* = \{y \in R^n : [c \in C] \Rightarrow [y \cdot c \leq 0]\}.$$

In geometric terms, C^* contains those vectors y that make an obtuse or right angle with every vector in C. Figure 17.3 depicts a subset of $\Re^2$ that is a closed convex cone and its polar cone C^*.

Figure 17.3. A closed convex cone C and its polar cone C^*.

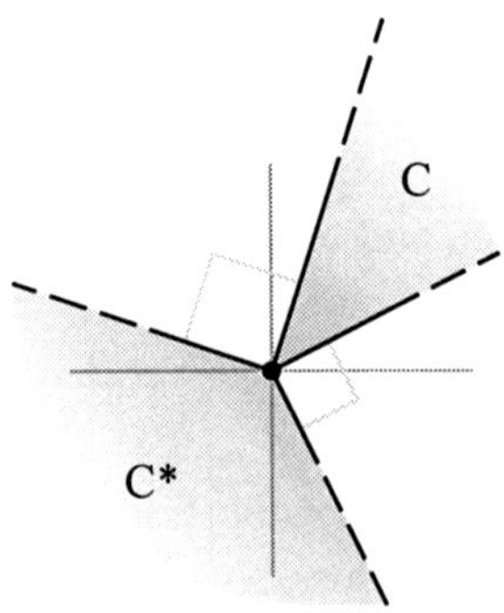

A duality theorem?

Figure 17.3 hints at a theorem. Every vector in C^* makes an obtuse or right angle with each vector in C. And every vector in C makes an obtuse or right angle with each vector in C^*. This connotes that if we begin with a closed convex cone C, take its polar cone C^*, and then take *its* polar cone $(C^*)^*$, we get C. That is correct! That this is so will soon be established. It is a corollary of the result that comes next.

A generalization

The main result of this section is a generalization of the theorem of the alternative in Chapter 12. Consider

Proposition 17.4 *(a theorem of the alternative).* Let the subset C of $\Re^n$ be a closed convex cone, and consider any vector b in $\Re^n$. Exactly one of the following alternatives occurs:

(a) The vector b is in C.

(b) There exists a vector y in C^* such that $y \cdot b > 0$.

A geometric perspective

Before proving Proposition 17.4, we pause to motivate it and to indicate how the vector y is chosen. Figure 17.4 displays a closed convex cone C and a vector b that is *not* in C. The extreme value theorem is used to identify a vector $\hat{c}$ in C that is closest to b, and then y is taken as $y = b - \hat{c}$. This vector y makes an acute angle with b.

Figure 17.4. Vectors $b \notin C$ and $y \in C^*$ have $b \cdot y > 0$.

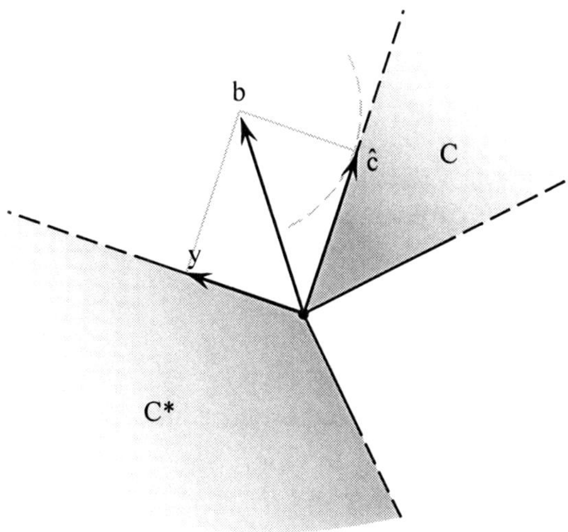

Proof of Proposition 17.4. First, suppose that (a) is satisfied, equivalently, that $b \in C$. By definition of C^*, each vector y in C^* has $y \cdot b \leq 0$, so (b) cannot hold. For the remainder of the proof, suppose that (a) is not satisfied, equivalently, that $b \notin C$. A four-step argument will show that part (b) is satisfied.

Step #1 of this argument will identify a vector in C that is "closest" to b. For each vector c in $\Re^n$, define the function f(c) by

$$(11) \qquad f(c) = (b - c) \cdot (b - c), \qquad \forall c \in \Re^n.$$

One can interpret $f(c) = ||b - c||^2$ as the "squared distance between b and c." The function f(c) defined by (11) is continuous. Aiming to use the extreme value theorem, pick any element $\bar{c}$ of C and define the set T by

$$(12) \qquad T = \{c \in C : f(c) \leq f(\bar{c})\}.$$

The intersection of two closed sets is a closed set. By hypothesis, C is closed, and T is closed because it is the intersection of the closed sets C and $D = \{c \in \Re^n : f(c) \leq f(\bar{c})\}$. The set T is bounded because D is bounded. The extreme value theorem guarantees the existence of a vector $\hat{c} \in T$ such that $f(\hat{c}) \leq f(c)$ for all $c \in T$. Hence, from (12),

$$(13) \qquad f(\hat{c}) \leq f(c) \qquad \forall c \in C.$$

Define the vector y by

$$(14) \qquad y = b - \hat{c}.$$

Step #2 of the proof will demonstrate that $y \in C^*$, equivalently, that $y \cdot c \leq 0$ for each $c \in C$. Consider any vector $c \in C$. It is immediate from (8) that the convex cone C contains $\hat{c} + \alpha c$ for every $\alpha \geq 0$. Thus, from (13)

$$(15) \qquad f(\hat{c}) \leq f(\hat{c} + \alpha c) \qquad \forall \alpha \geq 0.$$

Equations (11) and (14) give $f(\hat{c}) = (b - \hat{c}) \cdot (b - \hat{c}) = y \cdot y$ and $f(\hat{c} + \alpha c) = (y - \alpha c) \cdot (y - \alpha c)$. Substituting into (15) produces

$$(16) \qquad y \cdot y \leq (y - \alpha c) \cdot (y - \alpha c) = y \cdot y - 2\alpha c \cdot y + \alpha^2 c \cdot c.$$

In inequality (16), cancel the terms $y \cdot y$, then divide by α, and then let α decrease to zero to obtain $0 \leq -2c \cdot y$, equivalently, $0 \geq c \cdot y = y \cdot c$. This inequality holds for every $c \in C$, which shows that $y \in C^*$

Step #3 of the proof will demonstrate that $y \cdot \hat{c} = 0$. This is obvious if $\hat{c} = 0$. Suppose $\hat{c} \neq 0$. In this case, the cone C contains $c = \hat{c} + \alpha\hat{c}$ for all numbers $\alpha \geq -1$, and (13) and (15) give

$$(17) \qquad f(\hat{c}) \leq f(\hat{c} + \alpha\,\hat{c}) \qquad \forall \alpha \geq -1.$$

Step #2 with α decreasing to 0 gives $y \cdot \hat{c} \leq 0$. Repeating Step #2 with α increasing to zero reverses an inequality and gives $y \cdot \hat{c} \geq 0$. Hence, $y \cdot \hat{c} = 0$.

Step #4 will show that $y \cdot b$ is positive. Rewrite (14) as $b = y + \hat{c}$, so that $y \cdot b = y \cdot y + y \cdot \hat{c} = y \cdot y + 0 = y \cdot y$. Since $b \notin C$ and $\hat{c} \in C$, the vector $y = b - \hat{c}$ is nonzero, so $y \cdot y = ||y||^2$ is positive. In brief, $y \cdot b = y \cdot y > 0$, completing a proof. ■

The proof of Proposition 17.4 is lengthy, but it has only two themes. One theme is to use the extreme value theorem to identify a vector in C that is closest to b. The other is a "calculus trick," namely, to let α approach 0 in a way that gets rid of the quadratic term.

Farkas

Proposition 17.4 may remind you of a result from Chapter 12. That result appears here as

Proposition 17.5 *(Farkas).* Consider any $m \times n$ matrix A and any $m \times 1$ vector b. Exactly one of the following alternatives occurs:

(a) There exists an $n \times 1$ vector x such that $Ax = b$ and $x \geq 0$.

(b) There exists a $1 \times m$ vector y such that $yA \leq 0$ and $y\,b > 0$.

Proof. Given any $m \times n$ matrix A, define C by

$$C = \{Ax : x \in \Re^{n \times 1},\ x \geq 0\}.$$

This set C consists of nonnegative linear combinations of the columns of A, and C is a closed convex cone. Its polar cone C^* is easily seen to be

$$C^* = \{y \in R^{1\times m} : yA \le 0\}.$$

Thus, Proposition 17.5 is immediate from Proposition 17.4. ■

A pattern of inference

Proposition 17.5 (Farkas) had been proved in Chapter 12 as a corollary of the duality theorem of linear programming. A generalization has now been obtained from the extreme value theorem of analysis. Figure 17.5 records a pattern of inference. All but one of the logical implications in this figure has been verified. We have not shown that LP Duality can be obtained as a consequence of Farkas's theorem of the alternative. (See Problem 9 for an outline of that argument.)

Figure 17.5. A pattern of logical implication.

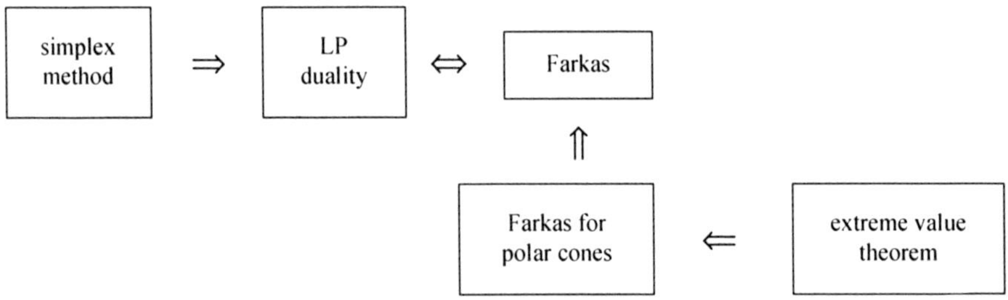

Evidently, starting with the extreme value theorem leads to deeper results than does starting with the simplex method.

5. A Duality Theorem

The duality theorem suggested by Figure 17.3 is now established as a direct consequence of Proposition 17.4.

Proposition 17.6. Let the subset C of $\Re^n$ be a closed convex cone. Then $C=(C^*)^*$.

Proof. It will first be established that $C \subseteq (C^*)^*$. Consider any vector $c \in C$. By definition of C^*,

$$c \cdot y \le 0 \qquad \forall\, y \in C^*.$$

This expression demonstrates that $c \in (C^*)^*$, thereby showing that $C \subseteq (C^*)^*$.

It remains to demonstrate that $C \supseteq (C^*)^*$. Consider any n-vector $b \notin C$. Proposition 17.4 guarantees the existence of a vector $y \in C^*$ such that $y \cdot b > 0$. This shows that $b \notin (C^*)^*$, hence that $C \supseteq (C^*)^*$, thereby completing a proof. ■

6. A Separating Hyperplane

Let a be a fixed nonzero n-vector, let β be a fixed number, and consider the sets

$$H = \{x \in \Re^n : a \cdot x = \beta\},$$

$$H^+ = \{x \in \Re^n : a \cdot x > \beta\},$$

$$H^- = \{x \in \Re^n : a \cdot x < \beta\}.$$

It is clear that these three sets are disjoint, that each of them is convex, that their union equals $\Re^n$. In addition, the set H is closed, and the sets H^+ and H^- are open. The set H is called a **hyperplane**, and the sets H^+ and H^- are called **open halfspaces**.

Illustration

Figure 17.6 exhibits a hyperplane in $\Re^2$. For the vector a=(2, 3) and the value β=6, it displays the hyperplane H given by

$$H = \{x : ax = \beta\} = \{(x_1, x_2) : 2x_1 + 3x_2 = 6\}.$$

Ask yourself where in Figure 17.6 the open halfspace H^+ lies.

In Figure 17.6 and in general, the hyperplane $H = \{x \in \Re^2 : a \cdot x = \beta\}$ is perpendicular (orthogonal) to the vector a. This is so because vectors x and y in H have $a \cdot x = \beta$ and $a \cdot y = \beta$, which implies $a \cdot (x - y) = \beta - \beta = 0$.

Figure 17.6. The vector a = (2, 3) and the hyperplane $H = \{x \in \Re^2 : a \cdot x = 6\}$.

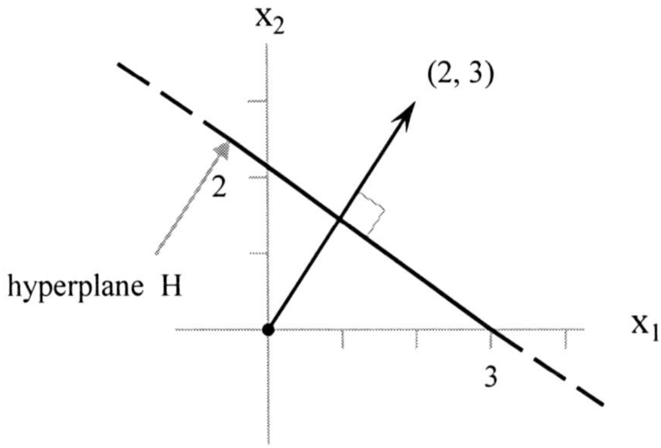

Separation

The closed sets S and T of n-vectors are said to be **separated** by a hyperplane H if the set S is contained in one of H's open halfspaces and if T is contained in the other. Some pairs of closed disjoint sets can be separated, and some pairs cannot.

If S is closed and convex and if T consists of single point that is not in S, they can be separated, as is suggested by Figure 17.7.

Figure 17.7. A closed convex set S, a vector $b \notin S$ and a separating hyperplane H.

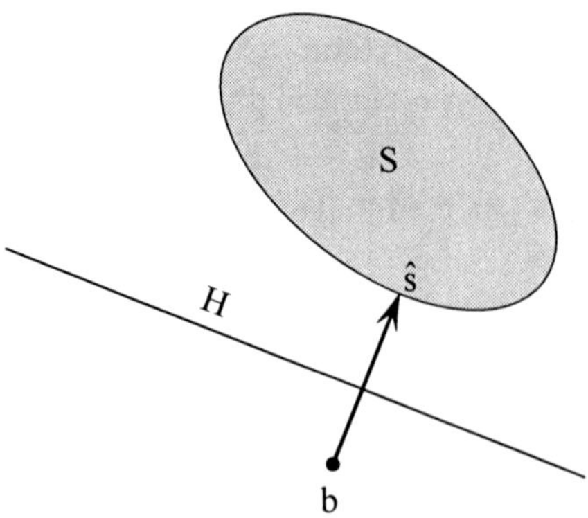

Proposition 17.7 *(a separating hyperplane).* Let S be a nonempty closed convex subset of $\Re^n$, and consider any n-vector $b \notin S$. There exists an n-vector a and a number β such that

$$(18) \qquad a \cdot b < \beta \qquad \text{and} \qquad a \cdot s > \beta \qquad \forall\, s \in S.$$

Outline of proof. As Figure 17.7 suggests, the proof of Proposition 17.7 is similar to that of Proposition 17.4 in that it begins with selection of a vector $\hat{s}$ in S that is closest to b. As before, the extreme value theorem shows that S contains a vector $\hat{s}$ having

$$(19) \qquad (\hat{s} - b) \cdot (\hat{s} - b) \le (s - b) \cdot (s - b) \qquad \forall s \in S.$$

Define the vector a by

$$(20) \qquad a = \hat{s} - b.$$

Consider any $s \in S$. Being convex, the set S contains $[(1 - \alpha)\hat{s} + \alpha s] = [\hat{s} + \alpha(s - \hat{s})]$ for each α between 0 and 1. Applying (19) with s replaced by $[\hat{s} + \alpha(s - \hat{s})]$ and then letting α decrease to zero yields

$$(21) \qquad a \cdot \hat{s} \le a \cdot s \qquad \forall s \in S.$$

Finally, since $\hat{s} \in S$ and $b \notin S$, the vector a defined by (20) is nonzero, and (20) gives $0 < a \cdot a = a \cdot \hat{s} - a \cdot b$, so

$$(22) \qquad a \cdot b < a \cdot \hat{s}.$$

Take $\beta = (a \cdot b + a \cdot \hat{s})/2$, and observe from (21) and (22) that (18) holds. ■

Proposition 17.7 is known as the **separating hyperplane theorem**. This proposition is not the most general result of its type. For instance, any pair of disjoint convex subsets of $\Re^n$ can be separated (See Problem 7.)

7. A Supporting Hyperplane

Let S be a convex subset of $\Re^n$; the vector x in S is said to be on the **boundary** of S if there exists no positive number ε such that S contains $B_\varepsilon(x)$. A famous corollary of the separating hyperplane theorem is presented as:

Proposition 17.8 *(a supporting hyperplane)*. Let S be a nonempty closed convex subset of $\Re^n$, and consider any vector $\hat{s}$ on the boundary of S. There exists a nonzero n-vector a such that

$$a \cdot \hat{s} = \min\{a \cdot s : s \in S\}. \tag{23}$$

Remark: You, the reader, are encouraged to draw the analog of Figure 17.7 that describes the supporting hyperplane.

Proof. Since $\hat{s}$ is on the boundary of S, there exists a sequence $\{b^m : m = 0, 1, \ldots\}$ of n-vectors that converge to $\hat{s}$ and none of which is in S. For each m, the separating hyperplane theorem applies to b^m, and (18) shows that there exists an n–vector a^m such that

$$a^m \cdot b^m < \inf\{a^m \cdot s : s \in S\} \qquad \text{for } m = 1, 2, \ldots. \tag{24}$$

Dividing (24) a^m by $||a^m||$ preserves the inequality, so (24) holds with $||a^m|| = 1$ for each m. Being bounded, this sequence has a convergent subsequence (Proposition 17.2); let a be its limit. Since b^m converges to $\hat{s}$, (24) guarantees $a \cdot \hat{s} \leq \inf\{a \cdot s : s \in S\}$, which establishes (23), completing a proof. ■

Proposition 17.8 is known as the **supporting hyperplane theorem**. It will play an important role in our discussion of convex functions.

8. Review

Nearly everything in this chapter is important. It is vital to understand the information presented here about the dot product, the norm, open and closed sets, limits, neighborhoods, and convergent sequences of vectors. The Extreme Value Theorem (Proposition 17.2) and the Supporting Hyperplane Theorem (Proposition 17.7) are among the most useful tools in real analysis.

This chapter is far from encyclopedic. One important topic that this chapter omits is the "implicit function theorem." It would be required for a more ambitious foray into nonlinear optimization than is found in Chapter 20.

9. Homework and Discussion Problems

1. Find the angle between the 3-vectors (1, 2, 3) and (2, –5, 1).

2. **Schwartz's inequality** is that each pair x and y of n-vectors whose entries are nonnegative satisfies

$$\left(\sum_{i=1}^{n} x_i\, y_i\right)^2 \le \left(\sum_{i=1}^{n} (x_i)^2\right)\left(\sum_{i=1}^{n} (y_i)^2\right).$$

   Supply a proof of this inequality. *Hint:* no computations are needed.

3. For any convex set S of n-vectors, define S^* by

$$S^* = \{y \in R^n : y \cdot s \le 0 \quad \forall s \in S\}$$

   (a) Is S^* closed? Is S^* a convex cone? Is $S \subseteq (S^*)^*$? Support your answers.

   (b) Which convex sets S have $S = (S^*)^*$. Why?

4. Draw the intersection of the positive orthant and the hyperplane $H = \{x : a \cdot x = 6\}$ with $a = (1, 2, 3)$.

5. Construct a convex set S of 2-vectors and a 2-vector b that is not in S for which no separating hyperplane exists.

6. Let S and T be convex subsets of n-vectors. Is the set $U = \{(s, t) : s \in S, t \in T\}$ convex? Support your answer.

7. **(separating hyperplane)**. Let S and T be disjoint closed convex sets of n-vectors. Show that there exists a hyperplane H such that S is contained in H^+ and T is contained in H^-. *Hints:* The preceding problem might help you to demonstrate that there exist $\hat{s}$ in S and $\hat{t}$ in T such $(\hat{s} - \hat{t}) \cdot (\hat{s} - \hat{t}) \le (s - t) \cdot (s - t)$ for all s in S and all t in T. Then mimic the proof of Proposition 17.7.

8. Draw a diagram that illustrates the supporting hyperplane theorem.

9. (*that Farkas implies LP Duality*). The data in the problem are the (familiar) $m \times n$ matrix A, the $m \times 1$ vector b and the $1 \times n$ vector c. Suppose that there do not exist an $n \times 1$ vector x and a $1 \times m$ vector y that satisfy

$$
\begin{array}{llr}
u: & Ax & \le b, \\
v: & & -yA \le -c, \\
\theta: & -cx + yb & \le 0, \\
& x \ge 0, & y \ge 0.
\end{array}
$$

(a) Show that must exist a $1 \times m$ vector u and a $n \times 1$ vector v that satisfy

(*) $\quad uA \ge 0, \quad u \ge 0, \quad Av \le 0, \quad v \ge 0, \quad ub < cv.$

Hint: Apply Farkas and then demonstrate that θ cannot be positive.

(b) Show that there cannot exist an $n \times 1$ vector x and a $1 \times m$ vector y such that

$Ax \le b, \quad x \ge 0, \quad yA \ge c, \quad y \ge 0.$

Hint: Farkas to (*).

(c) Use Farkas and weak duality to prove this theorem of the alternative: Either a linear program and its dual have the same optimal value or at least one of them is infeasible. *Hint:* This is immediate from part (b).

Chapter 18: Differentiation

1. Preview

The derivative of a function of one variable is familiar from college-level calculus. The derivative of a function of several variables plays an important role in nonlinear programming. Most college-level calculus books employ a particular definition of the derivative. This definition works for functions of one variable. It fails for functions of two or more variables. In this chapter, the standard definition is reviewed. Then a variant that generalizes is introduced. Its properties are explored.

Differentiation abounds with traps for the unwary. Many things that seem to be true turn out to be false. This chapter is sprinkled with examples that identify the pitfalls.

This chapter draws upon Chapter 17. Before tackling this chapter, you should be familiar with the norm, the dot product, the angle between two vectors, neighborhoods, and open sets.

E. V. Denardo, *Linear Programming and Generalizations*, International Series in Operations Research & Management Science 149,
DOI 10.1007/978-1-4419-6491-5_18,

2. A Definition of the Derivative

Let us begin with a definition from introductory calculus. A function f of one variable is said to be **differentiable at x** if f is defined in a neighborhood of x and if there exists a number y such that

(1) $$y = \lim_{\varepsilon \to 0} \left\{ \frac{f(x+\varepsilon) - f(x)}{\varepsilon} \right\},$$

where "$\varepsilon \to 0$" is code for any sequence of numbers that approach 0. In order for f to be differentiable at x, the *same* limit y must be obtained in (1) for every sequence of numbers that approaches zero, and y must be a number, rather than $+\infty$ or $-\infty$. If f is differentiable at x, the number y for which (1) holds is called the **derivative** of f(x) at x and is denoted $f'(x)$.

Can a function be continuous without being differentiable? Yes. The function $f(x) = \max\{0, x\}$ is continuous but is not differentiable at 0, for instance.

If a function is differentiable at x, must it be continuous at x? Yes. To verify that this is so, consider any function f that is differentiable at x and observe from (1) that

(2) $$f(x+\varepsilon) - f(x) = \frac{f(x+\varepsilon) - f(x)}{\varepsilon} \cdot \varepsilon \to f'(x) \cdot 0 = 0.$$

A function f of one variable is said to be **differentiable on the set S** if S is an open subset of $\Re$ and if f is differentiable at each point x in S.

A discontinuous derivative

A differentiable function can have derivative that is discontinuous. Witness

Example 18.1. A classic example of a function f with a discontinuous derivative is

(3) $$f(x) = x^2 \sin(1/x).$$

Let us examine how f(x) behaves as x approaches 0. Recall that the function sin(x) of x oscillates with a period of 2π, specifically, that every real number x has $\sin(x + 2\pi) = \sin(x)$. Recall also that sin(x) takes values between +1 and –1. Hence, as x approaches zero, sin(1/x) oscillates between +1 and –1 with increasing rapidity. The function f(x) given by (3) damps this oscillation by the factor x^2, which guarantees that f(x) is differentiable at 0 and that $f'(0) = 0$. Indeed, f(x) is differentiable everywhere, and the chain rule verifies that

$$f'(x) = \begin{cases} 0 & \text{for } x = 0 \\ 2x \sin(1/x) - \cos(1/x) & \text{for } x \neq 0 \end{cases},$$

which is not continuous at 0.

Rolle's theorem

Michel Rolle [1652-1719] was an early critic of calculus, much of which had seemed to him to be based on unsound reasoning. But he discovered a theorem that places much of calculus on a sound footing. This theorem now bears his name.

Proposition 18.1 *(Rolle's theorem).* Let the function f(x) of the variable x be continuous on the interval $a \le x \le b$, with $f(a)=f(b)=0$, and suppose that f(x) is differentiable on the interval $a < x < b$. Then there exists at least one number y that satisfies

$$a < y < b \quad \text{and} \quad f'(y) = 0. \tag{4}$$

Proof. Let's first suppose that this function f has $f(w)<0$ for at least one value of w that satisfies $a<w<b$. The set $S=\{x: a \le x \le b\}$ is closed and bounded, and f is continuous on S, so the Extreme Value theorem (Proposition 17.2) guarantees that there exists an element y of S that minimizes f over this interval, i.e.,

$$f(y) \le f(z) \quad \text{for } a \le z \le b.$$

Moreover, since $f(w)<0$ and $f(a)=f(b)=0$, it must be that y lies strictly between a and b. Taking ε positive but close to zero gives

$$\frac{f(y+\varepsilon) - f(y)}{\varepsilon} \ge 0,$$

and taking ε negative and close to zero gives

$$\frac{f(y+\varepsilon)-f(y)}{\varepsilon} \le 0.$$

By hypothesis, f is differentiable at y, so the inequalities that are displayed above guarantee $f'(y) = 0$.

If f(w) >0 for some number w between a and b, applying the prior argument to the function –f establishes the desired result. Finally, if f(w)=0 for every w between a and b, the function f has $f'(y) = 0$ for a<y<b. ■

Proposition 18.1 is known as **Rolle's theorem**. It and its proof are exquisitely simple. To appreciate them, you need only recall Example 18.1.

The mean value theorem

A famous corollary of Rolle's theorem is

Proposition 18.2 *(the Mean Value theorem).* Let the function f(x) of x be continuous for a≤x≤b, and let f(x) be differentiable on the interval a<x<b. Then there exists at least one number y that satisfies a<y<b and

$$f'(y) = \frac{f(b)-f(a)}{b-a}. \tag{5}$$

Proof. Consider the function g(x) given by

$$g(x) = f(x) - f(a) - \frac{x-a}{b-a} \cdot [f(b) - f(a)].$$

Since g(a)=0 and g(b)=0, Rolle's theorem applies to g, and $g'(y) = 0$ gives $f'(y) = [f(b) - f(a)/[b-a]$. as desired. ■

Proposition 18.2 is known as the **Mean Value theorem** of calculus. You are encouraged to draw a diagram that illustrates it.

3. A Better Definition of the Derivative

Expression (1) is the classic definition of the derivative of a function of one variable, but it has an important defect: It does not generalize to functions of several variables. A different (but equivalent) definition of the derivative is now presented. The function f of one variable is now said to be **differentiable** at x if f is defined in a neighborhood of x and if there exists a number y and such that

$$\text{(6)} \qquad \lim_{\varepsilon \to 0} \left\{ \frac{f(x+\varepsilon) - [f(x) + y \cdot \varepsilon]}{|\varepsilon|} \right\} = 0.$$

Whether the denominator in (6) equals ε or |ε| makes no difference; the ratio converges to 0 with ε as its denominator if and only if it converges to 0 with |ε| as its denominator. The number y for which (6) holds is (again) called the **derivative** of f at x, and this number y is denoted f′(x). It is easy to verify that the two definitions of differentiability are equivalent – that if either holds, so does the other, and with the same number y.

An interpretation

Figure 18.1 interprets expression (6) for the case of a function f that is differentiable at x. With x fixed, this figure plots the function f(x + ε) versus ε, and it also plots the linear function g(x + ε) whose value at x equals f(x) and whose slope equals f′(x).

Figure 18.1. A differentiable function f and a linear approximation g to it.

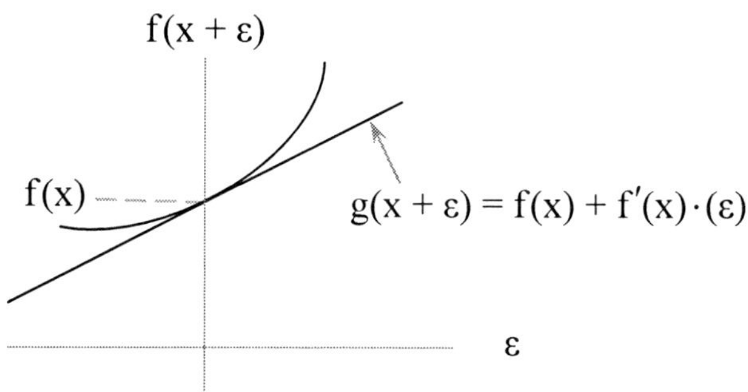

Expression (6) states that for small values of ε, the difference between f(x + ε) and g(x + ε) is small even when divided by |ε|. It is emphasized:

Differentiability, as defined by equation (6), states that the difference between f(x + ε) and the *linear approximation* [f(x) + y · ε] is so small that it approaches zero as ε approaches zero, even when divided by |ε|.

Evidently, a function of one variable is differentiable if it is well-approximated by a line. This hints at the general situation – that a function of two variables is differentiable if it is well-approximated by a plane, for instance.

A function of two or more variables

Expressions (6) motivates the definition of a derivative of a function of several variables. The real-valued function f of n variables is now said to be **differentiable** at the point x in $\Re^n$ if f is defined in a neighborhood of x and if there exists an n-vector y such that

$$\text{(7)} \qquad \lim_{||d||\to 0}\left\{\frac{f(x+d)-[f(x)+y\cdot d]}{||d||}\right\}=0.$$

In (7), the role of ε is played by the vector d, and the value that the function assigns to (x + d) is compared with the hyperplane whose slopes form the vector y. Evidently, to be differentiable is to be well-approximated by a hyperplane.

The limit

The limit in (7) must hold *no matter how* the norm of the vector d approaches zero. It is emphasized:

> For a function f of n variables to be differentiable at x, the ratio in (7) must converge to zero for *every* sequence $(d^1, d^2, \ldots, d^m, \ldots)$ of n-vectors that has $||d^m|| \to 0$.

Determining whether or not (7) holds can be onerous. A somewhat simpler test for differentiability will be provided later in this chapter.

It is easy to check that (6) and (7) coincide when n = 1. Later in this chapter, we will interpret y_i as the "partial derivative" of f(x) with respect to the variable x_i.

4. The Gradient

If the function f is differentiable at x, the unique vector y for which (7) holds is called the **gradient** of f at x and is denoted $\nabla f(x)$. To develop our understanding of the gradient, let us specialize (7) to the case in which the vector d approaches 0 in a particular "direction." To do so, we replace d by εd where d is a fixed n-vector and ε is a number that approaches 0.

Proposition 18.3. Suppose the function f of n variables is differentiable at x. There exists exactly one n-vector y for which

$$\text{(8)} \qquad y \cdot d = \lim_{\varepsilon \to 0} \left\{ \frac{f(x + \varepsilon d) - f(x)}{\varepsilon} \right\} \qquad \text{for all } d \in \Re^n,$$

and $y = \nabla f(x)$.

Proof. By hypothesis, f is differentiable at x. Set $y = \nabla f(x)$. Let d be any nonzero n-vector, apply (7) to $\hat{d} = \varepsilon d$ as $\varepsilon \to 0$, and observe that

$$\frac{f(x + \varepsilon d) - f(x) - \varepsilon y \cdot d}{||\varepsilon d||} \to 0 \qquad \text{as} \quad \varepsilon \to 0,$$

$$\frac{f(x + \varepsilon d) - f(x) - \varepsilon y \cdot d}{\varepsilon ||d||} \to 0 \qquad \text{as} \quad \varepsilon \to 0,$$

$$\frac{1}{||d||} \left[\frac{f(x + \varepsilon d) - f(x)}{\varepsilon} - y \cdot d \right] \to 0 \qquad \text{as} \quad \varepsilon \to 0,$$

which shows that (8) is satisfied by taking $y = \nabla f(x)$.

It remains to demonstrate that only $\nabla f(x)$ satisfies (8). Pick any i between 1 and n. Let e^i be the n-vector with 1 in its ith position and 0's elsewhere. Take $d = e^i$, and note from (8) that y_i must equal $\nabla f(x)_i$. ■

The right-hand side of (8) is familiar. It can be interpreted as the derivative of a function of one variable. It measures the rate of change of f as we move away from x in the fixed direction d.

A non-zero gradient

If the vector $\nabla f(x)$ is not zero, it determines both *rate of change* of f and the *direction of increase* of the function f.

Proposition 18.4. Suppose the function f of n variables is differentiable at x and that $\nabla f(x) \neq 0$. Let d be any n-vector having $||d|| = 1$. Then

(9) $$\lim_{\varepsilon\to 0}\left\{\frac{f(x+\varepsilon d)-f(x)}{\varepsilon}\right\} \leq ||\nabla f(x)||,$$

and (9) holds as an equality if and only if $d = \nabla f(x)/||\nabla f(x)||$.

Proof. Since $\nabla f(x)$ and d are nonzero n-vectors, the angle θ between them was shown in Chapter 17 to satisfy

(10) $$\cos(\theta) = \frac{\nabla f(x)\cdot d}{||\nabla f(x)||\;||d||}.$$

By hypothesis, $||d|| = 1$. Proposition 18.3 shows that the limit on the left-hand side of (9) equals $\nabla f(x)\cdot d$. Substituting gives

(11) $$\lim_{\varepsilon\to 0}\left\{\frac{f(x+\varepsilon d)-f(x)}{\varepsilon}\right\} = ||\nabla f(x)||\cos(\theta).$$

Since $\cos(\theta) \leq 1$, inequality (9) has been verified. The cosine of θ equals 1 if and only if the angle between $\nabla f(x)$ and d equals 0, and that occurs if and only if $d = \nabla f(x)/||\nabla f(x)||$, which completes a proof. ■

Proposition 18.4 identifies $\nabla f(x)$ as the direction of increase of f, and it identifies $||\nabla f(x)||$ as the rate of increase of f in that direction.

> If the gradient of a function is not zero, it *points uphill* (in the direction of increase) of that function.

This interpretation of the gradient will used again and again.

Gradients and extrema

If x maximizes a differentiable function f, it must be that $\nabla f(x) = 0$. Similarly, if x minimizes f, it must be that $\nabla f(x) = 0$. Can the gradient equal zero at points that are neither maxima nor minima? Yes. You may recall this example from high school.

Example 18.2. The function $f(x) = x^3$ is differentiable and has $\nabla f(0) = f'(0) = 0$, but f is neither maximized nor minimized at 0.

5. "Directional" Derivatives

For particular function f, the limit on the RHS of (8) may exist, and it may not exist. When the limit in

$$\lim_{\varepsilon\to 0}\left\{\frac{f(x+\varepsilon d)-f(x)}{\varepsilon}\right\}$$

exists and is finite, we call this limit the **bidirectional derivative** in the direction d. When the limit in

$$\lim_{\varepsilon\downarrow 0}\left\{\frac{f(x+\varepsilon d)-f(x)}{\varepsilon}\right\}$$

exists and is finite, we call this limit the **unidirectional derivative**, in the direction d. This terminology is *not* universally agreed upon. Let it be noted that:

- Some writers substitute **two-sided directional derivative** for bidirectional derivative and **one-sided directional derivative** for unidirectional derivative.
- Some writers use **directional derivative** in place of bidirectional derivative.

We avoid the phrase, "directional derivative." We do so because we will need to deal with convex functions; which can have unidirectional derivatives but not bidirectional derivatives.

Are bidirectional derivatives enough?

The test for differentiability is unwieldy. It requires us to check that the same limit is obtained in (7) no matter how the norm of the vector d approaches 0. Verifying (8) would be simpler because the limit is taken as the number ε approaches 0. This raises a question: If there exists a vector y that satisfies (8) for every direction d, must f be differentiable? Unfortunately, the answer is, "No."

Example 18.3. Consider the function f of two variables that has $f(0, 0)=0$ and has

$$(12) \qquad f(u, v) = \frac{2uv^3}{u^2 + v^6},$$

for all other pairs (u, v). Let us consider the behavior of this function in a neighborhood of (0, 0). For each number $v \neq 0$, we have $f(v^3, v) = 1$, so f is not continuous at (0, 0) and, for that reason, cannot be differentiable at (0, 0). On the other hand, an easy calculation verifies that this function has bidirectional derivative in each direction d at (0, 0), and these bidirectional derivatives equal 0. In other words, (8) holds with $y=0$.

6. Partial Derivatives

As was the case in Chapter 17, e^i denotes the n-vector that has 1 in its ith position and has 0's in all other positions. The real-valued function f of n variables is now said to have y_i as its **i^{th} partial derivative** at the point x in $\Re^n$ if f is defined in a neighborhood of x and if there exists a finite number y_i such that

$$(13) \qquad y_i = \lim_{\varepsilon \to 0} \left\{ \frac{f(x + \varepsilon e^i) - f(x)}{\varepsilon} \right\}.$$

This number y_i in (13) is familiar:

- y_i is the ordinary derivative of the function g(z) of the single variable z that is defined by $g(z) = f(x + e^i z)$.
- y_i is the bidirectional derivative of f in the direction e^i, evaluated at x.
- If f is differentiable at x, y_i is the ith entry in the gradient, $\nabla f(x)$.

It is emphasized:

If f is differentiable at x, its gradient $\nabla f(x)$ equals its vector of partial derivatives.

Needless to say, perhaps, a function can have partial derivatives without being differentiable.

The notation that is used to describe partial derivatives varies with the context. If f is thought of as a function of the n-vector x and if the ith partial derivative of f exists in a neighborhood, it and the value that it assigns to x are often denoted

$$\frac{\partial f}{\partial x_i} \quad \text{and} \quad \frac{\partial f}{\partial x_i}(x),$$

respectively. In the other hand, if f is regarded as a function of the (three) variables u, v and w and if its partial derivatives exist in a neighborhood of (u, v, w), then its partial derivative with respect to the second of these variables and value that this derivative assigns to the point (u, v, w) may be denoted by

$$\frac{\partial f}{\partial v} \quad \text{and} \quad \frac{\partial f}{\partial v}(u, v, w).$$

Are partial derivatives enough?

Consider a function that has partial derivatives. Must this function have bidirectional derivatives? No.

Example 18.4. Let the function f of two variables have $f(0, 0)=0$ and

$$f(x_1, x_2) = \frac{2x_1 x_2}{\sqrt{{x_1}^2 + {x_2}^2}},$$

otherwise. Clearly, $f(x_1, 0) = f(0, x_2) = 0$ so this function has partial derivatives at (0, 0), and they equal zero. Consider the direction $d=(1, 1)$. It is easy to check that $f(\varepsilon d) = f(-\varepsilon d) = |\varepsilon| \sqrt{2}$, so

$$\frac{f(\varepsilon d) - f(0)}{\varepsilon} = \begin{cases} +\sqrt{2} & \text{if } \varepsilon > 0 \\ -\sqrt{2} & \text{if } \varepsilon < 0 \end{cases} \qquad \text{with } d = (1, 1),$$

so equation (8) cannot hold, and f cannot have a bidirectional derivative at $x=0$ in the direction $d=(1, 1)$.

7. A Sufficient Condition

Is there any way to confirm that a function of several variables *is* differentiable, short of verifying that the limit in (7) holds no matter how the norm of d approaches 0? Yes, there is. Consider

Proposition 18.5. Let f map an open subset S of $\Re^n$ into $\Re$. The following are equivalent:

(a) The function f is differentiable on S.

(b) For $i = 1, \ldots, n$, the partial derivative $\dfrac{\partial f}{\partial x_i}$ exists and is continuous on S.

Proof. That (a) ⇒ (b) is obvious. We prove (b) ⇒ (a) for the special case in which S is a subset of $\Re^2$. This will reveal the general pattern of the proof with a modicum of notation.

Consider any 2–vector x in S. Since S is open, there exists a positive number r such that S contains the open ball $B_r(x)$ that is centered at x. The number r is fixed throughout this proof. By hypothesis, the partial derivatives exist and are continuous on $B_r(x)$. Consider any z in $B_r(x)$ and write $z = x + d$. We are guaranteed that $||d|| < r$. The identity,

$$\begin{aligned} f(x+d) - f(x) = {} & [f(x_1 + d_1, x_2) - f(x_1, x_2)] \\ & + [f(x_1 + d_1, x_2 + d_2) - f(x_1 + d_1, x_2)], \end{aligned}$$

parses $f(x+d) - f(x)$ into the sum of two terms, with only the 1st element of x varying in the 1st term and only the 2nd element of x varying in the 2nd term. The partial derivatives exist within $B_r(x)$, and the Mean Value theorem (Proposition 18.2) shows that there exist numbers α_1 and α_2 that lie strictly between 0 and 1 for which

$$\text{(14)} \qquad \begin{aligned} f(x+d) - f(x) = {} & d_1 \frac{\partial f}{\partial x_1}(x_1 + \alpha_1 d_1, x_2) \\ & + d_2 \frac{\partial f}{\partial x_1}(x_1 + d_1, x_2 + \alpha_2 d_2). \end{aligned}$$

Let ε be any positive number. The continuity of the partial derivatives on S guarantees that there exists a positive number δ that is a function of ε such that, for $i = 1, 2$,

$$\text{(15)} \qquad \left| \frac{\partial f}{\partial x_i}(z) - \frac{\partial f}{\partial x_i}(x) \right| < \varepsilon/2 \quad \forall \text{ z such that } ||z - x|| < \delta.$$

Define the 2-vector y by

$$\text{(16)} \qquad y_i = \frac{\partial f}{\partial x_i}(x) \qquad \text{for } i = 1, 2.$$

For each 2-vector d having $||d|| < \delta$, expressions (14)-(16) imply

$$(17) \qquad |f(x+d) - [f(x) + y \cdot d]| \le (|d_1| + |d_2|)(\varepsilon/2) \; < \; ||d|| \; \varepsilon.$$

Divide the above by $||d||$ and then let $\varepsilon \to 0$ to see that f is differentiable at x. This shows that f is differentiable on S. That the derivative is continuous on S is immediate from (16) and (17). ■

The key to Proposition 18.5 is to vary the coordinates one at a time and use the mean value theorem once per coordinate. Rolle to the rescue!

A function f is said to be **continuously differentiable** on an open set S if f is differentiable on S and if its gradient ∇f is continuous on S. Proposition 18.5 shows that a function is continuously differentiable if and only if its partial derivatives exist and are continuous on S.

8. Review

This chapter is focused on differentiation of a function of two or more variables. In order for a function to be differentiable at x, it must be defined in a neighborhood of x. This means that differentiation is defined with respect to open sets. As concerns differentiation, the key facts are:

- To be differentiable is to be well-approximated by a plane.
- A function is differentiable if it has partial derivatives *and* if they are continuous.
- The gradient of a differentiable function points in the direction of increase of that function, if it is not zero.
- Differentiation is rife with counterexamples:
  - A function can have partial derivatives without having bidirectional derivatives (Example 18.4).
  - A function can have bidirectional derivatives without being differentiable (Example 18.3).
  - A function can have a derivative that is discontinuous (Example 18.1).

This chapter is relatively brief. It presents information about differentiation that relates directly to nonlinear optimization. The test for differentiability that's given in Proposition 18.5 is less than fully satisfying because checking that the partial derivatives are continuous can be difficult. If the function is convex, there is a simpler test for differentiability, as we shall see in Chapter 19.

9. Homework and Discussion Problems

1. Does the Mean Value theorem apply to the function $f(x) = \sqrt{x}$ on the interval $0 \leq x \leq 4$? If so, draw a picture that illustrates it.

2. *(polar coordinates)* It can be convenient to express a function f(u, v) of two variables in terms of the *radius* $r = \sqrt{u^2 + v^2}$ and the *angle* θ = arc tan (v/u). Consider the function $g(r, \theta) = r \sin(2\theta)$. Does this function have partial derivatives at (0, 0)? Does it have bidirectional derivatives at (0, 0)? Support your answer.

3. *(polar coordinates, continued)* Consider the function $g(r, \theta) = r \sin(3\theta)$. Does this function have partial derivatives at (0, 0)? Does it have bidirectional derivatives at (0, 0)? Is it differentiable at (0, 0)? Support your answer.

4. Suppose that the function f of n variables is differentiable at x. Show that f is continuous at x. *Hint:* Mimic (2).

5. With the function f that is defined by (3), let g(x) = (1/2) f(x) + (1/4)f(x – 1) + (1/8) f(x – 1/2) + (1/16) f(x – 1/3) + (1/32) f(x – 2/3).

   a) Is g(x) is differentiable. Support your answer.

   b) For what values of x is $g'(x)$ discontinuous? Support your answer.

   c) Suppose (this is true) that the rational numbers can be placed in one-to-one correspondence with the positive integers. Does there exist a differentiable function h(x) whose derivative is discontinuous at every rational number? Support your answer.

6. Consider the function f of two variables that has f(0, 0)=0 and, for all other 2-vectors, has

$$f(u, v) = \frac{2uv^2}{u^2 + v^4}.$$

Is this function continuous at 0? Does it have bidirectional derivatives at 0? Is it differentiable at 0? Support your answer.

7. For what directions d does the function f given in Example 18.4 have bidirectional derivatives at $x = (0, 0)$? Support your answer.

8. Suppose that the bidirectional derivative $f(x{:}\ d)$ of f at x exists and is equal to 2.5. What can you say about $f(x{:}\ -d)$?

Chapter 19: Convex Functions

1. Preview

This chapter is focused on convex functions. The information in the first six sections is basic and is easy to master. Section 2 introduces the subject. Section 3 shows that each convex function lies on or below its "chords." Section 4 shows that each convex function satisfies "Jensen's inequality." Section 5 shows that a function is convex if and only if its "epigraph" is a convex set. Section 6 provides a variety of ways in which to determine whether or not a give function is convex.

Sections 7 through 11 describe the behavior of a convex function f at each point x in the "interior" of the set S on which it is convex. In particular:

E. V. Denardo, *Linear Programming and Generalizations*, International Series in Operations Research & Management Science 149,
DOI 10.1007/978-1-4419-6491-5_19,

- Proposition 19.9 shows that f is continuous at x.
- Proposition 19.10 shows that f has unidirectional derivatives at x.
- Proposition 19.11 shows that if f is differentiable at x, its gradient $\nabla f(x)$ is the vector of slopes of its "supporting hyperplane" at x.
- Proposition 19.12 shows that f has a "supporting hyperplane" at x even if it is not differentiable at x.
- Proposition 19.13 shows that f is differentiable at x if it has partial derivatives at x.

The set S on which a function is convex can have an empty interior. If it does, Propositions 19.9 through 19.13 seem to be vacuous. That is not so. Proposition 19.14 shows that each of these propositions holds when "interior" is replaced by "relative interior."

This chapter is sprinkled with examples of the pathologies that the analysis of convex functions must skirt.

A note of caution is appropriate. Propositions 19.9 through 19.14 have simple statements. Several of them have daunting proofs. Inclusion of these proofs provides you, the reader, with access to difficult material that plays a *minor* role in Chapter 20. An exception is Proposition 19.11. It plays a crucial role in Chapter 20. Its proof is straightforward and is well worth learning.

2. Introduction

Convex functions are closely related to convex sets. Let us recall from Chapter 17 that a subset S of $\Re^n$ is **convex** if S contains the line segment between every pair of n-vectors in S, that is, if

$$\alpha x + (1 - \alpha)y \in S \tag{1}$$

for each pair x and y of vectors in S and for every number α between 0 and 1.

A real-valued function f that is defined on a convex subset S of $\Re^n$ is said to be **convex on S** if the inequality

$$f[\alpha x + (1 - \alpha)y] \leq \alpha f(x) + (1 - \alpha)f(y) \tag{2}$$

holds for each pair x and y of vectors in S and for each number α that satisfies $0 \le \alpha \le 1$.

Geometric Insight

Figure 19.1 uses "chords" to provide geometric insight into convex functions. The horizontal axis marks the numbers a<b<c, and the vertical axis marks the values f(a), f(b) and f(c) that the convex function assigns to a, b and c. Figure 19.1 also exhibits two chords (line segments). One **chord** is the line segment that connects the pairs [a, f(a)] and [b, f(b)]. The other chord connects the pairs [b, f(b)] and [c, f(c)].

Figure 19.1 A convex function f of one variable and two of its chords.

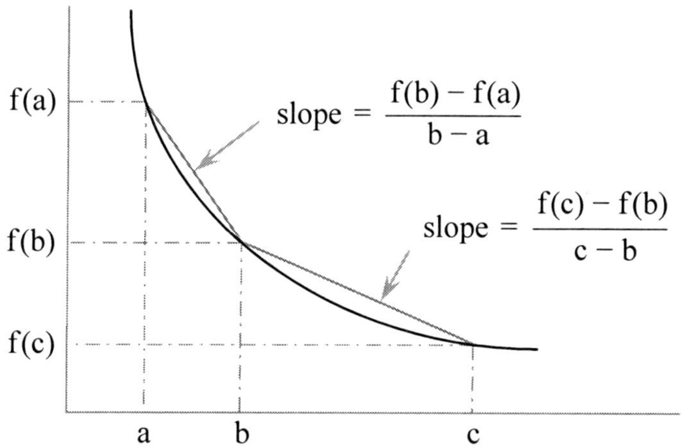

Not displayed in Figure 19.1 is the chord connecting the pairs [a, f(a)] and [c, f(c)]. This chord lies above the value that f assigns to each number x that lies strictly between a and c. Figure 19.1 suggests – correctly – that:

A convex function lies on or below its chords.

Inequality (2) need not hold strictly; a linear function is convex, for instance.

In Figure 19.1, the chord to the right has a higher (less negative) slope than the chord to the left. If a and b are close to each other, the slope of the chord that connects them approximates the derivative (if it exists) of f at, say (a + b)/2. This suggests – correctly, as we shall see – that:

> A differentiable function f of one variable is convex if and only if its derivative f'(x) cannot decrease as x increases.

If a function's first derivative can only increase, its second derivative – if it exists – must be nonnegative. In other words:

> A twice differentiable function f of one variable is convex if and only if its second derivative f''(x) is nonnegative.

Of the three properties that are highlighted above, the first is obvious, and the other two are verified in Proposition 19.5.

A mnemonic

It is handy to have a memory aid for convex functions. Here is a brief rhyme: *e to the x is convex.* This is so because the function $f(x) = e^x$ curves upward; this function equals its derivative, which increases as x increases.

Concave functions

A real-valued function f that is defined on a convex subset S of $\Re^n$ is said to be **concave** on S if the function –f is convex on S, equivalently, if the "≤" in (2) is replaced by "≥". Each property of a convex function becomes a property of a concave function when the requisite inequality is reversed. For instance, a concave function lies on or above its chords, Also, a differentiable function f of one variable is concave if its slope (derivative) f'(x) is nonincreasing.

Economic interpretation

Suppose f(x) measures the cost of acquiring x units of a good. If this cost function is convex, the marginal cost f(x + 1) – f(x) of acquiring one more unit can only go up as the quantity increases. Similarly, suppose g(x) measures the profit obtained producing x units of a good. If this profit function is concave, the marginal profit g(x + 1) – g(x) of producing one more unit can only go down as the quantity increases. Convex and concave functions are central to economic reasoning because they model increasing marginal cost and decreasing marginal profit.

Terminology

Within this book, a function f that assigns a real number to each vector x in a convex set S of n-vectors is said to be *convex on S* if f satisfies (2). As you explore the literature, you will find that some writers use different nomenclature: They extend to $\Re^n$ the domain of a function f that is convex on S by setting $f(z) = +\infty$ for each n-vector z that is not in S. In this book, functions whose values can be infinite are avoided.

3. Chords and Convexity

Let us begin by verifying a property of the chords of a convex function that is suggested by Figure 19.1, namely

Proposition 19.1. Let the function f be defined on a convex on a subset S of $\Re$. The following are equivalent:

(a) The function f is convex on S.

(b) For any numbers $a<b<c$ in S,

$$\frac{f(b) - f(a)}{b - a} \leq \frac{f(c) - f(b)}{c - b}. \tag{3}$$

Proof. First, suppose f is convex on S. The identity

$$b = \left(\frac{c - b}{c - a}\right) a + \left(\frac{b - a}{c - a}\right) c$$

couples with the convexity of f to give

$$f(b) \leq \left(\frac{c - b}{c - a}\right) f(a) + \left(\frac{b - a}{c - a}\right) f(c).$$

Multiplying the above inequality by the positive number $(c-a)=(c-b)+(b-a)$ results in the inequality $(c-b)\ [f(b)-f(a)] \le (b-a)\ [f(c)-f(b)]$, and dividing this inequality by the positive quantity $(c-b)\ (b-a)$ produces (3).

Now suppose part (b) is satisfied. Each step of the above argument is reversible, so f is convex on S. ■

In the interest of simplicity, Figure 19.1 and Proposition 19.1 have been cast in the context of a convex function of one variable. A similar property holds for a convex function of several variables, as is illustrated by Figure 19.2.

Figure 19.2 A convex function of several variables.

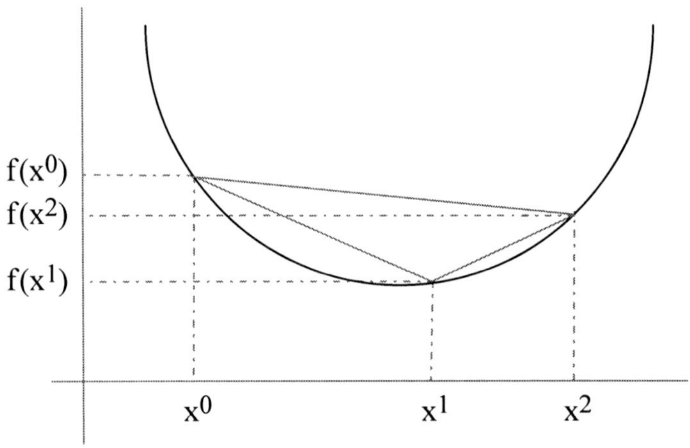

In Figure 19.2, the n-vectors x^0, x^1 and x^2 lie in the set S on which the function f is convex; the vector x^1 is reached by starting at x^0 and moving some positive number v of units in some direction d, and x^2 is reached by moving farther in the same direction, d.

Proposition 19.2. Let $S \subseteq \Re^n$ be a convex set, and let f be convex on S. Consider any n-vector d and positive numbers v and w such that S contains the n-vectors x^0, x^1 and x^2 given by

(4) $$x^1 = x^0 + vd, \qquad x^2 = x^0 + (v+w)\,d.$$

Then

$$\text{(5)}\qquad \frac{f(x^1) - f(x^0)}{v} \le \frac{f(x^2) - f(x^0)}{v + w} \le \frac{f(x^2) - f(x^1)}{w}.$$

Remarks: The inequalities in (5) mirror the relations between the slopes in Figure 19.2. The proof of Proposition 19.2 is similar to the proof of Proposition 19.1.

Proof. Equation (4) contains two expressions for d. Solving each expression for d produces

$$d = \frac{x^1 - x^0}{v} = \frac{x^2 - x^0}{v + w}.$$

Solving the above for x^1 in terms of x^0 and x^2 gives

$$x^1 = \frac{v}{v + w}x^2 + \frac{w}{v + w}x^0.$$

The numbers $v/(v + w)$ and $w/(v + w)$ are nonnegative, and they sum to 1, so the convexity of f on S gives

$$f(x^1) \le \frac{v}{v + w}f(x^2) + \frac{w}{v + w}f(x^0).$$

Multiply the above inequality by the positive number $(v + w)$ and rearrange the resulting inequality as $w[f(x^1) - f(x^0)] \le v[f(x^2) - f(x^1)]$, then divide by the product v w of the positive numbers v and w to obtain

$$\text{(6)}\qquad \frac{f(x^1) - f(x^0)}{v} \le \frac{f(x^2) - f(x^1)}{w},$$

which is one of the desired inequalities. Multiplying this inequality by the positive number w and then adding $f(x^1) - f(x^0)$ to both sides produces

$$\frac{w + v}{v}[f(x^1) - f(x^0)] \le f(x^2) - f(x^0).$$

Dividing the above by (v + w) gives

$$\frac{f(x^1) - f(x^0)}{v} \leq \frac{f(x^2) - f(x^0)}{v + w},$$

which is the second of the desired inequalities. To obtain the third, multiply (6) by v, then add $f(x^2) - f(x^1)$ to both sides and proceed as above. ■

Later in this chapter, Proposition 19.2 will be used to demonstrate that every convex function has unidirectional derivatives in the interior of its domain.

4. Jensen's Inequality

The definition of convexity requires that the value that a convex function f assigns to a convex combination of x and y cannot exceed the same convex combination of f(x) and f(y). A similar bound holds for the convex combination of three or more points.

Proposition 19.3 *(Jensen's Inequality).* Let $S \subseteq \Re^n$ be a convex set, let f be convex on S. For every finite set $\{x^1, x^2, \ldots, x^r\}$ vectors in S and for every set $\{\alpha_1, \alpha_2, \ldots, \alpha_r\}$ of nonnegative numbers that sum to 1,

$$(7) \qquad f(\alpha_1 x^1 + \alpha_2 x^2 + \cdots + \alpha_r x^r) \leq \alpha_1 f(x^1) + \alpha_2 f(x^2) + \cdots + \alpha_r f(x^r).$$

Proof. Equation (7) is trite for r = 1. When r = 2, it is true by definition of a convex function. For an inductive proof, suppose it is true for all convex combinations of r − 1 elements of S. Consider the convex combination of r vectors in S, as described in the hypothesis of Proposition 19.3. If $\alpha_r = 0$, the inductive hypothesis verifies (7). Suppose $\alpha_r > 0$ and write

$$\alpha_1 x^1 + \alpha_2 x^2 + \cdots + \alpha_r x^r = (1 - \alpha_r)\left[\frac{\alpha_1}{1 - \alpha_r} x^1 + \cdots + \frac{\alpha_{r-1}}{1 - \alpha_r} x^{r-1}\right] + \alpha_r x^r.$$

The term in brackets in the above equation is a convex combination of the vectors x^1 through x^{r-1}, hence is in the convex set S. The convexity of f gives

$$f[\alpha_1 x^1 + \alpha_2 x^2 + \cdots + \alpha_r x^r] \leq$$
$$(1-\alpha_r)\, f\left[\frac{\alpha_1}{1-\alpha_r}x^1 + \cdots + \frac{\alpha_{r-1}}{1-\alpha_r}x^{r-1}\right] + \alpha_r f(x^r),$$

and the inductive hypothesis completes a proof. ■

Proposition 19.3 is due to the Danish mathematician, Johan Jensen (1859-1925). It is known as **Jensen's inequality**. It is extremely simple, and it has a great many uses. This inequality is the key to two of the more challenging proofs in this chapter.

5. Epigraphs and Convexity

This section brings into view a relationship between convex functions and convex sets. Let S be a convex subset of $\Re^n$, and consider *any* function f that assigns a real number f(x) to each n-vector x in S. The **epigraph** of f is the set of all pairs (x, y) having $x \in S$ and $y \geq f(x)$. In brief, the epigraph of f is the subset T of $\Re^{n+1}$ given by

(8) $$T = \{(x, y) : x \in S, y \geq f(x)\}.$$

If the function f is convex on S, what can be said of its epigraph? To bring this question into view, consider

Example 19.1. The function f that is defined

(9) $$f(x) = \begin{cases} 0.5\,x & \text{for } 0 < x < 1 \\ 1 & \text{for } x = 1 \end{cases}$$

is convex on $S = \{x : 0 < x \leq 1\}$. The epigraph of this function is the convex subset T of $\Re^2$ that is depicted in Figure 19.3.

Figure 19.3 suggests – correctly – that the epigraph of a convex function is a convex set. The converse is true as well. Consider:

Proposition 19.4. Let $S \subseteq \Re^n$ be a convex set, and let f assign a real number f(x) to each element of S. The following are equivalent:

(a) The function f is convex on S.

(b) The epigraph of f is a convex set.

Figure 19.3 Epigraph T of the function f given by (9).

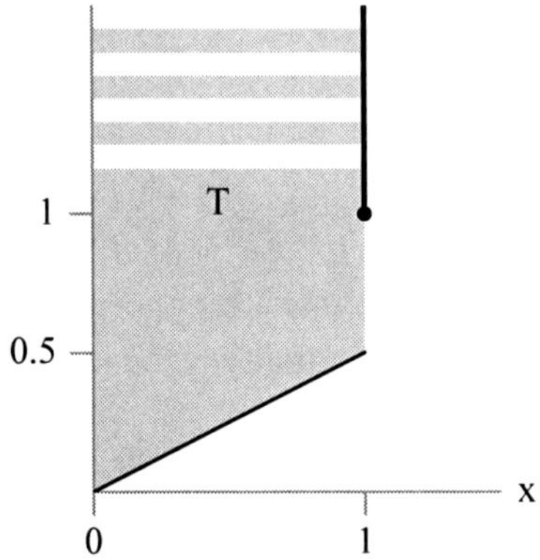

Proof. Omitted – it follows immediately from the definitions. ■

Proposition 19.4 may seem trite, but it is useful. It brings properties of convex sets to bear on the analysis of convex functions. Later in this chapter, the Supporting Hyperplane theorem (Proposition 17.8) will be used to demonstrate that a convex function has at least one support at each point in the "interior" of its domain.

6. Tests for Convexity

Described in this section are several ways in which to determine whether or not a particular function is convex.

Functions of one variable

Figure 19.1 exhibits a convex function of one variable and two of its chords. It's clear visually that as a chord shifts right-ward, its slope can only increase.

A function g of one variable is said to be **nondecreasing** on a subset S of $\Re$ if $g(a) \le g(b)$ for all members a and b of S that have $a < b$. The next proposition verifies two properties of convex functions that were highlighted earlier in this chapter.

Proposition 19.5. Let the real-valued function f be defined on an open convex subset S of $\Re$.

(a) Suppose f is differentiable on S. Then f is convex on S if and only if its derivative f' is nondecreasing on S.

(b) Suppose f is twice differentiable on S. Then f is convex on S if and only if its second derivative f' is nonnegative on S.

Proof. Part (b) is immediate from part (a).

To prove part (a), we first suppose that f' is nondecreasing on S. Consider any a and c in S and any b such that $a<b<c$. By hypothesis,

$$f(b) - f(a) = \int_a^b f'(z)dz \leq f'(b)(b - a),$$

$$f(c) - f(b) = \int_b^c f'(z)dz \geq f'(b)(c - b).$$

Divide the first inequality by $(b-a)$, divide the second by $(c-b)$ and then subtract to eliminate f'(b), obtaining

(10) $$\frac{f(b) - f(a)}{b - a} \leq \frac{f(c) - f(b)}{c - b}.$$

Proposition 19.1 and expression (10) show that f is convex on S.

Next, suppose that f is convex on S. For all triplets $a<b<c$ of elements of S, Proposition 19.1 shows that (10) holds. We let b decrease to a and conclude from (10) that $f'(a) \leq [f(c) - f(a)]/[c-a]$. Similarly, we let b increase to c and conclude from (10) that $[f(c) - f(a)]/[c-a] \leq f'(c)$. This shows that $f'(a) \leq f'(c)$, which completes a proof. ∎

Composites of convex functions

Listed below are properties of convex functions that follow directly from the definition.

Proposition 19.6. Let S be a convex subset of $\Re^n$. Then:

(a) For each number β, each n-vector a and each element $\hat{s}$ of S, the function

$$f(x) = \beta + a \cdot (x - \hat{s}) \qquad \forall\, x \in S$$

is convex on S.

(b) If f is convex on S and if β is a nonnegative number, the function

$$h(x) = \beta\, f(x) \qquad \forall\, x \in S$$

is convex on S.

(c) If f and g are is convex on S, the functions h and H given by

$$h(x) = f(x) + g(x) \qquad \forall\, x \in S$$

$$H(x) = \max\,\{f(x), g(x)\} \qquad \forall\, x \in S$$

are convex on S.

Proof. Immediate from the definition. ■

Part (a) of this proposition states that linear functions are convex. Part (b) states that convexity is preserved by multiplying a convex function by a nonnegative number. Part (c) states that the sum of two convex functions is convex and that the maximum of two convex functions is convex.

Example 19.2. From Proposition 19.5 and Proposition 19.6, we see that:

- The function $f(x) = x^3$ is convex on the set S of nonnegative numbers.
- The function $g(u, v) = \max\,\{-u, -v\}$ is convex on $\Re^2$.
- The function $f(x) = -\log(x)$ is convex on the set S of positive numbers.
- For fixed numbers a, b and c, the function $f(x) = ax^2 + bx + c$ is convex on $\Re$ if $a \geq 0$ and is concave on $\Re$ if $a \leq 0$.

Quadratic functions

A quadratic function of several variables is most easily described in matrix notation. Within this subsection, x is to be regarded as an $n \times 1$ vector. A function f of n variables is said to be **quadratic** if there exists an $n \times n$ matrix Q, a $1 \times n$ vector c and a number b such that

(11) $$f(x) = x^T Q x + cx + b.$$

Whether or not this function is convex depends on the matrix Q. An $n \times n$ matrix Q is said to be **positive semi-definite** if

(12) $$0 \leq d^T Q d = \sum_{i=1}^{n} \sum_{j=1}^{n} d_i Q_{ij} d_j \qquad \forall\, d \in \Re^{n \times 1}.$$

Condition (12) may seem to be difficult to verify, but it is not. Problem 10 suggests how to do so by a sequence of elementary row operations.

Proposition 19.7. The function f given by (11) is convex on $\Re^{n \times 1}$ if and only if the matrix Q is positive semi-definite.

Proof. The addend $cx + b$ in (11) is a linear function of x. Hence, the function f(x) given by (11) is convex if and only if $x^T Q x$ is a convex function of x. Convexity of $x^T Q x$ is equivalent to convexity on each line segment. To see whether or not $x^T Q x$ is convex, we fix any vectors x and y in $\Re^{n \times 1}$ and consider the function $g(\alpha)$ of the number α that is defined by

$$g(\alpha) = [(1 - \alpha)x + \alpha y]^T Q[(1 - \alpha)x + \alpha y].$$

Setting $d = y - x$ simplifies the formula for $g(\alpha)$ to

$$g(\alpha) = [x + \alpha d]^T Q[x + \alpha d].$$

Differentiating $g(\alpha)$ twice with respect to α results in

(13) $$g''(\alpha) = 2d^T Q d,$$

which is nonnegative if and only if Q is positive semi-definite. Thus, Proposition 19.7 is immediate from Part (b) of Proposition 19.5. ■

The Hessian

Let us consider a function f of n variables that is twice differentiable at each n-vector x in an open set. The $n \times n$ matrix H(x) given by

$$H(x)_{ij} = \frac{\partial^2 f}{\partial x_i \partial x_j}(x) \tag{14}$$

is known as the **Hessian** of f, evaluated at x.

Proposition 19.8. A twice-differentiable quadratic function f of n variables is convex on an open subset S of $\Re^{n\times 1}$ if and only if its Hessian H(x) is positive semi-definite at each x in S.

Proof. Exactly the same as for Proposition 19.7, but with (13) replaced by $g''(\alpha) = 2H(x)$. ■

Proposition 19.8 is handy if – and only if – it is relatively easy to determine whether or not the Hessian is positive semi-definite.

7. The Interior

The propositions in the prior sections have straight-forward proofs. Some of the propositions in the next few sections have difficult proofs. If those propositions were stated in their ultimate generality, their proofs would be even more daunting.

To ease the exposition, this material is presented in a setting that is somewhat restricted. Propositions 19.9 through 19.13 are established for points in the "interior" of a convex set. In Proposition 19.14, these results are shown to apply more generally, that is, to the points in the "relative interior" of a convex set.

Neighborhoods

Let us begin by reviewing the notion of a neighborhood. For each n-vector x and each positive number ε, the subset $B_\varepsilon(x)$ is defined by

$$B_\varepsilon(x) = \{y \in \Re^n : ||y - x|| < \varepsilon\}, \tag{15}$$

and each such set is called a **neighborhood** of x.

An n-vector x in a subset S of $\Re^n$ is said to be in the **interior** of S if there exists a positive number ε such that $B_\varepsilon(x) \subseteq S$. Similarly, an n-vector x in a subset S of $\Re^n$ is said to be on the **boundary** of S if x is not in the interior of S, equivalently, if every neighborhood of x contains at least one point that is not in S.

An empty interior

A convex set can contain many elements none of which are in its interior. Witness:

Example 19.3. Let $S = \{(u,\ v) \in \Re^2 : u + v = 1, u > 0, v > 0\}$. This set S is convex, but every vector x in S is in the boundary of S. The interior of S is empty.

The next few propositions describe properties that hold in the *interior* of a convex set. The interior may be empty, as it is in Example 19.3. When the interior is empty, these propositions are vacuous. Or so it seems. In Section 12, we will see how to apply these results to each point in the "relative interior" of a convex set. For Example 19.3, each vector in S is in the relative interior of S, incidentally.

8. Continuity

A function that is convex on the set S will soon be shown to be continuous in the interior of S.

The boundary

Example 19.1 exhibits a convex function that jumps upward at the boundary of the region on which it is convex. It may seem that a convex function can jump upward but not downward on the boundary. But consider

Example 19.4. Let $S = \{(u, v) \in \Re^2 : u > 0\} \cup \{(0, 0)\}$ and let the function f be defined by

$$f(u, v) = \begin{cases} v^2/u & \text{if } u > 0 \\ 0 & \text{if } u = v = 0 \end{cases}.$$

This set S is convex, and (0, 0) is the only point on its boundary. It is not hard to show (Problem 4 suggests how) that this function f is convex on S. Note that for any $u>0$ and any $k>0$, this function has $f(u, \sqrt{ku}) = k$, independent of u. This function jumps downward at (0, 0).

The interior

Proposition 19.9 (below) demonstrates that a function that is convex on S must be continuous on the interior of S.

Proposition 19.9 *(continuity).* With S as any convex subset of $\Re^n$, let f be convex on S. Then f is continuous at each point x in the interior of S.

Remark: Our proof of Proposition 19.9 is surprisingly long. It makes delicate use of Jensen's inequality. It earns a star. Skip it or skim it, at least on first reading.

Proof*. Consider any point x in the interior of S. There exists a positive number ε such that $B_\varepsilon(x) \subseteq S$. The proof has three main steps, each of which is illustrated by Figure 19.4.

Figure 19.4 A (shaded) simplex $A \subseteq B_\varepsilon(x)$.

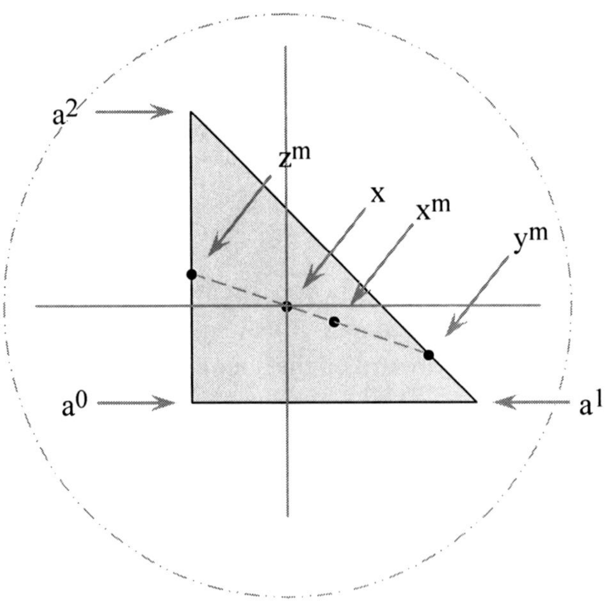

The first step will be to construct a simplex A in $\Re^n$ that has x in its interior and is contained in $B_\varepsilon(x)$. For i = 1, …, n, let e^i be the n-vector that has 1 in its ith position and has 0's in all other positions. Pick number $\beta>0$ that is small enough that $x+\beta e^i$ is in $B_\varepsilon(x)$ for each i. Let e be the n-vector each of whose entries equals 1, and set $a^0 = x - \beta e/n$. Set $a^i = x + \beta e^i$ for i = 1, 2, …, n. Define the set A as the set of all convex combinations of the vectors $a^0, a^1, \ldots, a^n$, so that

$$A = \left\{\sum_{i=0}^{n} \gamma_i a^i : \gamma \geq 0,\ \sum_{i=0}^{n} \gamma_i = 1\right\}$$

Figure 19.4 illustrates this construction for the case n = 2. Evidently, A is a convex subset of $B_\varepsilon(x)$, and x is in the interior of A. Define the constant K by

$$K = \max\{f(a^i) : i = 0, 1, \ldots, n\}.$$

The set A and the constant K are fixed throughout the remainder of the proof. Each vector y in A is a convex combination of a^0 through a^n, so Jensen's inequality (Proposition 19.3) guarantees

$$f(y) \leq K \qquad \forall\, y \in A. \tag{16}$$

Now consider any sequence $\{x^m : m = 1, 2, \ldots\}$ of n-vectors that converges to x. We must show that $f(x^m)$ converges to f(x). For m large enough, each of these n-vectors is in A. Renumber this sequence, if necessary, so that for each m the vector x^m is in the interior of A.

For the second step of the proof, consider any m for which $x^m \neq x$. This step places lower bounds on f(x) and on $f(x^m)$. With c as any number, consider the n-vector $x + c(x^m - x)$. For values of c that are close enough to zero, this vector is in A. For values of c that are sufficiently far from 0, this vector is not in A. (The dashed line segment in Figure 19.4 corresponds to the values of c for which this vector is in A.) Define λ_m and μ_m by

$$\lambda_m = \max\{c : x + c(x^m - x) \in A\},$$

$$\mu_m = \max\{c : x - c(x^m - x) \in A\}.$$

Define the n-vectors y^m and z^m by

(17) $$y^m = x + \lambda_m(x^m - x),$$

(18) $$z^m = x - \mu_m(x^m - x).$$

Figure 19.4 illustrates this construction. It is easy to verify that y^m and z^m are on the boundary of A, moreover, that $\lambda_m>1$ and $\mu_m>0$. Since λ_m exceeds 1, equation (17) lets x^m be expressed as the convex combination

(19) $$x^m = \frac{1}{\lambda_m}y^m + \frac{(\lambda_m - 1)}{\lambda_m}x$$

of y^m and x. Since $y^m \in A$, the convexity of f and (16) give

(20) $$f(x^m) \le \frac{1}{\lambda_m}K + \frac{(\lambda_m - 1)}{\lambda_m}f(x).$$

Similarly, since μ_m is positive, equation (18) lets x be expressed as the convex combination

(21) $$x = \frac{1}{(\mu_m + 1)}z^m + \frac{\mu_m}{(\mu_m + 1)}x^m$$

of z^m and x^m. Since $z^m \in A$, the convexity of f and (16) give

(22) $$f(x) \le \frac{1}{(\mu_m + 1)}K + \frac{\mu_m}{(\mu_m + 1)}f(x^m).$$

Inequalities (20) and (22) are the desired lower bounds on f(x) and $f(x^m)$.

The third major step of the proof is to let $m \to \infty$. Since $x^m \to x$ and Since y^m and z^m are on the boundary of A, equations (19) and (21) give $\lambda^m \to \infty$ and $\mu^m \to \infty$, so (20) and (22) give lin $\sup_{m\to\infty} f(x^m) \le f(x) \le$ lin $\inf_{m\to\infty} f(x^m)$. These inequalities show that $f(x^m) \to f(x)$, which completes a proof. ■

9. Unidirectional Derivatives

In this section, it is shown that a function that is convex on a set S must have unidirectional derivatives on the interior of S.

No derivative

Must a function that is convex on S be differentiable on the interior of S? Consider

Example 19.5. The function $f(x) = \max\{0, x\}$ is convex on $\Re$ but is not differentiable at 0.

The function f in Example 19.5 is convex, and it is differentiable, except at 0. Must the points at which such a function fails to be differentiable be isolated? Consider:

Example 19.6. Let $S = \{x \in \Re : 0 < x < 1\}$. The rational numbers (fractions) in S can be placed in one-to-one correspondence with the positive integers. In such a correspondence, let r(i) be the rational number that corresponds to the integer i, and consider the function f defined by

$$f(x) = \sum\nolimits_{i=1}^{\infty} (1/2)^i \cdot \max\{0, x - r(i)\}.$$

It is not difficult to show that f is increasing and convex on S, but that f fails to have a derivative at each rational number in S. It can also be shown that f has a derivative at each irrational number in S.

You may have observed that the functions in Examples 19.5 and 19.6 have "left" and "right" derivatives at each point in the interior of their domains.

The unidirectional derivative

"Unidirectional" and "bidirectional" derivatives were introduced in Chapter 18. For convenient reference, their definitions are reviewed here. Consider a function f whose domain is a subset S of $\Re^n$. With x as any vector in S and with d as any vector in $\Re^n$, let us suppose that the limit on the right-hand side of (23) exists and is finite.

$$f'_+(x, d) = \lim_{\varepsilon \downarrow 0} \frac{f(x + \varepsilon d) - f(x)}{\varepsilon}. \qquad (23)$$

If that occurs, $f'_+(x, d)$ is called the *unidirectional derivative* of f at x in the direction d. This definition requires that:

- The vector $(x + \varepsilon d)$ be in S for every positive number ε that is sufficiently close to 0.
- The same limit in (23) be obtained for every sequence of positive numbers that decreases to zero.
- This limit be a number, rather than $+\infty$ or $-\infty$.

The function f(x) in Example 19.5 is not differentiable at 0, but its unidirectional derivatives at 0 are easily seen to be

$$f'_+(0,\ d) = \begin{cases} d & \text{for } d \geq 0 \\ 0 & \text{for } d \leq 0 \end{cases}.$$

Bidirectional derivatives

In Chapter 18, the *bidirectional derivative* $f'(x, d)$ of f at x in the direction d was defined by the variant of (23) in which $\varepsilon \to 0$ replaces $\varepsilon \downarrow 0$. Thus, the bidirectional derivative has the more demanding requirement. A function can have a unidirectional derivative, without having a bidirectional derivative. It will soon be shown that a convex function has unidirectional derivatives on the interior of the region on which it is convex.

If the bidirectional derivative exists, it must satisfy

$$f'(x, d) = -f(x, -d).$$

The unidirectional derivatives do exist (Proposition 19.10, below), and they must satisfy

$$f'_+(x, d) \geq -f'_+(x, -d).$$

The boundary

Let S be a convex subset of $\Re^n$, let f be convex and continuous on S. Must the unidirectional derivative $f'_+(x, d)$ exist for a point x on the boundary of S and a direction d that points “into” S? Not necessarily. Consider

Example 19.7. Let $S = \{u \in \Re : -1 \leq u \leq +1\}$. The function.

$$f(u) = 1 - \sqrt{1 - u^2}$$

is plotted in Figure 19.5. For x=-1 and d=+1, the set S contains $x + \varepsilon\, d$ for all positive ε that are below 2. But $f'_+(-1, +1)$ does not exist because the ratio on the RHS of (23) approaches $-\infty$ as ε decreases to 0.

Figure 19.5 The convex function f in Example 19.7.

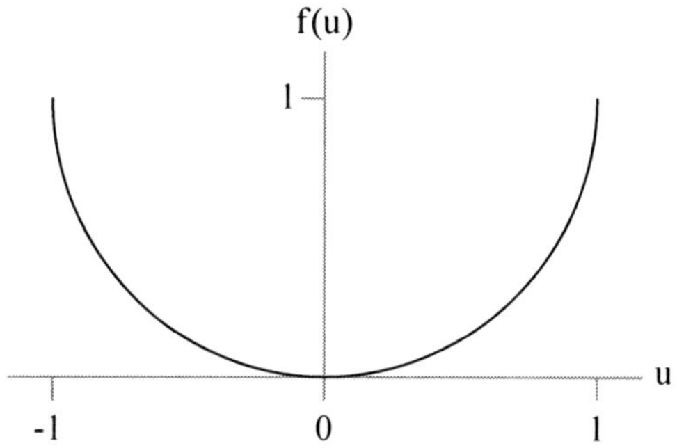

The interior

Evidently, if we want to guarantee the existence of unidirectional derivatives, we should avoid the boundary. Consider

Proposition 19.10 *(unidirectional derivatives).* Let the function f be convex on the convex subset S of $\Re^n$. Then $f'_+(x, d)$ exists for each n-vector x in the interior of S and each direction d in $\Re^n$.

Proof. By hypothesis, S contains some neighborhood of x. Proposition 19.2 with $x^0=x$ shows that the ratio on the RHS of (23) cannot increase as ε decreases. Proposition 19.2 with $x^1=x$ places a lower bound on this ratio. Thus, the completeness postulate of the set $\Re$ of real numbers shows that the limit on the RHS of (17) exists and is a real number. ∎

This proof of Proposition 19.10 is refreshingly simple; it rests squarely on Proposition 19.2.

10. Support of a Convex Function

Proposition 17.8 demonstrated that a convex set S has a supporting hyperplane H at each point x on its boundary. That proposition demonstrates that x is contained in a hyperplane H and that S is a subset of $H \cup H^+$.

The "support" of a convex function has a similar definition. Let the function f be convex on a convex subset S of $\Re^n$. This function is said to have a **support** at the n-vector x in S if there exists an n-vector d such that

$$f(y) \geq f(x) + d \cdot (y - x) \qquad \forall\, y \in S. \tag{24}$$

The expression on the RHS of (24) is linear in y, and (24) requires f(y) to be at least as large as the value that this linear expression assigns to y. The main result of this section is that a convex function has a support at each point x in the interior of its domain.

Illustration

Figure 19.6 presents a convex function and a support. This figure suggests (correctly, as we shall see) that if a convex function is differentiable at x, its support is unique, and (24) is satisfied if and only if $d = \nabla f(x)$.

Figure 19.6 A convex and function and a support.

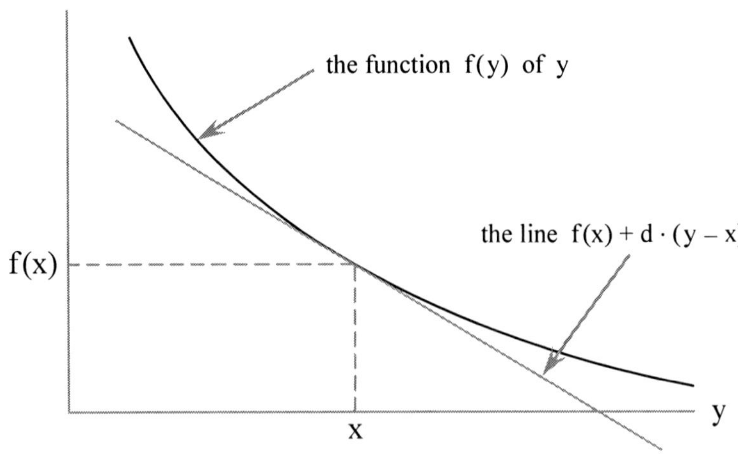

The boundary

The function f plotted in Figure 19.5 is continuous and convex on the set $S=\{x : -1 \leq x \leq +1\}$. It's clear, visually, that this function has a support at each number x that lies strictly between −1 and +1, but this function has no support at x=−1, and it has no support at x=+1.

Differentiable functions

To guarantee the existence of a support, we shall stay away from the boundary. Let us first suppose the function f of n variables is differentiable at a point x in the interior of its domain. Proposition 18.3 shows that its gradient $\nabla f(x)$ determines its directional derivatives, specifically, that

$$\text{(25)} \qquad \lim_{\varepsilon \to 0} \frac{f(x+\varepsilon d) - f(x)}{\varepsilon} = \nabla f(x) \cdot d \qquad \forall\, d \in \Re^n.$$

Consider

Proposition 19.11. Let the function f be convex on the subset S of $\Re^n$, and suppose that f is differentiable at the point x is in the interior of S. Then

$$\text{(26)} \qquad f(y) \geq f(x) + \nabla f(x) \cdot (y - x) \qquad \forall\, y \in S.$$

Proof. Since x and y are in the convex set S, the convex function f satisfies

$$f\left[(1-\varepsilon)x + \varepsilon y\right] \leq (1-\varepsilon)f(x) + \varepsilon f(y)$$

for all ε having $0<\varepsilon<1$. Divide the above inequality by the positive number ε and then rearrange it as

$$\frac{f[x+\varepsilon(y-x)] - f(x)}{\varepsilon} \leq f(y) - f(x).$$

Let ε approach 0, and note from (25) that the LHS of the above inequality approaches $\nabla f(x)\,(y - x)$. This completes a proof. ■

The proof of Proposition 19.11 is refreshingly straight-forward.

Trouble in the interior

A key part of the hypothesis of Proposition 19.11 is that the function f is differentiable at x. What if the function f is not differentiable at x? Proposition 19.10 guarantees that f has unidirectional derivatives at x. Do unidirectional derivatives in linearly independent directions determine a support? Consider

Example 19.8. The function f(u, v) = max {−2 u, −2 v} of two variables is convex on $\Re^2$ (Proposition 19.6 shows that the larger of two convex functions is convex). For any positive number ε,

$$f(u + \varepsilon, u) = f(u, u) = f(u, u + \varepsilon) = -2u.$$

Hence, with $e^1 = (1, 0)$ and $e^2 = (0, 1)$,

$$f'_+[(u, u), e^1] = f'_+[(u, u), e^2] = 0.$$

The unidirectional derivatives of f at (u, u) in the "forward" directions equal zero. And with d = 0, (24) cannot hold because it would require f(w, w) ≥ f(u, u) + 0, which is violated for every w > u.

In brief, the function f given in Example 19.8 does *not* lie on or above the plane that matches its value at x = (u, u) and whose slopes equal to the unidirectional derivatives $f'_+[x, e^1]$ and $f'_+[x, e^2]$, both of which equal zero.

An existential result

Proposition 19.12 (below) shows that function f that is convex on S has a support at each point x in the interior of S. The proof of Proposition 19.12 is starred, and for good reason.

Proposition 19.12. Let S be a convex subset of $\Re^n$, let f be convex on S, and let x be any n-vector in the interior of S. Then:

(a) There exists an n-vector d such that

(27) $$f(y) \geq f(x) + d \cdot (y - x) \qquad \forall\, y \in S.$$

(b) Furthermore, if the ith partial derivative of f exists at x, this partial derivative equals d_i.

Proof*. Since x is in the interior of S, there exists a positive number ε such that S contains every n-vector $\hat{x}$ having $||\hat{x} - x|| \le \varepsilon$. Consider the subset T of $\Re^{n+1}$ given by

$$T = \{(\hat{x}, y) : \hat{x} \in \Re^n, ||\hat{x} - x|| \le \varepsilon, y \ge f(\hat{x})\}.$$

Proposition 19.4 guarantees that T is a convex set. That T is closed is immediate from the fact that f is continuous (Proposition 19.9) on the interior of S. That the pair [x, f(x)] is on the boundary of T is evident from the fact that for every positive number δ the pair [x, f(x) – δ] is not in T. Thus, the Supporting Hyperplane theorem (Proposition 17.8) shows that T has a support at [x, f(x)]. This support identifies a pair (α, β) with these three properties: (i) α is an n-vector and β is a number, (ii) at least one of α and β are nonzero, and (iii)

$$(28) \qquad \alpha \cdot (\hat{x} - x) + \beta \cdot [y - f(x)] \ge 0 \qquad \forall \{(\hat{x}, y) \in T.$$

The fact that T contains each pair (x, y) with y>f(x) guarantees that β cannot be negative. Aiming for a contradiction, suppose β=0. In this case, the inequality in (28) reduces to $0 \le \alpha \cdot (\hat{x} - x)$ and (ii) guarantees that the vector α cannot equal 0. For each number δ that is sufficiently close to zero, the set T contains each pair$(\hat{x}, y)$ having $\hat{x} - x = \delta\alpha$ and $y = f(\hat{x})$. Premultiply $\hat{x} - x = \delta\alpha$ by α to obtain $0 \le \alpha \cdot (\hat{x} - x) = \delta\alpha \cdot \alpha$. Since α is not zero, $\alpha \cdot \alpha$ is positive, and the preceding inequality cannot hold for any negative value of δ, so the desired contradiction is established. Thus, (28) holds with β>0. Divide (28) by β, define the n-vector d by d = – α/β, and note from (28) that

$$(29) \qquad f(\hat{x}) - f(x) \ge d \cdot (\hat{x} - x) \qquad \text{whenever } ||\hat{x} - x|| \le \varepsilon.$$

Since f is convex on S, Proposition 19.2 shows that (29) remains true for all $\hat{x} \in S$. This proves part (a).

For part (b), suppose the ith partial derivative of f exists at x. Denote as e^i the n-vector having 1 in its ith position and 0's elsewhere. In (29), set $\hat{x} = x + \delta e^i$ to obtain $f(x + \delta e^i) - f(x) \ge d_i \delta$ for every number δ having $|\delta| \le \varepsilon$. For $\delta > 0$, divide the preceding inequality by δ and then let δ approach zero to obtain $f'(x, e^i) \ge d_i$. For $\delta > 0$, divide the preceding inequality by δ and then let δ approach zero to obtain $f'(x, e^i) \ge d_i$. For $\delta < 0$, divide the same inequality by δ and let δ approach zero to obtain $f'(x, e^i) \le d_i$. Hence, $f'(x, e^i) = d_i$, which completes a proof. ■

Part (a) is *existential*; it shows that a convex function has at least one support at each point x in the interior of its domain, but it does not show how to construct a support. Part (b) shows that a convex function that has partial derivatives at x has exactly one support at x, moreover, that this support has d equal to the vector of partial derivatives, evaluated at x.

11. Partial Derivatives and Convexity

In Chapter 18, we saw that a function can have partial derivatives without being differentiable. That is *not* true of convex functions. If a convex function has partial derivatives at x, it is differentiable at x. Witness

Proposition 19.13. Let S be a convex subset of $\Re^n$, and let f be convex on S. If f has partial derivatives at a point x in the interior of S, then f is differentiable at x.

Remark: The statement of Proposition 19.13 is simple. Our proof is not. It can be skimmed or skipped with no loss of continuity.

Proof*. By hypothesis, x lies in the interior of S. Part (a) of Proposition 19.12 shows that f has a support at x, and part (b) shows that f has only one support at x, indeed, that

$$f(y) \geq f(x) + z \cdot (y - x) \qquad \forall\, y \in S,$$

where z is the vector of partial derivatives of f, evaluated at x.

To establish the differentiability of f at x, we consider any sequence $\{d^1, d^2, \ldots, d^m, \ldots\}$ of nonzero n-vectors having $||d^m|| \to 0$. Substituting $x + d^m$ for y in the inequality that is displayed above gives

$$f(x + d^m) - f(x) - z \cdot d^m \geq 0.$$

This inequality is preserved if it is divided by $||d^m||$. Thus, a proof that f is differentiable at x can be completed by showing that

$$\text{(30)} \qquad \limsup_{m\to\infty} \frac{f(x + d^m) - f(x) - z \cdot d^m}{||d^m||} \leq 0.$$

Jensen's inequality will be used to verify (30). With d_i^m as the ith entry in d^m, we designate

$$|d^m| = \sum_{i=1}^{n} |d_i^m| \qquad \text{and} \quad \alpha_i^m = \frac{|d_i^m|}{|d^m|} \quad \text{for } i = 1, 2, \ldots, n.$$

Note that the sum over i of α_i^m equals 1. As usual, e^i denotes the n-vector having 1 in its ith position and 0's in all other positions.

To simplify the discussion, this paragraph is focused on a nonzero vector d^m all of whose entries are nonnegative. The fact that z_i is the partial derivative of f with respect to the ith variable, evaluated at x, guarantees

$$f(x + |d^m|e^i) - f(x) = z_i|d^m| + o(|d^m|), \tag{31}$$

where "o(ε)" is short for any function of a(ε) such that $a(\varepsilon)/\varepsilon \to 0$ as $\varepsilon \to 0$. Consider the identity

$$x + d^m = \sum_{i=1}^{n} \alpha_i(x + |d^m|e^i).$$

This identity, the convexity of f, and Jensen's inequality (Proposition 19.9) give

$$f(x + d^m) \le \sum_{i=1}^{n} \alpha_i f(x + |d^m|e^i),$$

and substituting (31) into the above gives.

$$f(x + d^m) \le \sum_{i=1}^{n} \alpha_i[z_i\,|d^m| + o(|d^m|) + f(x)].$$

Since d_i^m is nonnegative, we have $d_i^m = \alpha_i|d^m|$ and

$$\sum_{i=1}^{n} \alpha_i\, z_i\, |d^m| = \sum_{i=1}^{n} z_i\, d_i^m = z \cdot d^m,$$

so the preceding inequality yields

$$f(x + d^m) - f(x) - z \cdot d^m \le o(|d^m|).$$

For any nonzero vector d, the inequality $|d|/||d|| \leq \sqrt{n}$ holds because replacing any two non-equal entries of d by their average has no effect on $|d|$ but reduces $||d||$. Thus, dividing the inequality that is displayed above by $||d^m||$ yields

$$(32) \qquad \frac{f(x + d^m) - f(x) - z \cdot d^m}{||d^m||} \leq \frac{o(|d^m|)}{||d^m||} \leq o(\sqrt{n}).$$

Inequality (32) has been verified for $d^m > 0$. To verify it for any nonzero vector d^m, replace e^i by $-e^i$ throughout the preceding paragraph for those entries having $d_i^m < 0$. To verify (30), let $m \to \infty$ in (32). ■

Proposition 19.13 eases the task of determining whether or not a convex function is differentiable at x. If it has partial derivatives at x, it is. If it does not have partial derivatives at x, it isn't. This result remains true for a function that has bidirectional derivatives in any set of directions that form a basis for $\Re^n$. Virtually the same proof applies to that version, and it will prove useful when we deal with the "relative" interior.

12. The Relative Interior

Propositions 19.9 through 19.13 describe the behavior of a convex function in the interior of the convex set S. If the interior of S is empty, these propositions seem to be content-free. But that is not so. These results can easily be made to apply to each vector in the "relative interior" of a convex set. How to do so is the subject of this section.

A subspace

Until now, a convex set S of n-vectors and a neighborhood $B_\varepsilon(x)$ of the n vector x have has been viewed from the perspective of the vector space, $\Re^n$. They will soon be viewed from the perspective of a *subspace* of $\Re^n$. For any convex subset S of $\Re^n$, the set L(S) is defined by

$$(33) \qquad L(s) = \{\beta(x - y) : \beta \in \Re,\ x \in S,\ y \in S\}.$$

Thus, L(S) is obtained by taking the difference (x−y) of each pair of vectors in S and multiplying that difference by every real number β. An immediate consequence of the fact that S is convex is that:

- The subset L(S) of $\Re^n$ is a vector space.
- The set L(S) equals $\Re^n$ if and only if S has a non-empty interior.

Figure 19.7 illustrates L(S) for the convex set $S = \{(u,\ 1-u) : 0 < u < 1\}$ of all 2-vectors whose entries are positive numbers that sum to 1. The interior of S is empty. We will soon see that each vector in S is in its "relative interior."

Figure 19.7 A convex set S and its subspace L(S).

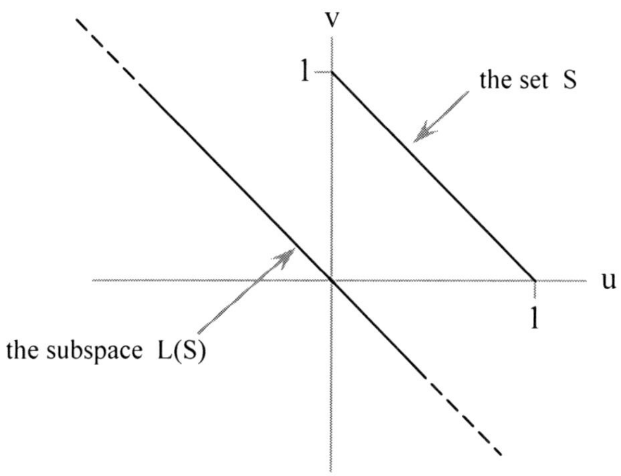

The sum of two sets

A bit of notation will prove handy. The **sum** of subsets S and T of $\Re^n$ is denoted S + T and is defined by

$$S + T = \{(x + y) \ :\ x \in S,\ y \in T\}.$$

In this context, the neighborhood $B_\varepsilon(x)$ relates to $B_\varepsilon(0)$ by

$$B_\varepsilon(x) = \{x\} + B_\varepsilon(0).$$

A new neighborhood system

A system of "relative neighborhoods" is now described. **The relative neighborhood** $B_\varepsilon^S(0)$ of 0 is defined by

$$B_\varepsilon^S(0) = B_\varepsilon(0) \cap L(S) \tag{34}$$

and the relative neighborhood $B_\varepsilon^S(0)$ is defined by

$$B_\varepsilon^S(x) = \{x\} + B_\varepsilon^S(0). \tag{35}$$

Evidently, $B_\varepsilon^S(0)$ is a proper subset of $B_\varepsilon(x)$ if L(S) is a proper subset of $\Re^n$.

The relative interior

An element x of a convex set S is now said to be in the **relative interior** of S if there exists a positive number ε such that $B_\varepsilon^S(x)$ is contained in S. Similarly, an element x of S is now said to be on the **relative boundary** of S if it is not in the relative interior of S. For the set S displayed in Figure 19.7, each member of S is a vector $(u, 1 - u)$ with $0 < u < 1$, and each such vector in S is in the relative interior of S. The relative interior of a convex set is the subject of:

Proposition 19.14. Consider a convex subset S of $\Re^n$ that contains at least two distinct n-vectors. Then:

(a) There exists a vector x in the relative interior of S.

(b) If x is in the relative interior of S and if y is in S, then $x + \alpha(y - x)$ is in the relative interior of S for every α such that $0 \le \alpha < 1$.

Proof. By hypothesis, S is a convex subset of $\Re^n$ that contains at least two elements. It follows that L(S) is a vector space whose dimension k is at least 1.

For part (a), we consider any pairs $\{x^1, z^1\}$ through $\{x^k, z^k\}$ of elements of S such that $0.5(x^1 + z^1)$ through $0.5(x^k + z^k)$ span L(S). The average of these k vectors is easily seen to be in the relative interior of S, which proves part (a).

Let the n-vectors v^1 through v^k be any basis for L(S).

For part (b), consider any vector x in the relative interior of S. For ε sufficiently close to 0, the set $B^S_\varepsilon(x)$ is in S, so nonzero numbers β^1 through β^k exist such that $x^i = x + \beta^i v^i \in B^S_\varepsilon(x)$ for i = 1, 2,..., k.

Consider any y in S and any number α such that $0 \le \alpha < 1$. Set $z = x + \alpha(y-x)$, set $z^i = x^i + \alpha(y - x^i)$ for i= 1, 2,..., k, and set $\lambda = (1-\alpha)\,\varepsilon$. For each i we have $z^i \in S$ because S is convex, and we have $z^i - z = (1-\alpha)(x^i - x)$, so $||z^i - z|| = (1-\alpha)||x^i - x|| \le \lambda$. This guarantees $z^i \in B^S_\lambda(z)$ for each i, hence that z is in the relative interior of S. ■

Thus, if a convex set S contains more than one vector, its relative interior is nonempty. And, if x is in the relative interior of S and y is in S, then each vector in the open line segment between x and y is also in the relative interior of S.

These results hold for *convex* sets. If a subset S of $\Re^n$ is not convex, L(S) need not be a vector space.

Generalizing prior results

To apply the principal results of Sections 8 through 11 to points x in the relative interior, we need only repeat the prior arguments with the vector space $\Re^n$ replaced by the subspace L(S) and with the neighborhood $B_\varepsilon(x)$ switched to $B^S_\varepsilon(x)$. The proof of Proposition 19.9 applies when $B_\varepsilon(x)$ is replaced by $B^S_\varepsilon(x)$. Proposition 19.10 holds as written for each direction d in L(S), rather than in $\Re^n$. Proposition 19.12 holds as written if the set T is required to lie in L(S) rather than in $\Re^n$.

13. Review

Propositions 19.1 through 19.7 describe the basic properties of convex functions and provide several ways in which to determine whether or not a particular function is convex. Propositions 19.8 through 19.14 probe the structure of convex functions. They demonstrate that every convex function is well-behaved in the relative interior of its domain: It is continuous, it has unidirectional derivatives, and it is differentiable if it has partial derivatives.

Of the earlier propositions, Jensen's inequality (Proposition 19.3) may be the most useful. It has been used repeatedly within this chapter, and other uses of it can be found within the chapter's homework and discussion problems. Of the later propositions, the most important may be that the function $g(y) = f(x) + \nabla f(x) \cdot (y - x)$ of y supports the function f at x if f is convex and differentiable. This fact will prove to be very handy in Chapter 20.

14. Homework and Discussion Problems

1. True or false: The epigraph of a convex function is a closed set.

2. On what set S is the function $f(x) = -\sqrt{x}$ convex? For what members of S does this function have a support, and what is that support?

3. Suppose that the functions f(x) and g(x) are convex and twice-differentiable on $\Re$, and that g is nondecreasing. Show that the function f[g(x)] is convex on $\Re$. (*Hint:* Differentiate f[g(x)] twice.)

4. This problem concerns Example 19.4 (on page 596).

   (a) Show that this function is convex on the interval between (0, 0) and any vector (u, v) in S.

   (b) Show that this function is convex on the interval between any two non-zero vectors in S. (*Hint*: compute its Hessian.)

5. (Unidirectional derivatives):

   (a) For Example 19.5, compute the sum $f_1^+(0,\ 1) + f_1^+(0,\ -1)$ of at the point 0 at which f is not differentiable.

   (b) For Example 19.6, compute the sum $f_1^+[r(i),\ 1] + f_1^+[r(i),\ -1]$ at the ith fraction r(i).

6. Show that the function $f(x) = e^x \log(x)$ is convex on $S = \{x : x \geq 1\}$.

7. Let $g(x) = -\log(x)$ and $h(x) = x^2$, and let $S = \{x : x > 0\}$. Support your answers to each of the following:

   (a) Is g convex on S?

   (b) Is h convex on $\Re$?

(c) Is the function $f(x)=g[h(x)]$ convex on S?

(d) Is the function $f(x)=h[g(x)]$ convex on S?

8. Suppose the functions f and g are convex on $\Re$, and suppose that these functions are twice differentiable. Under what circumstance is the function $h(x)=f[g(x)]$ convex on $\Re$? *Hint*: It might help to review the preceding problem.

9. *(classical uses of Jensen's inequality):*

(a) Is the function $g(x)=-\log(x)$ convex on $S=\{x : x>0\}$? If so, why?

(b) For each set $\{x_1, \ldots, x_n\}$ of positive numbers and each set $\{\alpha_1, \ldots, \alpha_n\}$ of nonnegative numbers that sum to 1, use part (a) to show that

$$x_1^{\alpha_1} \cdots x_n^{\alpha_n} \le \alpha_1 x_1 \cdots + \alpha_n x_n,$$

thereby verifying that the **geometric mean** does not exceed the arithmetic mean."

(c) With $p \ge 1$ as a constant, is the function $g(x)=x^p$ convex on $S=\{x : x \ge 0\}$? If so, why?

(d) With $p \ge 1$ as a constant, show that

$$(\alpha_1 x_1 \cdots + \alpha_n x_n)^p \le \alpha_1 (x_1)^p + \cdots + \alpha_n (x_n)^p$$

for each set $\{x_1, \ldots, x_n\}$ of positive numbers and each set $\{\alpha_1, \ldots, \alpha_n\}$ of nonnegative numbers that sum to 1. *Hint*: part (c) night help.

(e) With $p \ge 1$ as a constant, set $\alpha = 1/p$ and $\beta = 1-\alpha$. Show that

$$\sum_{i=1}^n w_i x_i \le \left(\sum_{i=1}^n w_i\right)^\beta \left(\sum_{i=1}^n w_i x_i{}^{1/\alpha}\right)^\alpha$$

for any sets $\{x_1, \ldots, x_n\}$ and $\{w_1, \ldots, w_n\}$ of positive numbers. *Hint*: In part (d), take $w_i/(\sum_{j=1}^n w_j)$.

(f) With constant α having $0<\alpha<1$ and with $\beta=1-\alpha$, prove **Hölder's inequality**, which is that

$$\sum_{i=1}^n y_i z_i \le \left(\sum_{i=1}^n y_i{}^{1/\beta}\right)^\beta \left(\sum_{i=1}^n z_i{}^{1/\alpha}\right)^\alpha$$

for any sets $\{y_1, \ldots, y_n\}$ and $\{z_1, \ldots, z_n\}$ of positive numbers.

10. (*quadratic functions*) This problem concerns the quadratic function $f(x) = x^T Q x$ where x is an $n \times 1$. vector and Q is a symmetric $n \times n$ matrix.

    (a) True or false: If Q were not symmetric, replacing Q by $0.5(Q + Q^T)$ would not effect f(x), so no generality is lost by the assumption that Q is symmetric.

    (b) For the symmetric 3×3 matrix Q whose entries are in cells B2:D4 of the spreadsheet that appears below, elementary row operations have produced a matrix L with 1's on the diagonal, with 0's above the diagonal, and with L Q given by cells L2:N4. Is this matrix L invertible? If so, what is its inverse? What sequence of elementary row operations transformed Q into L Q? Is the matrix LQL^T symmetric? Is LQL^T diagonal? If so, what entries are on its diagonal?

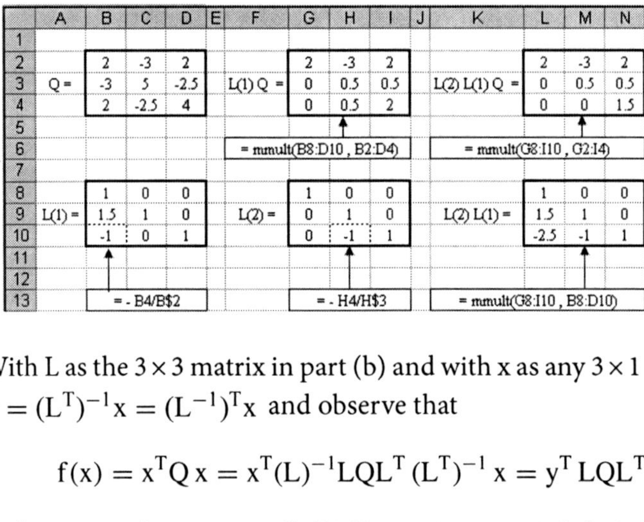

	A	B	C	D	E	F	G	H	I	J	K	L	M	N
1														
2		2	-3	2			2	-3	2			2	-3	2
3	Q =	-3	5	-2.5		L(1) Q =	0	0.5	0.5		L(2) L(1) Q =	0	0.5	0.5
4		2	-2.5	4			0	0.5	2			0	0	1.5
5														
6						= mmult(B8:D10 , B2:D4)					= mmult(G8:I10 , G2:I4)			
7														
8		1	0	0			1	0	0			1	0	0
9	L(1) =	1.5	1	0		L(2) =	0	1	0		L(2) L(1) =	1.5	1	0
10		-1	0	1			0	-1	1			-2.5	-1	1
11														
12														
13		= - B4/B$2					= - H4/H$3				= mmult(G8:I10 , B8:D10)			

    (c) With L as the 3×3 matrix in part (b) and with x as any 3×1 vector, set $y = (L^T)^{-1}x = (L^{-1})^T x$ and observe that

$$f(x) = x^T Q x = x^T (L)^{-1} LQL^T (L^T)^{-1} x = y^T LQL^T y.$$

    (d) Is the matrix Q given in cells B2:D4 positive semi-definite?

    (e) For the symmetric matrix in cells B2:D4, find the range on the value of Q_{21} (its current value equals –3) for which the matrix Q is positive semi-definite.

11. Ascertain whether or not the 4×4 matrix Q given by

$$Q = \begin{bmatrix} 1 & -2 & 3 & -4 \\ -2 & 6 & 4 & -4 \\ 3 & 4 & 62 & -51 \\ -4 & -4 & -51 & 239 \end{bmatrix}$$

is positive semi-definite. (*Hint*: mimic the recipe in the preceding problem.)

12. Can a matrix Q be positive semi-definite if $Q_{ii} < 0$ for some i? If not, why not?

13. Take $S \subseteq \Re^2$ as the intersection of surface of the unit circle and the positive orthant. Sketch the set L(S) that is defined by (33). Is it a vector space?

14. *(trivial supports)* Consider the convex subset S of $\Re^3$ that consists of each vector x that has $x_1^2 + x_2^2 \le 4$ and $x_3 = 1$.

    (a) Which points, if any, are in the interior of S?

    (b) Which points, if any, are in the relative interior of S?

    (c) Does there exist a plane that is a support of S at each point on its boundary? If so, what is it?

15. *(the closure)* Consider a function f that is convex on an open subset S of $\Re^n$. This function is continuous on S (Proposition 19.9). The **closure** of S is the set *cl*(S) that consists of S and its limit points. Let us attempt to extend f to *cl*(S) in a way that preserves its continuity. To do so, we must assign $f(s) = +\infty$ if s is a limit point of a sequence of elements of S whose f-values approach $+\infty$. Will we succeed? *Hint*: review Example 19.4.

Chapter 20: Nonlinear Programs

1. Preview

A nonlinear program differs from a linear program by allowing the objective and the constraints to be nonlinear. The fundamental questions for nonlinear programs are easy to pose:

- Is there an analog for nonlinear programs of conditions that determine whether or not a feasible solution to a linear program is optimal? If so, what is it?

E. V. Denardo, *Linear Programming and Generalizations*, International Series in Operations Research & Management Science 149,
DOI 10.1007/978-1-4419-6491-5_20, © Springer Science+Business Media, LLC 2011

- Is there an algorithm that computes optimal solutions to nonlinear programs quickly and reliably? If so, how does it work?

Neither of these questions has a simple answer:

- The Karush-Kuhn-Tucker (or KKT) conditions are an analog for nonlinear programs of conditions that characterize optimal solutions to linear programs.
- The KKT conditions are necessary and sufficient for a feasible solution to a nonlinear program to be a global optimum *if* the objective and constraints of the nonlinear program satisfy a "constraint qualification" that is presented in Section 5.
- The KKT conditions are shown to be necessary (but not sufficient) for a feasible solution to be a local optimum if the objective and constraints satisfy a different constraint qualification that is presented in Section 9.
- Several algorithms have been devised that do a good job of finding local or global optima to nonlinear programs. The generalized reduced gradient method (abbreviated GRG) is one of them. The GRG method is built upon the simplex method. It is implemented in Solver and in Premium Solver. These implementations work well if the functions are differentiable and if the derivatives are continuous.

The chapter begins with the presentation of the optimality conditions for a *linear* program in a format that becomes the KKT conditions when they are restated in the context of a nonlinear program. As noted above, the objective and constraints of a nonlinear program must be restricted if its optimal solution is to satisfy the KKT conditions. Any such restriction has long been known (somewhat inaccurately) as a constraint qualification. Examples are presented of the difficulties that constraint qualifications must rule out.

These examples are ruled out by a constraint qualification that is dubbed Hypothesis #1. For a nonlinear program that satisfies this hypothesis, a feasible solution is shown to be optimal if and only if it satisfies the KKT conditions.

A limitation of this hypothesis is then brought into view, and a less restrictive constraint qualification is introduced. If a nonlinear program satisfies that condition, the KKT conditions are seen to necessary for a feasible solution to be a local optimum.

No algorithm is known – or will ever be known – that solves all nonlinear programs efficiently. The GRG method seeks a local optimum. It works rather well. Whether it works can depend on how a problem is formulated. Tips for effective formulation are provided. A sketch is provided of the way in which the GRG method tackles a nonlinear program.

This chapter builds directly and indirectly on material in several earlier chapters. Prominent in this chapter's development are:

- The fact that a convex function lies on or above its supports (Proposition 19.11).
- The fact that a convex function is continuous on the interior of its domain (Proposition 19.9).
- The Duality Theorem of linear programming (Proposition 12.2).

2. Optimality Conditions for LPs

To prepare for a discussion of nonlinear programs, the optimality conditions for a linear program will be described in a way that can be generalized. This will be accomplished for a linear program that has been placed in the format of

Program 20.1. Maximize c x, subject to the constraints

$$Ax \le b,$$

$$x \in \Re^{n \times 1}.$$

The data in Program 20.1 are the $1 \times n$ vector c, the $m \times n$ matrix A, and the $m \times 1$ vector b. The decision variables form the $n \times 1$ vector x. The constraint $x \ge 0$ is omitted from Program 20.1. Any nonnegativity constraints on the decision variables are represented in Program 20.1 by rows of the constraint matrix.

Conditions that characterize an optimal solution to Program 20.1 were presented in Chapter 12. These conditions will now be stated in a way that suggests the optimality conditions for a nonlinear program.

Proposition 20.1. Let x^* be feasible for Program 20.1. The following are equivalent.

(a) The vector x^* is an optimal solution to Program 20.1.

(b) There exists a $1 \times m$ vector λ that satisfies

(1) $$c = \sum_{i=1}^{m} \lambda_i A^i,$$

(2) $$\lambda_i \geq 0 \qquad \text{for } i = 1, \ldots, m,$$

(3) $$\lambda_i [A^i x^* - b_i] = 0 \qquad \text{for } i = 1, \ldots, m.$$

Remark: This result and its proof are familiar. Expressions (1) and (2) are the constraints of the dual of Program 20.1, and (3) is complementary slackness.

Proof. In Chapter 12, we saw that the dual of Program 20.1 is the linear program:

Minimize λb, subject to the constraints

$$\lambda A = c, \quad \lambda \geq 0.$$

(a) ⇒ (b): Suppose x^* is optimal for Program 20.1. The Duality Theorem shows that there exists a row vector λ that satisfies (1) and (2) (which are the constraints of the dual linear program) and has $cx^* = \lambda b$. It remains to verify (3). By hypothesis, x^* satisfies $Ax \leq b$, so that $A x^* + s = b$ where the $m \times 1$ vector s satisfies $s \geq 0$. Premultiply the preceding equation by λ and use $\lambda A = c$ to obtain $c x^* + \lambda s = \lambda b$. Since $c x^* = \lambda b$, we have $0 = \lambda s = \lambda_1 s_1 + \cdots + \lambda_m s_m$. Each addend in this sum is nonnegative, so each addend must equal zero. Hence, if s_i is positive, it must be that λ_i equals 0. This verifies (3).

(b) ⇒ (a): Suppose x is feasible for Program 20.1 and that λ satisfies (1)-(3), hence that λ is feasible for the dual of Program 20.1. Multiply the constraint $A^i x \leq b_i$ by the nonnegative number λ_i and use (3) to get $\lambda_i A^i x = \lambda_i b_i$. Sum over i to obtain $\lambda A x = \lambda b$. Equation (1) is $\lambda A = c$, so we have $\lambda Ax = cx = \lambda b$. The Duality Theorem shows that x is optimal for Program 20.1. ■

In prior chapters, the variable that was complementary to the ith constraint of a linear program was called the *multiplier* for that constraint and was denoted y_i. The symbol λ_i suggests (correctly) that the variable that is complementary to the ith constraint of a nonlinear program will be called the *Lagrange multiplier* for that constraint.

3. Optimality Conditions for NLPs

Program 20.1 is a special case of a the nonlinear program that that appears below as

Program 20.2. Maximize f(x), subject to the constraints

$$g_i(x) \le 0 \qquad \text{for } i = 1, 2, \ldots, m,$$

$$x \in \Re^{n\times 1}.$$

In Program 20.2, f(x) and $g_1(x)$ through $g_m(x)$ are real-valued functions of the decision variables x_1 through x_n. To place Program 20.1 in the format of Program 20.2, set

(4) $$f(x) = cx,$$

(5) $$g_i(x) = A^i x - b_i \qquad \text{for } i = 1, 2, \cdots, m.$$

Here, as usual, A^i denotes the ith row of the matrix A.

Terminology

Some standard terminology is now adapted to Program 20.2. The n-vector x is said to be **feasible** for Program 20.2 if x satisfies $g_i(x) \le 0$ for i = 1, …, m. The **feasible region** for Program 20.2 consists of each n-vector x that is feasible for Program 20.2. The symbol S is reserved for the feasible region for Program 20.2, so that

$$(6) \qquad S = \{x \in \Re^n \; : \; g_i(x) \le 0 \text{ for } i = 1, \ldots, m\}.$$

A feasible solution x* for Program 20.2 is said to be a **global optimum** if

$$f(x^*) \ge f(x) \qquad \text{for every } x \in S.$$

Thus, a feasible solution x* is a global optimum if no feasible solution x has objective value f(x) that exceeds f(x*). Similarly, a feasible solution x* for Program 20.2 is said to be a **local optimum** if there exists a positive number ε such

$$f(x^*) \ge f(x) \qquad \text{for every } x \in S \cap B_\varepsilon(x^*).$$

Thus, a feasible solution x* is a local optimum if a positive number ε exists such that no feasible solution x whose distance from x* is below ε has objective value f(x) that exceeds f(x*).

A canonical form

A **nonlinear program** maximizes or minimizes a function of n real variables subject to finitely many constraints. Each constraint requires a function of these n variables to bear one of three relationships to the number 0; the function can be required to be ≤ 0, to be ≥ 0 or to be = 0. The usual tricks convert any nonlinear program into the format of Program 20.2. Thus, Program 20.2 is a canonical form for nonlinear programs.

The KKT conditions

Gradients are now used to express the optimality conditions for Program 20.1 in terms of the functions f(x) and $g_1(x)$ through $g_m(x)$. These functions are linear in x. Their gradients (vectors of partial derivatives) are

$$\nabla f(x) = c, \qquad \nabla g_i(x) = A^i \qquad \text{for } i = 1, \ldots, m.$$

When written in terms of gradients, equation (1), which is $c = \sum_{i=1}^m \lambda_i A^i$, becomes

$$(7) \qquad \nabla f(x) = \sum_{i=1}^m \lambda_i \nabla g_i(x).$$

Thus, with f and g_1 through g_m specified by (4) and (5), Proposition 20.1 shows that a feasible solution x to Program 20.1 is optimal if and only if there exist numbers λ_1 through λ_m that satisfy (7), (8) and (9), where

(8) $$\lambda_i \geq 0 \qquad \text{for } i = 1, \ldots, m,$$

(9) $$\lambda_i g_i(x) = 0 \qquad \text{for } i = 1, \ldots, m.$$

When expressed in terms of Program 20.2, rather than Program 20.1, equations (7)-(9) *are* the celebrated **KKT conditions** for nonlinear programs. "KKT" abbreviates the names of William Karush, Harold Kuhn, and Albert Tucker.

Nomenclature

In the context of Program 20.2, the numbers λ_1 through λ_m are called **Lagrange multipliers.** The Lagrange multiplier λ_i is said to be **complementary** to the constraint $g_i(x) \leq 0$. The constraint $g_i(x) \leq 0$ is said to be **binding** when it holds as an equation and to be **nonbinding** when it holds as a strict inequality. Expression (9) is the familiar **complementary slackness** condition; it states that if an inequality constraint is nonbinding (slack), its complementary multiplier must equal 0.

Interpreting the KKT conditions

The KKT conditions have a lovely interpretation. Equation (7) requires the gradient of the objective to equal a linear combination of the gradients of the constraints. Equation (8) requires the coefficients (Lagrange multipliers) to be nonnegative. Equation (9) requires a multiplier to equal 0 if its complementary constraint is nonbinding. In brief:

> A feasible solution to Program 20.2 satisfies the *KKT conditions* if and only if the gradient of its objective equals a nonnegative linear combination of the gradients of its binding constraints.

A qualification

One might hope that the analogue of Proposition 20.1 holds for nonlinear programs – that a feasible solution to Program 20.2 is optimal if and only if it and a vector of Lagrange multipliers satisfy the KKT conditions. That need not be true. It *is* true if the objective and constraints of Program 20.2 satisfy a "constraint qualification" that will be introduced shortly.

An illustration

The example in Figure 20.1 illustrates the KKT conditions. This example has three inequality constraints (m = 3) and two decision variables (n = 2). For each constraint, the line (possibly curved) on which $g_i(x) = 0$ is identified, and a "–" adjacent to this line identifies "side" of this line containing those vectors x that have $g_i(x) < 0$. The set S of feasible solutions is the region on the "–" side of all three lines.

Figure 20.1. A feasible region and a local optimum, y.

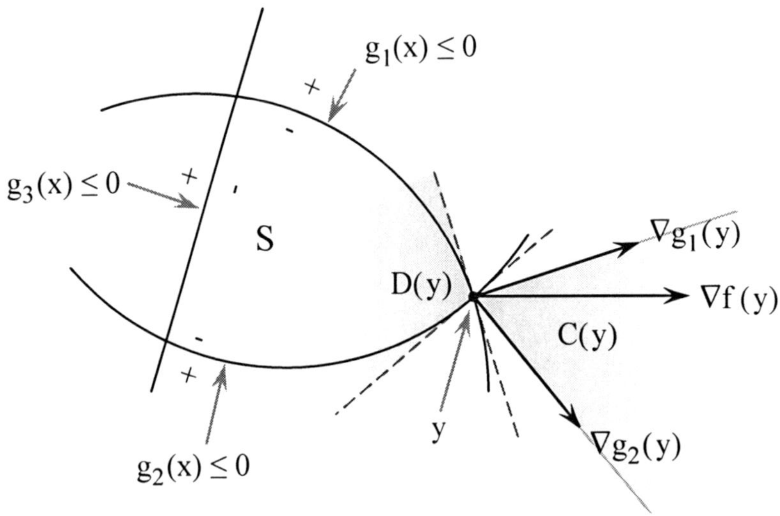

Figure 20.1 connotes – correctly – that each feasible solution y to Program 20.2 will be identified with a closed convex cone C(y) and with its polar cone D(y). The cone C(y) consists of all nonnegative linear combinations of the gradients of the constraints that are binding at y. Figure 20.1 depicts a feasible solution y for which the constraints $g_1(y) \leq 0$ and $g_2(y) \leq 0$ are binding and for which the constraint $g_3(y) \leq 0$ is not binding.

Figure 20.1 models an example in which the KKT conditions are satisfied at y and in which each feasible solution x other than y has $(x - y) \cdot \nabla f(y) < 0$. This is enough to guarantee that y is a local optimum.

4. The Need for a Constraint Qualification

If restrictions are not placed on the objective and the constraints of a nonlinear program, its optimal solution can fail to satisfy the KKT conditions. This section illustrates three difficulties that must be ruled out.

The vanishing gradient

The gradient of a differentiable function points uphill (in the direction of increase) *if* it is not zero. A difficulty can arise if an optimal solution x^* has a binding constraint whose gradient equals 0. Consider

Example 20.1. Maximize $\{x\}$, subject to $(x - 1)^3 \le 0$.

Example 20.1 falls into the format of Program 20.2 when we set $n = m = 1$ and

$$f(x) = x, \qquad g_1(x) = (x - 1)^3.$$

The optimal solution to Example 20.1 has $x^* = 1$, hence has

$$\nabla f(x^*) = 1 \quad \text{and} \quad \nabla g_1(x^*) = 3(x^* - 1)^2 = 0.$$

With $x^* = 1$, no number λ can satisfy $\nabla f(x^*) = \lambda \nabla g_1(x^*)$ because $\nabla g_1(1) = 0$ and $\nabla f(1) = 1$. The difficulty is that of the "vanishing gradient."

No interior

The optimal solution to a nonlinear program can fail to satisfy the KKT conditions if its feasible region S has no interior, as is illustrated by

Example 20.2. Maximize $\{x_2\}$, subject to

$$(x_1 - 1)^2 + (x_2)^2 \le 1,$$

$$(x_1 - 3)^2 + (x_2)^2 \le 1.$$

The first of these constraints keeps the pair (x_1, x_2) from lying outside the circle of radius 1 that is centered at (1, 0). The second constraint keeps the pair (x_1, x_2) from lying outside the circle of radius 1 that is centered at (3, 0). The only feasible solution is $x^* = (2, 0)$. Example 20.2 falls in the format of Program 20.2 when we take $n = 2$, $m = 2$ and define f, g_1 and g_2 by

$$f(x) = x_2, \quad g_1(x) = (x_1 - 1)^2 + (x_2)^2 - 1, \quad g_2(x) = (x_1 - 3)^2 + (x_2)^2 - 1.$$

Figure 20.2 records the result of casting Example 20.2 in the format of Program 20.2.

Figure 20.2. Example 20.2, in the format of Program 20.2.

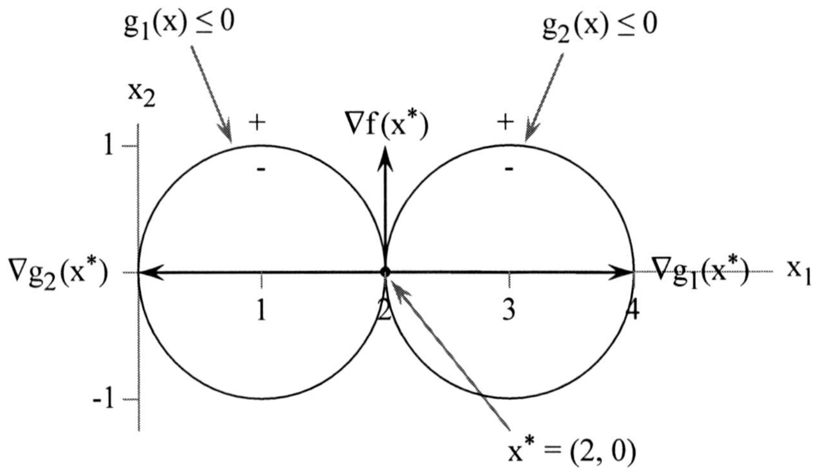

Note visually that $\nabla g_1(x^*)$ points to the right, that $\nabla g_2(x^*)$ points to the left and that $\nabla f(x^*)$ points toward the top of the page. This makes it impossible to express $\nabla f(x^*)$ as a linear combination of $\nabla g_1(x^*)$ and $\nabla g_2(x^*)$. Algebraically, we have $x^* = (2, 0)$, and

$$\nabla f(x^*) = (0, 1), \quad \nabla g_1(x^*) = (2, 0), \quad \nabla g_2(x^*) = (-2, 0),$$

for which reason no multipliers can satisfy (7). This difficulty can crop up when the feasible region has an empty interior.

A cusp

The optimal solution to a nonlinear program can occur at a cusp of its feasible region, and that presents the difficulty illustrated by

Example 20.3. Maximize $\{x_1\}$, subject to $x_2 \le (1 - x_1)^3$ and $x_2 \ge 0$.

If $x_1 > 1$, the RHS of the 1st constraint is negative, and 2nd constraint is violated. Hence, the unique optimal solution is $x^* = (1, 0)$. To place Example 20.3 in the format of Program 20.2, we take n = 2, m = 2, and

$$f(x) = x_1, \quad g_1(x) = x_2 - (1 - x_1)^3, \quad g_2(x) = -x_2.$$

The feasible solution $x^* = (1, 0)$ is optimal, but it has

$$\nabla f(x^*) = (1,\ 0), \quad \nabla g_1(x^*) = (0, 1), \quad \nabla g_2(x^*) = (0, -1),$$

so (7) cannot be satisfied. Figure 20.3 presents a visual record of this example.

Figure 20.3. Feasible region, optimal solution x*, and gradients for Example 20.3.

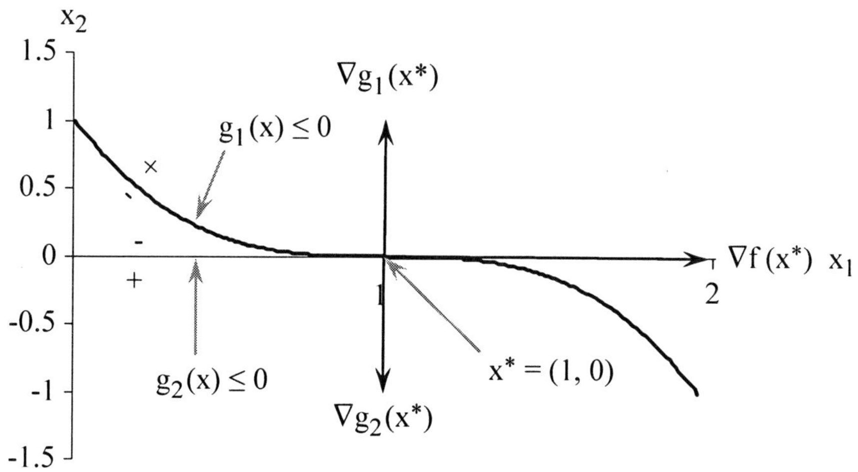

5. A Constraint Qualification

If the optimal solution to Program 20.2 is to satisfy the KKT conditions, restrictions must be placed on the functions f and g_1 through g_m. Any such restriction is known as a **constraint qualification**. Examples 20.1, 20.2 and 20.3 illustrate the pathologies that a constraint qualification needs to rule out.

The literature abounds with constraint qualifications, one of which will be presented shortly. For an instance of Program 20.2 that satisfies this constraint qualification, the following will be shown to be equivalent:

- A feasible solution x^* is a global optimum.
- A feasible solution x^* satisfies the KKT conditions.

Affine functions

Let us begin with a definition. A real-valued function g of n variables is said to be **affine** if there exists an n-vector a and a number b such that $g(x) = a \cdot x - b$ for each n-vector x. Here, as usual, the dot product $a \cdot x$ equals $\sum_{j=1}^{n} a_j\, x_j$ Affine functions are familiar from linear programming. Each constraint in a linear program requires an affine function g(x) to satisfy one of these three relations:

$$g(x) \le 0, \qquad g(x) \ge 0, \qquad g(x) = 0.$$

The up-coming constraint qualification distinguishes the affine constraints from the others. The set {1, 2, ..., m} is now partitioned into the sets L and N where

$$L = \{i \in \{1, 2, \ldots, m\} : g_i \text{ is affine}\},$$

$$N = \{1, 2, \ldots, m\} \backslash L.$$

Interpret N as the set consisting of those i for which the ith constraint is "genuinely nonlinear."

An hypothesis

Program 20.2 will soon be analyzed under the constraint qualification that is unimaginatively labeled

Hypothesis #1.

Part (a): The functions – f and g_1 through g_m are convex and differentiable on an open convex set T that contains S.

Part (b): There exists a feasible solution $\bar{x}$ to Program 20.2 that satisfies

$$g_i(\bar{x}) < 0 \qquad \text{for each } i \in N.$$

Part (b) requires that Program 20.2 has a feasible solution that satisfies each "genuinely nonlinear" constraint as a strict inequality. Let us see why Hypothesis #1 rules out Examples 20.1, 20.2 and 20.3:

- Examples 20.1 and 20.3 violate Part (a) because the function g_1 is not convex.
- Example 20.2 violates Part (b) because it has N = {1, 2}, and no feasible solution $\bar{x}$ satisfies both nonlinear constraints strictly.

Morton Slater

Very early in the history of nonlinear programming, Morton Slater introduced a constraint qualification that differs only slightly from Hypothesis #1 and which has been called the *Slater conditions* ever since. Slater did not require the functions – f and g_1 through g_m to be differentiable. The Slater conditions yield a weaker result that does Hypothesis #1. They are discussed in Section 12 of this chapter.

Appeal

Hypothesis #1 has considerable appeal. Reasons why this is so are listed below:

- It is often easy to check whether the functions $-f$ and g_1 through g_m are convex and differentiable.
- It is often easy to check that at least one feasible solution $\bar{x}$ satisfies each of the genuinely nonlinear constraints as a strict inequality.
- The objective function f models the case of decreasing marginal return.
- When the ith constraint measures the consumption of a particular resource, the convexity of g_i models increasing marginal consumption of that resource.
- For an instance of Program 20.2 that satisfies Hypothesis #1, the analog of Program 20.1 holds: A feasible solution is a global optimum if and only if it and a set of Lagrange multipliers satisfy the KKT conditions.

In brief, Hypothesis #1 can be easy to verify, it encompasses a useful class of models, and (as will soon be demonstrated) it allows the *global* optima to be characterized.

6. A Global Optimum

The implications of Hypothesis #1 are presented in a series of four propositions. The first of these propositions shows that the feasible region is convex.

Proposition 20.2. Consider an instance of Program 20.2 that satisfies Part (a) of Hypothesis #1. Its set S of feasible solutions is convex.

Proof. Consider n-vectors x and y in S, To demonstrate that S is convex, we need to show that the inequality

$$g_i[\alpha x + (1-\alpha)y] \le 0$$

holds for each number α between 0 and 1 and each i between 1 and m. The convexity of g_i on T and $S \subseteq T$ guarantee

$$g_i[\alpha x + (1-\alpha)y] \le \alpha g_i(x) + (1-\alpha) g_i(y).$$

The fact that x and y are in S guarantees $g_i(x) \le 0$ and $g_i(y) \le 0$, and the hypothesis includes $\alpha \ge 0$ and $(1-\alpha) \ge 0$. Thus, the right-hand side of the inequality that is displayed above cannot exceed 0, which completes a proof. ■

Sufficiency

This constraint qualification is now shown to suffice for a feasible solution that satisfies the KKT conditions to be a global optimum.

Proposition 20.3 *(sufficiency).* Consider an instance of Program 20.2 that satisfies Part (a) of Hypothesis #1. Suppose the n-vector x^* is feasible and that it and an m-vector λ satisfy the KKT conditions. Then x^* is a global optimum for Program 20.2.

Proof. Proposition 20.2 shows that the set S of feasible solutions to Program 20.2 is convex. The hypothesis of Proposition 20.3 is that x^* is feasible for Program 20.2 and that x^* and an m-vector λ satisfy (7)-(9).

Consider any feasible solution x for Program 20.2. For each i between 1 and m, we have $0 \le g_i(x)$ because x is feasible. By hypothesis, g_i is convex on S and is differentiable at x^*. A convex differentiable function lies on or above its supports (Proposition 19.11), which justifies the second inequality in

(10) $$0 \geq g_i(x) \geq g_i(x^*) + \nabla g_i(x^*) \cdot (x - x^*).$$

From (8), the Lagrange multiplier λ_i is nonnegative. Multiply (10) by λ_i. Expression (8) gives $\lambda_i \geq 0$, and expression (9) gives $\lambda_i g_i(x^*) = 0$, so

(11) $$0 \geq \lambda_i \nabla g_i(x^*) \cdot (x - x^*).$$

Sum this inequality over i, and note from (7) that $\nabla f(x^*) = \sum_{i=1}^{m} \lambda_i \nabla g_i(x^*)$, so

(12) $$0 \geq \nabla f(x^*) \cdot (x - x^*).$$

The function f is concave and is differentiable at x^*, so Proposition 19.11 also guarantees

(13) $$f(x) \leq f(x^*) + \nabla f(x^*) \cdot (x - x^*).$$

Inequalities (12) and (13) combine to give $f(x) \leq f(x^*) + 0 = f(x^*)$. This shows that x^* is a global optimum, completing a proof. ∎

Showing that the KKT conditions are sufficient for a feasible solution x^* to be a global optimum has been fairly straightforward. The main tool in the proof of Proposition 20.3 is the fact that a convex function lies on or above its supports.

Necessity

Let us suppose that x^* is an optimal solution to Program 20.2. It remains to show that there exist a set of multipliers that satisfy (7)-(9). That will be accomplished by a pair of propositions, both of whose proofs are starred. With x^* as an optimal solution to Program 20.2, the set E is now defined by

(14) $$E = \{i \in \{1, 2, \ldots, m\} : g_i(x^*) = 0\}.$$

This use of the letter E is mnemonic; E stands for the set of constraints that x^* satisfies as equalities.

Proposition 20.4. Suppose Part (a) of Hypothesis #1 is satisfied. Let x^* be a global optimum for Program 20.2, and let E be defined by (14). Then there exists no n-vector d such that

$$\begin{aligned} &\nabla f(x^*) \cdot d > 0, \\ &\nabla g_i(x^*) \cdot d < 0 \qquad \text{for each } i \in E \cap N, \\ &\nabla g_i(x^*) \cdot d \leq 0 \qquad \text{for each } i \in E \cap L. \end{aligned}$$

Proof*. Aiming for a contradiction, we suppose that such a vector d does exist. Since g_i is differentiable at x*, it has unidirectional derivatives at x*, and Proposition 18.3 gives

$$\lim_{\varepsilon \downarrow 0} \frac{g_i(x^* + \varepsilon d) - g_i(x^*)}{\varepsilon} = \nabla g_i(x^*) d.$$

It will be demonstrated that for all sufficiently small positive values of ε, the n-vector $(x^* + \varepsilon d)$ is feasible for Program 20.2.

First, consider any $i \in E \cap N$. By hypothesis, $\nabla g_i(x^*) \cdot d < 0$, so the equation that is displayed above guarantees $g_i(x^* + \varepsilon d) < g_i(x^*) = 0$ for each sufficiently small positive number ε.

Next, consider any $i \in E \cap L$. By hypothesis, $\nabla g_i(x^*) \cdot d \leq 0$. The fact that g_i is affine guarantees that for every $\varepsilon > 0, g_i(x^* + \varepsilon d) = g_i(x^*) + \varepsilon \nabla g_i(x^*) \cdot d = 0 + \varepsilon \nabla g_i(x^*) \cdot d \leq 0$.

It remains to consider any $j \notin E$. Since x* is in the interior of T, Proposition 19.9 shows that the function g_j is continuous at x*, so $g_j(x^* + \varepsilon d) < 0$ for all sufficiently small positive values of ε.

It has now been shown that $(x^* + \varepsilon d)$ is feasible for all sufficiently small positive values of ε. By hypothesis, $\nabla f(x^*) \cdot d > 0$, and the fact that f is differentiable at x* couples with Proposition 18.3 to give

$$\lim_{\varepsilon \downarrow 0} \frac{f(x^* + \varepsilon d) - f(x^*)}{\varepsilon} = \nabla f(x^*) \cdot d > 0.$$

This inequality shows that $f(x^* + \varepsilon d) > f(x^*)$ for all sufficiently small positive number ε. This contradicts the optimality of x*, which completes a proof. ∎

Proposition 20.4 prepares for the analysis of Program 20.3, below. It is a linear program; its decision variables are the number ε and the vector d. Program 20.3 is feasible because setting ε = 0 and d = 0 satisfies all of its constraints. Proposition 20.4 showed that this linear program can have no feasible solution in which ε > 0. Its optimal value z* must equal 0.

Program 20.3. z* = maximize {ε}, subject to the constraints

μ_0: $\varepsilon - \nabla f(x^*) \cdot d \leq 0,$

μ_i: $\varepsilon + \nabla g_i(x^*) \cdot d \leq 0$ for each $i \in E \cap N$,

μ_i: $\nabla g_i(x^*) \cdot d \leq 0$ for each $i \in E \cap L$.

The Duality Theorem of linear programming guarantees that the dual of Program 20.3 also has 0 as its optimal value. The dual must be feasible, which demonstrates existence of a solution to

(15) $$\mu_0 + \sum_{i \in E \cap N} \mu_i = 1,$$

(16) $$-\mu_0 \nabla f(x^*) + \sum_{i \in E} \mu_i \nabla g_i(x^*) = 0,$$

(17) $$u_0 \geq 0 \qquad \text{and} \qquad \mu_i \geq 0 \qquad \text{for each } i \in E.$$

Proposition 20.5. Suppose Hypothesis #1 is satisfied. Let x* be a global optimum for Program 20.2, and let E be defined by (14). Then (15)-(17) have a solution in which μ_0 is positive, and (7)-(9) are satisfied by setting

(18) $$\lambda_i = \begin{cases} \mu_i/\mu_0 & \text{for each } i \in E \\ 0 & \text{for each } i \notin E \end{cases}.$$

Proof*. Proposition 20.4 demonstrates the Program 20.3 has 0 as its optimal value, so the Duality Theorem guarantees that (15)-(17) have a solution.

Consider the case in which a solution to (15)-(17) has $\mu_0 > 0$. Recall that E is the set of constraints that are binding at x*, hence that dividing (16) by μ_0 shows that the gradient of the objective is a nonnegative linear combination of the gradients of the binding constraints, equivalently, that (7)-(9) hold.

Aiming for a contradiction, it is now assumed that (15)-(17) has a solution with $\mu_0 = 0$. In this case, (15) and (16) reduce to

(19) $$\sum_{i \in E \cap N} \mu_i = 1$$

(20) $$\sum_{i \in E} \mu_i \nabla g_i(x^*) = 0$$

Part (b) of Hypothesis #1 is that there exists a feasible solution $\bar{x}$ to Program 20.2 that satisfies $0 > g_i(\bar{x})$ for each $i \in N$.

Consider any $i \in E$. Since g_i is convex and differentiable at x*,

$$g_i(\bar{x}) \geq g_i(x^*) + \nabla g_i(x^*)(\bar{x} - x^*) \qquad \text{for each } i \in E.$$

If $i \in E \cap N$, we have $0 > g_i(\bar{x})$ and $g_i(x^*) = 0$, so the above inequality gives

$$0 > \nabla g_i(x^*)(\bar{x} - x^*) \qquad \text{for each } i \in E \cap N.$$

Alternatively, if $i \in E \cap L$, we have $0 \geq g_i(\bar{x})$ and $g_i(x^*) = 0$, so the same inequality gives

$$0 \geq \nabla g_i(x^*)(\bar{x} - x^*) \qquad \text{for each } i \in E \cap L.$$

Equation (19) guarantees that μ_i is positive for at least one $i \in E \cap N$. Multiply the ith displayed inequality by μ_i and then sum over each $i \in E$ to obtain

$$0 > \textstyle\sum_{i \in E} \mu_i \nabla g_i(x^*)(\bar{x} - x^*).$$

The above and (20) produce the contradiction $0 > 0$, which completes a proof. ∎

Recap

The proofs of Propositions 20.2 through 20.5 rely principally on the supporting hyperplane theorem for a convex function and the Duality Theorem of linear programming. In concert, these propositions prove

Proposition 20.6 *(characterization).* Let x^* be feasible for an instance of Program 20.2 that satisfies Hypothesis #1. The following are equivalent.

(a) The vector x^* is a global optimum for Program 20.2.

(b) There exists an m-vector λ such that x^* and λ satisfy the KKT conditions.

Thus, for nonlinear programs that satisfy Hypothesis #1, the KKT conditions are necessary and sufficient for a feasible solution to be optimal. For nonlinear programs that satisfy Hypothesis #1, Proposition 20.6 is the *exact* analogue of Proposition 20.1.

The KKT conditions are succinct because (7) is written in terms of gradients. It is actually a system of n equations, one per decision variable. The data in each equation are the partial derivatives of the objective and constraints with respect to its decision variable.

7. The Karush-Kuhn-Tucker Conditions

Equations (7)-(9) are the Karush-Kuhn-Tucker (KKT) for Program 20.2. Since Program 20.2 is a canonical form, the KKT conditions have been defined for every nonlinear program.

A recipe

The KKT conditions for a nonlinear program can be specified directly, however, without first forcing it into the format of Program 20.2. The cross-over table makes this possible. It determines the senses of the complementary constraints and multipliers exactly as it did for a linear program. In a linear program, the data in the constraint that is complementary to the decision variable x_j are its coefficients. More generally, in a *nonlinear* program, the data in the constraint that is complementary to the decision variable x_j are its partial derivatives. A recipe for the KKT conditions appears below:

- Each non-sign constraint in the nonlinear program is assigned a *complementary* decision variable, and each decision variable in the nonlinear program is assigned a *complementary* constraint.
- The senses of the complementary decision variables and constraints are determined by the cross-over table (Table 12.1 on page 383).
- The data in the constraint that is complementary to a particular decision variable are determined as follows:
  - Its RHS equals the partial derivative of the objective with respect to that decision variable.
  - Each addend on its LHS equals the product of (i) the partial derivative of a constraint with respect to that decision variable and (ii) the Lagrange multiplier that is complementary to that constraint.
- If an inequality constraint is not binding, its complementary variable must equal 0.

This recipe is wordy, but the procedure is familiar. It is *precisely* analogous to the scheme for taking the dual of a linear program and then invoking complementary slackness.

An example

To illustrate this recipe, we turn our attention to

Example 20.4. Minimize $\{e^x + y^2\}$, subject to

$$\begin{array}{ll} \lambda_1: & 4x - 3y = 6, \\ \lambda_2: & \sqrt{x} - 1y \geq 0.15, \\ & x \geq 0, \quad y \text{ is free.} \end{array}$$

Complementary variables

Example 20.4 is a minimization problem, so the cross-over table is read from right to left. This example has two non-sign constraints, which have been assigned the complementary variables λ_1 and λ_2. The first constraint is an equation, and the second constraint is a "≥" inequality. Reading row 5 and 4 of the cross-over table from right to left gives

$$\lambda_1 \text{ is free,}$$
$$\lambda_2 \geq 0.$$

Complementary constraints

Example 20.4 has two decision variables. The decision variable x is non-negative, so row 1 shows that its complementary constraint is a "≤" inequality. The decision variable y is free, so row 2 shows that its complementary constraint is an equation. The coefficients of the constraint that is complementary to x are found by differentiating the objective and constraints with respect to x, and that constraint is

$$x: \qquad 4\lambda_1 + 0.5x^{-0.5}\lambda_2 \leq e^x.$$

Similarly, the constraint that is complementary to y is

$$y: \quad -3\lambda_1 - 1\lambda_2 = 2y.$$

Complementary slackness

Complementary slackness states that if an inequality is not binding, its complementary variable must equal zero. Thus, for Example 20.4,

$$(x)(e^x - 4\lambda_1 - 0.5x^{-0.5}\lambda_2) = 0,$$
$$(\lambda_2)(0.15 - \sqrt{x} + 1y) = 0.$$

The KKT conditions that are obtained from this recipe are equivalent to those that would be obtained by forcing Example 20.4 into the format of Program 20.2 and then using (7)-(9). Proving that this is so would be cumbersome, elementary, and uninsightful. A proof is omitted.

8. Minimization

The major results in this chapter are presented in the context of a maximization problem. For convenient reference, these results are restated in the context of a minimization problem. Let us consider

Program 20.2^{MIN}. Minimize f(x), subject to the constraints

$$g_i(x) \geq 0 \qquad \text{for } i = 1, 2, \ldots, m,$$
$$x \in \Re^{n \times 1}.$$

The analogue for Hypothesis #1 for this minimization problem appears below as

Hypothesis #1^{MIN}.

Part (a): The functions f and $-g_1$ through $-g_m$ are convex and differentiable on a convex open set T that contains S.

Part (b): There exists a feasible solution $\bar{x}$ to Program 20.2^{MIN} that satisfies

$$g_i(\bar{x}) > 0 \qquad \text{for each } i \in N.$$

It is easy to check that Hypothesis #1^{MIN} becomes Hypothesis #1 when this minimization problem is converted into an equivalent maximization problem.

A feasible solution x to Program 20.2^{MIN} is said to satisfy the *KKT conditions* if there exists a vector λ such that

$$\nabla f(x) = \textstyle\sum_{i=1}^{m} \lambda_i \nabla g_i(x),$$

$$\lambda_i \geq 0 \qquad \text{for } i = 1, \ldots, m,$$

$$\lambda_i g_i(x) = 0 \qquad \text{for } i = 1, \ldots, m.$$

These KKT conditions can be verified by using the cross-over table or by converting Program 20.2^{MIN} into an equivalent maximization problem with "$\leq$" constraints. In brief:

> A feasible solution to Program 20.2^{MIN} satisfies the *KKT conditions* if and only if the gradient of its objective equals a nonnegative linear combination of the gradients of its binding constraints.

Evidently, the KKT conditions for Program 20.2^{MIN} are identical to those for Program 20.2.

9. A Local Optimum

Hypothesis #1 has a serious limitation. It cannot be satisfied by a nonlinear program that has at least one genuinely-nonlinear equality constraint. To illustrate this point, let us consider a nonlinear program that includes the constraint

$$g_3(x) = 0.$$

If the function g_3 is affine, replacing this constraint by the pair of inequalities,

$$g_3(x) \leq 0 \qquad \text{and} \qquad -g_3(x) \leq 0,$$

preserves Hypothesis #1. This replacement is equivalent to leaving the constraint $g_3(x) = 0$ in the model and allowing its multiplier λ_3 to be free (not constrained in sign), exactly as in linear programming.

On the other hand, if the function g_3 is not affine, replacing the constraint $g_3(x) = 0$ by the same pair of inequalities destroys Part (a) of Hypothesis #1 because it cannot be the case that the functions g_3 and $-g_3$ are both convex. Hypothesis #1 accommodates equality constraints *only if* they are affine.

A different nonlinear program

A constraint qualification that allows genuinely-nonlinear equality constraints will soon be introduced and discussed. This constraint qualification relates to a nonlinear program that is written in the format of

Program 20.4. Maximize f(x), subject to the constraints

$$\begin{aligned} \lambda_i : \quad & g_i(x) \le 0 && \text{for } i = 1, \ldots, r, \\ \lambda_i : \quad & g_i(x) = 0 && \text{for } i = r+1, \ldots, m, \\ & x \in \Re^n. \end{aligned}$$

Unlike Program 20.2, this formulation distinguishes between the inequality constraints and the equality constraints; the first r constraints are inequalities and the remaining m – r constraints are equations. It is allowed that r = 0, that r = m and even that r = m = 0.

The KKT conditions

Program 20.4 differs from Program 20.2 only in that m – r of its constraints are equations. From row 2 of the cross-over table, we see that the multipliers for those equations are free (unconstrained in sign). The KKT conditions for Program 20.4 are

(21) $$\nabla f(x) = \textstyle\sum_{i=1}^{m} \lambda_i \nabla g_i(x),$$

(22) $$\lambda_i \ge 0 \qquad \text{for } i = 1, \ldots, r.$$

(23) $$\lambda_i g_i(x) = 0 \qquad \text{for } i = 1, \ldots, r.$$

In brief, a feasible solution x to Program 20.4 satisfies the KKT conditions if it and an r-vector λ satisfy (21)-(23).

A different constraint qualification

The set S of feasible solutions to Program 20.4 consists of each n-vector x that satisfies its constraints. A constraint qualification for Program 20.4 appears below as

Hypothesis #2.

Part (a): The functions f and g_1 through g_m are differentiable on an open set T that contains S.

Part (b): The gradients of the constraints that are binding at each local optimum x* are linearly independent.

Requiring the gradients of the binding constraints to be linearly independent rules out the examples that were introduced earlier. In particular:

- Example 20.1 violates Part (b) because its optimal solution x^* has $g_1(x^*) = 0$ and $\nabla g_1(x^*) = 0$.
- Examples 20.2 and 20.3 violate Part (b) because the both examples have optimal solutions x^* that have $g_1(x^*) = g_2(x^*) = 0$ and $\nabla g_1(x^*) = -\nabla g_2(x^*)$.

Hypothesis #2 encompasses models that violate Hypothesis #1 because the functions – f and g_1 through g_m are no longer required to be convex.

Necessity

Suppose Hypothesis #2 is satisfied. Does each local optimum satisfy the KKT conditions? This question is answered in the affirmative by

Proposition 20.7 *(necessity).* Consider an instance of Program 20.4 that satisfies Hypothesis #2. Suppose that x^* is a local optimum for Program 20.4. Then there exists an m-vector vector λ such that x^* and λ satisfy the KKT conditions for Program 20.4.

An air-tight proof of Proposition 20.7 rests on the implicit function theorem and is omitted because it falls outside the scope of this text.

Sufficiency?

Suppose Hypothesis #2 is satisfied. If a feasible solution satisfies the KKT conditions, must it be a local maximum? No, as will be illustrated by

Example 20.5. Maximize f(x), subject to $g_1(x) = 0$, where

$$f(x) = \sqrt{3}x_1 + x_2 \quad \text{and} \quad g_1(x) = (x_1{}^2 + x_2{}^2 - 1).$$

The objective of Example 20.5 is linear. Its feasible solutions are the points (x_1, x_2) that lie on the circle of radius 1 that is centered at (0, 0). This example's gradients are

$$\nabla f(x) = (\sqrt{3},\ 1) \quad \text{and} \quad \nabla g_1(x) = (2x_1,\ 2x_2).$$

No point x on the unit circle has $\nabla g(x) = (0,\ 0)$, so Hypothesis #2 is satisfied.

A feasible solution for Example 20.5 satisfies the KKT conditions if there exist numbers x_1, x_2 and λ for which ${x_1}^2 + {x_2}^2 = 1$ and $\nabla f(x) = \lambda \nabla g(x)$. An easy computation verifies that these equations have two solutions, which are displayed below.

$$\lambda = 1, \qquad x_1 = \sqrt{3}/2, \qquad x_2 = 1/2$$
$$\lambda = -1, \qquad x_1 = -\sqrt{3}/2, \qquad x_2 = -1/2$$

One of these solutions is the point on the unit circle that maximizes f(x). The other is the point on the unit circle that minimizes f(x). Evidently, under Hypothesis #2, the KKT conditions are insufficient; they do not guarantee a local maximum.

10. A Bit of the History

The KKT conditions have a brilliant history. In the summer of 1950, they and a constraint qualification were first presented to the research community in a paper by Kuhn and Tucker.[1] That paper was instantly famous, and the conditions in it became known as the Kuhn-Tucker conditions. The constraint qualification that Kuhn and Tucker employed differs from Hypothesis #2. Their main result was akin to Proposition 20.7. It showed that their constraint qualification guarantees that each local optimum satisfies the KKT conditions. More than two decades elapsed before the research community became aware that William Karush had obtained *exactly* the same result in his unpublished 1939 master's thesis[2]. The Kuhn-Tucker conditions have hence (and aptly) been called the Karush-Kuhn-Tucker (or KKT) conditions.

Tucker

Albert William Tucker (1905-1995) earned his Ph. D. in mathematics from Princeton in 1932 and spent all but the first year of his academic career in Princeton's Mathematics Department. He chaired that department for

[1] Kuhn, H. K. and A. W. Tucker, "Nonlinear programming," *Proceedings of the Second Berkeley Symposium on Mathematical Statistics and Probability*, J. Neyman, editor, University of California Press, pp. 481-491, 1950.

[2] Karush, W, *Minima of functions of several variables with inequalities as side conditions*, M. Sc. Thesis, Department of Mathematics, University of Chicago, 1939.

nearly two decades – a particularly brilliant era, one in which he nurtured the careers of dozens of now-famous contributors to the mathematical underpinnings of operations research, game theory, and related areas.

Kuhn

Harold K. Kuhn (born in 1925) earned his Ph. D. in mathematics in 1950 at Princeton, where he had a long and distinguished career as a Professor of Mathematics. His work included fundamental contributions to nonlinear optimization, game theory and network flow.

Karush

William Karush (1917-1997) earned his Ph. D. in mathematics from the University of Chicago in 1942. During the war years, he participated in the Manhattan Project. After the war, he worked for Ramo-Wooldrige Corporation (now TRW), later became principal scientist at System Development Corporation in Santa Monica, and a Professor of Mathematics at California State University, Northbridge.

It was a book by Takayama[3] that alerted the research community to Karush's work. Prior to its publication in 1974, Richard Bellman and a few others were aware that the "Kuhn-Tucker" conditions and constraint qualification were due to Karush, but that fact was not widely known. Karush spent a lifetime in research, but he did not feel that was important to inform the community that his work anticipated that of Kuhn and Tucker.

John

By 1948, Fritz John[4] had obtained a weakened form of the KKT conditions in which $\nabla f(x^*)$ is replaced by $\lambda_0 \nabla f(x^*)$, where λ_0 must be nonnegative, but can equal 0. John's paper omits the constraint qualification that is shared by the work of Karush and of Kuhn and Tucker.

John (1910-1994) earned his Ph. D. in mathematics in 1934 at Göttingen. Like many others, he emigrated from Germany to the United States early

---

[3] Takayama, A., *Mathematical Economics*, Drysdale Press, Hinsdale, Illinois, 1974.

[4] John, F., Extremum problems with inequalities as subsidiary conditions, *Studies and essays presented to Richard Courant on his 60th birthday*, Interscience, New York, pp. 187-204, 1948.

in the Hitler era. John was a professor of mathematics at the University of Kentucky from 1935-1946 and at New York University thereafter, except for the war years, 1943-45, during which he worked at the Aberdeen Proving Ground.

Slater

Hypothesis #1, when relaxed to allow the functions to be nondifferentiable, is due to Morton L. Slater and is known as the **Slater conditions**. They appear in his Cowles Commission discussion paper,[5] which was written only a few months after the work of Kuhn and Tucker. The Slater conditions are discussed in Section 12 of this chapter.

A personal reminiscence

Readers who wish to learn more about the origins of nonlinear programming and its relationship to the work of Lagrange and Euler are referred to a personal reminiscence by a pioneer, Harold W. Kuhn[6].

11. Getting Results with the GRG Solver

Solver and Premium Solver implement the **Generalized Reduced Gradient** method, which is abbreviated as the **GRG** method. It finds solutions to systems of equations and inequalities that can be linear or nonlinear. It also finds solutions to nonlinear programs. It is designed to do these things quickly. Is it guaranteed to work? No! A search for an algorithm that works well on all nonlinear programs is akin to a quest for a philosopher's stone. No such thing exits.

Discussed in this section are a few tips that can help you to obtain good results with the GRG method. These tips are presented in the context of a nonlinear program, but some of them apply to nonlinear systems as well.

---

[5] Slater, M., "Lagrange multipliers revisited: a contribution to nonlinear programming," *Cowles Commission Discussion Paper*, Mathematics 403, November, 1950.

[6] Kuhn, H. K., "Nonlinear programming: a historical note," *A history of mathematical programming: A collection of personal reminiscences*, J. K. Lenstra, Alexander H. K. Rinnooy Kan, and Alexander Schrijver, eds., Elsevier Sci., Amsterdam, pp. 82-96, 1991.

What the GRG method seeks

When the GRG method is applied to a nonlinear program, it seeks a local optimum. It stops when it finds a local optimum. The local optimum that the GRG method finds may satisfy the KKT conditions, and it may not. In Example 20.3, the optimum occurs at a cusp, which does not satisfy the KKT conditions, but the GRG method finds it anyhow. In addition, if the GRG method is initiated at a solution to the KKT conditions that is not a local optimum, the GRG method is very likely to improve on it. It is emphasized:

> The GRG method seeks a local optimum, which may or may not satisfy the KKT conditions.

Strive for convexity

A nonlinear program is said to be **convex** if it can be written in the format of Program 20.2, with functions $-f$ and g_1 through g_m that are convex on an open set T that includes the set S of feasible solutions. A minimization problem is convex if its objective function f(x) is convex and if its constraints can be written in the format $g_1(x) \geq 0$ through $g_m(x) \geq 0$, where the functions g_1 through g_m are concave on a set T that includes S. The GRG code works best when the NLP is convex. If you are having trouble solving a nonconvex problem, it can pay to use a convex approximation to it.

Strive for continuous derivatives

Solver and Premium Solver are equipped with versions of the GRG method that differentiate "numerically." This means that it approximates each partial derivative by evaluating the function at closely spaced values. As might be expected, this works best when the functions are differentiable *and* when the derivatives are continuous.

Try to start close

There is an important difference between the way in which the simplex method and the GRG method are executed. When you use the simplex method, Solver and Premium Solver *ignore* whatever trial values you have placed in the changing cells. When you use the GRG method, Solver and Premium Solver *begin* with the values that you have placed in the changing cells. For this reason, the GRG method is more likely to work if you start with reasonable values of the decision variables in the changing cells. It can also pay to experiment – initialize the GRG method several times, with different values in the changing cells. It is emphasized:

Try to initialize the GRG method with reasonable values in the changing cells. If necessary, experiment.

Try the "Multistart" feature

The GRG method in Premium Solver is equipped with a "Multistart" feature that can help find solutions to nonconvex optimization problems and to nonconvex equation systems. This feature is on the "Options" menu in Premium Solver. Try it if you are encountering difficulty.

Avoid discontinuous functions

If you use functions that are continuous but not differentiable, you may get lucky. You can even get lucky if you use a discontinuous function. Using a discontinuous function is *not recommended!* Use a binary variable instead. Solver and Premium Solver are equipped to tackle nonlinear systems some of whose variables are explicitly required to be integer-valued.

If a problem includes integer-valued variables, strive for a formulation whose constraints and objective would be linear if the integrality conditions were removed. That will enable you to use the "Standard LP Simplex" code, which works very well.

A quirk

The GRG code has a quirk. It may attempt to evaluate a function for a value that lies outside of the range on which the function is defined. It can attempt to compute log(x) for a value of x that is negative, for instance. Including the constraint $x \geq 0$ does not keep this from occurring. Its occurrence can bring Excel to a halt. Two ways around this quirk are presented below.

This will not occur if you start "close enough." Place a positive lower bound K on the value of those variables whose logarithms are being computed, and solve the problem repeatedly, gradually reducing K to 0. Initialize each iteration with the optimal solution for a somewhat higher value of K. This tactic can avoid logarithms of negative numbers.

A slicker way is to use Excel's "ISERROR" function. Suppose that the objective of a nonlinear program is to maximize the expression

(24) $$\sum_{j=1}^{n} c_j \ln (x_j),$$

where c_1 through c_n are positive constants and x_1 through x_n are decision variables whose values must be nonnegative. To use this "slick" method:

- Enter expression (24) in a cell, say, cell B3.
- Enter the function = **IF(ISERROR(B3), – 1000000, B3)** in a different cell, say, cell B4. Cell B4 will record an objective value of –1,000,000 if the logarithm of a negative number had been taken.
- Ask Solver or Premium Solver to maximize the value in cell B4.

As mentioned earlier, every method for solving nonlinear systems or nonlinear programs will fail on occasion. The GRG method works rather well, and the tips that are mentioned in this section can help it to work a bit better.

12. Sketch of the GRG Method*

The GRG method has a great deal in common with the simplex method. A sketch of the GRG method is presented in this starred section. This sketch is focused on its use to solve a nonlinear program. It seeks a local optimum. It parses the problem of finding a local optimum into a sequence of "line searches," each of which optimizes the objective over a half-line or an interval.

Line search

A **line search** is initialized with a feasible solution x to the nonlinear program and with an **improving direction** d, namely, an n-vector d such that $x + \varepsilon d$ remains feasible and has objective value $f(x + \varepsilon d)$ that improves on $f(x)$ for all positive values of ε that are sufficiently close to zero. The line search finds the value θ for which $f(x + \theta d)$ is best (largest in a maximization problem), without violating any of the constraints that were nonbinding at x.

Having solved the line search, the GRG method corrects the vector $x + \theta d$ to account for any curvature in the set of solutions to the constraints that were binding at x. It then iterates by finding a new improving direction, executing a new line search, and so forth. How it accomplishes these steps will be explored in a series of examples.

No binding constraints

If a feasible solution x to a maximization program has no binding constraints, it seems natural to execute a line search in the "uphill" direction $d = \nabla f(x)$. To see what this accomplishes, we consider

A naïve start (for a maximization problem):

1. Begin with a feasible solution x for which no constraints are binding.
2. Select the direction $d = \nabla f(x)$.
3. With these values of x and d, find the value of θ that maximizes $f(x + \theta d)$ subject to the constraints of the nonlinear program.
4. Replace x by $(x + \theta d)$. If no constraints are binding at (the new vector) x, go to Step 2. Otherwise, do something else.

How the "naïve start" gets its name will be exposed in the context of

Example 20.6. Maximize $f(x_1, x_2) = \ln(x_1) - 0.5(x_2)^2$, subject to $x_1 \leq 100$.

The optimal solution to Example 20.6 is easily seen to be $x_1 = 100$ and $x_2 = 0$. For Example 20.6, the gradient (vector of partial derivatives) is given by

$$\nabla f(x) = (1/x_1, -x_2).$$

A zigzag

For Example 20.6, let's initiate the naïve start with the feasible solution x = (1, 1), for which $\nabla f(x) = (1, -1)$. For its first line search, this algorithm takes $d = (1, -1)$, so

$$f(x + \theta\, d) = f(1 + \theta, 1 - \theta) = \ln(1 + \theta) - 0.5(1 - \theta)^2.$$

Differentiation verifies that f(x + θd) is maximized at $\theta = \sqrt{2}$. The first iteration replaces (1, 1) by $(1 + \sqrt{2},\ 1 - \sqrt{2})$.

The constraint continues to be nonbinding. Again, this algorithm takes $d = \nabla f(x)$. Having maximized in the direction $d = (1, -1)$, the direction in which we next maximize must be *perpendicular* to $(1, -1)$; a **zigzag** has commenced. Figure 20.4 displays the path taken by the first 10 iterations of the "naïve start."

Figure 20.4. The naïve start.

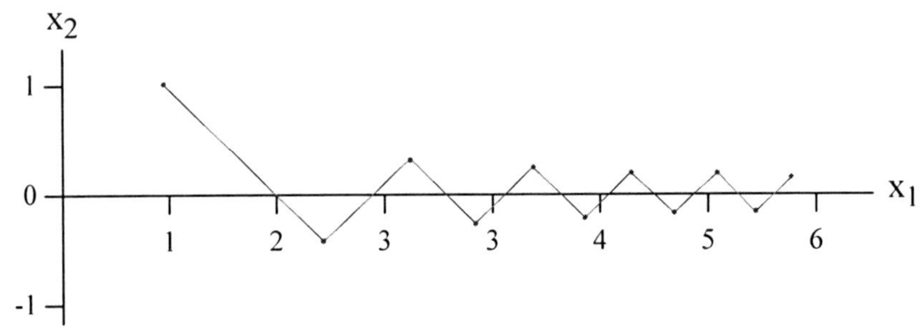

This performance is dismal. The odd iterations move in the direction $(1, -1)$. The even iterations move in the direction $(1, 1)$. Each iteration moves a shorter distance than the last. An enormous number of iterations will be needed before the constraint $x_1 \le 100$ becomes binding. Zigzagging needs to be fixed.

Attenuation

There are several ways in which to attenuate the zigzags. Solver and Premium Solver use one of them. Table 20.1 reports the result of applying Solver to Example 20.6. Its first line search proceeds exactly as does the naïve start. Subsequent iterations *correct* for the zigzag. The constraint $x_1 \le 100$ becomes binding at the 7th iteration, and the optimal solution, $(x_1, x_2) = (100, 0)$, is reached at the 8th iteration.

Table 20.1. Application of Solver to Example 20.6.

iteration	x_1	x_2	$f(x_1, x_2)$
0	1.000	1.000	-0.500
1	2.429	-0.429	0.795
2	4.559	0.512	1.386
3	9.247	-0.511	2.094
4	17.983	0.496	2.766
5	36.243	-0.514	3.458
6	70.528	0.496	4.133
7	100.000	0.080	4.602
8	100.000	0.000	4.605

Zigzagging begins whenever a line search fails to change the set of binding constraints. The *Generalized* Reduced Gradient method picks its improving direction d so as to attenuate the zigzags.

A nonlinear objective and linear constraints

The GRG method builds upon the simplex method. To indicate how, we begin with an optimization problem whose constraints are linear and whose objective is not, namely

Program 20.5. Maximize f(x), subject to A x = b and x ≥ 0.

The decision variables in Program 20.5 form the $n \times 1$ vector x. Its data are the $m \times n$ matrix A, the $m \times 1$ vector b, and the function f(x).

Let us begin with a vector x that is feasible, so A x = b and x ≥ 0. Each line search attempts to move a positive amount θ in a direction d that preserves feasibility, so it must be that

$$A(x + \theta d) = b.$$

Since A x = b, the direction d must satisfy the *homogeneous* system,

$$Ad = 0.$$

The line search moves an amount θ in the direction d that satisfies A d = 0. This line search maximizes $f(x + \theta d)$ while keeping $x + \theta d \geq 0$.

An illustration

An example will help us to understand how the GRG method selects an improving direction, d. Let us particularize Program 20.5 by taking

$$f(x) = 40x_1 - 10(x_1)^2 + 30x_2 - 20(x_2)^2 + 20x_3 - 30(x_3)^2 + 10x_4 - 5(x_4)^2,$$

$$A = \begin{bmatrix} 1 & 1 & 1 & 1 \\ 6 & 4 & 2 & 1 \end{bmatrix} \quad \text{and} \quad b = \begin{bmatrix} 3 \\ 12 \end{bmatrix}.$$

The decision variables in this example are x_1 through x_4. Let us initialize the GRG method with the 4×1 vector x that is given by

$$x = \begin{bmatrix} 1 & 1 & 1 & 0 \end{bmatrix}^T.$$

This vector x is easily seen to be feasible. It has

$$\nabla f(x) = \begin{bmatrix} 20 & -10 & -40 & 0 \end{bmatrix}.$$

Interpret the entries in $\nabla f(x)$ as marginal contributions. Decreasing x_2 by θ increases the objective by approximately 10 θ, for instance.

As noted earlier, each direction d that preserves feasibility must satisfy the homogeneous equation A d = 0. Given a feasible solution x, a direction d that satisfies A d = 0 and improves the objective will be found by pivoting on coefficients in columns having $x_j > 0$ so as to create a basic variable for each row other than the topmost of

$$(25) \qquad \begin{bmatrix} \nabla f(x) \\ A \end{bmatrix} = \begin{bmatrix} 20 & -10 & -40 & 0 \\ 1 & 1 & 1 & 1 \\ 6 & 4 & 2 & 1 \end{bmatrix}.$$

Pivoting

The feasible solution $x = \begin{bmatrix} 1 & 1 & 1 & 0 \end{bmatrix}^T$ equates x_1, x_2 and x_3 to positive values, but the matrix A has only two rows, so there is a *choice* as to the columns that are to become basic. Let's pivot on the coefficient of x_1 in the 1st row of A and on the coefficient of x_3 in the 2nd row of A. These two pivots transform the tableau on the RHS of (25) into

$$(26) \qquad \begin{bmatrix} 0 & 0 & 0 & 55 \\ 1 & 0.5 & 0 & -0.25 \\ 0 & 0.5 & 1 & 1.25 \end{bmatrix}.$$

Search direction

The entries in the top row of (26) play the role of *reduced costs*. Evidently, perturbing $x = \begin{bmatrix} 1 & 1 & 1 & 0 \end{bmatrix}^T$ by setting variable x_4 equal to θ changes the objective by approximately 55 θ when the values of the variables x_1 and x_3 whose columns have become basic are adjusted to preserve a solution to the equation system. The changes ∇x_1 and ∇x_3 that must occur in the values of these variables are found by placing the homogeneous system whose LHS is given by (26) (and whose RHS consists of 0's) in dictionary format;

$$\Delta x_1 = 0.25\theta,$$

$$\Delta x_3 = -1.25\theta.$$

Evidently, the line search is to occur in the direction d given by

$$d = \begin{bmatrix}0.25 & 0 & -1.25 & 1\end{bmatrix}^T.$$

This line search finds the value of θ that maximizes $f(x + \theta d)$ while keeping $x + \theta d \geq 0$. The optimal value of θ equals 0.8, at which point x_3 decreases to 0. This line search results in the feasible solution

$$x = [1.2 \quad 1 \quad 0 \quad 0.8]^T,$$

whose gradient $\nabla f(x)$ is given by

$$\nabla f(x) = [16 \quad -10 \quad 20 \quad 2]. \qquad (27)$$

The next pivot

The variable x_3 that had been basic now equals 0. The variable x_4 that had been nonbasic now equals 0.8. Replacing the top row of (26) by (27) and then pivoting so as to keep x_1 basic for the 1st constraint and to make x_4 basic for the 2nd constraint produces the tableau

$$\begin{bmatrix} 0 & -20.4 & 15.2 & 0 \\ 1 & 0.6 & 0.2 & 0 \\ 0 & 0.4 & 0.8 & 1 \end{bmatrix}. \qquad (28)$$

The current feasible solution has $x_2 = 1.2$, which is positive. The reduced costs (top- row coefficients) in (28) show that the next line search will reduce the nonbasic variable x_2 (its reduced cost is negative) and increase the nonbasic variable x_3 (its reduced cost is positive). The direction d in which this search occurs will adjust the values of the basic variable so as to preserve a solution to the homogeneous equation A d = 0. This direction d will satisfy

$$d_2 = -20.4,$$

$$d_3 = 15.2,$$

$$d_1 = -[0.6\,d_2 + 0.2\,d_3] = 9.2,$$

$$d_4 = -[0.4\,d_2 + 0.8\,d_3] = -4.0.$$

Thus, the next line search will be initiated with $x = \begin{bmatrix}1.2 & 1 & 0 & 0.8\end{bmatrix}^T$, and it will occur in the direction $d = \begin{bmatrix}9.2 & -20.4 & 15.2 & -4.0\end{bmatrix}^T$.

Program 20.5 revisited

The ideas that were just introduced are now adapted to Program 20.5 itself. Each iteration begins with a vector x that satisfies $A\,x = b$ and $x \geq 0$. Barring degeneracy, x has at least m positive entries (one per row of A), and x may have more than m positive entries. The direction d in which the next line search will occur is selected as follows:

1. Given this vector x, pivot to create a basic variable for each row but the topmost of the tableau

(29) $$\begin{bmatrix} \nabla f(x) \\ A \end{bmatrix},$$

but do not pivot on any entry in any column j for which $x_j = 0$. Denote as β the set of columns on which pivots occur. (If x has more than m positive elements, there is choice as to β.) The tableau that results from these pivots is denoted

(30) $$\begin{bmatrix} \bar{c}(x) \\ \bar{A} \end{bmatrix}.$$

For each j, the number $\bar{c}(x)_j$ is the *reduced cost* of x_j.

2. The search direction d is selected by this rule:

- If $x_k = 0$, then $d_k = \max\{0, \bar{c}(x)_k\}$.
- If $x_k > 0$ and $k \notin \beta$, then $d_k = \bar{c}(x)_k$.
- If x_k has been made basic for row i, then

(31) $$d_k = -\sum\nolimits_{j \notin \beta} \bar{A}_{ij}\, d_j$$

To interpret Step 2, we call x_k *active* if $k \in \beta$ and *inactive* if $k \notin \beta$. Barring degeneracy, each of the active variables is positive. Some of the inactive variables may be positive. The reduced costs determine d_k for each inactive variable. If an inactive variable x_k is positive, then it can its value can be increased or decreased, and $d_k = \bar{c}(x)_k$. If an inactive variable x_k is zero, then it can only be increased, and d_k equals the larger of 0 and $\bar{c}(x)_k$. Finally, if x_k is active, then d_k is determined from the dictionary, using (31). The direction d that is selected by Step 2 is known as the **reduced gradient**. The reduced gradient is determined by x *and* by the set β of columns that have been made basic.

Deja vu

This procedure is strikingly reminiscent of the simplex method. Step 1 pivots to create a basis, thereby transforming

$$\begin{bmatrix} \nabla f(x) \\ A \end{bmatrix} \quad \text{into} \quad \begin{bmatrix} \bar{c}(x) \\ \bar{A} \end{bmatrix}.$$

The directions d in which x can be perturbed must satisfy $\bar{A}\, d = 0$. The reduced costs determine d_j for each column j that has not been made basic. Placing the equation $\bar{A}\, d = 0$ in dictionary format determines d_j for each column that has been made basic.

The ensuing line search will find the value of θ that maximizes $f(x + \theta d)$ while keeping $x + \theta d \geq 0$. The usual ratios determine the largest number ρ for which $x + \rho d \geq 0$. If θ is less than ρ, a zigzag has commenced, and it will need to be attenuated.

Nonlinear constraints

To discuss the GRG method in its full generality, we turn our attention to a nonlinear program that has been cast in the format of

Program 20.6. Maximize f(x), subject to A(x) = b and x ≥ 0.

Program 20.6 generalizes Program 20.5 by replacing the matrix product A x by the vector-valued function A(x) of x. For i = 1, ..., m, the ith entry in the function A(x) is denoted $A^i(x)$. Let us denote as $\nabla A(x)$ the $m \times n$ matrix whose ij^{th} entry equals the partial derivative of the function $A^i(x)$ with respect to x_j. The reduced gradient d is selected exactly as in the preceding section, but with $\nabla A(x)$ replacing A in (29).

With nonlinear constraints, the vector $x + \theta d$ that results from the line search is very likely to violate $A(x + \theta d) = b$. When that occurs, a correction is needed. Methods that implement such corrections lie well beyond the scope of this discussion. The "G" in GRG owes its existence, in the main, to the way in which corrections are made.

Our sketch of the GRG method has been far from complete. Not a word has been said about how it finds a feasible solution x to a nonlinear program, for instance.

13. The Slater Conditions*

Morton Slater's constraint qualification is presented in the context of Program 20.2. This constraint qualification differs from Hypothesis #1 in two ways, one of which is minor. The minor difference is that Slater required the existence of a vector $\bar{x}$ that satisfies $g_i(\bar{x}) < 0$ for each i. That is easily relaxed. It is enough to require that the genuinely-nonlinear constraints hold strictly, i.e., that at least one feasible solution $\bar{x}$ satisfies $g_i(\bar{x}) < 0$ for each $i \in N$. The major difference is that Slater did not require that the functions f and g_1 through g_m be differentiable. That difference leads to a more subtle analysis and a weaker conclusion.

For current purposes, the **Slater conditions** are identical to Hypothesis #1, except that the functions f and g_1 through g_m need not be differentiable. When these functions are not differentiable, they do not have gradients, and equation (7) cannot hold as stated. The Slater conditions do require the function –f and g_1 through g_m to be convex on an open set T that includes each feasible solution x* to Program 20.2. These functions do have supports at x* (Proposition 19.12). Thus, for each n-vector x* that is feasible for Program 20.2, there exist n-vectors $a^0, a^1, \ldots, a^m$ such that

(32) $$f(x) \le f(x^*) + a^0 \cdot (x - x^*), \quad \text{for each } x \in S,$$

(33) $$g_i(x) \ge g_i\left(x^*\right) + a^i \cdot \left(x - x^*\right) \quad \text{for each } x \in S.$$

The dependence of a^0 through a^m on x* has been suppressed to simplify the notation. The vectors $a^0, a^1, \ldots, a^m$ that satisfy (32) and (33) need not be unique. If g_i is differentiable at x*, then a^i is unique, and conversely.

The KKT conditions

With supports substituted for gradients, the KKT conditions for Program 20.2 become

(34) $$a^0 = \sum_{i=1}^{m} \lambda_i a^i,$$

(35) $$\lambda_i \geq 0 \qquad \text{for } i = 1, 2, \ldots, m,$$

(36) $$\lambda_i g_i(x^*) = 0 \qquad \text{for } i = 1, 2, \ldots, m$$

where a^0 satisfies (32) and a^1 through a^m satisfy (33).

Sufficiency

A demonstration that the KKT conditions are sufficient follows the exactly same pattern that it did under Hypothesis #1.

Proposition 20.8 *(sufficiency)*. Suppose that x^* is feasible for an instance of Program 20.2 that satisfies Part (a) of the Slater Conditions. If a set $\{a^0, a^1, \ldots, a^m\}$ of vectors and a set $\{\lambda^0, \lambda^1, \ldots, \lambda^m\}$ of scalars satisfy (32)-(36), then x^* is a global optimum.

Proof. Proposition 20.2 holds as written because its proof does not use differentiability. Proposition 20.3 holds when (10) is replaced by (33), when (7) is replaced by (34), and when (11) is replaced by (32). ■

Necessity?

As concerns necessity, the ambiguity in a^0 through a^m leads to a more delicate analysis. To suggest why, consider

Example 20.7. Maximize $-x_2$, subject to $|x_1| \leq x_2$.

Setting $f(x_1, x_2) = -x_2$ and $g_1(x_1, x_2) = |x_1| - x_2$ places Example 20.7 in the format of Program 20.2. Figure 20.5 graphs Example 20.7. Its feasible region S consists of all pairs (x_1, x_2) having $x_2 \geq |x_1|$. Its unique optimal solution is $x^* = (0, 0)$, and $\nabla f(0, 0) = (0, -1)$. The function g_1 is not differentiable at (0, 0), and inequality (33) is satisfied by many vectors a^1, including $a^1 = (1, -2)$. With $a^1 = (1, -2)$, no scalar λ_1 can satisfy (34) because $\nabla f(0, 0)$ points straight down, and a^1 does not.

Figure 20.5. The optimal solution to Example 20.7.

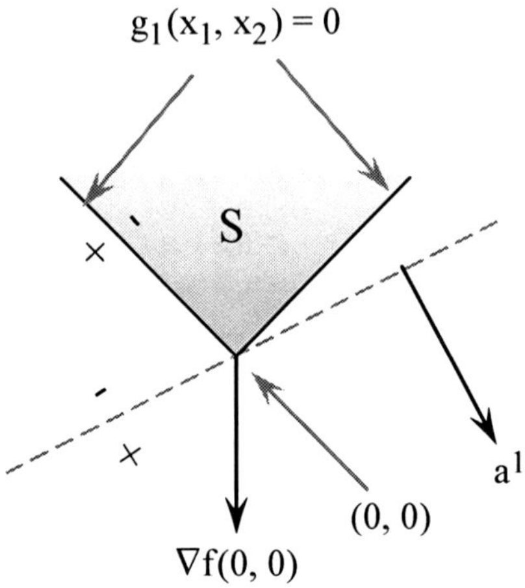

Necessity

Figure 20.5 indicates that with arbitrary supports, an optimal solution can violate the KKT conditions. Figure 20.5 does leave open the possibility that an optimal solution has supports that satisfy the KKT conditions.

Proposition 20.9 *(necessity).* Suppose that x^* is optimal for an instance of Program 20.2 that satisfies the Slater Conditions. Then for every n-vector a^0 that satisfies (32), there exist n-vectors $\{a^1, a^2, \ldots, a^m\}$ and numbers $\{\lambda_1, \lambda_2, \ldots, \lambda_m\}$ that satisfy (33)–(36).

Proof of Proposition 20.9 is omitted. Optimization with functions that are not differentiable is a difficult subject that falls well beyond the scope of this book. The statement of Proposition 20.9 is included because it exhibits a use of Part (b) of the Slater conditions.

14. Review

Hypothesis #1 guarantees that a feasible solution is a global optimum if and only if it satisfies the KKT conditions. This hypothesis does not accommodate equality constraints that are genuinely nonlinear. A second constraint qualification allows such constraints, but it produces a weaker result, namely, that each local optimum satisfies the KKT conditions.

The GRG method seeks a local optimum. It executes a sequence of line searches. The direction in which each line search occurs is found by employing linear approximations to the binding constraints. The direction is guided by the reduced gradient, but in a way that attenuates zigzags.

15. Homework and Discussion Problems

1. For the example illustrated in Figure 20.1, suppose x is feasible and has no binding constraints. Argue that x is not a local maximum if $\nabla f(x) \neq 0$.

2. For the example illustrated in Figure 20.1, suppose x is feasible, that only the constraint $g_3(x)$ is binding, and that the function g_3 is affine. Argue that x is not a local maximum if $\nabla f(x)$ is not a nonnegative multiple of $\nabla g_3(x)$.

3. For example illustrated in Figure 20.1, y is feasible, and every feasible solution $x \neq y$ has $(x - y) \cdot \nabla f(y) < 0$. Demonstrate that y is a local maximum.

4. Draw the analogue of Figure 20.1 for a nonlinear program that is cast in the format of Program 20.2^{MIN}. Interpret the KKT conditions at a feasible solution y to this nonlinear program.

5. For Program 20.2, suppose that the functions f and g_1 through g_m are differentiable. Let x^* be feasible, and suppose every feasible solution other than x^* has $(x - x^*) \cdot \nabla f(x^*) < 0$. Show that x^* is a local maximum.

6. Use Solver to maximize $f(x, y, z) = x\,y\,z$, subject to

$$4xy + 3xz + 2yz \leq 72,$$

$$x \geq 0, \quad y \geq 0, \quad z \geq 0.$$

   Then write the KKT conditions for the same optimization problem, and solve them analytically. Do you get the same solution?

7. The data in the optimization problem that appears below are the positive numbers a_1 through a_m and the positive numbers b_1 through b_m. What is its optimal solution? Why?

$$\text{Minimize } \sum_{j=1}^{m} a_j(x_j)^2, \text{ subject to}$$
$$\sum_{j=1}^{m} b_j x_j = 100, x_j \geq 0 \text{ for } j = 1, 2, \ldots, n.$$

8. Let S be the set of $n \times 1$ vectors x such that $A\,x = b$. (There are no sign restrictions on x.) Let c be any $1 \times n$ vector, let Q be any symmetric $n \times n$ matrix, and consider the problem of minimizing $f(x) =$ $f(x) = cx + \frac{1}{2}x^TQx$ subject to $x \in S$. Suppose that $x^* \in S$ is a local minimum. Consider any $x \in S$. Set $d = x - x^*$. (The fact that d depends on x has been suppressed to simplify the notation.)

   (a) Show that x^* is a global minimum. *Big hint:* Do parts (b)-f) first.

   (b) Is $A\,d = 0$?

   (c) Is $(x^* + \varepsilon d) \in S$ for every real number ε?

   (d) Is the function $\varphi(\varepsilon) = f(x^* + \varepsilon\, d) - f(x^*)$ of ε quadratic? If so, what are its coefficients?

   (e) Does $c\,d + d^TQx^*$ equal 0? If so, why?

   (f) Is d^TQd nonnegative? If so, why?

9. In Program 20.2, suppose that the functions $-f$ and g_1 through g_m are convex on an open set that includes the set S of feasible solutions, as defined by (6). The functions $-f$ and g_1 through g_m need not be differentiable. Justify your answers to parts (a)-(c).

   (a) Is S a convex set? Is S a closed set?

   (b) Suppose the vector x^* in S is a local maximum. Is x^* a global maximum? *Hint:* Suppose x is in S, and write down what you know about the value taken by $f[(1 - \varepsilon)x^* + \varepsilon x]$ for all sufficiently small positive values of ε.

10. A slight variant of the linear program that was used in Chapter 4 to introduce the simplex method is as follows: Maximize $(2\,x + 3\,y)$, subject to the six constraints

$$x - 6 \leq 0, \quad (x + y - 7)^3 \leq 0, \quad 2y - 9 \leq 0,$$
$$-x + 3y - 9 \leq 0, \quad -x \leq 0, \quad -y \leq 0.$$

Exhibit its feasible region and solve it graphically. Does its optimal solution satisfy the KKT conditions? If not, why not?

11. Use the GRG method to find an optimal solution to Example 20.3 (on page 626). Did it work? If so, does the solution that it finds satisfy the KKT conditions?

12. Prove the following: Part (a) of Hypothesis #1 guarantees that a local maximum for Problem 2 is a global maximum.

13. Suppose that x* is a local maximum for Program 20.2 and Hypothesis #1 is satisfied, except that the functions – f and g_1 through g_m are not differentiable. Show that x* is a global maximum.

14. This problem concerns Example 20.4.

    (a) Show that this NLP satisfies Hypothesis 1^{MIN}.

    (b) Use Solver or Premium Solver to find an optimal solution to it. Obtain a sensitivity report.

    (c) Verify that the KKT conditions are satisfied.

15. The data in the nonlinear program that appears below are the $m \times n$ matrix A, the $m \times 1$ vector b, the $1 \times n$ vector c and the symmetric $n \times n$ matrix Q. Write down the KKT conditions for this nonlinear program.

$$z_* = \min\left\{cx + \frac{1}{2}x^T Q x\right\}, \quad \text{subject to } Ax = b, \quad x \geq 0.$$

16. In system (25) with x equal to $\begin{bmatrix} 1 & 1 & 1 & 0 \end{bmatrix}^T$, as in the text, do as follows:

    (a) Pivot to make x_1 basic for the 1st row of A and to make x_2 basic for the 2nd row of A, so that $\beta = \{1, 2\}$ rather than $\{1, 3\}$.

    (b) With reference to this (new) basis, find the reduced gradient d.

    (c) Execute a line search in this direction d. Specify the feasible solution that results from this line search.

(d) True or false: In an iteration of the GRG method, the set β of columns is made basic has no effect on the feasible solution that results from the line search.

17. On pages 649-651, the GRG method "pivoted" from the feasible solution $x = [1 \quad 1 \quad 1 \quad 0]^T$ to the feasible solution $x = [1.2 \quad 1 \quad 0 \quad 0.8]^T$. Describe and execute the next iteration.

18. The data in NLP #1 and NLP #2 (below) are the $m \times n$ matrix A, the $m \times 1$ vector b and the $n \times 1$ vector c. Set $S = \{x \in \Re^{n\times 1} : Ax = b,\ x \geq 0\}$. Assume that the numbers c_1 through c_n are positive, that S is bounded, and that S contains at least one vector x each of whose entries is positive.

NLP #1: Minimize y b, subject to

$$Ax = b, \quad x \geq 0, \quad y A_j \geq c_j/x_j \quad \text{for } j = 1, 2, \ldots, n.$$

NLP #2. Maximize $\left\{\sum_{j=1}^{n} c_j \ln(x_j)\right\}$, subject to

$$Ax = b, \quad x \geq 0.$$

(a) Show that every feasible solution to NLP#1 has $y\,b \geq \sum_{j=1}^{n} c_j$.

(b) Show that there exists a positive number ε such that the optimal solution to NLP #2 is guaranteed to satisfy $x_j > \varepsilon$ for j = 1, 2, …, n. Does the variant of NMP #2 that includes these positive lower bounds satisfy Hypothesis #1? If so, write down its KKT conditions.

(c) Use part (b) to show that NLP #1 has an optimal solution and that each of its optimal solutions:

- has $y\,b = \sum_{j=1}^{n} c_j$,

- has $y\,A_j = c_j/x_j$ for each j,

- has the same vector x.

19. *(critical path with workforce allocation).* The tasks in a project correspond to the arcs in a directed acyclic network. This network has exactly one node α at which no arcs terminate and exactly on node ω from which no arcs emanate. Nodes α and ω represent the start and end of the project. Each arc (i, j) represents a task and has a positive datum c_{ij}, which equals

the number of weeks needed to complete this task if the entire workforce is devoted to it. If a fraction x_{ij} of the workforce is assigned to task (i, j), its completion time equals c_{ij}/x_{ij}. Work on each task (i, j) can begin as soon as work on every task (k, i) has been completed. The problem is to allocate the workforce to tasks so as to minimize the time needed to complete the project.

(a) Build a model of this workforce allocation model akin to NLP #1 of the preceding problem. *Hint:* Let the (node-arc incidence) matrix A have one row per node and one column per arc; the column A_{ij} that corresponds to arc (i, j) has –1 in row i, +1 in row j, and 0's in all other rows.

(b) Show that the minimum project completion time equals $\sum_{i,j} c_{ij}$ weeks, show that all tasks are critical (delaying the start of any task would increase the project completion time), and show how to find the unique optimal allocation x of the workforce to tasks.

Note: Problems 18 and 19 draw upon the paper, "A nonlinear allocation problem," by E. V. Denardo, A. J. Hoffman, T. MacKensie and W. R. Polleyblank, *IBM J. Res. Dev.*, vol. 36, pp. 301-306, 1994.

Index

E. V. Denardo, *Linear Programming and Generalizations,* International Series in Operations Research & Management Science 149,
DOI 10.1007/978-1-4419-6491-5, © Springer Science+Business Media, LLC 2011

CPSIA information can be obtained at www.ICGtesting.com
Printed in the USA
LVOW072200260613

340449LV00007B/112/P

9 781441 964908